MF-64

GEOLOGY OF CLAYS

GEOLOGY OF CLAYS

Weathering • Sedimentology • Geochemistry

by

GEORGES MILLOT

Professor of Geology
Dean of the Faculty of Sciences
University of Strasbourg, France

Translated by

W. R. FARRAND
Department of Geology and Mineralogy
University of Michigan, Ann Arbor, U.S.A

HÉLÈNE PAQUET
Centre de Sédimentologie et de Géochimie
de la Surface, Strasbourg, France

SPRINGER-VERLAG, NEW YORK • HEIDELBERG • BERLIN
MASSON ET Cie, PARIS
CHAPMAN & HALL, LONDON

1970

Traduction en langue anglaise de l'ouvrage du Pr G. MILLOT *Géologie des Argiles altérations, sédimentologie, géochimie*, publié par Masson et Cie, éditeurs.

"All rights reserved. No part of this book may be translated or reproduced without the written permission of Masson et Cie."

© Masson et Cie, Paris, 1970
Library of Congress Card Number 76-113458.
\ Standard Book Number 412 10050 9/69

Printed in Belgium

Title No. 1671

PREFACE

Geology, in common with other fundamental scientific disciplines, is diversifying. Today's geologist has to specialize, and the study of hydrogeology is as far removed from metallogenetics as both are from paleontology. Despite this proliferation, every geologist must retain a broad view of the Earth sciences, keep in touch with their advances and call upon the neighboring specialty as his work demands. Again, he has to be able to switch from one specialty to another. Indeed, assuming the professional career of scientists and engineers to span some forty years, we can be certain that in future most of them will have to change their specialty in the course of it, due to rapidly changing technology, new needs and unforeseen situations. Thus the future belongs to those nations or groups with the foresight to train key men of broad culture, intellectual flexibility and ready adaptability. This means that secondary and higher education must avoid the pitfalls of early specialization which, paradoxically, is the answer regularly proposed by the prophets of doom. It means, too, that when the time comes to specialize the bright boys must be given the proper tools. Among these tools, books—though no substitute for practical work—are indispensable. This is one of the main reasons why scientific and technical books have enjoyed a post-war boom. It is also why I decided to write a book, after 20 years of research into the geochemistry of the Earth's surface. Clays being the most important ingredient, I have called it The Geology of Clays.

I dedicate this book, first, to the memory of Jacques DE LAPPARENT, *Professor at the University of Strasbourg from 1919 to 1947, who died prematurely in 1948. I consider him to be the founder of the geochemistry of the formation of clays. He discovered the main features of the mechanisms effecting the genesis and evolution of clays. I was fortunate enough to work with him and I can still recall the sense of illumination our conversations gave me. His published work records his penetrating insights from which we are only gradually extracting the full meaning, either because we did not appreciate their profundity until recently, or because we were dazzled by his charm and picturesque turns of phrase.*

Next, I dedicate this book to Professor C. W. CORRENS, *of Göttingen and Professor* R. E. GRIM, *of Urbana, the respective leaders of the German and American Schools. Their students today hold sought-after positions in all parts of the world. Both schools have made a substantial contribution to research on clays and have not ceased to have a direct influence on it for some thirty years. To these two breeds of geologists and their leaders, I express my grateful thanks for the kindness and friendship shown to their French colleagues.*

The aim of the book is to present an up-to-date compilation of work done throughout the world on the origin and development of these phyllitic silicates.

My 1949 work was more restricted in scope, being basically concerned with sedimentary rocks, moreover, it has long been out of print. Many of its arguments are dated and I am glad to have the opportunity of correcting them myself.

I was at that time unduly impressed by the variations in some argillaceous rocks, due to the very sharply defined genetic environment I had elected to study; I therefore overemphasized the neoformation of clays. This phenomenon does exist, but it is not nearly as general as I supposed. In fact, the legacy of detrital clay minerals is responsible for many aggregations of phyllites in both recent and older sediments. Moreover, between neoformation and inheritance there occurs transformation of clay minerals, involving both the degradation of clays during the weathering of rock and their reconstitution by aggradation during sedimentation. This is the current view.

I thus intend to bring my earlier work up to date, at the same time extending it to include the clays of the whole surface of the Earth's crust, from their formation by weathering to their disappearance due to metamorphism. So the specialist has to touch upon many disciplines: sedimentology and geology are augmented by chapters on pedology, oceanography, geography, petrography and geochemistry. This makes his work harder and carries the risk of error, yet it is vital in tracing the evolution of silicates in the surface portion of the geochemical cycle.

The prime duty of a textbook is to cover the literature. This is a familiar problem: the volume of the literature swells from day to day and it is hard to see how the scientist of the future will keep up with it. Thanks to an excellent library and with the assistance of my friends, I was able to consult two thousand references in the course of two summer vacations. I have cited only one thousand in this book. I cannot claim that the sifting process guarantees freedom from errors and omissions. My choice of examples, a special knowledge of certain topics or authors, my ignorance of Slav and oriental languages, even the arrangement of my lists have led me to include some references rather than others. However, barring accidents, I think I may say that anything with a bearing on the history of clays in the widest sense is quoted here, the main work on the bibliography having been done in the summer of 1962.

Unquestionably, one thousand references make for a weighty book, but it is impossible to provide full documentation in a slim one. Today it is no longer acceptable to research in a field of any extent without knowing exactly what has been done already. This need arises not only from the risk of repeating work all over again, or reporting with a flourish some detail published ten years before—this would not matter too much, although such points have to be watched. More serious is the fear of the research worker that he might not make use of existing means for handling known material, although the risk of this happening increases daily, and some paths have to be trodden many times before the route is confirmed. The essential point, however, is this: we should not apply scientific consideration unless to the whole body of observations and measurements at present known. We have no right to rely solely upon our own results and our own points of view when judging broad issues. The volume of work done earlier or done by other people may seem burdensome, but it has to be taken into account in constructing arguments in the natural sciences. Documentation is not an obsession or a disease of the natural and human sciences: it is a duty and a moral obligation.

A further obligation in a textbook is to present a considered view of the subject. In the present case the author has devised a system to account for the various mechanisms by which clays are formed, their types of development and their progressive disappearance. Some stages in this geochemical evolution are fairly well known, while others are still obscure. Thus, the areas where further research is most needed may be pointed out.

This process of reflection on the genesis and evolution of clays has produced conflicting opinions in some countries. Thus, if you live on a huge delta, you tend to think all clays are detrital because this is true of modern clays in your country. But it is not true of clays at all times and in all countries. Naturally, this gives rise to disputes, and these are reviewed in my book. Alternatively, we could study the phenomenon of scientific controversy for its own sake. In the natural sciences, including the human sciences, medicine and sociology, individual interpretation still plays a large part in research. And each of us, whether he knows it or not, is the product of an environment, a history, a culture and, not least, his own character type. In current jargon, he is "situated". What he proposes is not the Truth but an aspect of truth discernible from his situation—"his" truth, in fact. It has recently been shown that the character of a philosopher influences his choice of system as much if not more than his times, his environment or his persuasion. In this view, the personality of a philosopher is his real "situation". As Gaston BERGER demonstrates, this is what makes the study of personality so enthralling. This study not only seeks to identify and classify personality characteristics, it aspires to interpret the writings of philosophers or politicians in the light of the author's personality. "One can envisage setting up an objective critique of systems which would make due allowance for variation of character and diversity of points of view without sacrificing what is absolute in Truth ([1])."

Each of us could apply such studies to his scientific discipline. Thus, an analysis of geological disputes—fortunately moderate and extremely rare—by personality characteristics might provide valuable insights and even some amusement.

But the fun would be short-lived, I fear. While a few data can be interpreted differently according to point of view, school, or observer, the majority, coded, recorded and sorted by modern data-processing methods, leave no room for a subjective interpretation of the facts. And so, as the natural and human sciences yield to the computer, there will be less and less to argue about. There will be fewer opportunities for diverse judgments, so that scientific controversies will die out. The only remaining haven for the born controversialist in a climate so dangerously anodyne will be to argue about how to use computers which, to some extent, give out only what is put into them. We can expect some spectacular clashes on this front.

The present book sets forth the opinions of various authors in all tranquillity,

---

([1]) BERGER, G. (1962), *L'homme moderne et son éducation*. Presses Universitaires, p. 23.

just as they were printed in the scientific journals. The foregoing remarks explain this lack of passion. Often the strongest oppositions can be seen as simply mirroring differences in situation, all of equal credibility.

SOME FALLACIES. — *When, after long toil and patient debate, specialists achieve solutions which are universally accepted, there still remains one task to be done: these discoveries must be incorporated into basic teaching and become general knowledge. With a subject which is quite new, this work of popularization is not difficult. Today every schoolboy knows all about atomic reactors and the orbits of spaceships. But with an established subject, we run up against out-of-date precepts which are still taught, although they are wrong, because of the great authority conveyed by force of habit. This certainly happens in geology. And, even in geology, these neat aphorisms are dangerous for those without sufficient background knowledge.*

The specialists are working to root out such fallacies, the most famous example being "man is descended from the ape". Here, in the field which concerns us, are a few fallacies we could well do without:

The weathering mechanism of feldspars is kaolinization. This is wrong. The weathering mechanism of feldspars is hydrolysis. Kaolinization is not inevitable and it is a secondary mechanism occurring only under special conditions.

The intensity of weathering in humid tropical or equatorial climates is due to the high CO_2 content of the rain. This is wrong. The intensity of weathering is caused by heavy rainfall and high temperature.

Lateritic hardpan results when iron rises by capillarity under the influence of evaporation. This is wrong. Hardpan is a concretion within a permeable material at the level where the water table circulates horizontally.

Silica in natural waters is in the colloidal state. This is wrong. Silica in natural surface waters is in true solution. This refers to the water in the hydrosphere and not to hydrothermal phenomena.

Cherts and silicifications are formed by the flocculation of colloidal silica. This is wrong, as the silica in natural waters is not in the colloidal state. A mechanism has been proposed of crystal growth from solutions as in other chemical sediments.

It is my hope that the text which follows will not make this list any longer because of any mistakes it may contain.

ACKNOWLEDGEMENTS. — *I should like to thank my publishers, Masson et Cie, who have done so much for scientific books, especially in geology, for the confidence they have shown in me by accepting this book.*

Observations made over 20 years and the subsequent discussion of them in the laboratory are the outcome both of our own studies and those communicated to us by many others. The method is always the same: study the samples, correlate measurements, meet to evaluate results, discuss with the specialist who raised the question. I should like to thank all our clients—geologists, geographers, pedologists, agronomists, petroleum geologists, ceramists, civil engineers and oceanographers—

who by submitting their problems to us have helped to enlarge our field of knowledge.

It is a labor of love rather than a duty to thank those who, over the years, have assured the day-to-day running of the laboratory by issuing regular reports of results, perfecting methods, training new staff and entertaining visitors. Marthe BONIFAS, Jacques LUCAS, Thérèse CAMEZ *and* Hélène PAQUET *have been my daily companions over the years and it is to them I owe the efficiency and harmony of my scientific life. What is more, the work we shared blossomed into a friendship which is a daily source of strength in a somewhat harassed existence.*

We and the laboratory staff as a whole have not worked in isolation but have enjoyed the encouragement and support of many, among whom I should like to mention Georges KULBICKI, *of the Centre de Recherches de la SNPA at Pau,* Bernard KUBLER, *of the Laboratoire Central de la CFP at Bordeaux, who sustained a scientific dialogue with us and gave us help and inspiration. Also* Raymond WEY, *Director of the Laboratoire de Physicochimie des Sols in our Faculty, whose activities so happily complemented ours. Further removed, there are my teachers and friends at other universities, especially Paris where I was a student. I must mention my teacher,* M. ROUBAULT, *who is coming to Strasbourg to direct the Centre de Recherches de Sédimentologie et de Géochimie de la Surface, thus reinforcing doubly the links forged between us over 20 years. I must mention, too, my colleagues at Strasbourg who have supported us in our work, in particular* M. GOLDSZTAUB *and* M. SAUCIER, *and the deans of the Faculty,* M. MARESQUELLE *and* M. VIVIEN.

Finally, I owe a great deal to all my colleagues, coworkers and friends at the Institut de Géologie and at the Service de la Carte Géologique d'Alsace et de Lorraine. Each of them carries on simultaneously the jobs of training the young people entrusted to our care, helping to run a large scientific institution and doing or supporting research in the various sectors of our activities. I herewith thank each one personally, whatever his job, for doing it with care, speed and courtesy; I can only express my gratitude by saying: I am at your service.

Georges MILLOT.
January 1964.

TRANSLATORS' PREFACE

No two languages have 100 percent equivalency of terms, although in the mid-20th century there is increasing internationalization of scientific terminology, thanks to the rapid exchange of information between countries and hemispheres.

Nevertheless, one of the biggest problems for any translator remains that of finding word-for-word equivalents for technical terms in order to translate accurately and to compose sentences economically in the language of the translation. If one must replace a single noun (e.g., marne*) by a longer noun-phrase (e.g., a more-or-less argillaceous limestone), then many awkward situations will arise concerning style and sentence structure in the translation, not to mention the impossibility of forming adjectives from such noun-phrases. Therefore, many translators justifiably retain a foreign word in the translation, either as a foreign word and italicized (e.g.,* schistes lustrés*) or incorporated as a normal word (e.g.,* marne*) in the language of the translation. Although we believe that such cases should be kept to a minimum, we have found this procedure necessary in a few cases.*

In geology many descriptive terms are endemic to a given language. These terms are in many cases colloquial, archaic or regional terms that have been applied to natural phenomena long before modern science incorporated them. The following problems of translation arose somewhat frequently in this book:

— marne, *the French word for a rather soft rock ranging between argillaceous limestone and calcareous mudstone.* Marne *is translated as "marl" in some cases, but this only leads to greater confusion because "marl" in English seems to have a wider range of meanings than* marne *in French. "Marl" is very commonly applied to Recent lacustrine, highly calcareous muds in the central U.S.A., but to incoherent sands on the Coastal Plain, while in Britain "marl" is used in a manner very similar to* marne. *It seems that the only way to avoid confusion for some readers is to retain* marne *and even to incorporate it into the geological vocabulary in English.*

— lessivage *(literally "leaching") is used by French pedologists to designate a complex of phenomena involved in removing substance from the upper part of the soil profile. It includes* both *the mechanical movement of clay particles, not chemically broken down,* and *the removal of ions in solution as the result of hydrolysis within the soil profile. Clearly this phenomenon is much more comprehensive than "leaching", as that word is commonly applied by American pedologists, which involves simply the transfer of alkalis, mainly* $CaCO_3$, *by chemical solution.* Lessivage *is more nearly equivalent to the entire process of eluviation*. To avoid confusion, however, we have decided to retain lessivage throughout.

— arène *is used by French geologists to designate the sandsized residue of the grain-by-grain mechanical disintegration of granular rocks, such as granite. It implies much more than simply "sand", and the only common English equivalent is "gruss" (from the German* Grus*).* Arène *is retained here to avoid further complicating the international vocabulary.*

— cuirasse *is French for a breast-plate of armor and has been adopted by French pedologists for the armor-like crust that forms upon exposure of a lateritic soil profile.* Cuirassement *is the action of forming such a crust. Since "lateritic crust" and "encrustation" are quite common terms for this phenomenon,* cuirasse *and* cuirassement *have not been used in the translation, although they appear in a few places in parentheses in order to clarify the kind of crust being discussed.*

— schistes lustrés *are the fine-grained, lustrous, low-grade metamorphic phyllites or schists of the Alpine geosyncline. General practice in English is to retain this term which has great specificity.*

— meulières *are silicified freshwater limestones, commonly utilized as millstones. Other than the geologically vague term "millstone" there is no English equivalent.*

William R. FARRAND,
Ann Arbor

Hélène PAQUET.
Strasbourg

CONTENTS

PREFACE V

TRANSLATORS' PREFACE XI

CHAPTER I. — *The clay minerals* 1

— **Kaolinite group:** Kaolinite. — Dickite and nacrite. — Halloysite. — Disordered kaolinites (fireclay minerals). — Serpentines and homeotypes of kaolinite . 2

— **Mica group:** Micas. — Pyrophyllite and talc. — Illites. — Polytypes of muscovite. — Glauconite and "glauconie". 5

— **Montmorillonite group:** Dioctahedral "montmorillonite-beidellite" series. — Dioctahedral "nontronite-beidellite" series. — Trioctahedral "stevensite-saponite" series 10

— **Chlorite group:** True chlorites. — Chlorites of iron ores. — Expanding chlorites. — Dioctahedral chlorites 12

— **Vermiculite group:** The mineral vermiculite. — The vermiculite of clays . . . 13

— **Mixed-layer clay minerals:** Mixed-layer clay minerals. — The types of interstratification. — The main mixed-layer minerals. — Identification of mixed-layer minerals 15

— **Attapulgite, sepiolite and palygorskite group:** Attapulgites. — Sepiolites. — Nomenclatural problems: FERSMANN's palygorskite series. — Attapulgite and sepiolite structures. — Fibrous facies and chemical composition. . . . 18

— **Iron and aluminum oxides and hydroxides:** Natural iron hydroxides and oxides. — Limonite. — Aluminum hydroxides and oxides. — Mixed oxides. 20

— **Conclusion: The values and dangers of nomenclature** 23

CHAPTER II. — *Argillaceous rocks* 29

— **History of research on clay geology:** Before 1930: clay geology by chemical methods. — From 1930 to 1950: intervention of X-ray diffraction. — From 1950 to 1962: modern works 29

— **Classification and texture:** Classification. — Texture 34

— **Argillaceous rocks of hydrothermal and volcanic origin:** Clays of hydrothermal deposits. — Argillaceous rocks resulting from volcanic materials (bentonites) 40

CHAPTER III. — *Geochemistry of ions in the hydrosphere* 49

— **The behavior of ions in water:** Hypothesis of GOLDSCHMIDT: ionic potential. — Interpretation of WICKMANN (1945), according to BERNAL and MEGAW (1935). — The radius of hydrated ions in water. — Behavior of alkalis in surface geochemistry. — Conclusion and classification. 49

— **Solubility of silica and alumina:** History. — Solubility of silica. — Solubility of alumina. — Comparison of the solubilities of silica and alumina. — Influence of ions present in solution. — Geochemical consequences . . . 55

— **Hydrolysis of silicates:** Hydration of feldspar. — Cation exchange between water and crystal. — Attack on the silicate framework. — Summary on the hydrolysis of silicates. — Conclusion 62

CHAPTER IV. — *The place of clays in the geochemical cycle* 68
  — **Geochemistry of the normal sedimentary series** 68
  — **Geodynamic interpretation of the normal geochemical series of sediments** . . 71
  — **Positive and negative sedimentary sequences** 73
  — **The evolutionary series of pedological origin** 76
  — **Confrontation and general view on sedimentary series** 78
  — **The place of clays in the surficial part of the geochemical cycle** 82

CHAPTER V. — *Weathering and soil clays* 84
  — **Weathering mechanisms:** Freeze and thaw. Cryoclasticism. — Variations of temperature. — Humidity. Hydration and effect of salt crystallization. — Oxidation and reduction. — Hydrolysis and its variations. — Biochemical activity 89
  — **Arenization:** History prior to weathering. — Petrographic data on the weathering of feldspathic rocks. — Arenization and geochemistry of arenization. — Variations depending on climates. — Experimental data on arenization and rock disintegration. — The clay minerals of *arènes* and of the beginnings of weathering. — Conclusion: the myth of the kaolinization of feldspars. . 92
  — **The soil clays:** History. — Non-evolved soils: cold countries and deserts. — Soils under an Atlantic climate. — The ferruginous soils of warm climate. — Calcimorphic soils. — Lateritic soils. — Lateritic crusts (*cuirasses*). — Bauxites 99
— **Conclusions** 133

CHAPTER VI. — *Clays of continental sediments* 135
  — **Glacial deposits** 136
  — **Eolian deposits** 137
  — **Deposits of rivers and estuaries** 139
  — **The siderolithic facies:** Definition and history. — The siderolithic Eocene of northern Aquitaine. — The siderolithic facies in Europe. — The siderolithic facies in Africa. — The siderolithic facies in the U.S.A. 142
  — **The great detrital red-bed series** 154
  — **Coal measures and Tonstein:** The coal measures. — Tonstein 162
  — **Alkaline lakes in plains:** Clays of present-day lakes. — Oligocene lacustrine clays in France. — Lakes with attapulgite and sepiolite 169
  — **Analcimolites** 177

CHAPTER VII. — *Clays of marine sediments* 178
  — **Present-day marine sediments:** The facts. — Interpretation 178
  — **The detrital series: from sandstones to shales and argillites:** Basal sandstone series. — Graywackes. — Flysch and molasse. — Alternating series . . . 186

- **The marno-carbonate series: from argillites to limestones:** Early works: the clay fraction of limestones. — Continuing inventory of limestones and marnes. — Variations of clays with paleogeography and facies. — Study of the alternating argillo-calcareous series. — Diagenesis in limestone   193

- **Alkaline chemical sedimentation with attapulgite, sepiolite and montmorillonite:** Marine attapulgite and sepiolite. — The clay fraction of chalks. — Origin of attapulgite and montmorillonite in sediments. — Origin of the elements necessary for this neoformation. — Geology of landscapes involving attapulgitic and montmorillonitic sedimentation. — Conclusion: Sedimentary sequence and geochemical sequence   198

- **Glauconitic sediments:** Varieties of occurrence of glauconie in sediments. — Mineralogical varieties. — The conditions of genesis   204

- **The clay minerals of iron ores:** Mineralogical identifications. — Conditions of formation — Origin of the iron   210

- **The hypersaline facies:** General survey of the hypersaline facies. — Clay minerals of hypersaline facies   215

- **Conclusions on clays of marine sediments**   234

CHAPTER VIII. — *Evolution of the clay fraction in some great sedimentary series*   235

- **The Cambro-Ordovician sandstones of the Central Sahara:** Regional situation. — The geological section. — Petrography of the lower coarse series. — Petrography of the upper series. — Reconstruction of the history of the sandstones of Hassi-Messaoud and Hassi el Gassi   235

- **The Carboniferous of the central U.S.A.:** The American Carboniferous. — Study of argillaceous sediments of Oklahoma. — Study of argillaceous sediments of Illinois. — Study of sediments associated with fireclays in Missouri   248

- **Triassic argillaceous sedimentation in Morocco and in France:** The Triassic of Morocco. — The Triassic of the Jura. — The Triassic of Lorraine and of the Paris Basin. — Comparison with neighboring basins   257

- **Clay minerals in the Tertiary basins of West Africa:** The eastern Sudan basin. — The basin of Senegal-Mauritania. — The basins of Dahomey-Togo and of the Ivory Coast. — Age of the attapulgite- and sepiolite-bearing formations in West Africa. — Attapulgite-bearing facies and associated facies: antagonism and paragenesis. — Sedimentary sequence of the alumino-magnesian clay minerals. — Geochemical sequence. — The geochemical cycle in the African Tertiary basins   262

- **Overall view of the evolution of clays in the great sedimentary series:** Diagenesis in Cambrian sandstones of the Sahara. — Heritage in the Carboniferous series of the U.S.A. — Transformation in Triassic clays of Eurafrica. — Neoformation in the Eocene of West Africa   274

CHAPTER IX. — *Silicifications, flint and growth of crystals*   277

- **Chemical data: Silica in natural waters:** Silica is a mineral macro-molecule. — Geochemical consequences   279

- **Mineralogical data. Chalcedony and opal:** Chalcedony. — Opal. — Geological consequences   283

- **Geological and petrographic data:** Climatic silicification. — Flint, cherts and silicifications of sedimentary rocks   293

- **Silicification and growth of crystals:** Quartzification. — Silicification into chalcedony. — Silicification into opal. — Formation of silica gels in nature. — Importance of the factors of silicification 296
- **Conclusions** 300

CHAPTER X. — *Genesis of clay minerals: Inheritance and transformation* 302
- **Mechanical inheritance:** Clays inherited by weathering horizons. — Clays inherited by soils. — Clays inherited by sediments. — Inherited clays subjected to diagenesis. — The conditions of inheritance : stability and instability 303
- **Transformation:** Definition. — Degradation. — Aggradation 306
- **Conclusions: Inheritance and transformations by addition and subtraction** . . 322

CHAPTER XI. — *Genesis of clays: Neoformation and syntheses* 323
- **Neoformation in nature:** Neoformation in weathered products and soils. — Neoformation in sediments. — Neoformation in the domain of diagenesis. — Summary on the neoformation of clay minerals in nature 325
- **Laboratory syntheses:** Syntheses of clay minerals by CAILLÈRE, HÉNIN, ROBICHET and ESQUEVIN. — Syntheses by PÉDRO based on solutions from weathering. — Studies of gels and syntheses by FRIPIAT, GASTUCHE and DE KIMPE. — Synthesis of sepiolite and kaolinite by WEY and SIFFERT. — Syntheses at high temperature 336
- **Neoformations and syntheses by subtraction and addition: conclusions** . . . 347

CHAPTER XII. — *Superficial geochemistry and the silicate cycle* 357
- **Geochemistry of clays:** Places and mechanisms of the evolution of clays. — The two major pathways of surficial geochemical evolution of silicates. — Geochemical evolution by subtraction in an environment of lessivage. — Geochemical evolution by addition in confined environments 355
- **Geochemistry of the constituent elements of clays:** Alkaline ions: Na and K. — Alkaline-earth ions: Ca and Mg. — Iron and aluminum. — Aluminum and silicon. — Silicon and other cations in neoformations. — Disorder, order and crystal size 363
- **Geochemistry of landscapes and of sedimentary environments:** Landscapes of physical erosion. Detrital sedimentation. — Moderate tectonics. Moderate weathering. Detrital sedimentation with transformations. — Null tectonics. Strong weathering. Chemical sedimentation. — The reciprocal. Reconstruction of genetic environments 370
- **The silicate cycle:** Metamorphism of clay minerals. — Convergence and divergence of the geochemical cycles of elements 381

Bibliography 389

Index 427

CHAPTER I

THE CLAY MINERALS

Three excellent books about clay mineralogy have been written in recent years. They are the following:

— *Die silicatischen Tonminerale* by K. JASMUND. — First edition in 1951. Second edition in 1955.

— *X-ray Identification and Crystal Structure of Clay Minerals*, a work edited by C. W. BRINDLEY (1951) in collaboration with the best specialists. A new edition has been presented by G. BROWN (1961).

— *Clay Mineralogy*, the treatise of R. E. GRIM (1953).

A book on clay mineralogy, *Minéralogie des Argiles*, by Miss CAILLÈRE and Mr. HÉNIN, was published in French at the same time as the present volume ([1]). Accordingly, discussion of the methods of studying clay minerals and their mineralogy is not included in this book.

Since 1955 many studies and discussions at different meetings of the "Comité International pour l'Etude des Argiles" (CIPEA, now AIPEA) have been concerned with the nomenclature and classification of clay minerals. Several tables and propositions have been made by BROWN (1955), HÉNIN (1956), CAILLÈRE and HÉNIN (1956, in Mexico; 1959), MACKENZIE (1959), FRANK-KAMENETSKY (1960) and CAILLÈRE (1960). For reference the tables of CAILLÈRE and HÉNIN are presented at the end of this chapter (Tables I to VI).

The first chapter will be simply a schematic and pedagogical introduction to the mineral constituents of the clays we are going to study. This is necessary for two reasons.

First, I must present the language, the mineralogical vocabulary that I use, and the rudiments which are indispensable to the naturalist for understanding the history of these silicates.

Second, it is necessary to emphasize that our mineralogical knowledge evolves rapidly. The rationale presented in this book on clays is based on the identification of clay minerals. Even assuming this rationale is valid, it will be ephemeral if our ideas about clay minerals change. Therefore, it is essential to describe present-day concepts in this field.

([1]) Masson et Cie, publishers.

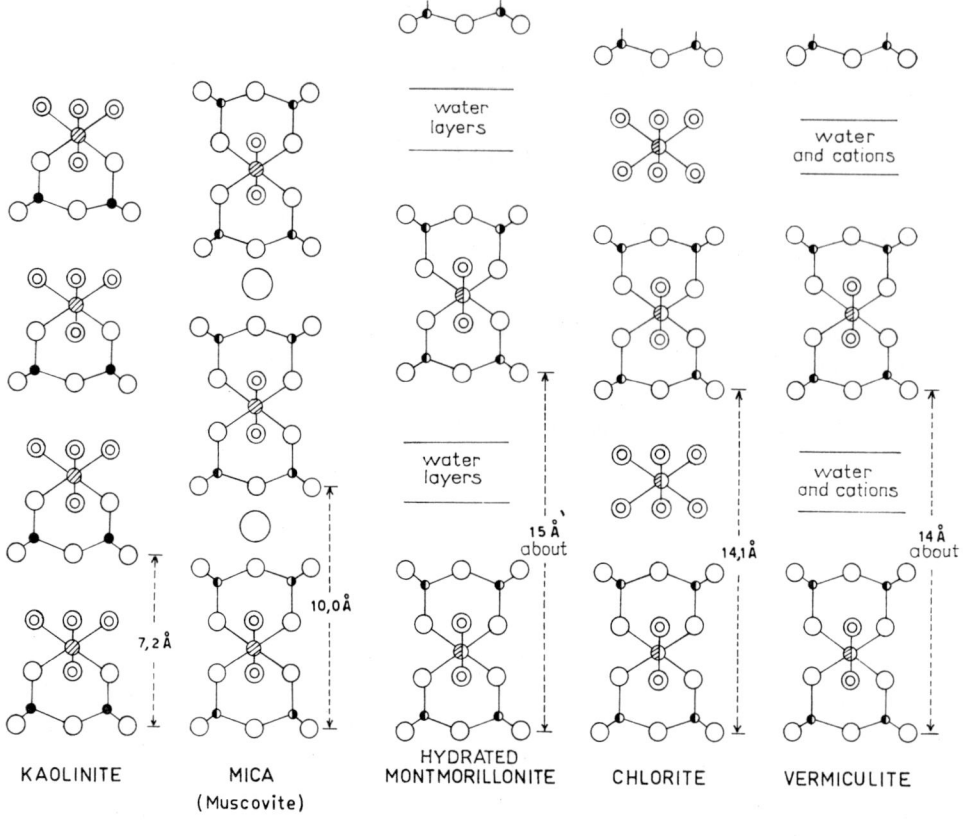

FIG. 1. — *Schematic representation of the structure of the principal clay minerals* (after BRINDLEY, 1951).

I. — KAOLINITE GROUP

1° *Kaolinite*

The term kaolin originates from the name of a Chinese hill where this product was extracted for centuries. It appears that the term kaolinite was first used in scientific vocabulary by JOHNSON and BLAKE (1867).

The structure of kaolinite was established by GRÜNER (1932) and revised by BRINDLEY and ROBINSON (1946). It is represented on Figure 1.

Like all the clay minerals, kaolinite is a phyllosilicate, i.e., a layer silicate. Every crystal flake is composed of a stacked arrangement of layers. The study of the individual layer defines the mineral.

Every layer is an association of two different sheets, named the tetrahedral sheet and the octahedral sheet. The tetrahedral sheet is so called because it is formed by the association of tetrahedra arranged in a plane. The four tips of the tetrahedra are occupied by oxygen ions and their center by a silicon ion which shares its four positive charges with the four oxygen ions of the tips. The octahedral sheet is composed of octahedra arranged in a plane. The six tips of the octahedra are occupied by oxygen ions or hydroxyl groups and their center by an aluminum ion. The association of a silica tetrahedral sheet with an alumina octahedral sheet forms one layer of kaolinite.

The structural formula of kaolinite is

$$(OH)_4 Al_2 Si_2 O_5.$$

This formula shows that the layer is electrically neutral; the fourteen positive charges corresponding to the aluminum and silicon ions are balanced by the fourteen negative ones corresponding to the oxygen ions and hydroxyl groups.

The regular stacking of these layers defines kaolinite, and the periodicity of this stacking is about 7 Å. X-ray diffraction patterns are characterized by strong 7.1 and 3.58 Å reflections, which disappear upon heating above 450° (LUCAS and JEHL, 1961).

Electron microscopy shows hexagonal crystal flakes generally well defined. Sometimes, however, the crystals are small and poorly formed, although the stacking is not disordered (OBERLIN et al., 1961).

2° *Dickite and nacrite*

The term dickite was proposed by ROSS and KERR (1931) to name what was previously called "Dick minerals". The word nacrite was introduced into the literature by BRONGNIART (1807) and confirmed by ROSS and KERR (1931). These minerals are rare and usually associated with hydrothermal deposits.

These two minerals have a chemical composition identical with that of kaolinite. They differ only in the structure of their lattice. Whereas kaolinite is identified as a triclinic mineral, dickite and nacrite are both monoclinic, although the latter is almost orthorhombic, with $\beta = 90°20' \pm 10'$.

3° **Halloysite**

The term halloysite was proposed by BERTHIER (1826). Halloysite is composed of kaolinite layers separated from each other by sheets of water (ROSS and KERR, 1934; MEHMEL, 1935; HENDRICKS, 1938; BRINDLEY et al., 1948). The basal spacing is about 10 Å, but by heating the water can be driven out, reducing the spacing to that of kaolinite at 7.2 Å. This dehydrated halloysite is called metahalloysite. In electron micrographs halloysite shows a tubular nature, owing to the rolling-up of layers. Deposits with halloysite are hydrothermal or in some instances sedimentary.

4° Disordered kaolinites (fireclay minerals)

Many natural kaolinites identified in refractory earths turn out to be composed of layers of kaolinite type, but the stacking of the layers is disordered. Shifts parallel to the b axis take place between the layers. Changes in the X-ray diffraction pattern brought about by this irregular stacking lead to the characterization of these structures.

At the London congress in 1948 several names for these minerals were proposed, such as "mellorite" by BRINDLEY and "livesite" by GRIMSHAW and ROBERTS. Some authors are still in favor of a mineral name, as ROBERTS (1958) is for livesite. Others think that the degree of disorder is too variable and that a group name is preferable. We hear commonly "fireclay minerals" or "fireclays", but BRINDLEY (in BROWN, 1961) not liking this rather unsuitable language suggests using simply "disordered kaolinites".

5° Serpentines and homeotypes of kaolinite

Definition. — The mineral in serpentines is a magnesian hydrosilicate whose structure is similar to that of kaolinite. Its chemical formula is inferred from that of kaolinite by a substitution of 3 Mg for 2 Al in the octahedral sheet. It is the trioctahedral magnesian homeotype of kaolinite. This requires some explanation.

Isomorphy. Isotypy. Homeotypy. — The substitution of 3 Mg for 2 Al is sometimes called isomorphous substitution. The term is well understandable but improper. In reality, the laws of isomorphism require not only the similarity of the crystal lattice but also the possibility of syncrystallization. Therefore, kaolinite and serpentine, or muscovite and phlogopite, are not isomorphous.

Isotypic substitution and isotypy have also been mentioned. Here again, some explanation is necessary. Isotypy (BILLIET, 1944; WINKLER, 1955) does not require the possibility of syncrystallization. On the contrary, it calls for the same group of symmetry, but the substitution of 3 Mg for 2 Al changes the symmetry. There remains a case of less exacting typy called homeotypy (WINKLER, 1955). So, if muscovite and paragonite are isotypes, muscovite and phlogopite, as well as kaolinite and serpentine, are homeotypes. Regarding the fine-grained clay minerals, the more prudent notion of homeotypy must be preferred to that of isotypy, since there still are important unknowns with respect to their crystal chemistry.

Dioctahedry. Trioctahedry (STEVENS, 1942-1945). — Kaolinite is characterized by 2 Al in its octahedral sheet; thus, it is called *dioctahedral*. In contrast, the mineral of serpentines contains 3 Mg in its octahedral sheet and it is called *trioctahedral*.

Variations in the serpentine minerals. — Serpentines show a great diversity of petrographic facies, which have been described by LACROIX (1893-1895). Minerals of serpentines have different habits as observable by electron microscopy and sometimes in macroscopic samples. The diversity of these habits corresponds to variations of order and disorder in the stacking of the type layer, which is the magnesian homeotype of kaolinite:

$$(OH)_4 Mg_3 Si_2 O_5.$$

At the present time there is a special name for each variety, such as chrysotile, ortho-serpentine, antigorite (BRINDLEY, in BROWN, 1961). The fibrous habit is called chrysotile. BRAGG and WARREN (1930) proposed a double chain structure, analogous to that of amphiboles. GRÜNER (1937) showed that it was difficult to explain chrysotile by such a structure and proposed a layer structure for antigorite. It was J. DE LAPPARENT (1938) who pointed out that chrysotile had a layer structure, but that the poor alignment of the magnesian sheet upon the silica sheet "twisted the structure during its growth like a paper ribbon gets twisted when coated with glue upon one of its faces". Actually, the tubular habit of chrysotile has been observed by electron microscope in 1950 (BATES, SAND and MINK), and the structure of chrysotile was established definitively by WARREN in 1942.

The other serpentine minerals, antigorites and ortho-serpentines, are composed of the same elementary layers, but with different degrees of disorder in their arrangement (ZUSSMAN, BRINDLEY and COMER, 1957).

Other homeotypes of kaolinite. — The ferrous homeotype of kaolinite is greenalite. The nickeliferous homeotypes (garnierite, noumeite, nepouite) are found in the serpentines of New Caledonia. Furthermore, substitutions can also occur in the tetrahedral sheet, which ceases to be entirely siliceous. Such are cronstedtite (ferrous and ferric) and *berthiérine* (alumino-ferrous). The latter mineral is called chamosite by Anglo-Saxons and will be studied in connection with the clay minerals of iron ores, in which it is frequent.

II. — MICA GROUP

1° *Micas*

The structure of micas was determined by MAUGUIN (1928) and PAULING (1930), and can be described as follows:

a) An octahedral sheet between two tetrahedral sheets;

b) The octahedral sheet contains varied hexa-coordinate ions which determine the diversity of micas;

c) Only three out of every four tetrahedra of the tetrahedral sheet are occupied by Si; the fourth tetrahedron being occupied by Al. This results in a deficit of charge for that sheet;

d) This deficit is filled by large ions, generally K, which occur between unit layers and strengthen their bonding. These layers connected by K ions cannot slip past each other. This bonding is weak enough, however, to allow an easy cleavage of micas.

The resulting structure is illustrated in figure 1. The basal spacing is about 10 Å, which is the basic periodicity of all the minerals, argillaceous or not, which are constructed according to the mica type.

The type of the dioctahedral series is muscovite:

$$(OH)_2 Al_2 Si_3 AlO_{10} . \\ | \\ K$$

The sodic equivalent, or isotype, of muscovite is paragonite:

$$(OH)_2 Al_2 Si_3 AlO_{10} . \\ | \\ Na$$

In many rocks imperfect varieties of muscovite are commonly observed by petrographers. These varieties have been named sericite, damourite, phengite, pinite, etc. These minerals occur in metamorphic schists (*schistes lustrés*, greenstone schists, mica schists) and hydrothermal deposits, as well as in alteration products of feldspars and many silicates. These minerals correspond to structures undergoing growth or reorganization, processes which result in chemical and structural disorders in comparison with well crystallized muscovite.

The type of the trioctahedral series is phlogopite:

$$(OH)_2 Mg_3 Si_3 AlO_{10} . \\ | \\ K$$

In biotite, the octahedral sheet contains Mg and Fe^{2+} and accessorily Fe^{3+} and Al:

$$(OH)_2 (Mg, Fe^{2+})_3 Si_3 AlO_{10} . \\ | \\ K$$

The varieties of chemical composition in the mica group have been studied (WINCHELL, 1927). The series of sericites, rich in aluminum and magnesium, and the series of dioctahedral micas were analysed according to crystallochemical laws by SCHALLER (1950) and FOSTER (1956).

2° *Pyrophyllite and talc*

Pyrophyllite and talc are also composed of layers at approximately 10 Å. The exact basal spacing, according to GRÜNER (1934), is 9.16 and 9.3 Å. There is no substitution of aluminum for tetrahedral silicon, and therefore no alkali between the layers. This explains the ease with which talc layers slip past each other. The schematic structural formulas are the following:

Pyrophyllite . . . $(OH)_2 Al_2 \; Si_4 O_{10}$

Talc . . . . . $(OH)_2 Mg_3 Si_4 O_{10}$

As far as I know, these minerals have been found only rarely in sediments. Pyrophyllite is reported three times by ROUGE (1954) from the basal Gothlandian in the northern Sahara (borehole of El Golea), by KULBICKI and ESQUEVIN (1961) in the basal Devonian of Ougarta, and finally by LAFOND (1961) in the alluvium of the Vilaine, in which it is detrital. This pyrophyllite is a diagenetic product within lower Devonian shales and sandstones in the Laval syncline (DUNOYER DE SEGONZAC and MILLOT, 1962). Talc is mentioned four times in the oolitic limestones of African Infracambrian (BIGOTTE, BONIFAS, MILLOT, 1957; MILLOT and PALAUSI, 1959), in the German Zechstein (FÜCHTBAUER and GOLDSCHMIDT, 1959), and in the modern sediments of the Florida coasts (WEAVER, 1959).

These minerals, which are rare in hydrosphere, occur in hydrothermal and even in metamorphic rocks, where their presence poses problems quite different from those concerning sedimentary rocks.

3° Illites

The term illite was proposed (GRIM, BRAY and BRADLEY, 1937) as a group name after the State of Illinois, U.S.A. Originally this word designated the whole group of clay minerals with a structure similar to that of micas.

Chemical characteristics. — The first chemical analyses have already shown that less than one silicon out of four was replaced by an aluminium ion, resulting in a proportionate diminution of alkali ions between layers. A partial replacement of the Al octahedral ions by Mg, Fe^{2+} and Fe^{3+} ions is frequent. It is still not known to what degree these substitutions are possible, all the more so since one is never certain about the purity of the clay fraction analysed.

Structural formula and demonstration of disorder in illites. — The structural formula is a variable one, of the following type:

$$(OH)_2 Al_2 Si_{4-x} Al_x O_{10} \cdot \underset{K_x}{|} \qquad 0.5 < x < 0.75$$

In fact, it is difficult to determine the structural formula of illites (or hydromicas) from the results of chemical analyses (BROWN and NORRISH, 1952).

One method of calculating the formula involves neglecting the amount of water, the removal of which is provoked by heating above 105 °C. The calculation is based on the ten theoretical oxygens and the two theoretical hydroxyl ions. A formula is obtained wherein cations are apportioned between these twenty two negative charges, but there is no assurance that hydroxyls are not more numerous. This process is used by HENDRICKS and ROSS (1941) and by GRIM and BRADLEY (in BRADLEY, 1951).

Another process, used by BRAMMAL, LEECH and BANNISTER (1937), consists of taking into account the removal of water above 105 °C. This results in a formula with less than ten oxygens and more than two hydroxyls, the total of the negative charges remaining equal to twenty two. This formula is rather different from that of an ideal mica.

BROWN and NORRISH (1952) showed that the hypothesis of replacement of the interlayer

K ions by H_3O^+ oxonium ions resulted in more satisfying structural formulas, such as those previously assumed. We arrive at a very general formula of following type:

$$(OH)_2 (Al, Ti, Mg, Fe)_2 (Si_{4-x}Al_x)_{10}$$
$$|$$
$$(Ca, Na, K, H_3O)_x .$$

Presently illites are considered on the whole to be fine-grained micas in which potassium ions show a deficit and are replaced by water. This water is detected by chemical analyses, and some authors consider that it occurs in the form of oxonium or hydronium ions. According to the degree of the deficiency, layers are no longer as completely bound as in a theoretical mica, and the expansion capabilities correspond to the degree of hydration. There is a discrete disordered interstratification according to a mechanism which will be described below.

Illites are the most common clay minerals in nature. These small micas no longer have or do not yet have their interlayer spacings saturated by K ions, a condition which disturbs the arrangement of their octahedral and tetrahedral sheets in comparison with the ideal mica. Chapters XI and XII will discuss the mechanisms of transformation which explain this beginning of disorder, of opening and of interstratification in the structure of micas of clay size, i.e., smaller than 2 μ.

Isotypes. — If potassium between the layers of illite is largely replaced by sodium, we have brammalite (BANNISTER, 1943). If octahedral aluminum is largely replaced by ferric iron, the mineral is called glauconite.

History of nomenclature. — In the past these minerals structured according to the muscovite model were anticipated and they received diverse names. Worthy of mention are bravaisite (MALLARD, 1878), hydromuscovite (JOHNSTONE, 1889), hydromica (GALPIN, 1912), *potash-bearing clay* (ROSS and KERR, 1931), *sericite-like mineral* (GRIM, 1935), and *Glimmerton* (ENDELL, HOFMANN and MÄGDEFRAU, 1935).

GRIM, BRAY and BRADLEY (1937) wondered whether it was necessary to create the new term illite in place of the previous names. Hydromuscovite and hydromica seem to relate the difference only to the water content; "sericite-like mineral" is vague in as much as sericite itself is vague; *Glimmerton* is very significant, but it has not been used.

Bravaisite, defined as early as 1878 by MALLARD, would have considerable precedence. Such was the opinion of several authors, especially FLEISHER (1943), ROSS (1945) and RIVIÈRE (1946). However, GRIM and ROWLAND (1942), NAGELSCHMIDT (1944) and BRADLEY (1945) proved that bravaisite was a mixed-layer mineral of illite and montmorillonite. GRIM and BRADLEY (1948) insisted upon this point and redefined the illite family.

The term illite, proposed by GRIM, BRAY and BRADLEY, is a convenient name for the population of clay minerals structured according to the mica type. Here we are concerned with the clay minerals most commonly found in sedimentary rocks and soils; it is necessary to try to understand their origin and evolution.

4° Polytypes of muscovite

In a population as large as that of the illites every criterion for discriminating varieties is a very important help in analysis. Recent works have led to the definition and study of two types of variations: interstratifications, which will be considered later on, and the polytype varieties. These latter variations have been studied in the muscovite group, and they are also found among the muscovites of clay size, the illites.

There are several polytype varieties of muscovite, and they have been described and enumerated by SMITH and YODER (1956). They are distinguishable from each other by the mode of stacking of their component sheets. Polytypism is induced by a displacement of the oxygen sheets in the octahedral zone by an amount equal to $a/3$. Because of the hexagonal symmetry of the planar network at this level, the displacement can occur in six directions; thus six polytypes are theoretically possible. In fact, only three polytypes of muscovite have been found in nature. They are represented by a numeral which gives the number of layers (1, 2, or 3) in the unit cell and by a letter indicating the crystal system: M (monoclinic) or T (triclinic).

These three polytypes are the following: 1 M, 2 M_1 (the subscript indicates one of the two possibilities of the 2 M type) and 3 T, the latter being much rarer than the first two. Certain authors immediately thought they could prove that these varieties in the illite group had a genetic implication: the 2 M types being detrital and the 1 M ones newly formed. This is premature and must be preceded by a serious inventory.

When these polytype variations are taken into account, as well as the great richness of interstratifications, it can be seen that the systematics of this group become difficult and complex. But there is a good chance that it will allow us to understand better the story of the micaceous minerals in nature.

5° Glauconite and "glauconie"

The term *glauconie* was proposed by BRONGNIART (1823) as a result of the chemical analyses of BERTHIER (1820-1821). The word glauconite was proposed by KEFERSTEIN (1828).

The mineralogical structure was established by GRÜNER (1935) who showed it similar to that of micas. MÄGDEFRAU and HOFMANN (1937) pointed out its relationship with sericite. On the basis of many analyses HENDRICKS and ROSS (1941) presented a structural formula showing that there was no difference between glauconite and celadonite, which occurs in the weathered products of basic lavas. The structural formula is close to that of illite:

$$(OH)_2 (Fe^{3+}, Al, Fe^{2+}, Mg)_2 (Si_{4-x} Al_x) O_{10} \cdot \underset{K_x}{|} \qquad 0.5 < x < 1$$

The average chemical formula obtained by HENDRICKS and ROSS (1941) is:

$$(OH)_2 (K, Ca_{1/2}, Na)_{0.84} (Al_{0.47}, Fe^{3+}_{0.97}, Fe^{2+}_{0.19}, Mg_{0.40}) (Si_{3.65} Al_{0.35}) O_{10} \cdot$$

SABATIER (1949) took up the mineralogical study again, examined the behavior of the water by differential thermal analysis and showed by X-ray analysis that there was disorder in the stacking of the layers. BURST (1956-1958) pointed out the variety of glauconies and that they are not all constituted like the mineral of GRÜNER. He distinguishes four categories. Utilizing the mica polytypes of YODER and EUGSTER (1954) as the first two, these four categories are:

a) The first group corresponds to the dioctahedral layer mineral of GRÜNER (1935) called glauconite. The reflections are sharp and symmetrical; it is a regular 1 M structure;

b) The second group is also monomineralic and micaceous, but the peaks are low and asymmetrical. The peak falls off hesitatingly and slowly on the side of the large diffraction angles. The study of the X-ray patterns shows that it is a 1 Md structure (the monoclinic stacking is disordered, thus d);

c) The third group is composed of mixed layers in which montmorillonite alternates in a rather random manner with a 10 Å mineral;

d) The fourth group presents mixtures of minerals among which BURST determined in different cases illite, montmorillonite, chlorite or varied mixed layers.

JUNG (1954) and KELLER (1958) presented evidence for ferriferous illites or glauconitic micas which lie half-way between illite and *glauconie* with respect to the ferric iron content.

Thus, it is obvious that the *glauconies* represent a veritable population; the studies of petrography and genesis must take that into account. Otherwise, our nomenclature is ambiguous. The word *glauconie* in French represents thus in some cases a well defined clay mineral, the ferric homeotype of illite, and in other situations products of mixed or varied mineralogical nature, which have in common a green color. Accordingly I propose to distinguish three terms, *glauconie*, glauconite and glauconitite.

— **Glauconie** will be useful to geologists; it will designate the green products in the form of grains, trails or accordion-folds, mineralogical study having to be done. *Glauconie* would be a facies name;

— **Glauconite,** as defined by mineralogists, would correspond to the ferric mineral, a homeotype of illite;

— And **glauconitite,** of use to petrographers, would designate a rock whose cardinal and predominant mineral would be glauconite.

III. — THE MONTMORILLONITE GROUP

The term montmorillonite was proposed by DAMOUR and SALVETAT (1847) and was named after Montmorillon (Vienne, France). Montmorillonites are minerals close to micas, but the bonds between the layers are weakened. Thus water in variable quantity can enter between the unit layers causing the stacking periodicity to have a variable value, in many cases close to 14 Å. The permutation of exchangeable ions, heat treatment, and the action of polyalcohols cause this periodicity to vary between 10 to 20 Å. Such reproducible reactions form the basis of modern determinative methods.

The structure of montmorillonite was established by HOFMANN, ENDELL and WILM (1933). The study of the varieties within this group dates from the fundamental work of ROSS and HENDRICKS (1943-1944); these varieties, based on the substitution of ions in the structure, are very numerous (see fig. 1).

Dioctahedral "montmorillonite-beidellite" series. — At the "montmorillonite" pole of this series, there is no substitution by aluminum for tetrahedral silicon, but a partial replacement of octahedral aluminum by magnesium. This results in a charge deficit, that is balanced by exchangeable interlayer ions. Any kind of cation may occur in the interlayer position, but it is frequently Na or Ca under natural conditions. An example of montmorillonite:

$$(OH)_2 (Al_{1.67} Mg_{0.33}) Si_4 O_{10} \cdot \underset{Na_{0.33}}{|}$$

At the "beidellite" pole, the replacement of silicon by aluminum occurs in the tetrahedral sheets. An example of beidellite:

$$(OH)_2 (Al_{1.46}, Fe^{3+}_{0.50}, Mg_{0.04}) (Si_{3.64} Al_{0.36}) O_{10} \cdot \underset{Na_{0.36}}{|}$$

All the intermediate types exist between the two poles of this series.

The dioctahedral "nontronite-beidellite" series. — Ferric iron can replace aluminum in a sensitive manner in all the stages of the previous series. Thus we arrive at the "nontronite-beidellite" series, which comprises the ferric homeotypes of the "montmorillonite-beidellite" series. An example of structural formula:

$$(OH)_2 (Fe^{3+}_{1.61}, Al_{0.39}) (Si_{3.67} Al_{0.33}) O_{10} \cdot \underset{K_{0.33}}{|}$$

The trioctahedral "stevensite-saponite" series. — Stevensite is the magnesian homeotype of montmorillonite, such as talc is for pyrophyllite. The theoretical formula would then be:

$$(OH)_2 Mg_3 (Si_4 O_{10}) \cdot$$

But substitutions in the octahedral and tetrahedral sheets can occur. The partial replacement of Mg by Li gives hectorite, and the replacement of Si by Al results in saponite. Ghassoulite, which I had defined (MILLOT, 1954) as the magnesian pole of the series being unaware of the existence of stevensite, is in fact intermediate between stevensite and hectorite because of a small amount of Li (FAUST, HATHAWAY and MILLOT, 1959).

The formulas of the minerals just mentioned are the following:

Stevensite $\quad (OH)_2 Mg_{2.92} (Si_4 O_{10}) \cdot Na_{0.16}$

Saponite $\quad (OH)_2 Mg_3 (Si_{3.67} Al_{0.33} O_{10}) \cdot Na_{0.33}$

Hectorite $\quad (OH)_2 Mg_{2.67} Li_{0.33} (Si_4 O_{10}) \cdot Na_{0.33}$

Ghassoulite (Faust, Hathaway and Millot, 1959)

$$(OH, F)_2 (Mg_{2.57}, Fe^{2+}_{0.05}, Al_{0.08}, Li_{0.10}) (Si_{3.98} Al_{0.02} O_{10}) \cdot \left[Na_{0.13}, K_{0.04} \left(\frac{Ca}{2}\right)_{0.02}, \left(\frac{Mg}{2}\right)_{0.12} \right]$$

IV. — THE CHLORITE GROUP

The term chlorite was used by Werner in the last century to designate green foliated minerals rich in ferrous iron. There exists a great variety of these minerals because of the numerous homeotype substitutions which can occur in their structure. Chlorites have been well known for a long time in crystalline schists, in hydrothermal rocks and in the alteration products of many silicates. Among the clay minerals their identification has been much more recent; they have been identified in modern sedimentary deposits (Correns, 1937; Grim et al., 1949), in soils (MacEwan, 1948) and in sedimentary rocks (Millot, 1949; Eckhardt, 1958). Some sedimentary deposits are even composed principally of chlorite (Millot, 1954).

True chlorites. — Mineralogical knowledge of this group was acquired through the works of Orcel (1927), Mauguin (1930), Pauling (1930), MacMurchy (1934) and with greater precision through the works of Robinson and Brindley (1949). The structure (fig. 1) consists of alternate sheets of trioctahedral mica bonded by sheets with a brucite structure. The basal spacing is about 14 Å, but here, in contrast to the montmorillonites, it is fixed. The schematic formula is:

$$(OH)_2 (Mg_{3-y}, Al_y) (Si_{4-x} Al_x) O_{10} \cdot (Mg_{3-z}, Al_z) (OH)_6$$

However, substitutions in the tetrahedral and micaceous octahedral sheets, as well as in the brucitic sheet, are extremely varied. Based on the quality and quantity of these substitutions, a classification of many varieties can be set up. Some such classifications are presented by Orcel, Caillère and Hénin (1950), and by Brindley and Gillery (1956).

Chlorites of iron ores (7 Å minerals). — The problem of the iron ore chlorites will be examined in detail later with regard to the study of these ore deposits. However, it can be said here that, in comparison to the true chlorites, iron ores contain many foliated minerals which looked like chlorites under the microscope, but which turned out to be 7 Å clay minerals (BRINDLEY, 1949; ORCEL, HÉNIN and CAILLÈRE, 1949; HARDER, 1951). They belong to the kaolinite group, but are close to serpentine. The French call them berthierines and the Anglo-Saxons chamosites.

Expanding chlorites. — A mineral that was able to expand like a montmorillonite when treated with polyalcohols but remained stable after heating was described from the marls of the English Keuper by HONEYBORNE (1951) and by STEPHEN and MACEWAN (1951).

This mineral is considered to be a chlorite in which the incomplete brucitic sheet forms pillars between the mica layers. Thus the ionic attraction force between the mica sheets is weak enough to allow the entrance of two sheets of glycol or glycerol, whereas the pillars resist the collapse of the structure upon heating (BRINDLEY in BROWN, 1961).

These expanding chlorites were also found in iron ores (CAILLÈRE and HÉNIN, 1952), and they occur either as distinct individuals or interstratified with true chlorites in Triassic marls (LIPPMANN, 1954; MARTIN VIVALDI and MACEWAN, 1960; LUCAS and BRONNER, 1961).

Dioctahedral chlorites. — BRINDLEY and GILLERY (1956) predicted the dioctahedral chlorites in their classification. Their stacking is based on a micaceous layer of the muscovite type rather than a layer of the phlogopite type. The following structure is obtained:

$$(OH)_2 (Al_{2-y}, Mg_y) (Si_{4-x}Al_x) O_{10} .$$
$$(Mg_{3-z}, Al_z) (OH)_6$$

During the transformation of these minerals, such as we see in nature, the occurrence of such products appears very likely in soils (BROWN and JACKSON, 1958) and in sediments (CAILLÈRE, HÉNIN and POBEGUIN, 1962).

V. — VERMICULITE GROUP

The mineral vermiculite. — Vermiculite is a mineral resembling mica that, when heated, has an appearance of little worms. It has been considered for a long time as a variety of trioctahedral mica. GRÜNER (1934) determined its structure and showed it to be composed of mica layers separated by water molecules. HENDRICKS and JEFFERSON (1938) point out that the structure (fig. 1) is not neutral, and that it is balanced by interlayer ions, most commonly Mg and sometimes Ca. The structural formula is:

$$(OH)_2 (Mg, Fe)_3 (Si_{4-x} Al_x) O_{10}$$

$$(Mg, Ca)_{x/2}.$$

The study of the behavior of vermiculite in X-ray diffraction was carried out in minute detail by BARSHAD (1948, 1949, 1950) and by WALKER (1949, 1951, 1957, 1958). Vermiculites have a basal spacing of 14 Å, but this spacing is not fixed. The permutation of interlayer cations makes this basal spacing vary from 14 to 10 Å, according to the cations involved: about 14 Å for Mg and Ca; about 12 Å for Ba, Li and Na; and about 10 Å for NH_4, K, Rb, Cs. Thus, the action of polyalcohols, glycerol and ethyleneglycol differs according to samples and to interlayer ions present. The most constant criterion for identification was sought by WALKER (1957-1958); it can be said that the great majority of vermiculites treated with Mg yield a basal spacing of 14.3 Å under the action of glycerol. This is the primary distinction in contrast to the minerals of the montmorillonite group.

The vermiculite of clays. — All the previous results were obtained on mineral samples of large size, well characterized as trioctahedral vermiculites according to the definition of GRÜNER. It is now necessary to see what occurs in the domain of small sizes, i.e., that of clays. More than ten years ago minerals with characteristics resembling those of vermiculite were discovered to be common or predominant constituents in natural clays. They were first found in soils by MACEWAN (1948), JACKSON et al. (1952), BROWN (1953), HATHAWAY (1955), and then by very many authors. They were discovered in sedimentary rocks by MILLOT (1949), and after that in a great number of sediments and sedimentary rocks.

What exactly are these vermiculites in clays? They are defined as such by experimentalists only because of their behavior in X-ray diffraction. They are 14 Å minerals which are not chlorites because their basal spacing collapses to 10 Å after heating, and they are not montmorillonites because they do not expand beyond about 14 Å upon saturation by Mg and the action of glycerol. Chemical study is not possible because these minerals occur in mixtures. But X-ray analysis shows that the vermiculite of soils and sediments is dioctahedral as well as trioctahedral (BROWN, 1953; HATHAWAY, 1955; WALKER, 1957; MARTIN VIVALDI and SANCHEZ CAMAZANO, 1961), contrary to macroscopic vermiculites.

Consequently we face here an extension of language. The "clay vermiculites" are dioctahedral or trioctahedral micaceous structures in which bonding between the layers is weaker than in micas and chlorites, but stronger than in montmorillonites. It is now certain that all the intermediate forms occur. Today, as well as in 1949, it is my opinion that we are arbitrarily separating under different names some structures which pass by gradual transition into others. All depends on the degree of freedom realized between layers of the mica type.

VI. — MIXED-LAYER CLAY MINERALS

Mixed-layer clay minerals. — Mixed-layer clay minerals are clay minerals in which different kinds of layers alternate with each other.

The existence of such minerals was discovered by GRÜNER (1934) and HENDRICKS and JEFFERSON (1938) in connection with the study of vermiculite, as well as by NAGELSCHMIDT (1944) working on bravaisite (illite-montmorillonite mixed-layer). Since then many studies on these mixed structures were carried out mainly in three directions:

a) The theoretical study of the problem and especially the calculation of the effects of the interstratification on X-ray diffraction. The main works are those of HENDRICKS and TELLER (1942), BRADLEY (1945), MERING (1949), MACEWAN (1949, 1956) and MACEWAN, RUIZ AMIL and BROWN (in BROWN, 1961);

b) The improvement of the practical methods for the recognition and the inventory of the possible kinds of interstratification. These are the fruits of the labor of many authors, including those given above and BRINDLEY (1951), BRADLEY (1953) and BRADLEY and WEAVER (1956);

c) The distribution of mixed-layer minerals in soils, weathering products, sedimentary rocks, and hydrothermal clays, aspects which will be studied throughout the chapters of this book.

The types of interstratification. — MACEWAN (1949) showed that mixed-layer structures are of three types:

a) Regular. — Layers of different types, called for instance A and B, alternate according to a specific law. The most simple law will be of the form ABABAB... Some such regular mixed layers have been given mineral names;

b) Irregular. — No law governs the alternation, the layers being randomly interstratified;

c) Segregation of alternating packets, being themselves in some cases interstratified. It is a very complex case, difficult to study.

The main mixed-layer minerals. — Theoretically, all kinds of layers can interstratify with each other. Accordingly, BRINDLEY and GILLERY (1953) described a chlorite-kaolinite mixed layer and BRADLEY (1950) a regular vermiculite-pyrophyllite mixed layer, which was called "rectorite." Likewise, many mixed layers including biotite are well known (RUTHRUFF, 1941), such as biotite-chlorite or biotite-vermiculite which occurs in weathering products and is called "hydrobiotite". But biotite is fragile in hydrosphere, so that in soils and sediments the majority of layers are of the four following types: illite, montmorillonite, vermiculite, chlorite. These four types of layers can, indeed, interstratify simultaneously, but at the present time we cannot analyze such a case correctly. Mixed layers involving three or four constituents are being studied by several authors such as VANDERSTAPPEN and CORNIL (1958) and

JONAS and BROWN (1959), but interstratifications of two types of layers are by far the most common and retain the attention of many present-day investigators.

By interstratification of illite, chlorite, montmorillonite, and vermiculite layers two by two, the following six types of interstratification are obtained:

Illite-montmorillonite:	I-M	Montmorillonite-vermiculite:	M-V
Illite-vermiculite:	I-V	Montmorillonite-chlorite:	M-C
Illite-chlorite:	I-C	Vermiculite-chlorite:	V-C

If the existence of swelling chlorite layers is taken into account, as considered by MARTIN VIVALDI and MACEWAN (1960), as well as that of 12 Å montmorillonite, which is rather common in nature, the varieties are multiplied. Each case will be studied.

Specific names were given to some mixed layers. In addition to the names rectorite and hydrobiotite noted above, the term bravaisite (MALLARD, 1878) can be mentioned. It has been shown that bravaisite, which seems to have priority over illite, was an illite-montmorillonite mixed layer. The name corrensite was similarly proposed by LIPPMANN (1954) for the mixed layer identified by him as a chlorite-montmorillonite mixed layer (1959) and by BRADLEY and WEAVER (1956) as a chlorite-vermiculite mixed layer. MARTIN VIVALDI and MACEWAN (1960) prefer to keep the name chlorite-swelling chlorite. However, it seems that such specific names are no more convenient than the identification of the component layers. The variety of the stackings and the variety of the proportions and types of stacking would yield an inextricable nomenclature. However, it is agreed (BROWN, 1955) that the only regular mixed layers should receive names.

Identification of mixed-layer minerals. — The complete identification of mixed-layer clay minerals would necessitate several steps:

— Identification of the type of the two associated layers;

— Determination of the regularity of interstratification;

— Determination of the proportion of each of the two constituents;

— Determination of the manner of interstratification: AB AB AB or ABB ABB ABB...

Such efforts encounter theoretical and practical difficulties, almost insurmountable in concrete cases. In everyday work one must be satisfied for the time being to determine the types of interstratified layers and to estimate the regularity of stacking and the approximate proportion of the two components.

The method of identification is based on X-ray diffraction analysis in the following way:

a) The three usual tests are carried out by X-ray diffraction on well oriented flakes of clay:

Test of the natural or normal clay	N
Test after heating to 490°	Ch
Test after action of glycerol on the flake . . .	G

b) The interpretation is based on the basal spacings obtained for the single minerals. These spacings, given in Å, are the following:

	N (normal)	Ch (heating)	G (glycerol)
Illite	10	10	10
Chlorite	14	14	14
Montmorillonite	14	10	17.7
Vermiculite	14	10	14

c) If these single minerals were mechanically mixed as separate particles, the patterns would give the diffraction reflections of the two minerals present at the same time.

d) Regular mixed layers do not behave as mixtures, but as true minerals whose basal spacing is equal to the sum of those of the two interstratified layers. A new periodicity is introduced. For instance, corrensite, the chlorite-montmorillonite mixed

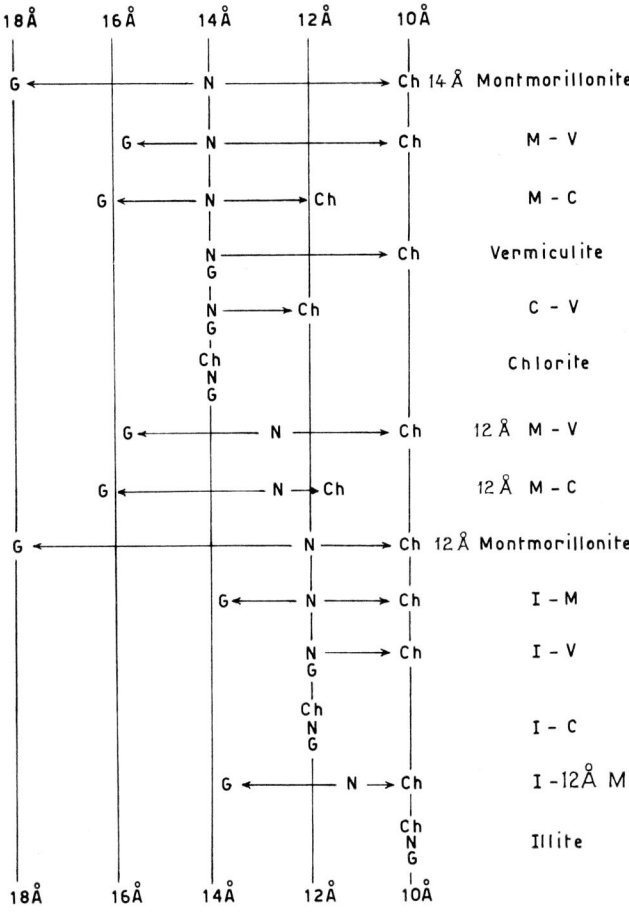

FIG. 2. — *Schematic representation of the behavior of the reflections corresponding to basal spacing during X-ray diffraction after different treatments* (after LUCAS, CAMEZ and MILLOT, 1959).
G: Treatment with glycerol; N: Natural sample; Ch: Heating at 490°.

layer which occurs commonly in sediments shows a basal spacing equal to $14 + 14 = 28$ Å. The sequences of the diffraction reflection are regular, as follows:

Normal	28	14	9.3	7	
Heating	24	12		8	6
Glycerol	32	16	10.6	8	5.3

e) The irregular mixed layers have no periodicity in their lattice. Their sequence is not regular. In fact, if widely separated reflections of each of the two component minerals are not affected, the reflections which are close to each other do interfere. A single reflection is obtained by interference of very close diffracted rays; this is called a maximum. Every structure shows a succession of maxima peculiar to it, resulting in a non-linear series of reflections. A great number of examples have been given by specialists (MacEwan, Ruiz Amil and Brown, in Brown, 1961).

f) A first approach to the identification of mixed-layer structures is thus possible. As a result, guides to identification have been set up such as those tables presented by Lucas, Camez and Millot (1959), Martin Vivaldi and MacEwan (1960) and Warshaw and Roy (1961). The simplified table of Lucas, Camez and Millot is presented in figure 2.

VII. — ATTAPULGITE, SEPIOLITE AND PALYGORSKITE GROUP

Attapulgites. — These minerals have been well known for a long time as constituents of asbestos board, mountain wood or mountain cork. As early as 1807 Brongniart classified them under the name of *asbestes subériformes*. Later they were recognized in fibrous bundles in the asbestos facies, then in earthy masses, or in lacustrine or marine sedimentary layers, outcrops of which have a foliated or papyraceous aspect.

Attapulgites are alumino-magnesian silicates in which aluminum and magnesium are present in nearly equal proportions. One characteristic is their fibrous habit, sometimes visible to the naked eye in the attapulgites of the cardboard-like or asbestiform facies. Electron microscopy shows that attapulgites of sedimentary rocks also occur in long fibers.

Jacques de Lapparent (1935) named "attapulgite" the constituent of fullers' earths of Attapulgus, Georgia (U.S.A.) and of Mormoiron (France). Longchambon (1936) showed that these attapulgites belonged to Fersmann's palygorskite series.

The main diffraction reflections of attapulgite are the following:

10.5; 6.44; 5.42; 4.48; 3.24; 2.15

Sepiolites. — Sepiolites have also been well known for a long time and were first named "Meerschaum" (sea foam) by Werner in 1789. Afterwards the term magnesite was used, but it was reserved later in some languages for magnesium carbonate. The term sepiolite is used by everyone today. Like attapulgites, these minerals occur either with the fibrous facies in mountain woods, or in hydrothermal veins, or with the earthy facies in the lacustrine or marine sedimentary rocks. The electron microscope (micrograph of MATHIEU-SICAUD, in MILLOT, 1949) has shown that the fibrous *habit* persists in the sedimentary sepiolites with an earthy aspect (fig. 3).

Sepiolite is a magnesian silicate in which aluminum plays an often very modest role; it constitutes the magnesian pole of FERSMANN's palygorskite series.

FIG. 3. — *Electron photomicrograph of particles 1 μ and smaller of sepiolite extracted from the argillite of Salinelles, Gard, France* (MILLOT, 1949; photo by MATHIEU-SICAUD).

The main diffraction reflections of sepiolites are the following:

12.10;   7.6;   5.05;   4.5;   4.30;   3.75;   3.33

Nomenclatural problems: Fersmann's palygorskite series. — FERSMANN (1908, 1913) defined the palygorskite series as an isomorphous series between a known magnesian pole, sepiolite, and an unknown aluminous pole, which he called para-montmorillonite. Palygorskite appears thus to be the name for a group which includes, between its two poles, intermediate members, such as lassalite and pilolite, the usage of which has been lost.

As a matter of fact, modern mineralogical work shows that this continuous series between the magnesian pole and the aluminous pole does not exist. What does exist are two types of minerals with different structures; the first one, alumino-magnesian with various isotypic substitutions, is attapulgite; the second one, principally magnesian, is sepiolite.

Concerning the choice of nomenclature, two positions are possible: One can consider that attapulgites represent all the essential variety of FERSMANN's palygorskite

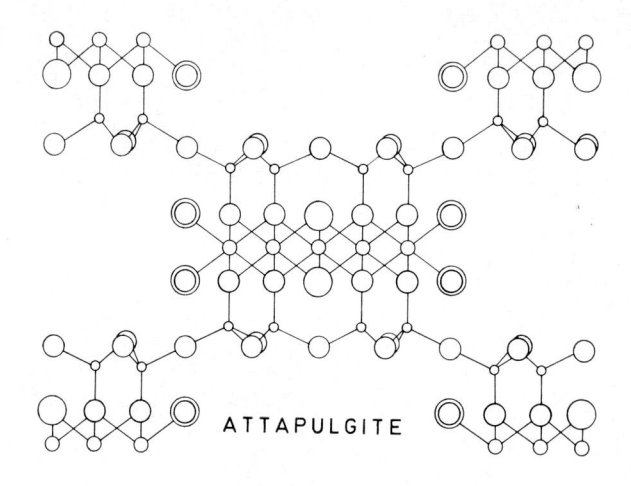

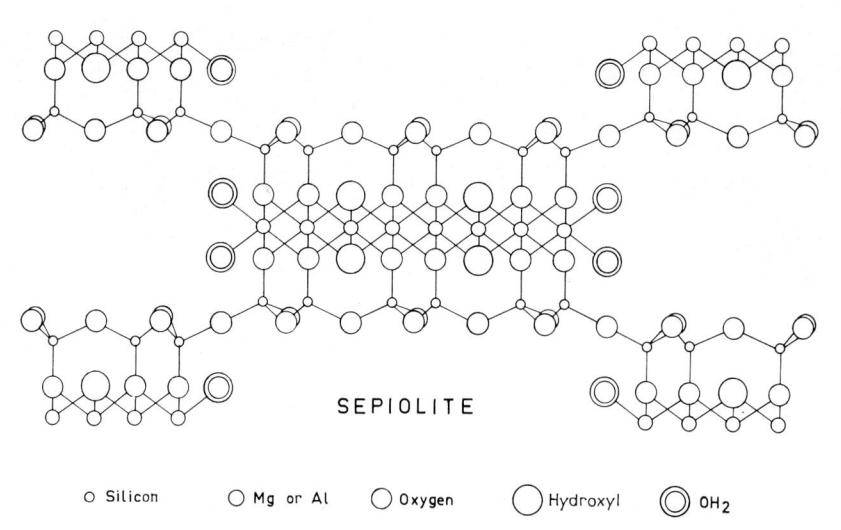

FIG. 4. — *Schematic structure of sepiolite* (after BRAUNER and PREISINGER, 1956) *and of attapulgite* (after BRADLEY, 1940).

series; in this case, the term palygorskite has priority over the term attapulgite. Or it can be considered that the term palygorskite designates an ensemble in which are found the formerly known sepiolites as well as the attapulgites of J. DE LAPPARENT, in which case palygorskite remains as a family name covering the two groups, whose structures are, moreover, not identical. The latter attitude will be chosen here.

Attapulgite and sepiolite structures. — For sepiolite LONGCHAMBON proposed in 1937 an amphibole chain structure of Si_4O_{11}. Taking account of the fibrous development, J. DE LAPPARENT (1938) considered that the fibrous mineral facies does not require the amphibole structure, as he had shown for chrysotile, and he proposed a layered structure of Si_4O_{10}. For attapulgite BRADLEY (1940) suggested a structure of long hollow bricks; later an analogous structure for sepiolite was proposed (NAGY and BRADLEY, 1955). Further improvements have been put forward by BRAUNER and PREISINGER (1956) and BRINDLEY (1959).

The outline of these different structures is presented on figure 4. It is not a question of fibers, but of laths elongated parallel to the c axis. These laths are composed of elementary alternating ribbons, every ribbon having a mica structure. These ribbons comprise five octahedral positions in the case of attapulgite (BRADLEY, 1940) and eight in the case of sepiolite (BRAUNER and PREISINGER, 1956). In the channels that remain open between the ribbons water molecules with a zeolitic behavior can enter.

The proposed structural formulas are the following:

— Attapulgite BRADLEY (1940):

$$(OH_2)_4 (OH)_2 Mg_5 Si_8 O_{20} \; 4 H_2O$$

with the possibility of replacement of Mg by Al.

— Sepiolite NAGY and BRADLEY (1955):

$$(OH_2)_4 (OH)_6 Mg_9 Si_{12} O_{30} \; 6 H_2O$$

— Sepiolite BRAUNER and PREISINGER (1956):

$$(OH_2)_4 (OH)_4 Mg_8 Si_{12} O_{30} \; 8 H_2O$$

Fibrous facies and chemical composition. — Attapulgites and sepiolites have chemical compositions rather close to those of some members of the montmorillonite family. However, their facies is very special because of their development in long rods. It has been already pointed out that magnesian montmorillonites of hectorite type have a facies which develops broad, flat ribbons. It is an important question why the habit of these alumino-magnesian silicates is sometimes lamellar, sometimes fibrous.

MARTIN VIVALDI and CANO RUIZ (1956) have studied this question. They concluded that attapulgites and sepiolites were intermediate between the dioctahedral and trioctahedral series. The dioctahedral montmorillonites develop planar lattices as does the trioctahedral series of saponite. But attapulgites and sepiolites would be in an intermediate state in which the octahedral positions are neither completely occupied (3/3) as in saponite, nor occupied in a 2/3 ratio as in the dioctahedral series. This intermediate situation corresponds to the lath facies and is probably the basis for it.

The proposal is interesting and would explain how crystals in process of neoformation assume an elongate or lamellar facies, according to the composition of the growth environment.

VIII. — IRON AND ALUMINUM OXIDES AND HYDROXIDES

Iron and aluminum oxides and hydroxides occur in a great number of weathering products, in soils, and even in rocks rich in clay. Thus it is necessary to be on the alert. Knowledge of these minerals will be particularly useful in the study of laterites and bauxites.

Natural iron hydroxides and oxides. — The works of GAUBERT (1925), BÖHM (1928) and GOLDSZTAUB (1935) have increased our knowledge of natural iron hydroxides and oxides considerably, through the use of X-rays.

Previous descriptions noted a great number of products, which received the names limonite, turgite, stilpnosiderite, xanthosiderite, hydrohematite, hydrogoethite, and so on. It was natural since chemical analysis could furnish only varied percentages of iron oxide and water content, leading to this variety of proposals.

Today, it must be considered that there are only four mineral species, besides magnetite:

2 hydroxides : goethite (α FeOOH) and lepidocrocite (γ FeOOH)
and 2 oxides : hematite (α Fe$_2$O$_3$) and maghemite (γ Fe$_2$O$_3$)

Criteria for the discrimination of these species can be found in treatises, especially in those which treat clay minerals, such as those of BRINDLEY (1951) and GRIM (1953). X-ray diffraction, differential thermal analysis and thermomagnetic analysis are utilized.

The main diffraction reflections are the following:

Goethite (α FeOOH):	4.15;	2.67;	2.43;	1.70	
Lepidocrocite (γ FeOOH):	6.25;	3.28;	2.46;	1.93	
Hematite (α Fe$_2$O$_3$):	3.67;	2.68;	2.50;	2.19;	1.83; 1.68
Maghemite (γ Fe$_2$O$_3$):	2.93;	2.50;	2.07;	1.60;	1.47

Limonite. — Limonite, a very widespread name in geological literature, is not a mineral but a rock. It consists of a cryptocrystalline aggregate of ferric oxides. Often, it turns out to be goethite alone, but in some cases it is a mixture involving goethite, lepidocrocite, hematite and other clay minerals, aluminous or not. When observed under the microscope, limonite has a misleading aspect; the crystallites are so small that an apparent isotropy occurs. This is really a statistical isotropy by compensation, the same as that which prevails in cryptocrystalline aggregates.

Attempts were also made to establish distinctions by color. Thus it was hoped that we would be able to estimate the quantity of water present. In reality, it is not a question of structural water, but of intermediate water which occurs between the crystallites. The overall color depends on the size of these crystallites and on their arrangement with respect to the water of imbibition; this leads to extremely varied surface effects which are in no way characteristic.

Aluminum hydroxides and oxides. — The most common hydroxide in nature is gibbsite or hydrargillite: $\gamma(AlOH)_3$. The polymorphic variety is bayerite, $\alpha Al(OH)_3$, which has never been found with certainty in natural products, although it must be emphasized that it is almost impossible to identify bayerite when it is mixed with gibbsite.

Two other hydroxides are well known: boehmite, $\gamma AlOOH$ (BÖHM, 1925; J. DE LAPPARENT, 1927) and diaspore, $\alpha AlOOH$. The former is much more common than the latter.

Lastly, the only natural oxide is corundum αAl_2O_3, which belongs to the metamorphic domain.

The main diffraction reflections are the following (BROWN, 1961):

Corundum:	3.48;	2.55;	2.38;	2.08;	1.74;	1.60;	1.40;	1.37
Gibbsite:	4.85;	4.37;	4.31;	2.45;	2.38;	2.04		
Boehmite:	6.11;	3.16;	2.35;	1.86				
Diaspore:	4.71;	3.98;	2.55;	2.31;	2.13;	2.07;	1.63	

Mixed oxides. — FORESTIER and CHAUDRON (1925) showed that beginning with a precipitate of iron and aluminum hydroxides it was possible, by annealing, to obtain oxides of the $(Fe_{2-x} Al_x)O_3$ type; such oxides have been obtained at low temperature by CAILLÈRE, GATINEAU and HÉNIN (1960). Alumina is dissimulated in the lattice of hematite in a proportion which does not exceed 0.25 with an accompanying weak deformation of this lattice.

At the other extreme, the study of some bauxites of the Hérault (France) revealed the presence of ferriferous boehmite (CAILLÈRE and POBEGUIN, 1961) in which iron can be equal to aluminum without any variation in the reticular spacings. Such boehmite has been experimentally obtained (CAILLÈRE and HÉNIN, 1961) without any modification of the boehmite lattice. Thus, we come to certain minerals, which can be written by means of dualistic formula:

$$(Al_{2-x}Fe_x)O_3, H_2O \qquad (\text{where } x \leqslant 1).$$

These experiments are of great interest when one knows the role played by sesquioxides in laterites and bauxites.

IX. — CONCLUSION :
THE VALUES AND DANGERS OF NOMENCLATURE

This has been a very brief exposé of our knowledge about clay minerals. In spite of the obviously pedagogical character of this summary, one may get an idea of the various possibilities offered by mineralogic determinations, as expressed in the nomenclature. This variety permits us to make numerous distinctions within the "natural clays", soils, sediments and other materials. Actually this constitutes the first step for naturalists, namely, to identify correctly the simple or mixed-layer clay minerals within a mixture. Although there is very little to be seen under the microscope, we end with a nomenclature almost as varied as that resulting from the study of deep-seated siliceous rocks.

This nomenclature is, however, still deceiving in part; it presents some dangers. In reality, a soil profile carefully studied can comprise in its clay mineral fraction the following succession, from the bottom to top:

chlorite — (chlorite-vermiculite) — vermiculite — (vermiculite-montmorillonite) — montmorillonite.

We see on paper a series of terms. They are generally the results of X-ray diffraction measurements. But the steps within this succession cannot have been isolated because they are mixed with other minerals of the fine fraction. What chlorite is involved? Is it the same as in crystalline schists? Which vermiculite is present? Dioctahedral or trioctahedral? What is its chemical composition? Besides its X-ray diagrams does it resemble macroscopic vermiculite? And this montmorillonite generally observed in the finest fractions, how does it resemble the montmorillonites well known to the mineralogists?

There are as many problems as there are questions. The precise succession of names masks a reality which is complicated in two ways:

The first complication rests in the fact that the terms utilized designate products which probably have only certain characteristics in common with the minerals whose names they bear. Thus the precise language of the mineralogist is transposed to some fractions of a mixture from which it is difficult to assemble several criteria for identification. It is this difficulty which explains and makes legitimate the application of particular names to clays. Illite is, everything considered, nothing but a fine muscovite; but the inventory of its irregularities continues to grow, and for that reason the distinctions between the terms is without a doubt necessary.

The second complication comes from the fact that a succession of several distinct terms, assuming that they are appropriate, encompasses a reality which is more nearly continuous, with unnoticeable transitions; in other words, it is an evolution.

It is in this way that nomenclature is a precious necessity but at the same time a trap. It is natural to utilize to the maximum the analytical methods which are at our disposal and to designate objects by means of words. Still we must realize that these words are without doubt applied to a reality with more nuances than the rigorous definitions of the words would lead us to believe. And we must not break up the continuity of a natural evolution simply because we have only a small number of words available.

It is certain that the species which we know well by means of mineralogical study are more stable stages than those of products undergoing evolution. Being more stable, they are more probable and thus more common. *We must in reality guard against two dangers: that of drying up a natural evolution by means of the jerky mechanism of going from one species to the next, and that of diluting the concept of a mineral species into a continuous succession in which no distinct steps are discernible.*

In utilizing the present nomenclature, which becomes more precise each year, we shall try to preserve the understanding of real evolution. It is tenuous and difficult to document. Such is the subject of this book on clays.

TABLES OF CLAY MINERAL CLASSIFICATION

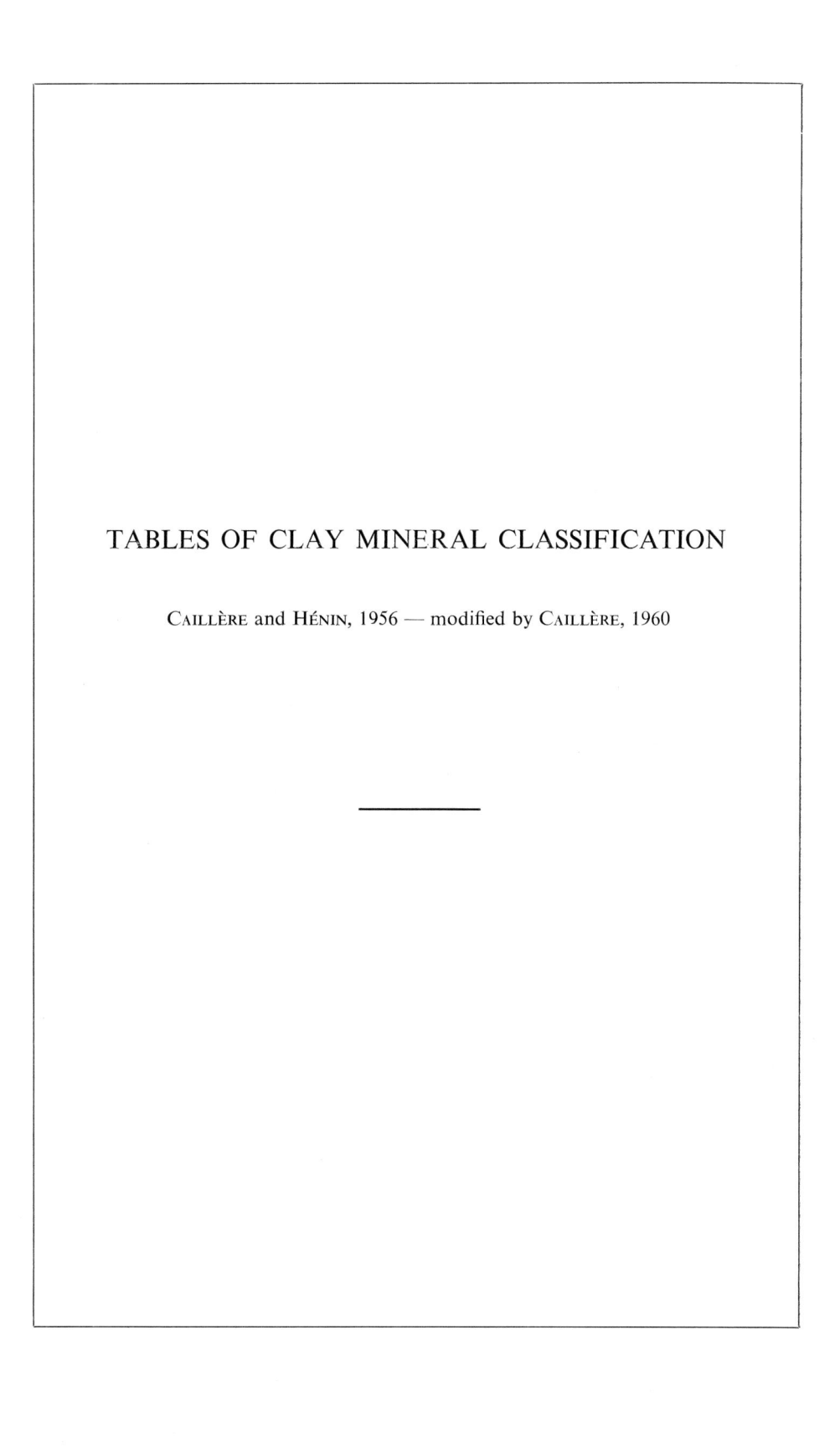

Caillère and Hénin, 1956 — modified by Caillère, 1960

TABLE I. — GENERAL CLASSIFICATION

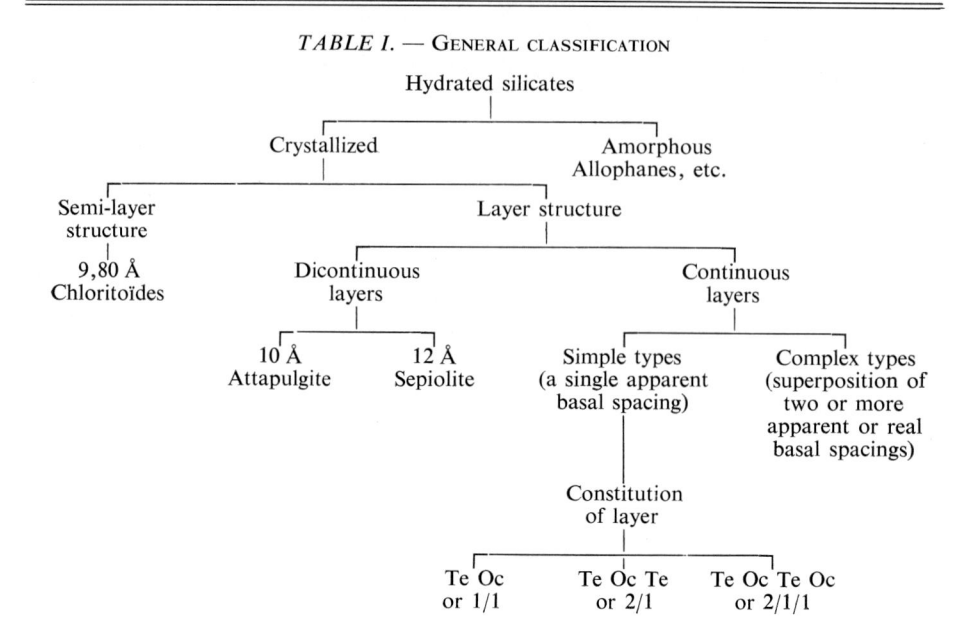

```
                        Hydrated silicates
                               |
              ┌────────────────┴────────────────┐
          Crystallized                       Amorphous
              |                              Allophanes, etc.
   ┌──────────┴──────────────────────┐
Semi-layer                      Layer structure
structure                            |
   |                    ┌────────────┴────────────┐
 9,80 Å             Dicontinuous              Continuous
 Chloritoïdes         layers                    layers
              ┌─────────┴─────────┐     ┌─────────┴─────────────┐
           10 Å                 12 Å   Simple types        Complex types
        Attapulgite           Sepiolite (a single apparent  (superposition of
                                        basal spacing)      two or more
                                              |             apparent or real
                                       Constitution         basal spacings)
                                         of layer
                              ┌───────────────┼───────────────┐
                           Te Oc          Te Oc Te         Te Oc Te Oc
                           or 1/1         or 2/1           or 2/1/1
```

TABLE II. — KAOLINITES AND SERPENTINES
Layer of Te-Oc or 1/1 type

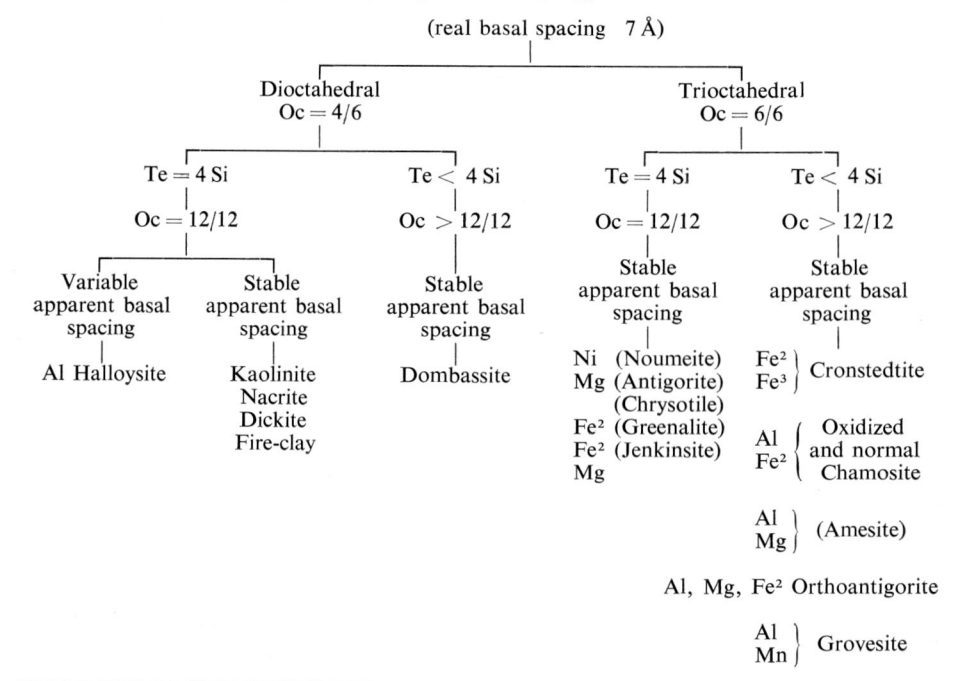

(real basal spacing 7 Å)

```
                           ┌──────────────────────┴──────────────────────┐
                    Dioctahedral                                   Trioctahedral
                    Oc = 4/6                                       Oc = 6/6
            ┌───────────┴───────────┐                    ┌───────────┴───────────┐
        Te = 4 Si                Te < 4 Si            Te = 4 Si               Te < 4 Si
            |                        |                     |                       |
       Oc = 12/12               Oc > 12/12             Oc = 12/12              Oc > 12/12
   ┌────────┴────────┐               |                     |                       |
Variable          Stable           Stable                Stable                  Stable
apparent basal    apparent basal   apparent basal        apparent basal          apparent basal
spacing           spacing          spacing               spacing                 spacing
   |                |                |                     |                       |
Al Halloysite    Kaolinite         Dombassite           Ni  (Noumeite)         Fe²  ⎫ Cronstedtite
                 Nacrite                                Mg  (Antigorite)       Fe³  ⎭
                 Dickite                                    (Chrysotile)
                 Fire-clay                              Fe² (Greenalite)       Al  ⎧ Oxidized
                                                        Fe² (Jenkinsite)       Fe² ⎨ and normal
                                                        Mg                         ⎩ Chamosite

                                                                               Al  ⎫ (Amesite)
                                                                               Mg  ⎭

                                                        Al, Mg, Fe² Orthoantigorite

                                                                               Al  ⎫ Grovesite
                                                                               Mn  ⎭
```

TABLE III. — MICAS AND DIOCTAHEDRAL MONTMORILLONITES
Layer of Te-Oc-Te (2/1) *type*
(real basal spacing: 10 Å)

Dioctahedral Oc = 4/6

	Te = 8 Si		Te < 8 Si		
	Stable apparent basal spacing	Variable apparent basal spacing	Variable apparent basal spacing		Stable apparent basal spacing
	Oc = 12/12	Oc < 12/12	Oc = 12/12	Oc > 12/12	Oc = 12/12
Al	Pyrophyllite	Montmorillonite	Beidellite	Vermiculite ?	K Illite (Muscovite) Sericite Damourite / Na Brammalite (Paragonite)
Al Cr	—	—	Wolchonskoïte		Chromocre (Fuchsite)
Fe^3	—	—	Nontronite	—	K (Glauconite)

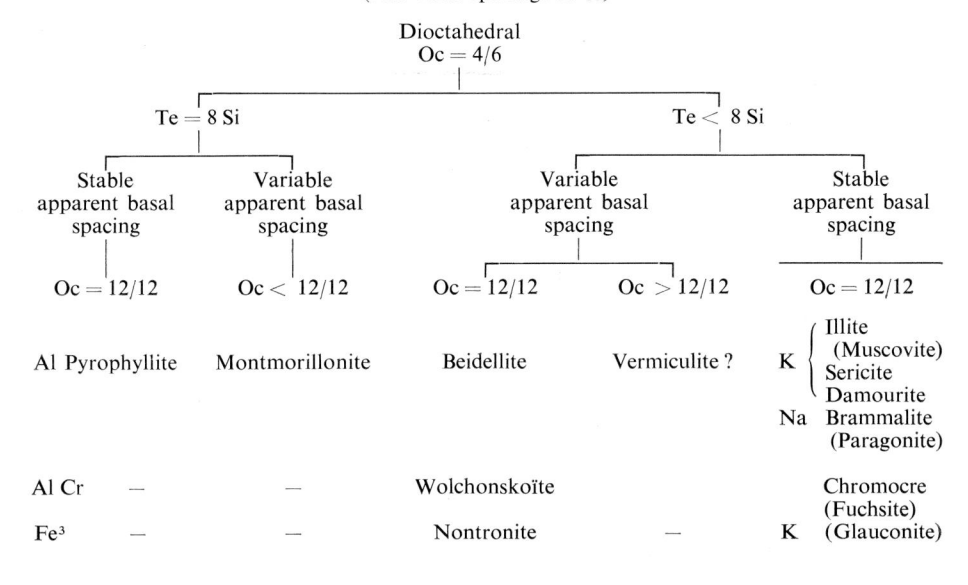

TABLE IV. — MICAS AND TRIOCTAHEDRAL MONTMORILLONITES
Layer of Te-Oc-Te (2/1) *type*
(real basal spacing: 10 Å)

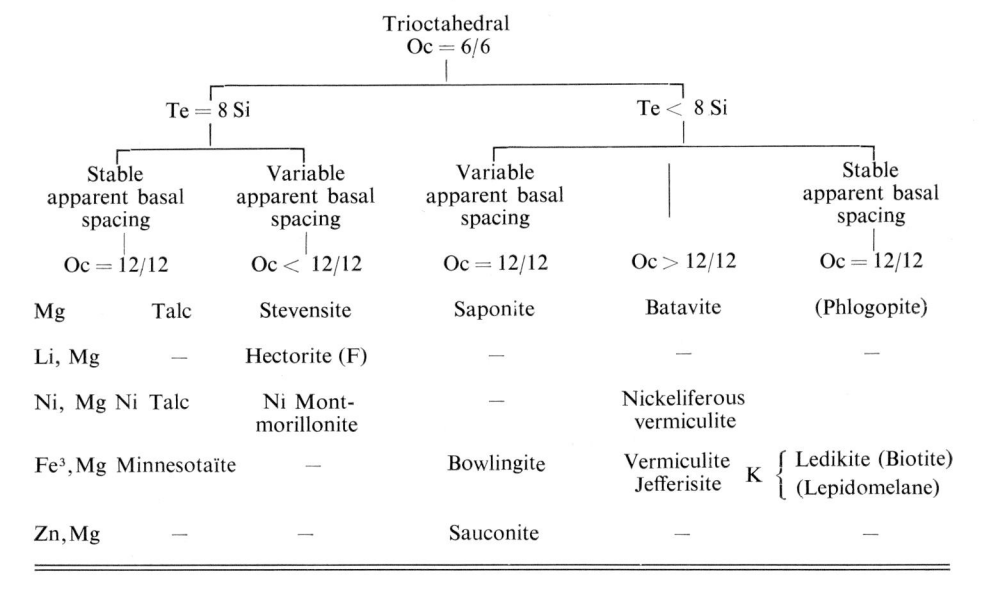

Trioctahedral Oc = 6/6

	Te = 8 Si		Te < 8 Si		
	Stable apparent basal spacing	Variable apparent basal spacing	Variable apparent basal spacing		Stable apparent basal spacing
	Oc = 12/12	Oc < 12/12	Oc = 12/12	Oc > 12/12	Oc = 12/12
Mg	Talc	Stevensite	Saponite	Batavite	(Phlogopite)
Li, Mg	—	Hectorite (F)	—	—	—
Ni, Mg	Ni Talc	Ni Montmorillonite	—	Nickeliferous vermiculite	—
Fe^3, Mg	Minnesotaïte	—	Bowlingite	Vermiculite Jefferisite	K { Ledikite (Biotite) (Lepidomelane)
Zn, Mg	—	—	Sauconite	—	—

TABLE V. — CHLORITES

14 Å

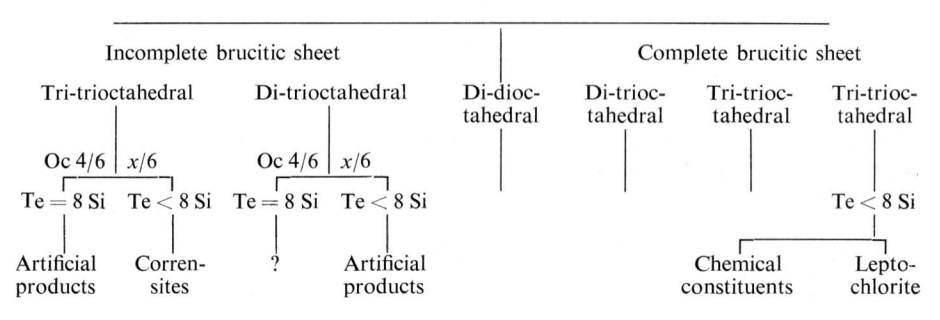

Incomplete brucitic sheet				Complete brucitic sheet			
Tri-trioctahedral		Di-trioctahedral		Di-dioc-tahedral	Di-trioc-tahedral	Tri-trioc-tahedral	Tri-trioc-tahedral
Oc 4/6	x/6	Oc 4/6	x/6				
Te = 8 Si	Te < 8 Si	Te = 8 Si	Te < 8 Si				Te < 8 Si
Artificial products	Corrensites	?	Artificial products			Chemical constituents	Leptochlorite

TABLE VI. — MIXED-LAYER COMPLEXES

Interstratified types

	10 Å type	Mixed 10 - 14 Å types				14 Å types
	10 Å stable	10 Å variable	10 Å variable	14 Å stable	14 Å variable
	10 Å variable	14 Å variable	14 Å stable	10 Å stable	14 Å stable
Irregular stacking	Hydromicas					
Regular stacking.	{ Rectorite { Allevardite					

14 - 7 Å types		Diverse types
14 Å stable	14 Å variable	Kaolinite - Silica
7 Å stable	7 Å stable	Anauxite

CHAPTER II

ARGILLACEOUS ROCKS

I. — HISTORY OF RESEARCH ON CLAY GEOLOGY

It is indispensable to give a simple historical review of the progress realized in the study of the clay geology and especially in the understanding of the genesis of clay minerals. This history is intimately linked with that of clay mineralogy which perfects the tools of the trade and the knowledge of clay minerals. But it depends also on the progress made by chemists in the chemistry of clay components, by pedologists in the evolution of soils and weathering, by metallogenists studying hydrothermal kaolinization and sericitization. The main turning points in the history of clay mineralogy would fall differently from those of clay geology because some time is needed before discoveries in the experimental sciences are assimilated and used by the natural sciences.

Therefore, this historical essay will bear only on the geological point of view.

1° Before 1930: clay geology by chemical methods

In 1894 LE CHÂTELIER analysed the insoluble residues extracted from limestones used for cement and hydraulic lime. His chemical analyses showed that three of the residues studied could be compared to some aluminum silicates, ill-defined at that time, called montmorillonite, confolensite, steargilite, cimolite. On the contrary, the fourth one was similar to the bravaisite of MALLARD (1878). Likewise, VOGT (1906), studying the composition of clays, identified kaolinite in clay used as refractories, but he found in marnes a clay fraction of quite a different nature; it seemed composed of fragments of magnesian minerals (biotite, chlorite, and others). LAVEZARD (1906) studied the calcareous clays of Brunoy and Montmirail in the neighborhood of Paris; he found a clay fraction in which the SiO_2/Al_2O_3 ratio is greater than 2 and in which magnesia, lime and potash occur. WÜLFING (1900) analyzed Keuper marnes and showed that there was only a little kaolinite. He thought that these marnes were constituted mainly of an alkaline chlorite and of pilolite, one of the members of FERSMANN's palygorskite series (1908-1913); here we find again potash and magnesia silicates. MERRILL (1902) in the U.S.A. reached the same conclusions.

In 1914 LE CHÂTELIER, writing his fundamental treatise on silica and silicates, said that the terms "clayey" limestones or calcareous "clays" are often but wrongly used to designate marnes such as those of the Oxfordian of the Paris Basin. "In fact, the siliceous matter of marnes, which can be easily isolated by the action of a weak acid such as acetic acid, bears no relation to the composition of kaolinite." This assertion, amazing today, is explained when one realizes that, for the scientists at the beginning of the century, only that which was kaolinitic was considered to be clay. In fact, kaolinite was the only well known and defined clay mineral; moreover, the kaolinite earths were the purest, the most appreciated and the best studied because of their use as ceramic and refractory earths. After the first world war Jacques DE LAPPARENT (1923) wrote in his treatise that, if these are grounds for thinking that the clay fraction of marnes is made partially of kaolinite and halloysite, there are also reasons for believing that it is constituted, and often largely, of sericite and chlorite.

In 1925 the work of THIÉBAUT was published and some important results were obtained:

a) The marine and lagoonal sedimentary deposits of the Paris Basin contain an alumino-silicate of iron and magnesium, similar to bravaisite;

b) The greater part of these sediments does not contain "clay" (we would say today no "kaolinite");

c) The refractory clays contain kaolinite and are not marine deposits;

d) Bravaisite, like glauconite, was formed by action of sea water on terrigenous deposits. According to my knowledge, it is in the work of THIÉBAUT (1925) that, for the first time, there is an attempt to find a relation between the composition of an argillaceous rock, its origin and depositional environment. Herein we see the origin of a kind of diptych, in which marine sedimentation would give rise to minerals similar to bravaisite, whereas in fresh-water deposits kaolinitic refractory clays would occur. This work will have a long influence.

Thus, the first studies on clay geology, conducted by means of chemical methods alone, result in making distinctions between the different constituents of different facies and in breaking the ground for studies of genesis.

2° *From 1930 to 1950: intervention of X-ray diffraction*

The following twenty years were marked principally by the introduction into geological studies of the mineralogical technique of X-ray diffraction. Clays, which by definition are fine-grained minerals, remained difficult to study because the microscope was not powerful enough and because chemistry provided good information only on pure minerals, whereas the majority of argillaceous rocks are complex mixtures.

Ross and KERR (1931, 1931, 1934) made a careful study of the kaolinite group, in which the SiO_2/Al_2O_3 ratio was equal to 2, and contrasted it to the montmorillonite and beidellite group, in which SiO_2/Al_2O_3 can vary from 2 to 9. In addition to these two groups, Ross and KERR brought to light a still ill-defined group of potash-bearing clay minerals. The German, French and American schools were coming around to greater precision on this third group in the definition of the "sericite-like minerals" of GRIM (1935), of the *Glimmerton* of ENDELL, HOFMANN, MÄGDEFRAU (1935), of the bravaisite of J. DE LAPPARENT (1937). Finally the term "illite" of GRIM, BRAY and BRADLEY (1937) was attained. This mineral, related to micas, turns out to be the constituent of many sedimentary rocks and the most common clay mineral in nature.

Thanks to these better established systematics, the inventory progressed quickly and in three directions. The refractory argillaceous rocks, used in ceramics, were confirmed to be essentially kaolinitic in nature. Bentonites, bleaching earths, fuller's earths or smectic clays appeared to be almost always constituted of montmorillonite. Common clays, brick or pot earths, and the clays of the major part of sedimentary series are composed mainly of illite.

Generalizations on the origin of clay minerals are not very numerous because the time of investigators is filled with inventorying, and it is difficult for them, with so few data, to make generalizations. However, J. DE LAPPARENT (1937, 1937), with a striking insight, outlined the major principles dominating all the recent works. At the second World Petroleum Congress in 1937, he summarized his thought in the following way: kaolinitic deposits have accumulated in sedimentary basins close to regions in which tropical climate and vigorous vegetation favor the formation of kaolinite during the weathering process. On the contrary, the ensemble of minerals whose structure is mainly based on the micaceous lattice, characterized by a basal spacing of about 10 Å, can only be formed in saline environments. This ensemble, which, for that author, includes bravaisite, glauconite, attapulgite, sepiolite and montmorillonite, requires basins rich in calcic and magnesian salts. On the other hand, the montmorillonite of bentonites would require the action of a hydromagnesian environment on an original volcanic material. Using thermodynamic relations, URBAIN (1942) showed that clayey sediments could be authigenic, i.e., could have been fully developed in a sedimentary environment that would lead to a certain kind of neoformation rather than to another.

In 1945 Ross emphasized the preponderant role which would be played by bases, from the geological point of view. Even today what he wrote twenty years ago can be read again with great profit. Kaolinite is born in acid environments under the influence of either humic acids or the oxidation of sulfides; but it can result also from strong leaching (lessivage). The two processes tend to eliminate bases from the system, and it seems obvious that these processes, more than the acidity itself, rule the neoformation of kaolinitic minerals. This amounts to saying that kaolinitic minerals are more easily formed in systems characterized by a minimum of elements other than silica, alumina and water. Clay minerals resulting from continental weathering would react with potassium upon reaching the sea, thus explaining the frequency of illite in marine sedimentary rocks. Here one sees the argument extended beyond the characteristics of climate and environment to the analysis of the chemical composition of the genetic milieu. In 1946 RIVIÈRE proposed a genetic classification: the weathering of granular, acidic rocks under the influence of acid waters results in clays which are essentially kaolinitic. These clays are found in secondary sedimentary deposits in which physico-chemical conditions are similar, being characterized by the acidity of environment. The weathering of glassy rocks would result in rocks with montmorillonites. "The ordinary weathering of most rocks under common conditions of pH" would lead to illites and bravaisites. Lastly, lagoonal facies would be characterized by rocks containing palygorskites and sepiolites.

In 1947, by the study of the substitutions in lattices, GRIM showed how clay minerals were able to undergo important "diagenetic" transformations and, consequently, to be indicators of sedimentation environment. In 1949, MILLOT described the coincidences occurring between clay minerals and the genetic conditions of argillaceous sedimentary rocks. His hypotheses were the following. "The silico-aluminous material brought into sedimentary basins will yield sedimentary rocks which will differ according to the conditions prevailing in the different environments. In acidic lakes and strongly leached fluvio-lacustrine deposits, sedimentary rocks are kaolinitic. In calcareous lakes the three-layered micaceous lattice is predominant, and in the alkaline lakes, poor in potash but rich in magnesia, montmorillonite and in some cases attapulgite and sepiolite occur. Lagoonal deposits, either continental or connected to the sea, yield a clay fraction in which the three-layered lattice—illite, chlorite, montmorillonite—is almost the only one present. Marine sedimentary rocks are variable; in general, micaceous minerals dominate over kaolinite, and among these micaceous minerals, there is especially illite, containing potassium." These general conceptions attribute to the environment of sedimentation and to the diverse chemical characteristics of its composition (pH, kinds and amounts of cations, SiO_2/Al_2O_3 ratio) a determinant role in argillaceous sedimentation, thus underestimating the importance of inert mechanically transported detritus.

Before considering the intervention of new methods and new knowledge, let us summarize the aspect of this period from 1930 to 1950. The inventory was carried on with greater certainty, thanks to X-ray diffraction; and information accumulated rapidly. Four major types of sediments were definitively established, and extensive hypotheses were constructed to account for the coincidences observed in nature.

3° *1950-1962: modern works*

The last decade of study has been very productive. It is characterized principally by the following features:

— The multiplication of laboratories, investigators and quantitative results;

— General discussion about "detrital inheritage or authigenic neoformation";

— The introduction of the recognition of mixed-layer types;

— Modern studies by pedologists and oceanographers on the clays of soils and present-day sediments;

— Syntheses of clay minerals at room temperature.

Let us consider these different features in order:

— The multiplication of results is due to the multiplication of laboratories and investigators. Previously there were only some tens of laboratories studying clay minerals; today there are several hundreds. There were hardly more than a hundred investigators; today there are several thousands. Measurements on sedimentary deposits and rocks were counted by hundreds; today our laboratory, like many others, has made more than 12,000 analyses. Confrontation of previous hypotheses by such a considerable mass of documents was inevitable.

— Many years have been occupied by the famous discussion of the problem of *detrital clays - neoformed clays*, also called *inheritance and neoformation*. The great number of measurements effected in the most diverse sedimentary deposits and environments of sedimentation showed that the great hypotheses of the early days were very schematic. It appeared very quickly that it was not a matter of making a choice; rather both of these processes exist: there are detrital clay minerals, which are, moreover, the most important, and there are neoformed clay minerals (MILLOT, 1953; GRIM, 1953). *But these two terms enclose reality in a framework that is too limited;* clay minerals transformed *during sedimentation can exist also;* this is the "diagenesis" of American authors (GRIM, 1951-1953). Material which is altered by continental weathering is capable of being transformed or reconstructed into new clay minerals under the influence of environmental conditions. This is a third way in which detrital or inherited minerals can be modified chemically. These mechanisms are often weak and difficult to demonstrate, however. Therefore, some authors, such as RIVIÈRE (1953) and WEAVER (1958, 1958), played down the importance of the latter by multiplying the examples in which heritage is the only or predominant process. On the contrary, other investigators, such as GRIM (1953, 1954, 1958), MILLOT (1953, 1957, 1958) and KELLER (1953, 1956) insisted on favorable cases in which transformation and neoformation were clearly manifest. Today we know that all these mechanisms exist, and we can combine their effects; however, there remains the problem of distinguishing among them in each case.

— Fundamental contributions were brought to this discussion and to all the works on clay geology by *newly acquired knowledge about mixed layers* which was placed into the hands of naturalists. The possibilities of interstratification and degrees of interstratification are so numerous that there suddenly became available an infinity of cases and of valuable variations by means of which distinctions, correlations, etc., could be made. Moreover, mixed layers turned out to be stages of degradation, then of aggradation of clay minerals. A very accurate tool was placed at our disposal for the analysis of genesis.

— The degrading *transformations* were the first to be well known, thanks to the works on soil clays by the Scottish school (MACEWAN, 1948; WALKER, 1949; BROWN, 1953 and MITCHELL, 1955) and by the American school of JACKSON (1948, 1952, 1959) and his numerous collaborators. One can grasp all the stages of transformation leading from original illites and chlorites to open minerals or to weathering products. This is the degradation of layer silicates.

— Reciprocally, oceanographic works, begun by GRIM, DIETZ and BRADLEY (1949) and GRIM and JOHNS (1954) and carried on by many investigators, showed crystalline *transformations* of clays between fresh-water, estuarine and marine environments. This amounts to the partial reconstruction of the layer silicates from their debris. Recently, these results have been applied to paleo-oceanography. Not only are the transformed minerals typical of different facies, but also transformations can be observed within sedimentary basins from the shore to the open-sea. Thus, SMOOT (1960) pointed out the evolution of mixed layers into illite within the Mississippian basin of Illinois (U.S.A.). LUCAS and BRONNER (1961) and LUCAS (1962) were able to show the transitions from degraded illites to corrensites and trioctahedral magnesian chlorites in the Triassic basins of France and Morocco.

— Lastly, the works of chemists on the solubility of silica and *the experiments of clay mineral synthesis at low temperature* allow the geologist to look once again at siliceous and argillaceous neoformations. We will examine this aspect carefully, especially the beautiful works by specialists of synthesis.

— Thus these twelve years of study, with all their activity and the intervention of new techniques, allow one today to delve more deeply into the understanding of the geology of clays. This is the object of this book.

II. — CLASSIFICATION AND TEXTURE

1° *Classification*

In 1949 I made propositions about the nomenclature of argillaceous rocks and about their classification. These propositions were followed by some authors, especially by CAROZZI (1953), and by my colleagues HILLY (1957) and BERNARD (1958), who were willing to study the question and to advise me on this difficult problem of nomenclature.

Classification of detrital rocks. — Two parallel classifications relating to the size of the reworked elements were proposed for detrital rocks. The one is based on Latin etymologies, the other on Greek.

Latin roots (GRABAU, 1904) ([1])		*Greek roots* (CORRENS, 1939)
Rudites	Coarse	Psephites
Arenites	Medium	Psammites
Lutites	Fine	Pelites

The limits of size were fixed by GRABAU : 2 mm between rudites and arenites, 50 μ between arenites and lutites. This second limit was modified by NIGGLI (1938) and placed at 64 μ, because of the common size of sieve meshes.

The nomenclature based on Greek etymology is difficult to use in France. In fact, the term psammite has a very precise meaning, confirmed by CAYEUX (1929), that is in general use: it designates a micaceous sandstone in which the bedding is emphasized by highly micaceous layers that facilitate cleavage. Moreover, in French pelite certainly has the meaning of a detrital rock which is fine grained but essentially quartzose; the mineralogical nature becomes confused with the question of size.

If one develops the classification of GRABAU, which is used more and more and which is accepted both by the Committee on Sedimentation (1936) and by the Chambre syndicale du Pétrole in France (1961), argillaceous rocks will thus be classified among the lutites.

([1]) *American Geologist* is the journal which was changed into *Economic Geology* in 1905. The data of GRABAU are contained in PETTIJOHN (1956).

Detrital clays and clays of chemical origin. — The above classification gives the impression that clays are always detrital sediments. Although this is often the case, there exist also neoformed clays of chemical origin and intermediate clays, called transformed. In truth, GRABAU had provided that sedimentary rocks of chemical origin be named "pulverites", which represented the same granulometric class as lutites. But this term fell into disuse, and it seems that, for clays, this is of no importance. In any case, at the moment of their formation, clays are particulate; the determination of detrital or chemical origin is often doubtful, often mixed, and always the result of patient work in the laboratory. Moreover, I remain faithful to the old rule of the ancient naturalists, according to which *a classification must not be primarily genetic because it runs the risk of being overthrown whenever genetic interpretations change.*

We will thus consider that argillaceous rocks are lutites.

Nomenclature of clays within the lutite class. — This nomenclature is presented in Table VII (MILLOT, 1949; modified).

TABLE VII. — CLASSIFICATION OF ARGILLACEOUS ROCKS

	Homogeneous texture	*Layered texture*	*Deformations*
Essentially ARGILLACEOUS rocks	Claystone, mudstone or ARGILLITE	SHALE (layered argillite)	Phyllite Slate SCHIST
ARGILLACEOUS CARBONATE rocks	Calcareous claystone MARNE Dolomitic marne Argillaceous limestone Argillaceous dolomite Sandy marne	Marny shale	Schistose marne Calc-schist
ARGILLACEOUS SILICEOUS rocks	SANDY CLAYSTONE	Sandy shale	Arenaceous schist Quartzophyllite

Some explanation of these terms is necessary.

Argillite. — The word *argilite* in French (claystone in English) is used in the sense in which I defined it (MILLOT, 1949). It designates a common rock composed for the most part of one or several clay minerals and without any notable bedding. It is to this group of rocks that the name "clay" is commonly given, and that usage is quite significant. But the word "clay" designates also weathering products, muds, the clay minerals themselves, and material of volcanic or hydrothermal origin. It seems that, when one wants to indicate with a single word that one is speaking of a rock, one could say "argillite". The same reasoning has been used for serpentinite, and I propose the same thing for glauconitite. The English vocabulary is richer than ours. The word "clay" will translate the French term *argile* and the word "mud" the French term *boue*; thus, the argillaceous rocks that are called *argilites* in French can be translated as "claystone" or "mudstone". Unfortunately, in English "argillite" has variable usage, but generally is applied to a clay which was indurated during incipient metamorphism without cleavage or parting (TWENHOFEL, 1939; FLAWN, 1953; PETTIJOHN, 1956). This coincidence is annoying, but the French language does not have words for either claystone or mudstone, and the use of *argilite* is already wide-spread.

Shale. — "Shale" means layered argillite. This bedding must be understood as that which is principally due to the sedimentation process itself. These rocks are very common in sedimentary series, and French does not have a term to designate them. Therefore, I proposed (MILLOT, 1949) to introduce simply the word "shale" into the French geological language. To make themselves understood, the French must use the term *schiste*, which evokes immediately the idea of schistosity acquired either by mechanical deformation or during an early stage of metamorphism. This is bothersome. On the other hand, the word "shale" which is short and without any historical variation in meaning, remains very convenient.

Schist. — Every rock foliated by mechanical stresses is a schist. Such schistose parting is independent of bedding, which is the result of stratification and which is typical of shales. It is obvious that the two, bedding and foliation, can coincide. This schistosity can be acquired for purely tectonic reasons, far from the action of metamorphism. However, since metamorphic transformations also are the result of deformation, metamorphic rocks are schistose. All rocks are able to take on a mechanical schistosity, but argillaceous rocks are the most affected because they are composed of layered minerals. The great schistose series are sometimes called "phyllites". When the schistosity is very fine, one speaks of slaty schists, and when beautiful planar laminae can be quarried, one is dealing with slates.

Marne. — Between claystones and carbonate rocks, all the intermediate stages exist. In the common language as well as in the geological vocabulary, a calcareous claystone is called *marne*. Many people try to codify this language by means of quantitative criteria. I maintain that this is superfluous and that it suffices to conform to the common usage. Some have proposed *marne* for rocks with more than 35 % calcium carbonate and *argillaceous limestone* for rocks with more than 65 %: these numbers are without doubt 33.3 % and 66.6 % rounded off. Others have pointed out that effervescence occurs readily with a carbonate content of 13 to 15 % and proposed that this value should mark the limit of marne, but it must be admitted that this depends on the grain size and on the kind of carbonate present.

It is better to be more modest. The language of natural science always fluctuates between numerical exactitude and the talent of being comprehensible. I prefer the second path. Whenever effervescence is evident, we will use "marne"; before that point, "argillite" will be used. Thereafter, so long as the consistency of the rock is more clayey than calcareous, "marne" will be used; then, when the rock is hard and consolidated, we will speak of "argillaceous limestone", and this modification occurs with quite varied amounts of clay.

The terms "argillaceous marnes", "marny clays" and "marny limestones" are pleonasms, but the latter term (*calcaires marneux* in French) is in wide usage and everybody understands it. On the contrary, the carbonate fraction is commonly dolomitic and the locutions "dolomitic clay", "dolomitic marne" and "argillaceous dolomite" are very useful.

If use of the word "shale" spreads in French, the terms "marny shale" and "dolomitic shale" will be welcome. As the rock acquires schistosity, it advances to the stage of "schistose marne" and "calc-schist".

Sandy argillite. — All intermediate types exist between argillites (claystones) and sands and sandstones. One can use the terms "sandy argillite", "sandy shale", "clayey sand", and "clayey sandstone". The international classification specifies as silt, among lutites ($<64\,\mu$), that fraction included between 64 and $4\,\mu$. This can be useful not only from the point of view of size, but also because in this size range, which is commonly greater than that of the majority of clay minerals, rocks are mainly quartzose. Thus, to speak of silt, siltite and siltstone evokes not only the larger sizes of the lutite category or, if one prefers, the sizes greater than those of clays, but also the very siliceous petrographic nature. It is to such rocks as these that French usage commonly gives the name "pelite". Certainly for the French the word pelite designates a rock finer-grained than sandstone and coarser and more quartzose than argillite; included are the micro-sandstones (*micro-grès*) of Lucas (1942) and the *pélites* of Hilly (1957). The acquisition of a mechanical schistosity transforms them into arenaceous schists, schistose sandstones and quartzophyllites.

TABLE VIII. — Some varieties of argillaceous rocks

Mineralogical varieties	Kaolin Saliferous clay Saline marne Gypsiferous marne	Glauconitic marne Micaceous marne	Ferruginous clay Clayey gaize
Facies varieties	Variegated clay Versicolored clay Banded marne Varved clay Plastic clay	Platy marne Papyraceous clay Paper schist Oolitic marne Nodular marne	Lateritic clay Flint clay *Lehm*
Organic varieties	Foraminiferal marne Sponge-bearing marne Fish-bearing shale		Bituminous shale Carbonaceous shale Black shale (*ampélite*)
Varieties according to utilization	Fire clay	Tile marne Clayey cement stone	Bentonite Fuller's earth Absorbent clay Swelling clay Smectite

Varieties. — According to the advice of CAYEUX, language must retain great freedom in order to characterize rocks by adjectives which depend on accessory minerals, cements, organic matter, organisms, tradition and uses. Concerning tradition, I will mention the term *ampélite* (black shale, in English) which designates carbonaceous or organic shales, rich in pyrite. Formerly, peasants scattered such shale in vineyards [*ampes* = (grape) vines], the oxidation of pyrites improving the soil. This simple word takes account of many characteristics, it remains evocative and must be kept; everything should not yield to the rigor of numbers, which are no easier to keep in mind than language of the ancients.

Table VIII presents some of the varieties.

Conclusion. — Under the influence of CAYEUX (1929) who dominates French sedimentary petrography, the nomenclature of sedimentary rocks has maintained its simplicity and adaptability. Let us remain faithful to this principle and, if greater precision is necessary, it can always be rendered by supplementary numerals.

2° *Texture*

The polarizing microscope can be used in two ways to study "clays". As a tool of mineralogist it serves for the study of the clay mineral itself, either in context or after its isolation. The main works of this sort were effected by CORRENS (1929), MARSHALL (1935), URBAIN (1937), CORRENS and PILLER (1954). As a tool of the petrographer, the microscope allows one to study the rock itself.

PREPARATION OF THIN SECTIONS. — The cutting of thin sections requires special techniques perfected for this purpose (MEYER, 1946; KULBICKI, 1949; DEBYSER, 1957; CAVANAUGH, 1960; DURAND, 1960; TOURTELOT, 1961). The principle consists of impregnating the rock with a product which makes it cohesive without modifying its texture. Such products are Canada balsam, silicates, and plastics. Each laboratory refines its own techniques with respect to a series of products.

Texture normal and parallel to stratification (MILLOT, 1949). — Variations of texture are infinite; there is no code by which to designate it, and it would be regrettable if our vocabulary, which has remained flexible, would become rigid. However, the varieties of the terms used do not always designate the same kind of relations and this often comes from the fact that texture can be studied in two ways. Therefore we will distinguish:

— texture normal to stratification;

— texture parallel to stratification.

1° **Textures normal to stratification.** — Textures can be qualified by a great number of terms, according to the desires of the observer:

Homogeneous, if stratification of bedding is not apparent;

Oriented, if it is obvious that clay minerals have a preferential arrangement, due to sedimentation, diagenesis or any other cause;

Layered, if the rock appears to be composed of successive layers;

Cyclic or sequential, if there is an alternation of facies in sedimentary microsequences, as in varved clays, banded marnes and many other sedimentary rocks. Very beautiful examples from modern sediments are given by DEBYSER (1959);

Microlenticular (URBAIN, 1937), if these cycles are so local that they seem unconformable at the scale of the sample or thin section.

2° **Textures parallel to stratification.** — The clay fraction can be masked by other rock constituents, and in that case observation is the same as that of any common sedimentary rock.

The clay fraction can be visible, and usually it is much more apparent in the thin edges of thin sections. In such cases one finds a series of textures which are defined in the works of BERTRAND and LANQUINE (1919, 1922) and URBAIN (1937).

The texture will be:

Crystalline, if well-individualized flakes form the matrix;

Cryptocrystalline, if the crystallinity is still appreciable but in a confused manner, with fuzzy edges at the limits of small birefringent zones;

Hidden, invisible or *amorphous,* if the clayey matrix seems isotropic; the term "amorphous" is not recommended, however, because the matter is crystalline, although very small in size; one is the victim of statistical isotropy by compensation.

Cryptocrystalline textures can involve the following peculiarities:

Fissured, reticulated and *bulls-eye-form* (URBAIN, 1937), according to the arrangement of minerals at the edges of fissures;

Meshed and *flocky* (MILLOT, 1949); the meshed texture occurs when the crystallites are entangled after the fashion of the microscopic facies of antigorite. The frequently observed flocky texture results from the organization of the clayey material in less birefringent circular masses surrounded either by more birefringent clay minerals, by calcite crystals or by various impurities.

Flow texture or *moiré* (KULBICKI, 1953) shown by numerous subtle optical effects seen in some cases.

Textures distinguished by VIKOULOVA (1943) in STRAKHOV (1957) repeat or complement the terms above: *spotted, marbled, oriented, reticulate, banded, stratified, micro-stratified, floating, "eyed", fluidal, porous, massive.*

The essential is to observe and to describe the observations clearly; in so doing, one sees that acquisitions obtained independently confirm one another.

Textures of clays in carbonate and siliceous rocks. Cements. — The following chapters will show that argillaceous fraction of rocks is of great interest, even in the interior of rocks, where it constitutes only a subordinate or accessory fraction. Briefly, one can predict if the clay minerals are inert within a material which is evolving, that they will be the evidence of stages in former evolution. On the contrary, if it is the argillaceous material which is being transformed vis-à-vis a fixed environment, it will give information about the stages of recent evolution.

— Here we come to the problem of argillaceous cements. A careful systematic treatment is given by STRAKHOV (1957), who describes the following textures: micro-aggregated, lamellar, pellicular, radiating, vermiculiform, lepidoblastic and in chips.

— In sands and sandstones, clay minerals of diagenesis are the sensitive part of rock, in contrast to the quartz grains which play a passive role in the history of the rock. Plates I and II show examples of the successive evolution of clay minerals within a sandstone.

— In limestones and dolomites, epigenesis and evolution are not less common and lead to the understanding of the succession of events. If oolites or concretionary structures exist, clay minerals participate in various stages of their formation. Rather close to this category is that of iron-ore, or oolitic ferruginous rocks: the evolution of clay minerals of chamositic and chloritic types allows one to reconstruct a portion of the early or late stages of diagenesis.

— In crystalline rocks, volcanic, metamorphic or abyssal, microscopic examination obviously gives information about metamorphic or retromorphic evolution of clay minerals and also about hydrothermal and atmospheric alteration.

— For the petrographic study of the clay fraction of sedimentary rocks, books on sedimentary petrography (CAROZZI, 1960) are invaluable guides. Geochemical sedimentologists have the duty to emphasize that the improvement of physical and chemical methods must *in no case dispense with microscopic studies*. Direct visual observation of textures often shows the direction of the evolution of which the "finer" methods can rather blindly give only the stages.

III. — ARGILLACEOUS ROCKS OF HYDROTHERMAL AND VOLCANIC ORIGIN

There are two great ensembles of argillaceous rocks that are related to emanations from depth: they are the clays of hydrothermal deposits and the clays transformed from volcanic rocks. The former concern the domain of veined ore deposits, the latter that of volcanism. The literature in these two domains is considerable, involving several hundreds of titles. Otherwise, the problem posed by these hydrothermal clays and the rules governing their genesis are the same as those which concern the clays resulting from atmospheric weathering and sediments. Therefore, without treating the entirety and variety of these questions, I shall present here a simplified summary for the purpose of allowing comparisons. Furthermore it would not be appropriate to neglect, in this book, formations which are born under conditions peculiar or different from the others, but which are nevertheless quite "true clays".

1° *Clays of hydrothermal deposits*

Definition. — In the vicinity of hydrothermal veins, alteration extending laterally from a few meters to a few tens of meters is common. This alteration transforms the surrounding rocks into clay minerals or related minerals: kaolinite, sericite, montmorillonite, chlorite, pyrophyllite, alunite, serpentine, talc and even palygorskite. In some cases the alteration, instead of being disposed in symmetrical bands on both sides of veins, extends through greater volumes of rock and transforms complete massifs.

Kaolin deposits. — Hydrothermal kaolin deposits are the result of the transformation *en masse* of crystalline rocks, metamorphic or volcanic, under the influence of deep hydrothermal solutions. From the genetic point of view, they are completely different from kaolin resulting from atmospheric weathering and sedimentary kaolin.

— The large deposits in Brittany were studied by NICOLAS (1956). At Plémet, the rock having undergone kaolinization is a quartz diorite. At Ploemeur, it is a monzonitic granite. The study of the quartz revealed that solutions of about 150 to 300 °C had circulated in these rocks. When solutions are hot, zones of alteration different from kaolinization and anterior to it can be discerned. Kaolinization involves an important loss in silica and an almost complete disappearance of alkalis.

— The deposit of St Austell (Cornwall, Great Britain) is one of the most important kaolin deposits of Great Britain, according to L. G. BROWN (1953). After the primary consolidation of the granite, there was a long period of deformation including the opening of fissures and intrusions of pegmatites, aplites and felsites, which have the same composition as granite. Then, a pneumatolytic phase occurs, which leads to three forms of alteration, tourmalinization, greisenization, and kaolinization. Tourmalinization develops tourmaline and quartz at the expense of feldspars and mica. Sericitization forms white mica and quartz at the expense of feldspars. Kaolinization yields kaolinite, sericite and quartz at the expense of feldspars.

— The deposit of Geisenheim (Rheingau, Germany) is formed from a quartz keratophyre massif, the vault of which it occupies. The alteration is of hydrothermal origin (BEHNE and HOENES, 1955), involving a sericitization phase followed by a kaolinization phase acting upon the sericite.

Supergene and hypergene origin. — Specialists have for a long time been hesitant about the origin of the large deposits of kaolin, and supergene origin has often been invoked. There were two principal reasons for this: first, kaolin deposits resulting from weathering exist (KONTA and POUBA, 1961), and distinctions are not always easy. From another point of view, earlier authors emphasize that some deposits wedge out at depth (DE LAUNAY, 1913). Modern studies on successive alteration, zonation of alteration, relations to hypogene fissures, and geological thermometry, allow one to assert that kaolin deposits are certainly hydrothermal. It must be added that hydrothermal action produces a disorganization of feldspars that makes them much more vulnerable to weathering, so that in many cases the latter is superimposed upon hydrothermal alterations.

Moreover, it will be noticed that the large deposits of kaolin are formed in acidic rocks, poor in plagioclases or having previously undergone sericitization. The calcium content is very low, rarely equal to 1 %. It generally decreases in kaolinized zones. This is very important for genetic considerations. In any case, the responsible hydrothermal vapors are considered to be acid by metallogenists, according to the study of mineral paragenesis.

Walls of hydrothermal vein deposits. — The walls of ore-deposit veins include in many cases rocks altered to clayey products through a thickness of several meters. These phenomena have been studied by metallogenists, and general views are given by LINDGREEN (1933), LOVERING (1941, 1950) and KERR (1955).

The zonal disposition of alteration most common in granites and related rocks is the following (LOVERING, 1950) (fig. 5):

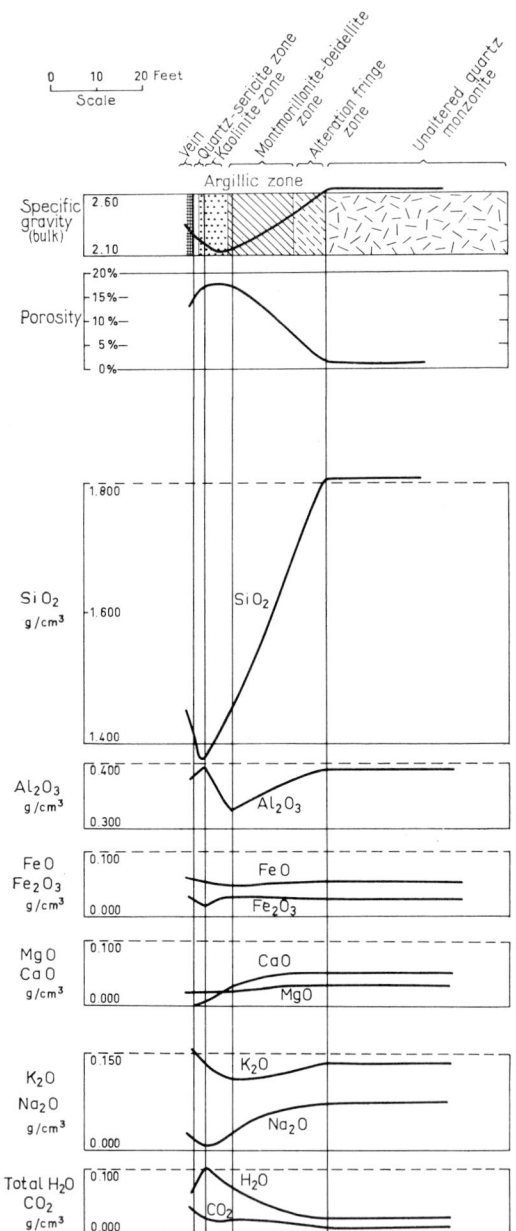

1. *The vein* itself.

2. *The siliceous envelope*, largely converted to quartz, with or without orthoclase and sericite.

3. *The sericitized zone*, in which the country rock is transformed into sericite, but with substantial amounts of hypogene quartz and kaolin residue.

4. *A zone of alteration properly called argillaceous*, with a rather abrupt transition; this zone includes clays rich in alumina, such as kaolinite and dickite, associated locally with rutile, leucoxene and alunite.

5. *A montmorillonite - beidellite zone* which becomes more marked as one moves away from the mineralizing vein and as the intensity of alteration decreases.

6. *An allophane-rich fringe*, which contains poorly organized siliceous gels and sporadic mineralization into chlorite, epidote, montmorillonite, carbonates and sericite; the zone of chloritization can be thicker in some cases.

FIG. 5. — *Hydrothermal alteration zone in quartz monzonites, Boulder County tungsten district. Mineralogical, physical and chemical changes* (after LOVERING, 1950).

Field studies have been compared with studies on hydrothermal synthesis. The temperatures of transformations were estimated by means of geological thermometers (KONTA, 1961). The geochemistry of these alterations has been reconstructed on many examples, which are varied since the mechanisms depend on temperature, on the kind of rocks subject to alteration and on the nature of the hydrothermal fluids. Each alteration zone is younger than its neighbor on the side away from the vein which nourished it. The acid solutions (HCl, HF, H_2S) work by leaching throughout surrounding rocks. Cations and silica are leached from rocks and put into circulation. In zones of strong leaching kaolinite appears, and where leaching is weaker, minerals of chlorite group and montmorillonite occur. The process is gradational because the circulation of solutions follows gradients of temperature, of leaching and of acidity. At the termination of alteration, cations leached in the vein induce a final sericitization of its walls.

Progressive transformations. — The utilization of modern knowledge on mixed layers allows one to study progressive transformations during hydrothermal alteration. Some such works have been carried out by Japanese investigators (SUDO et al., 1962), who, thereby, have been able to show evidence in epithermal deposits for the following transitions:

— Wall of a pyrite-galena deposit: transition from a montmorillonite zone to a zone of chlorite-illite;

— Wall of a pyrite deposit: transition from a pyrophyllite zone to a pyrophyllite-diaspore zone;

— Wall of a pyrite deposit: transition from a kaolinite zone (kaolinite, dickite, nacrite, alunite) to a pyrophyllite zone.

In every case regular and irregular mixed layers were identified, and the authors reach a kind of rule governing the transition from one clay mineral to another. In an environment where mineral A occurred, but which becomes favorable for the neo-formation of mineral B, an evolution takes place, which can be summarized as follows: mineral A $\longrightarrow$ modification of A by interstratification $\longrightarrow$ mixed-layer AB $\longrightarrow$ modification of B $\longrightarrow$ mineral B.

Supergene and hypogene origin. — Intervention of weathering raises the same problems for hydrothermal deposits as for kaolin deposits. In fact, the routes provided by ore-bearing veins are favorable to the percolation of meteoric waters. Moreover, in the oxidation zone these meteoric waters oxidize sulfides and induce sulfuric leachings, which produce strong alterations of supergene origin. There is a superimposition of two kinds of alterations.

In some cases alterations which are observed in the walls of veins can be attributed entirely to surficial percolation. This interpretation is supported by the recent presentation by TCHOUKOV (1961) of research carried out in Russia, whose main point is the following. The clay minerals of ore-bearing deposits are mainly of supergene origin. This is true not only for the clay minerals of oxidized zone, but also for the clay minerals occurring below the zone of ore oxidation. Thus, in the deposit of Kounrad in Kazakstan, the 50-meter thick oxidized zone contains mainly halloysite, but below, reaching to a depth of 210 meters in the tectonized zones, the principal mineral is montmorillonite. Lastly, alteration clays of ore deposits are often opalized by a strong surficial silicification, which decreases with depth. This calls our attention to the ease with which clay minerals are transformed into opal; this results from climatic weathering which, in these regions, dates from late Cretaceous or early Tertiary time.

Magnesian clay minerals of hydrothermal origin. — Description of the mineralogy and petrography of magnesian clay minerals produced by hydrothermal or metamorphic action would require a whole book. It is quite impossible to undertake such a task here. Let us say only that all the large families of clay minerals are involved in these processes: serpentines, talc, micas, montmorillonites, vermiculites, chlorites, mixed layers, and even palygorskites. The structural interpretation of variations between the different magnesian clay minerals has been given by BRADLEY (1955). Serpentines are magnesian clay minerals that can occur as voluminous bodies in geological formations, or in veins or as discreet alterations within rocks. The petrographic and genetic study of serpentines would be a digression from our purpose; the reader may refer to the good, specialized books by BENSON (1918), CAILLÈRE (1936) and PAVLOVITCH (1937). Serpentines are, on the whole, the product of transformation of former magnesian rocks; these can be peridotites, as is most commonly the case, or calco-magnesian rocks with amphibole and pyroxene. These transformation phenomena are varied and can be spread in time from the period contemporaneous with the emplacement of basic magmas, where they result from the pneumatolytic action of the vapor phase, down to the tardive hydrothermal activity occurring at low pressure and temperature. In every case it is observed that serpentines would be more stable than the magnesian minerals in crystalline rocks under conditions of shallow depth, where water comes into play. From the geochemical point of view, there would be some analogy between the kaolinite-aluminous silicates couple and that of the serpentine-magnesian silicates. Nevertheless, the neoformation of serpentines does not commonly belong to the realm of weathering, but to that of hydrothermal alterations. We shall see that the development of talc is rare in diagenetic processes; its domain is located in the zone of metamorphic and hydrothermal activity; basic information can be found in LACROIX (1893-1913). Magnesian micas, mainly phlogopite and biotite, belong both to the metamorphic domain and to the hydrothermal domain. Vermiculite is not only a product resulting from the hydrothermal alteration of biotite and phlogopite, but also a micaceous structure that forms in the reaction borders between intrusives and magnesian rocks, as shown by the studies of VARLEY (1952), AGARD et al. (1953), CLABAUGH and BARNES (1959) in Texas, and WIMMENAUER (1959) on the Kaiserstuhl in the Black Forest (Germany). Innumerable studies show also neoformations of magnesian clay minerals of the montmorillonite and chlorite groups under hydrothermal conditions: nontronites, saponites, hectorites, sauconites, mixed layers, chlorites, vermiculites, talc, etc. Also to be mentioned are the works of Italian investigators ALIETTI and GALLITELLI, a list of which is given in GALLITELLI (1961).

Palygorskites, attapulgites and sepiolites are also produced by hydrothermal action in quite variable conditions, within rocks as well as in hydrothermal veins. Many examples are given by FERSMANN (1913), CAILLÈRE (1936, 1951, 1952) and STEPHEN (1954).

From the geochemical point of view, we shall only retain the following very general rule; clay minerals, and moreover, layer minerals as a whole group, are typical of the hydrosphere; they are the stable forms of silicates in the hydrosphere, whereas at depth structures occur as chains, tridimensional lattices or isolated tetrahedra. But the hydrosphere has two faces: the external face, which is subject to "ordinary" temperatures and pressures, and the deep one, which is under the influence of hydrothermal conditions.

2° *Argillaceous rocks resulting from volcanic material (bentonites)*

Volcanic extrusive rocks contain in many cases an important amount of glass. This glass is more vulnerable to hydrothermal actions than crystallized silicates and commonly gives rise to major formations of argillaceous rocks.

The transformation of volcanic glass into argillaceous rocks can occur in three very different ways, however:

1. By hydrothermal action of volcanic fumaroles;
2. By transformation of volcanic ash and tuff within the aqueous environment of sedimentary basins;
3. By weathering of volcanic ash and tuff.

These three possible types of evolution of glassy or partially glassy volcanic rocks must not be confused, although it is not always easy to distinguish them in the field. The term bentonite was created by KNIGHT to designate a deposit of a highly colloidal clay, in the Cretaceous sediments at Fort Benton, Wyoming (U.S.A.). Subsequently it was shown that these clays were composed essentially of montmorillonite and that they constituted the product of *in situ* evolution of volcanic ash. ROSS and SHANNON (1926) defined bentonites as the product of *in situ* transformation of volcanic ash. But this restriction has been weakened for commercial reasons and today any clay with developed colloidal and adsorbant properties is called bentonite in everyday language, even if it is not composed of montmorillonite.

It is important that geologists and mineralogists restrict the term bentonite, whenever it is used—and it is sometimes convenient—to montmorillonite-bearing rocks. It happens that they are of volcanic origin in many cases. Thus, a kind of symmetry is established with the term kaolin.

The true kaolins are products of hydrothermal alteration of silico-aluminous rocks.

The true bentonites are products of *in situ* alteration of volcanic material.

But sedimentary kaolin and weathering kaolin exist, as well as sedimentary bentonites and weathering bentonites. This entails the inevitable broadening of the terms.

Hydrothermal action on volcanic material. — The example of the bentonites of Lalla Maghnia (Oran, Algeria) is particularly typical (SADRAN, MILLOT, BONIFAS, 1955). There are several volcanic outcrops in the region; they are partly altered to montmorillonite, transforming the rock, whether it be obsidian or rhyolite. Transformation reaches some of the beds which are concentric, or on the contrary diagonal to the structure of volcanic plug, then it reaches the core of the plug itself. The lateral volcanic breccias are also transformed; the cement and the clastic elements are in some cases of different color, but equally transformed into soapstone. Sedimentary formations in the vicinity of the plugs include montmorillonite beds and rhyolitic gravel horizons. All these facts show that the hydrothermal alteration was produced by volcanic activity that transformed lavas along the fissures through which fumarolic emanations escaped. The transformed materials were reworked within the basin and deposited as fine particles; on the other hand, rhyolitic rocks gave rise to gravelly levels. Autochthonous or allochthonous, the bentonites of Lalla Maghnia are the evolution products of volcanic glass under the influence of deep-seated hydrothermal activity.

A large number of deposits are of this type; these are the true bentonite deposits of Ross and Shannon (1926). They can be the transformation products of lavas or breccias, but more frequently of tuff and of ash. Most commonly the transformed volcanic glass is acidic with a composition of rhyolite, trachyte, phonolite or more seldom andesite.

Transformation of volcanic ash in sedimentary basins. — A second type of complete transformation of volcanic ash occurs when it is deposited in a sedimentary basin, a lake or the sea. The volcanic glass will evolve very quickly in the aqueous environment, as is observed today in the pyroclastic deposits of the Mediterranean, near Vesuvius and Sicily (Norin, 1953; Muller, 1961; Grim and Vernet, 1961).

The example of the montmorillonite of Camp Berteaux in eastern Morocco is typical (Bourcart, 1936; J. de Lapparent, 1937, 1945; Urbain, 1941; Jeannette, 1952). Layers of smectic clays are interstratified in a Miocene lacustrine or lagoono-lacustrine series of brown, green and blue marnes and beds of volcanic ash. The entire transition is observed between cinerites and smectic clays, within which are found non-transformed biotite flakes. These bentonites are the transformation product of glassy volcanic ash in the Miocene lagoono-lacustrine basin of Camp Berteaux.

Schlocker and Van Horn (1958) described an interesting case among the many examples known in America. In the vicinity of Denver, Colorado, they studied bentonites resulting from the alteration of volcanic ash in a Pliocene lake. They discovered two types of alteration: one yields montmorillonite, the other gives rise to abundant illite, irregularly interstratified with a small amount of montmorillonite. This alteration into illite is considered to be determined by the sedimentary environment. Likewise, other examples are known in which normal bentonites, i.e., composed of montmorillonite evolve within the sedimentary environment. This is the case of the Ordovician potash-bearing bentonites of Pennsylvania (Weaver, 1953) and of Scandinavia (Byström, 1954, 1957). If these bentonites are potash-bearing, it is because they are of marine origin and have fixed potassium between the montmorillonite layers.

Thus we have seen bentonites which are at the same time volcanic and sedimentary and which are the products of the evolution of volcanic ash in lakes or in the sea. We shall also see that there are strictly sedimentary bentonites, which are the products of the authigenic neoformation of montmorillonite in some alkaline and siliceous basins.

Weathering of pyroclastic formations. — Volcanic materials, like all the others, are subjected to weathering. But pyroclastic materials, such as ash and tuff have two supplementary characteristics which make them a particularly vulnerable material: they are permeable and glassy.

In these materials alterations will thus be rapid and strong. SUDO (1954) showed in Japan that moderate weathering is sufficient for the evolution of ash into montmorillonite, which is different from hydrothermal bentonite; by stronger weathering, however, halloysite is reached. The study of this kaolinization of glassy volcanic ash is very interesting because it shows that volcanic glass goes through an un-organized stage of allophane type. At first composed of very fine particles (0.02 μ), these colloids join together into rounded grains (0.2-0.5 μ). Gradually fibers of halloysite are formed, and then larger crystals of halloysite.

KELLER (1952-1956) showed that at alkaline pH and in a closed environment, volcanic material can evolve into zeolites, especially analcime, or into montmorillonite. By acid leaching they give rise to kaolinite, as do the other silicates, and to alumina when drainage is strong. Here we have all the possibilities, and the works of ZANS (1952) on the weathering of Jamaican cinerites into bauxites showed very well that the ultimate stage can be reached.

GASTUCHE and DE KIMPE (1961) studied the evolution of cinerites on Nyamuragira volcano in northern Kivu (Congo). Around the volcano the ash is coarse; it becomes finer and finer as one moves away. For a given lapse of time, weathering is much more rapid in the fine ash than in the coarse. The amount of CaO decreases as one moves away from the volcano. At a distance of about 20 kilometers from Nyamuragira, the coarse ash, around pH 7, gives rise to a highly charged and very hydrated gel; the corresponding X-ray patterns show the rough outlines of 14, 10 and 7 Å reflections indicating the appearance of three-layered minerals. At a distance of about 50 kilometers from the volcano, at about pH 5, a mixture of halloysite and kaolinite occurs, whereas at a distance of about 75 kilometers the clay fraction is composed of disordered kaolinite. These successive stages point out the gradual evolutions that are possible from the same glassy volcanic products.

Conclusions

1° *Hydrothermal fluids of deep-seated origin* have the capability of transforming silicate rocks into argillaceous ones.

2° When the emanations are acid, and the leaching *per ascensum* is strong, complete leaching of cations results in the deposits of hydrothermal kaolins and the kaolinitic wall rocks of ore-bearing veins.

3° When leaching is less strong, or for reasons of distance, temperature, porosity and composition, the evolution leads chiefly to montmorillonite, chlorite and sericite. The alteration products of the glassy volcanic material are a special case of these actions: these are bentonites in the strict and original meaning of word.

4° *The deposits of true kaolin* are hydrothermal, but the term was extended to deposits of sedimentary kaolin and kaolin resulting from atmospheric weathering. The former are reworked deposits, the latter the product of strong weathering under a tropical climate.

5° *The deposits of true bentonite* are volcanic and hydrothermal, but the term was extended to deposits of sedimentary bentonite and bentonite resulting from atmospheric weathering.

6° Sedimentary bentonites can be:
either the product of reworking bentonite which originated from autochthonous volcanic alteration;
or the product of the transformation of volcanic glass which is unstable in the aqueous environment of sedimentary basins: this is a typical example of subaquatic alteration called halmyrolysis by HUMMEL (1922);
or the product of a sedimentary neoformation, which has nothing to do with either volcanic rocks or with the deep-seated hydrothermal phenomena, and which forms a part of neogenesis in alkaline and siliceous environments;
or the reworked product of the weathering of volcanic ash.

7° Atmospheric weathering of volcanic ash, moreover, is not limited to alteration to montmorillonite. It can assume other aspects according to the influence of the weathering environment and arrive at the stage of kaolinite and bauxite; this will be studied in chapter V with the weathering of rocks.

8° Clays, being surface materials, are especially well developed in the alteration and sedimentation phenomena. But hydrothermal veins and volcanism are also phenomena of the surface of the crust, in which water intervenes. Therefore there is an interference between these two types of aqueous environments, the supergene and the hypogene, and it is for this reason that the limits of these phenomena must be precisely defined.

CHAPTER III

GEOCHEMISTRY OF IONS IN THE HYDROSPHERE

Clays belong to the surface of Earth's crust because metamorphism at depth transforms them very quickly into silicates of entirely different sizes and behavior. The geochemistry of clays thus is part of the geochemistry of surface where water plays a leading role. Three aspects of the geochemistry of hydrosphere will be presented in this chapter.

1° In mineralogy and petrography one is accustomed to reasoning on the basis of the behavior of ions in minerals, and especially in silicates. But when these ions are released by weathering and when they migrate in natural solutions, the rules governing their behavior in water are quite different.

2° Among the ions of prime importance for our purpose silicon and aluminum are to be put in first place, since clays are in most cases silico-aluminous structures. The chemistry of silica and alumina in solution has seen great improvements which must be taken into account.

3° Lastly, natural solutions have characteristics which depend on the dissolved elements they contain. These result from the weathering of rocks and minerals, and chiefly of silicates, which constitute at least nine-tenths of Earth's crust. It will be useful to recall the processes of silicate alteration.

These three aspects of the geochemistry of surface will be entitled:

I. *The behavior of ions in water;*
II. *Solubility of silica and alumina;*
III. *Hydrolysis of silicates.*

I. — THE BEHAVIOR OF IONS IN WATER

1° *Hypothesis of Goldschmidt: ionic potential*

GOLDSCHMIDT (1934, 1937, 1945), the founder of modern geochemistry, investigated the rules which lead to variations in sedimentary rocks as a function of their constituent elements. GOLDSCHMIDT used a function, defined by CARTLEDGE (1928), which is called the ionic potential. The ionic potential is the quotient of charge Z and ionic radius r: $Q = Z/r$.

GOLDSCHMIDT shows the principal ions found in sediments on a graph of the ionic radii as a function of the charge. The graph allows one to separate the elements into three groups (fig. 6.):

Group 1. Ions with a weak ionic potential which remain in true solution up to high values of pH; examples: Na, K, Ca, Mg.

Group 2. Ions with an intermediate ionic potential which precipitate by hydrolysis in the form of hydroxides at still low alkaline pH values; examples: Al, Fe, Si.

Group 3. Ions with a strong ionic potential which form complex anions with oxygen and commonly result once more in true ionic solutions; examples: P_2O_5, SO_3, CO_3, etc.

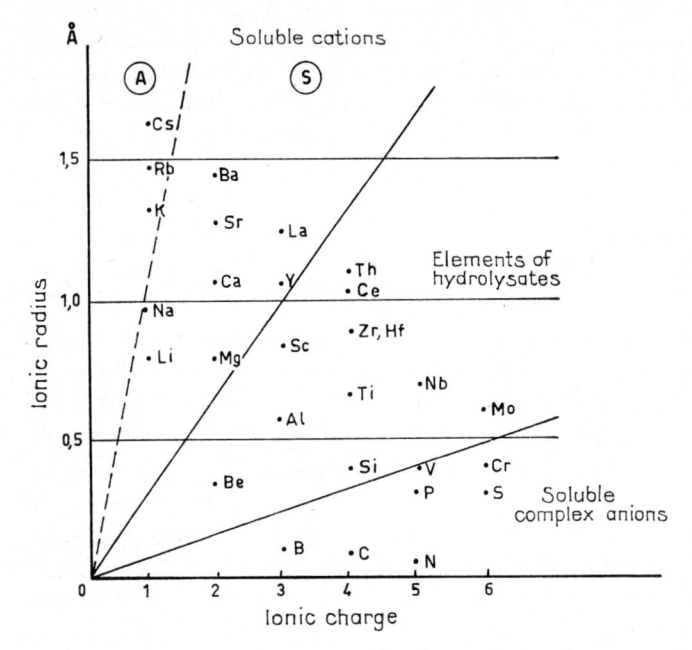

FIG. 6. — *Distribution of elements according to ionic potential* (after GOLDSCHMIDT, 1954, modified to distinguish the Antistokes —A— and Stokes —S— soluble cations).

GOLDSCHMIDT himself admitted in 1934 that the physical interpretation of this function was not easy to understand; later works have thrown light on this point.

2° Interpretation of Wickmann (1945), according to Bernal and Megaw (1935)

WICKMANN, anxious to give a better interpretation of the former considerations, used the works of BERNAL and MEGAW on the hydrogen ion-hydroxyl ion bond.

In the hydroxyl ion, hydrogen is bound to oxygen in such a manner that it does not occupy any space. This hydrogen is situated within the limits of the oxygen radius. So constituted, the $(OH)^-$ ion must have a cylindrical symmetry resulting in a simple dipole.

If an increasing and divergent electrical field is applied to such an ion, one will expect this ion to pass through three stages:

1° In the first case, where polarization is weak, it retains its cylindrical polar symmetry and increases simply its dipole moment.

2° In the second case, as polarization increases, the hydrogen atom is separated farther from the oxygen.

3° In the third case, with a still stronger polarization, the energy necessary to separate the hydrogen from the oxygen becomes weak. In these conditions, the H^+ ion is almost released and can effect a migration toward a neighboring oxygen.

The concrete effects on the hydroxyl ion of the application of such increasing fields by ions are the following:

1° The ions of large diameter (Na, K, Ca, Mg) and with only one or two charges spread them over a broad surface. They exert on hydroxyl ions only a weak field. Examples: $Na-OH$, $Ca-(OH)_2$. These ions result in alkaline solutions.

2° The ions of intermediate diameter (Al, Fe) and whose charge is 2 or 3 exert an action of only intermediate intensity on hydroxyl ions. We have here the amphoteric hydroxides which are able to react as acids or as bases, according to their environment. Moreover, their stability in solution is weak. Examples: $Al(OH)_3$, $Fe(OH)_3$, etc.

3° The ions of small diameter (B, P, S ...) and with a very high charge (3, 5, 6) develop an important electrical field over their surface. The hydrogen of the hydroxyl ion is released and becomes an acid hydrogen, whereas the oxygen is annexed and goes into complex anions. The ions B, P, S, thus, do not exist in solution, but are found within the structures of the anions BO_3, PO_4, SO_4, etc.

Today it is thus obvious that the ionic potential is a simple mode of expression of the power of polarization.

3° *The radius of hydrated ions in water*

The reasoning of GOLDSCHMIDT is based upon the ionic radius as it is measured and calculated in crystals with a predominantly ionic bonding. When ions leave the silicate lattice to enter into solution, the ionic bonds are broken and the field is exerted on water molecules, which can be represented as small dipoles. In other terms, the cations are polarizing and the water dipoles are polarizable. The latter cover the surface of ions, and the cations are hydrated. One can immediately deduce that their apparent and measurable physical diameter has changed because it now comprises that of the ion increased by the thickness of the water of hydration.

In point of fact, the majority of the monatomic ions are hydrated in water. DARMOIS (1946) made a review of the methods of measuring the radius of the hydrated ion. Different methods give different results, according to the property used for the measurement. DARMOIS proposed a simple method that uses the conductibility and the density of solutions; Miss SUTRA (1946, 1949) has applied it methodically.

The conductibility of strong electrolytic solutions depends on the mobility of ions, which decreases as the concentration increases; it is only when dilution is infinite that mobility is a characteristic property. One is thus led to extrapolate conductibilities. Therefrom one can determine the size of the ion in solution by assuming that the hydrodynamic braking obeys STOKES' law: $F = 6\pi\eta r v$ (η: viscosity; r: radius; v: velocity). This necessitates a series of conditions: the ion is spherical, the velocity is uniform, the equations of movement are linear, the symmetry of the phenomenon is axial, the liquid adheres to the sphere, and the initial state is at rest. In addition, the demonstration is based on the hydrodynamics of continuous media; for instance, in water the sphere must have a radius large enough for the liquid to be considered continuous. All these conditions must be fulfilled if STOKES' law is to be applied.

Certain ions satisfy correctly these requirements and follow STOKES' law. Other ions do not; they are called "antistokes". The results obtained for the monatomic cations which interest us here are presented in Table IX. In the same table are given the STOKES' radii of SUTRA (1946) and the ionic radii provided by GOLDSCHMIDT et al. (1926), by PAULING (1927) and by AHRENS (1952). The values of GOLDSCHMIDT and PAULING can be found in GOLDSCHMIDT (1954).

It can be seen that the majority of the ions have a STOKES' radius much larger than the radius measured in crystals. This results from hydration, which is the very process

TABLE IX. — COMPARISON BETWEEN IONIC RADII IN WATER AND IONIC RADII IN CRYSTAL STRUCTURES

	Stokes' radii in water	Ionic radii in crystals with heteropolar or ionic bond		
	SUTRA (1946)	GOLDSCHMIDT (1926)	PAULING (1927)	AHRENS (1952)
Al	4.57	0.57	0.50	0.51
Cr^{3+}	4.09	0.64		0.63
Fr^{3+}	4.02	0.67		0.64
Mg	3.45	0.78	0.65	0.66
Fe^{2+}	3.42	0.83	0.75	0.74
Co^{2+}	3.42	0.82	0.72	0.72
Ni	3.42	0.78	0.69	0.69
Mn^{+2}	3.42	0.91	0.80	0.80
Ca	3.07	1.06	0.99	0.99
Sr	3.07	1.27	1.13	1.12
Ba	2.87	1.43	1.35	1.34
Li	2.36	0.70	0.60	0.68
Na	1.83	0.98	0.95	0.97
K	1.24 ⎫	1.33	1.33	1.33
Rb	1.20 ⎬ ANTISTOKES	1.49	1.48	1.47
Cs	1.19 ⎭	1.65	1.69	1.67

by which they increase up to the sizes necessary for the application of the law. In fact, when an ion is introduced into water, an important polarization occurs because of the polar character of water and continues up to the point of the annexation of n water molecules, to which is attributed a diameter of 1.38 Å. It is probably the sum "dry ion" $+ n\, H_2O$ which constitutes the STOKES' radius.

On the contrary, there are ions which, because they are not hydrated, remain too small to follow STOKES' law. These "antistokes" are among the monatomic cations: K, Rb, Cs. It is thought that these monovalent ions develop over their large surface a field too weak to annex water molecules; from another point of view, they remain too small for water to constitute a continuous medium for them, and they "fall" into the holes in the water structure. These effects prevent them from obeying correctly the braking law of STOKES.

4° *Behavior of alkalis in surface geochemistry*

The outstanding behavior of potassium in comparison to that of sodium in the hydrosphere has been noted for a long time (NOLL, 1931; URBAIN, 1933; GOLDSCHMIDT, 1937; HARVEY, 1949). In order to give an account of this geochemical behavior, I have used the results of SUTRA (MILLOT, 1949).

The K/Na ratio is equal to 2.8 for schists, 3.3 for sandstones, 7.7 for limestones, it is close to 3 for the ensemble of sediments. It can be said therefore, that sediments, whatever they may be, adsorb potassium preferentially.

On the other hand, the K/Na ratio is equal to 1/10 in fresh water, and to 1/28.5 in sea water. One sees that in the course of continental weathering sodium turns out to be much more mobile than potassium and dominates the latter in natural solutions. Potassium is stored up and conserved in a preferential way, on the average, in weathering products. In order to account for the sodic character of sea water vis-à-vis the potassic character of sedimentary rocks, one must call upon neoformation of potash minerals, we will come to this point later, and also upon a gigantic preferential adsorption.

The underlying reason for these phenomena, if one accepts the apparent STOKES' volumes, is the following: the radius of K in solution is equal to about 2/3 of that of Na. The surface areas are in a 4/9 ratio and the volumes in a 8/27 ratio, i.e., just slightly less than 1/3. The sodium ion is three times more voluminous than the potassium ion.

Whatever the complex mechanism of the ion adsorption in water may be with respect to the quite varied particles in sediments, this adsorption is in inverse ratio to a power of the volume; and the potassium will be adsorbed and withdrawn from solutions much more strongly than sodium.

Rubidium and cesium have a behavior analogous to that of potassium, but are, in point of fact, adsorbed and concentrated much more strongly in marine sedimentary rocks (GOLDSCHMIDT, 1937). The lack of hydration explains their behavior.

In addition, it should be noted that the introduction of exchangeable bases into minerals with variable basal spacing, such as montmorillonite or vermiculite, induces variation of this basal spacing (MERING, 1946; BARSHAD, 1948). When potassium is introduced as the exchangeable base, the basal spacing of the mineral remains near 10 Å; however, this basal spacing reaches 12, then 14 Å, when Na, Ca, Mg are introduced; this increase is due to the hydration of these ions. The basal spacing remains fixed at 10 Å in the case of Rb and Cs. All this must be related to the hydration of ions.

Thus, potassium and sodium have quite different behaviors in the hydrosphere, whereas they are so frequently similar and associated in crystalline systems, wherein their large size allows them to fill the holes in the crystalline structures.

WELBY (1958), approaching this problem, used different values for the radii of the hydrated alkaline ions: Li, 3.65; Na, 2.80; K, 1.20; Rb, 1.80; Cs, 1.67 (identical with the crystalline radius). But the relative results are the same. The ratio of the hydrated radii of K and Na is equal to 1.8/2.8, i.e., about 2/3, and the volumes are in a ratio of about 1/3. Even if the values differ, the ratios are corroborated and the differential action of the two elements must be the same. Some examples of that will be examined later.

Conclusion and classification

It is obvious that the comparison of the ionic radius of an ion in a crystal lattice, where the ionic bonds are predominant, with the apparent radius of this ion in solution, is a rather formal exercise, more especially since the methods of measurement are different. But it has been established that the behavior of the ions in crystals is partly controlled by their size, which determines their coordination. It has likewise been established that the laws of stability, migration, adsorption and evolution of ions in solutions depend also on their hydration. It is for these reasons that, if comparisons between volumes are made, one arrives at informative results.

In a crystal, the higher the electrical field developed over the surface of an ion, the smaller will be the space which belongs to the ion vis-à-vis its neighbors, the space which is called the ionic radius.

In water, the stronger the field developed over the surface of an ion, the greater the capacity for hydration, and the greater its apparent radius in the solvent will be.

Thus, aluminum in an ionic crystal has a crystalline radius which is about one-ninth of its STOKES' radius in solution; this means that its apparent volume in water is 730 times greater than the space available for it in a silicate. On the contrary, the potassium ion, the largest of the common cations in silicates, does not change its volume in solution. In a crystal it was almost 70 times larger than aluminum, but in water it will behave as if 10 times smaller.

For the geologist who approaches the geochemistry of the hydrosphere, the comparison between these different reactions is extremely illuminating, as we shall be able to verify. A single physical cause produces different effects, according to the medium in which these elements evolve. *The crystal medium and the aqueous medium oppose each other in an important way; the transition from the first to the second during weathering and from the second to the first during crystal growth occurs with a change of comportment that must be not neglected.*

Therefore, according to the point of view of hydrosphere geochemistry, ions in solution must be classified in four great divisions, which are the following:

First group: A. Non-hydrated ions (antistokes): The ions with a very weak ionic potential representing the non-hydrated cations, e.g., K, Rb, Cs.

First group: B. Hydrated ions (stokes): The ions with a weak ionic potential representing the soluble hydrated cations, e.g., Ca, Mg, Ba.

Second group: The ions with an intermediate ionic potential constituting the amphoteric hydroxides, easy to precipitate, e.g., Al, Fe, Cr, Si.

Third group: The ions with a strong ionic potential constituting the soluble acid hydroxides, e.g., B, C, S, P.

This classification can be presented on the graph of GOLDSCHMIDT, modified to show the non-hydrated monatomic cations (fig. 6, p. 50).

II. — SOLUBILITY OF SILICA AND ALUMINA

1° History

The study of clays and silicates in the hydrosphere requires knowledge of the behavior of silica in water. If silicates can be both formed and destroyed in the hydrosphere, then there must exist silica-silicate equilibria which are indispensable to define.

In geology we have been caught in a kind of tenacious tradition disseminated throughout the decades since the beginning of century. This tradition considers that the silica of natural waters is in a colloidal state. It dates from about 1900 when brilliant works on the colloidal state and on colloidal silica were carried out by chemists. Geologists, in difficulty to explain the genesis of flint and chert and silicifications, jumped at this new possibility: they inferred from it that silicifications were the product of the flocculation of natural colloidal silica.

The first warning vis-à-vis this tradition was provided by CORRENS and his collaborators (1938, 1939, 1940). They showed that during weathering silica and alumina passed into ionic solution and could nourish neoformations. Very important was also the work of ROY (1945), which I already used in the problems of clay genesis (MILLOT, 1949). ROY presented to geologists a comparative review of the geological and chemical literature on this important question. Since the time of KAHLENBERG and LINCOLN (1898) geologists have not ceased considering that the silica of natural solutions was colloidal, and this is still often taught today. On the contrary, chemists, owing to experimental work, evolved in a divergent manner.

By the weight method, HARMAN (1925, 1926 and 1927) had already demonstrated the basic errors of KAHLENBERG and LINCOLN. Likewise CORRENS (1941) was able to establish by gravimetric method a curve of the solubility of silica in water as a function of pH, but it seems that the presence of a small fraction of colloidal micelles modified the plot.

Today one uses the colorimetric method, which was suggested by the precursors JOLLES and NEURATH (1898), but which was introduced as the current method by DIENERT and WANDENBULCKE (1923). By constant improvements we come to the modern works.

2° Solubility of silica

The parallel works of American and Japanese researchers provide our present-day documentation on solubility of silica: ALEXANDER (1953); GOTO, OKURA and KAYAMA (1953); ALEXANDER, HESTON and ILER (1954); ILER (1955); KRAUSKOPF (1956); WHITE, MURATA and BRANNOCK (1956); OKAMOTO, OKURA and GOTO (1957).

Solubility of amorphous silica. — The solubility of silica is regulated by the laws of polymerization and depolymerization. As long as the total amount of silica in water is less than 100-140 ppm (parts per million) at 25 °C, silica is dispersed, at equilibrium, in the form of *monomolecules* of monosilicic acid: $Si(OH)_4$; this is a true solution.

This solubility of silica in water is *practically independent* of pH, as long as pH remains lower than 9. Above pH 9 monosilicic acid dissociates, and the true solubility of silica increases (fig. 7).

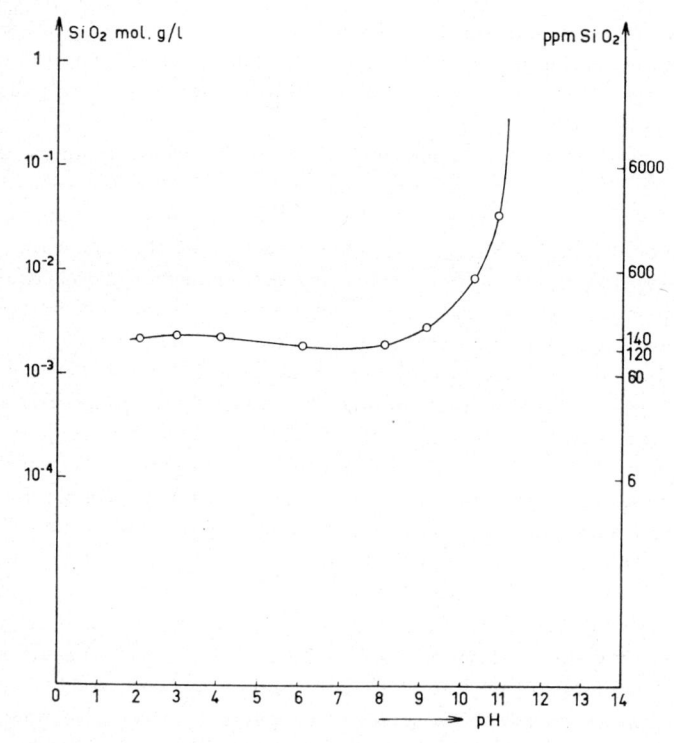

FIG. 7. — *Solubility of silica as a function of pH* (after KRAUSKOPF, 1956, 1959).

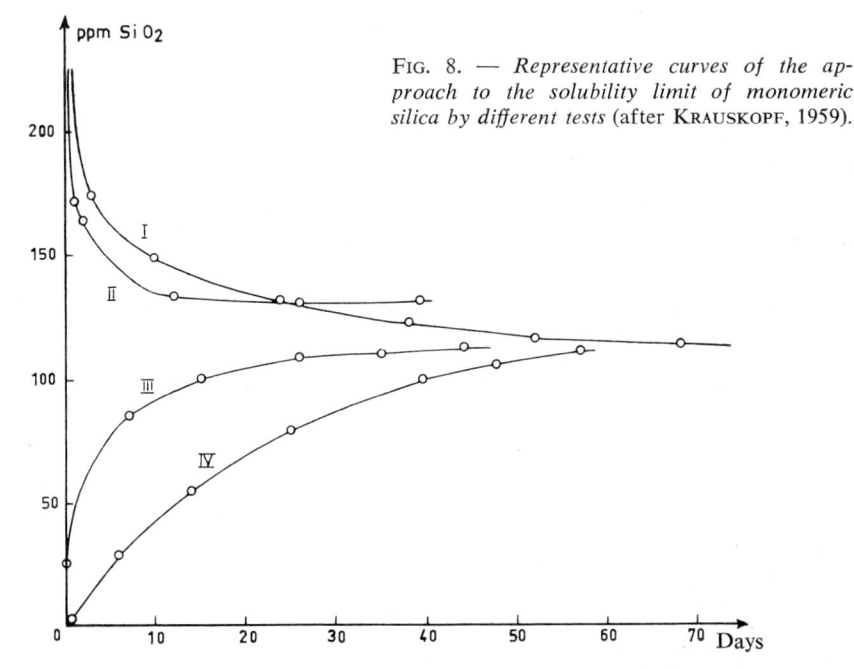

FIG. 8. — *Representative curves of the approach to the solubility limit of monomeric silica by different tests* (after KRAUSKOPF, 1959).

Curve I: Water from a hot spring was boiled to dissolve the maximum of silica. This water is cooled and its evolution studied.

pH close to 8.
Total amount of silica: 320 ppm.
Amount of monomeric silica:
— initial amount: 284 ppm
— after 40 days : 120 ppm.

Curve II: A Na_2SiO_3 solution is acidified by HCl. Its evolution is studied.
pH = 8.
Total amount of silica: 975 ppm.
Amount of monomeric silica:
— initial amount: 544 ppm
— after 40 days: 130 ppm.

Curve III: A concentrated Na_2SiO_3 solution acidified by HCl is allowed to age (silica becomes polymerized), then it is diluted, and one studies its evolution.

pH = 8.
Total amount of silica : 187 ppm.
Amount of monomeric silica:
— initial amount: 25 ppm
— after 40 days: 120 ppm.

Remark: The concentrated and aged solution contains polymers and about 120 ppm of monomeric silica. By dilution the amount of monomeric silica is decreased to 25 ppm; it will increase at the expense of the polymers present.

Curve IV: Silica gel in the presence of distilled water. pH = 5; the initial amount of monomeric silica is zero.

If the total amount of silica in water goes beyond 140 ppm (at 25 °C and at pH less than 9), the excess silica will be found in the form of condensed molecules, or polymers, along side the monomers of $Si(OH)_4$. The presence of these polymers characterizes the colloidal solutions of silica. The process is reversible; if solutions are diluted, and if the total amount of silica falls below 120 ppm, depolymerization leads to a true solution: amorphous silica is dissolved.

All this concerns solutions in equilibrium and we must insist on the fact that *these equilibria are established slowly*. A fresh solution can be lacking in polymers and contain an amount of monomeric silica greater than 140 ppm. But then it evolves towards equilibrium by formation of polymers until there remains only 140 ppm in true solution. The contrary is verified in the same manner (fig. 8).

Solubility of the other forms of silica. — The solubility of the other forms of silica is lower, although measurements are still not very numerous (KRAUSKOPF, 1959). However, partially dehydrated silica gel has the same solubility as amorphous silica. As for quartz, measurements are extremely difficult because of the slowness with which equilibrium is established and because of the necessity to avoid impurities, especially aluminum. The measurements carried out and thermodynamic calculations show a solubility of 7 to 14 ppm at 25 °C.

The same difficulties are encountered for opal and chalcedony. A trial measurement of solution from a diatomite was performed, but equilibrium did not seem to be reached at the end of two years (WHITE, in KRAUSKOPF, 1959); the measured solubilities were 22 to 34 ppm. For chalcedony, values were similar to those of quartz. Recent

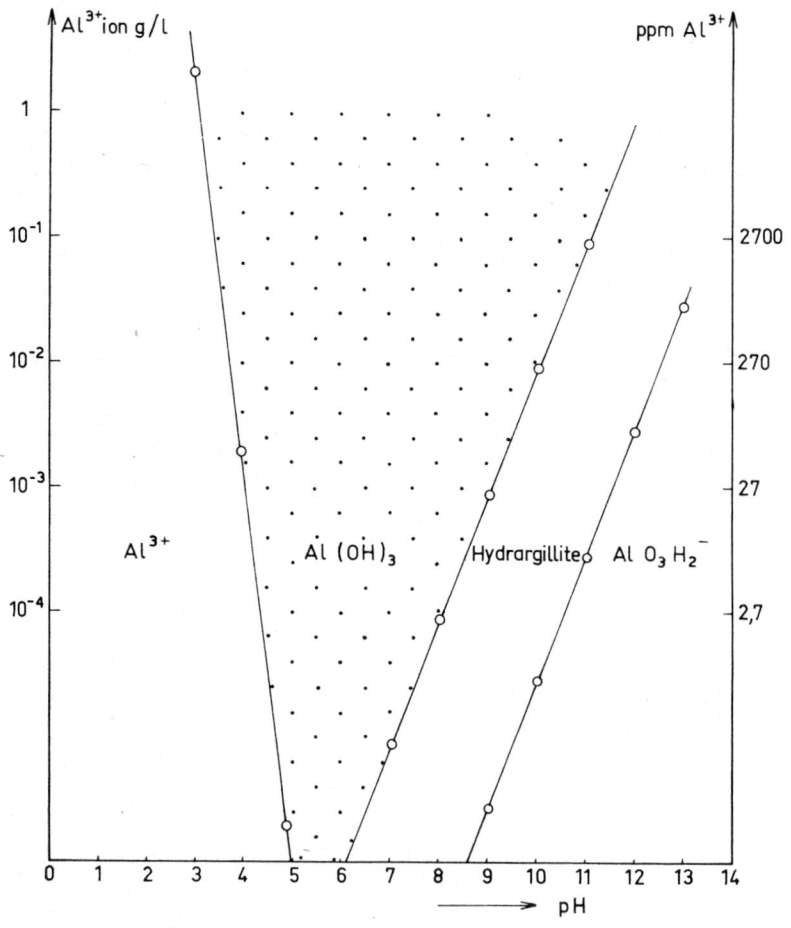

FIG. 9. — *Solubility of alumina as a function of pH* (after WEY, 1962).

measurements were made by WEY and SIFFERT (1961); solubilities of three samples of quartz, cristobalite and opal are shown in figure 61 in comparison with that of amorphous silica. Slight differences appear between the three crystalline varieties of silica, but they remain less than 20 ppm at the end of twenty days.

It can be assumed that solubilities of the natural forms of silica vary with the structure and the degree of crystallization, but the essential point is that *they are less than the solubility of amorphous silica*.

3° Solubility of alumina

Hydrated alumina, $Al(OH)_3$, is an amphoteric hydroxide. It is soluble in an acid medium, yielding Al^{3+} cations, and it is also soluble in an alkaline medium in the form of $AlO_3H_2^-$ anions. The pH values for which Al^{3+} and $Al(OH)_3$, on the one hand, and $Al(OH)_3$ and AlO_3H_2, on the other, can be calculated if the solubility product of alumina in equilibrium either with the cation, or with the anion, is known (CHARLOT, 1949). Figure 9 reproduces the variation of these dissolution pH values as a function of the ionic concentration. The dotted portion represents the insolubility field of the hydroxide when its degree of organization is low: amorphous alumina, pseudoboehmite. If the precipitate ages, it generally undergoes a crystalline reorganization and its solubility decreases. This is especially obvious with respect to bases; the insolubility field expands towards the higher pH values. A third straight line represents this dissolution limit of hydrargillite or gibbsite.

4° Comparison of the solubilities of silica and alumina

Figure 10 superposes the curve of alumina solubility upon the curve of silica solubility as unanimously accepted today.

The relative behaviors of the two oxides are immediately apparent:

— In an acid medium, alumina is more soluble than silica;

— In a neutral medium alumina is insoluble, whereas silica retains its solubility;

— In an alkaline medium the two solubilities meet and increase together.

These comparisons are very instructive and will serve as a guide for us in the interpretation of natural phenomena. However, it is necessary to exercise caution and to examine the influence of other ions present in solutions upon the solubilities of silica and alumina.

5° Influence of ions present in solutions

The solubility of monosilicic acid in water is *not affected* by the presence of ions in solution in water, and especially not by the *ions in sea water*. This eliminates entirely the hypotheses about the precipitation of silica under the influence of saline waters. OKAMOTO et al. (1957) pointed out that the only Al^{3+} ion, the abundance of which,

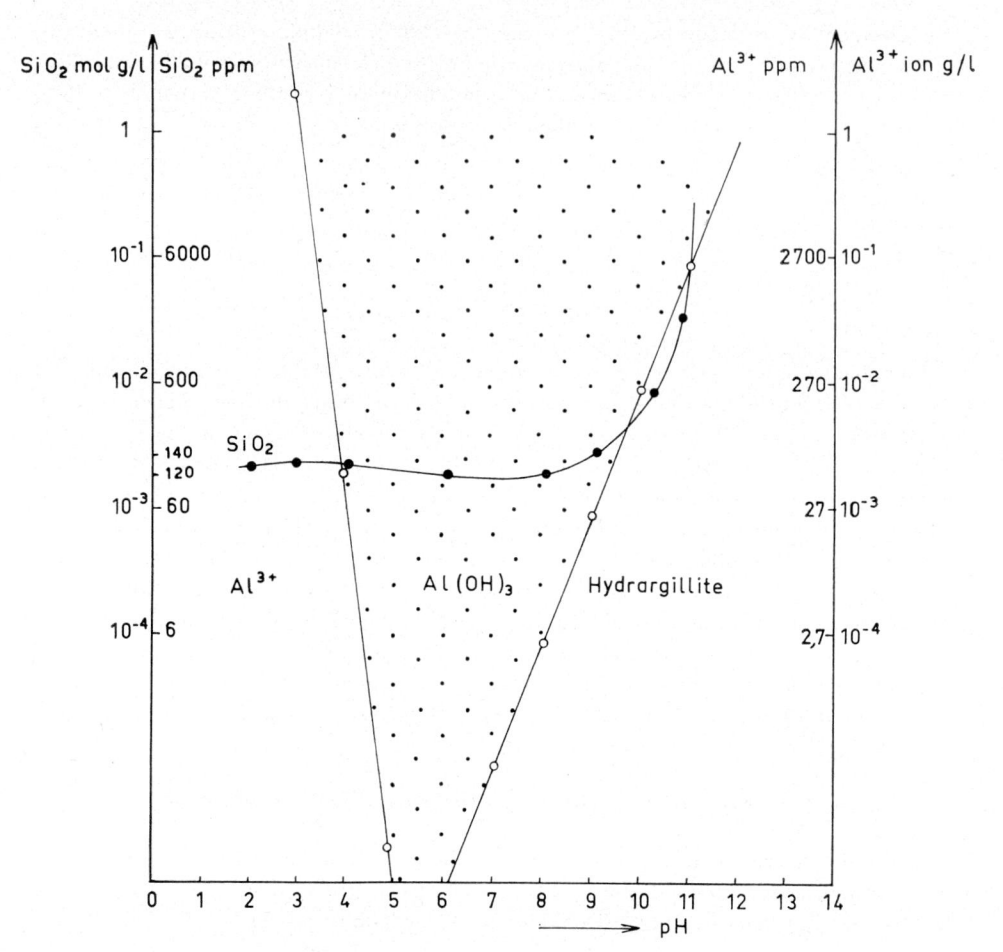

FIG. 10. — *Superposed curves of the solubility of silica and of alumina as functions of pH* (after KRAUSKOPF, 1956 and WEY, 1962).

however, is very small in sea water, is able to reduce the solubility of silica in water. An amount of 20 ppm of Al^{3+} reduces the solubility of silica to 15 ppm at pH values of 8 to 9; this is a considerable influence.

WEY and SIFFERT (1962) made a systematic study of the action of Al^{3+} on a saturated solution of silica. Moreover, they showed that the Mg^{2+} ion also acted under special conditions. The examination of the curves of WEY and SIFFERT is very instructive.

Figure 11 shows the variation in the concentration of SiO_2 and Al^{3+} as a function of pH, in a solution saturated with monomeric silica and containing Al^{3+} introduced in the form of sulfate. Between pH 5 and pH 10.5, a field in which alumina is scarcely soluble, the solubility of silica decreases strongly. At pH 8.5, 20 ppm of silica and less than 1 ppm of Al^{3+} remain in solution. It can be seen that the concentration of silica in the silica-alumina co-precipitate increases when NaCl is present in the solution.

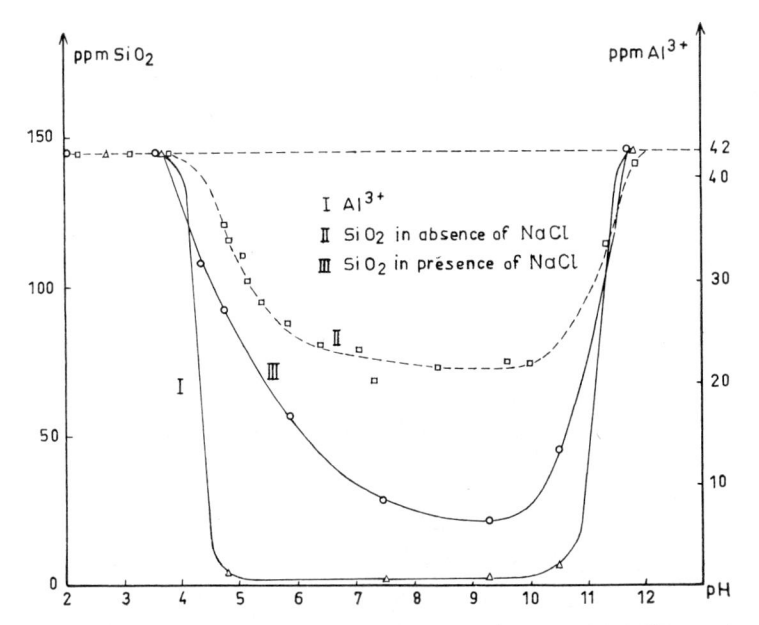

FIG. 11. — *Variation as a function of pH of the amount of SiO_2 and Al^{3+} of a solution saturated with monomeric silica and containing Al^{3+} introduced in the form of sulfate* (after WEY and SIFFERT, 1961).

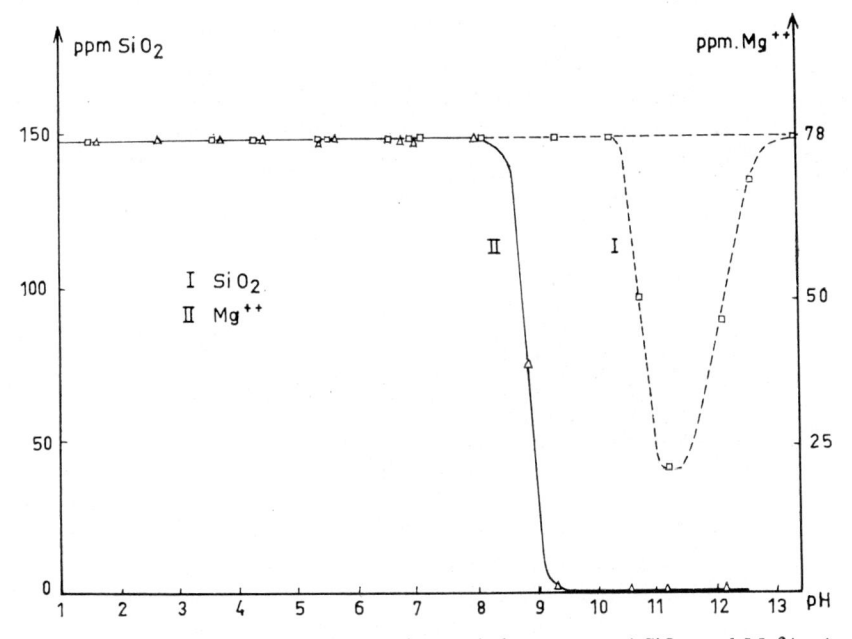

FIG. 12. — *Variation as a function of pH of the amount of SiO_2 and Mg^{2+} of a solution saturated with monomeric silica and containing an equimolecular quantity of Mg^{2+} introduced in the form of chloride* (after WEY and SIFFERT, 1961).

Figure 12 shows the variation in the amount of Mg^{2+} as a function of pH, of a solution saturated with monomeric silica and containing an equimolecular quantity of Mg, introduced in the form of $MgCl_2$. It is obvious that the solubility of silica decreases rapidly in the narrow pH range between 10 and 12. In this field, which is beyond natural pH conditions, the precipitation of magnesia accompanies that of silica. SIFFERT (1962) showed more recently that this common precipitation shifts toward pH 7 for higher concentrations of Mg^{2+}.

The method used in industry for the last 20 years to purify hot water which contains silica are based on this property of the mixed precipitation of magnesia and silica.

6° Geochemical consequences

These new chemical data on the solubility of silica and on the influence of ions upon the solubility of silica and alumina have considerable repercussion in geochemistry. They are the basis for a complete revision of our geological logic.

The following chapters of this book will allow us to illustrate that importance, but let us here evoke only a few important points.

— The alteration of silicates will be governed by these physico-chemical data; thus it will be possible to give interpretations of the different possibilities offered by nature.

— The neoformation of clays is directly dependant upon these rules, as confirmed by the experiments on synthesis: this will be investigated in Chapter XI.

— The problem of silicification is completely renewed. In actuality, the concentration of silica in natural solutions (fresh water, sea water, connate water of sedimentary rocks, soil solution) is clearly lower than 120-140 ppm. The silica of natural waters is not in colloidal solution, but in true solution. Thus, it is not possible to call upon flocculation, and a quite different approach will be investigated in Chapter IX.

III. — HYDROLYSIS OF SILICATES

No crystal is insoluble. An element which is called insoluble in common language is simply very slightly soluble. No matter what crystal is put into water, there will be ions migrating from the crystal surface into the aqueous medium. This includes the silicates: in contact with water they are altered by hydrolysis. A very old experiment shows that the powder of a silicate rock, left in a container with distilled water, transmits an alkaline reaction to the latter. If water percolates through a vertical tube filled with granite powder, a pH of 9.6 can be measured in the collected solution. These phenomena are interpreted as being the results of hydrolytic reactions between the crystal ions and the solution. Silicates can be considered as salts of weak acids and strong bases; the reaction of hydrolysis is thus alkaline.

These hydrolytic decompositions, or, more briefly, hydrolyses of silicates, were investigated by numerous and important studies performed in most cases on feldspars.

Certainly, the latter are by far the most abundant in silicate rocks. As early as 1938, CORRENS and ENGELHARDT showed that *feldspars release all their constituents in the ionic state*, and they studied the dynamics of this alteration under different conditions of pH. Numerous experiments have been performed since that time; they will be cited in Chapter V on weathering and soil clays.

It is often necessary to visualize natural processes for oneself. These visualizations are partly hypothetical, but they guide our observations and experiments which allow, in their turn, modification or specification of the starting scheme. For that reason I shall present here some views on the mechanism of feldspar hydrolysis that will lead us to the notion of what is essential in the process, and also to foresee variations of its intensity.

1° **Hydration of feldspar**

The relations which take place between the crystal surface and water have been approached in two schematic ways.

FREDERICKSON (1951) made comparisons between the structure of water, such as it was imagined by BERNAL and FOWLER (1933), and the structure of albite, as an example; then he considered a kind of arrangement of the adjacent layer of water onto the crystal. On the surface of albite there are exposed oxygens whose negative valences are not entirely satisfied. Water adheres to the crystal surface like a film, and the positive poles of water molecules saturate the negative surface offered at the contact. The negative poles of water face outward, where they will attract the positive poles of the following layer, and so on. As one moves farther away from the crystal, the organization of water layers decreases. On the contrary, in contact with albite, the first layer of water would have a structure rather similar to that of ice and would be arranged with respect to the albite surface in a manner similar to an epitaxy. Then begins the exchange between the Na^+ cations of the albite and the H^+ ions of the solution.

DEVORE (1956) did not think that the hydration of the feldspar surface by means of the association between water molecules and unsaturated oxygen ions could weaken the bonds between the alkaline ions and the crystal structure. He supposed a permutation between the $(OH)^-$ hydroxyls of the solution and the oxygens of the Si-O-Si bonds. This permutation would be possible because the Si-OH bond would be stronger than the Si-O bond. When several Si-O-Si bonds are thus converted into Si-OH bonds, the alkaline ions are more weakly bound and available for exchange with the H^+ ions.

Sometimes it has been considered that the schemes on alteration of FREDERICKSON and of DEVORE are mutually exclusive. In fact, these two schemes can work very well together. Furthermore, they constitute only a first step, one which precedes the release of cations from the crystal by exchange with the H^+ ions of solution. Here begins the true attack.

2° **Cation exchange between water and crystal**

Whatever mechanism may be envisaged, the feldspar surface is in fact hydrated. Two media are in contact; one might say that they are at grips with each other. Their chemical compositions are very different, however; there is no sodium in water, and there is no hydrogen ion in the albite lattice. It is here that alteration begins under the influence of the H$^+$ ions. These hydrogen ions have an extremely small diameter for a monovalent charge. They will exert a strong field on the oxygen ions of the crystals. These oxygen ions will have the tendency to be bound to the hydrogen ions rather than the sodium ions, whose ionic potential is weaker (equal monovalent charge, but much larger diameter). The sodium ions are released and the surface of the albite lattice is mechanically distorted. The concentration of Na is high in the albite lattice and zero in the water film outside the crystal. The Na ions will migrate through the channels of the distorted lattice. The reaction thus begun will continue. As for the crystal, lattice is distorted, and there are holes in it; thus, the surface area increases and water can easily enter the structure. As for the liquid, the Na$^+$ ions move into layers of water in which the concentration of sodium is low. This migration is facilitated by thermal agitation, and if water is circulating, Na$^+$ is exported. The renewed supply of hydrogen ions is assured by rotation of molecules and thermal agitation (FREDERICKSON, 1951).

The same reasoning can be applied to calco-sodic and potassic feldspars (NASH and MARSHALL, 1956). Hydrolysis thus amounts to a base exchange phenomenon by means of extraction of cations in the lattice. The Na, K and Ca ions are released first because of their weak ionic potentials.

By means of experiments on the reactions of muscovite and adularia with water at low temperature, GARRELS and HOWARD (1959) determined the thermodynamic conditions of hydrolysis. It is experimentally shown that muscovite is more stable than potassic feldspar at 20 °C during alteration.

3° *Attack on the silicate framework*

To alter a feldspar or other silicate, it does not suffice simply to release alkaline or alkaline earth cations from its lattice. Tetrahedra that contain the Si and Al ions must also be attacked. Many pertinent experiments have been performed under different conditions.

Experiments carried out at moderate temperatures and with continuous leaching are easy enough to transpose into the natural environment. Information thereon was given to us first by CORRENS and ENGELHARDT as early as 1938. In the attack on orthoclase by continuous leaching, silica and alumina are released, but in proportions which are variable according to the pH conditions. *There is more alumina than silica in an acid medium, more silica than alumina in a neutral medium, and there are equal quantities of both elements in an alkaline medium.*

A glance at the compared curves of silica and alumina presented in figure 10 explains these facts. The release of silica and alumina depends on their solubility at different pH values. This was completely confirmed by the elegant experiments of PÉDRO (1960, 1961); in a neutral medium, silica is leached preferentially.

Experiments at high temperature and in a confined medium are more difficult to transpose into natural phenomena. In fact, solutions of finite volume become saturated, that which does not happen in nature. Moreover, the kinetics of the reactions change as a function of temperature. Lastly, the hydration of ions varies with temperature (WILHELM, 1962). It is thus necessary to distinguish between that which is transposable to natural phenomena and that which is not. LAGACHE, WYART and SABATIER (1960, 1961) performed experiments on the dissolution of albite and adularia at 200 °C in pure water and in water charged with CO_2. Elements do not go into solution in the proportions existing in the crystals. Alumina is almost insoluble in an acid as well as in an alkaline medium. Alcalis are easily put into solution, but their dissolution diminishes with increasing concentration of the solution. The solubility of silica is equal to 150 mg/l for albite and 110 mg/l for adularia, but saturation is also reached. *These experiments confirm that during the weathering of feldspars alumina accumulates in the solid state because its solubility is so slight between* pH 4 *and* 9 *or* 10. In addition, they show that the attack occurs in the manner of a true chemical reaction, the products of which go into solution or accumulate as an insoluble fraction. Finally, the accumulation of the dissolved products in the water works like a brake in the experimental attack, but not in nature where the environment is leached and where the weathering solutions are renewed.

In another way, MURATA (1942) studied the attack on silicates by strong acids. Some of these reactions yield insoluble aggregates of tiny silicate fragments. Others result in siliceous gels. Thus, the attack is much stronger in the latter case, and it is shown that in this case the Al/Si ratio is about 2/3. In the opposite case, the minerals are richer in silica and more resistant against acid attack.

This would indicate that the "extraction effect", studied with regard to alkaline and alkaline earth ions and which applies also to silicon and aluminum, is facilitated by the abundance of the latter, i.e., by the abundance of the Si-O-Al bonds which are weaker than the Si-O-Si bonds. When the Si-O-Al bonds are numerous, the crystal is pulverized and only the smallest fragments of silicate lattice remain, constituting the gel. This can be transposed at once into nature by the intermediary of PÉDRO's experiments (1958). During his attempts at experimental weathering, PÉDRO described the micro-division of crystals that resulted in amorphous units of globular aspect which he could observe by means of the electron microscope. Such globules have been observed in soils by SUDO (1954), AOMINE and YOSHINAGA (1955) and FIELDES (1955).

The previous paragraph, concerning the solubility of silica in water, provides the basis for the prediction that at equilibrium, after the necessary lapse of time, these polymers of large size will also go into solution. Thus one reaches the complete dissolution of the most vulnerable silicates.

In conclusion, the attack on tetrahedral frameworks is readily pursued, and the abundance of tetrahedral aluminum facilitates it. Thereafter, the release of silica and alumina into solutions is controlled by the solubility laws of these elements as a function of environmental conditions.

4° *Summary on the hydrolysis of silicates*

It is not necessary to emphasize how hypothetical and provisional our schematic views are. However, an overall view of the phenomenon can be achieved.

Common experience shows that silicates are not equally vulnerable, and minerals have been listed in order of increasing alterability. Then, attempts were made to understand the basis for these differences in terms of silicate crystallochemistry. Several approaches are possible:

a) IONIZATION POTENTIAL. — The ionization potential is the energy needed to transform an element into an ion in a given valence state. Each valence state has a corresponding ionization potential. This energy is measured in volts and can be obtained very accurately by spectroscopic methods. A list of ionization potentials is given by AHRENS (1952). The weaker the ionization potential of the ions contained in silicates, the easier it will be to put them into solution. Furthermore, the more abundant these elements are, the more vulnerable the crystalline framework will be.

b) STABILITY OF COORDINATION POLYHEDRA OF ELEMENTS IN CRYSTALS. — A second approach involves consideration of the stability of the coordination polyhedra which make up the silicates. These polyhedra, whose arrangement is governed by the laws of PAULING, are all the more stable as the number of their apexes is small and, consequently, as the cations which occupy their center have a high valence and a small diameter. Thus, it can be predicted that silicates composed of polyhedra with numerous apexes will be more vulnerable than those composed principally of tetrahedra.

c) THE RELATIVE ABUNDANCE OF TETRAHEDRA. — The same order of vulnerability is found again in the appreciation of the number of tetrahedra within a crystalline framework. When tetrahedra, which are the most stable polyhedra, are numerous, structure is less alterable. In this context, the considerations of FAIRBAIRN (1943) on the "*packing index*" should be mentioned.

These different approaches, summarized by ANDREATTA (1950), are interesting, *but they are not independent*. On the contrary, they are different modes of expression of a single reality, that announced by the founders of crystallochemistry, GOLDSCHMIDT and PAULING.

— Ions of small diameter and having a high ionization potential enter into polyhedra with few apexes that are resistant to hydrolysis.

— Ions of large diameter and having a weak ionization potential enter into bulky and vulnerable polyhedra.

Here is the essential point, and it is not necessary to resort to epiphenomena. Thus, the ionization potential does not play a direct role in alteration because the silicate constituents are already in an ionic state within crystalline structures. However the problem may be approached, one comes back to simple, practical considerations based on causes and not on effects or appearances.

— *Alterability of silicates depends on the abundance of ions of large diameter, small charge, weak ionic potential, involved in bulky polyhedra with relative weak bonds.*

— Moreover, alteration by hydrolysis will always progress most rapidly among the bulky and vulnerable polyhedra and only later will attack the small polyhedra. The extraction of ions here evoked is progressive. It takes place first in polyhedra with

12, 10 and 8 apexes occupied by K, Na, Ca. It is then pursued in octahedra with 6 apexes occupied by Fe, Mg, Al, and in the tetrahedra occupied by Al and Si, *the solubility of the "occupants" governing the speed of their release.* Contrary to that which is often stated, the solubility of the alumina of silicates under the common conditions of pH is less than that of silica.

Conclusion

The hydrolysis of silicates, all told, amounts to a phenomenon of base exchange between two media, water and crystal. The renewed supply of hydrogen ions allow the release of cations to continue. Being released, these cations diffuse outward into the water, where concentrations are low.

The silicate will be all the more vulnerable as it is richer in cations, since the siliceous framework will be easily broken down into very small fragments that can be transported away or dissolved.

But weathering will be all the stronger as the renewal and the amount of hydrogen ions increases. Here appears the influence of climates conducive to lessivage and of organic acids. We shall consider these factors again in the study of weathering and soil clays.

CHAPTER IV

THE PLACE OF CLAYS IN THE GEOCHEMICAL CYCLE

The reconstruction of the path and the behavior of chemical elements in the course of evolution of the earth's crust is fascinating work. The "geochemical cycle" is the evolutionary succession that occurs during the following stages: weathering, transportation, sedimentation, diagenesis, metamorphism, and genesis of crystalline rocks. Clays belong within the first four stages of this cycle, these being surficial stages. In fact, after that, they are transformed by metamorphism and granitization. Clays have thus a birth and a death, or better, several births and deaths during the geochemical cycle. In order to situate correctly the place of these births and the place of clays in the succession of events, it is necessary to study the main geochemical characteristics of sedimentary rock series. This study will be subdivided as follows:

I. *Geochemistry of normal sedimentary series.*

II. *Geodynamic interpretation of this series.*

III. *The positive and negative sedimentary sequences.*

IV. *The evolutionary series of pedological origin.*

V. *Confrontation and general view on sedimentary series.*

VI. *The place of clays in the surficial part of the geochemical cycle.*

I. — GEOCHEMISTRY OF THE NORMAL SEDIMENTARY SERIES
Goldschmidt (1937, 1945)

Sediments, piled up in a sedimentary series, succeed one another in an order which is not indifferent. To see a succession of gravels, sands, then "fines", and lastly limestones and hypersaline facies is a century-old observation of geology. This succession occurs in many sedimentary series and is commonly called the "sedimentary cycle", a term somewhat inappropriate, but traditional. Since it has been now agreed upon that the term "cycle" can be applied only to a succession which comes back to its starting point, we shall say "series".

These sedimentary series have been considered from the stratigraphic, paleogeographical and historical points of view; the geochemical approach is extremely instructive. It was GOLDSCHMIDT (1934, 1937, 1945) who treated this series in *geochemical terms*. He called it the "geochemical cycle of the sediments", but we shall call it the "geochemistry of the normal sedimentary series". I reproduce here his proposal verbatim; the succession of stages is the following:

1° **Insoluble residues,** such as sands and sandstones, with resistant minerals such as quartz and zircon.

2° **Hydrolysates,** such as bauxites, clays and shales, with hydrated oxides and aluminum hydroxides.

3° **Oxidates,** such as a great number of sedimentary ores of iron and manganese with Fe^{3+} and Mn^{3+} or Mn^{4+} oxides and hydroxides.

4° **Reduction sediments,** such as coal, bituminous sediments, sedimentary sulfides and sulfur.

5° **Carbonate sediments,** such as limestones and dolomites.

6° **Evaporite sediments** involving common salts, chlorides, sulfates and borates of the alkaline and alkaline-earth metals, as well as carbonates and, exceptionally, nitrates.

GOLDSCHMIDT (1945) then showed how each of these groups serves as a preferential shelter for some concentrations of elements.

In fact, group 1 concentrates, besides quartz which is its main constituent, minerals such as zircon, tourmaline, monazite, garnets, ilmenite, as well as gold and diamond; but this is a question of mechanical concentration owing to their resistance or hardness.

Group 2 brings together the companions of aluminum (Ga, Ti), those of magnesium (Ni, Sc, Sn, Nb, Ta, Sb and rare earth elements), and lastly those of potassium (Rb, Cs).

In group 3: P, V, Pb, Zn, Sb, As, Ge, Se, Mo and W along with iron and manganese are concentrated. This explains the natural purification of many toxic elements in marine water.

In group 4: As, V, Cu, Zn, In and Ag would also be found along with iron and sulfur.

Finally, in group 5 and 6 are found, besides Ca and Mg, important amounts of Sr, Ba, as well as the halogens (F, Cl, Br, I) and the large acid anions (SO_4, BO_4, etc.).

On the whole, and this is the main point, GOLDSCHMIDT saw in the sedimentary series the result of a gigantic semi-quantitative chemical analysis, which separates in successive lots the unsorted materials from the continents. Comparing the associated elements, GOLDSCHMIDT pointed out that similar behavior is related to similar ionic potentials. It was in the consideration of these cycles that he discovered the application of ionic potentials to the surficial geochemistry.

Discussion and modernization of Goldschmidt's geochemical series

The striking proposals of GOLDSCHMIDT are thirty years old. Furthermore, their author was much more chemist than geologist. For these reasons, this typical succession is deserving of nuances of interpretation at the present time and needs to be updated. This will be done point by point in order to give to this comprehensive overview a new visage which is in accord with present-day knowledge.

Insoluble residues. — This category remains as is without comment.

Hydrolysates. — Three observations are to be made. First, the image evoked by this term is misleading. If, in fact, the majority of the "hydrolysate elements" defined by GOLDSCHMIDT himself are grouped together here, it does not mean that clays form by precipitation of these elements from solutions. This does happen, but the clays which are situated at this place in the normal geochemical series are generally detrital.

Next, it is not possible today to group bauxites together with sedimentary clays. It is well known that they are ancient tropical ironstone crusts, belonging to a previous phase of emergence.

Finally, the term "aluminum hydrosilicate" has been abandoned; we have taken the habit of saying "clay minerals". Under these conditions one is hesitant to retain the term "hydrolysate". This was the term chosen by GOLDSCHMIDT and it evokes the elements grouped together here. Some authors say "colloids", but clays, even though sometimes of colloidal size, are minerals. One could say simply "clays", but we will see that clays occur at various points of the sedimentary geochemical series. It seems that there is no confusion and that the term "hydrolysate" can be retained without giving it causal significance and with the knowledge that this is a question of fine-grained detrital sedimentation.

Oxidates. — This category remains as a witness to the frequent occurrence of iron ores between clays and limestones, or alternating with them.

Reduction sediments. — This group is heterogeneous and its place here is very debatable. I propose to eliminate it.

In reality, coals correspond to periods of emergence and characterize the end of a series rather than an intermediate stage. We shall come across them again in reworked continental products. On the other hand, bituminous shales are ubiquitous. They only give evidence of reducing conditions, which can affect all facies. The bituminous character is accessory to diverse stages of the geochemical series, but it does not constitute one of these stages by itself. Likewise, the occurrence of sulfides is typical of the proliferation of sulfate-reducing bacteria under anaerobic conditions; this can be found in clays as well as in carbonates and evaporites.

Carbonate sediments. — This category remains without any modification.

Evaporite sediments. — The only comment necessary concerns the term itself. It is becoming to our geological inheritance to situate the birthplace of evaporites in kinds of immense natural salt marshes. Recent authors have been able to show that many of these facies, including the most important ones, such as the Keuper, were not subaerial lagoonal deposits, but submarine concentrations. A discussion on these important points has been presented by SLOSS (1953) and recently in French by RICOUR (1960).

The term "evaporite" evokes the same landscape and the same mechanism. Even if evaporation takes part in the concentration of natural solutions, it seems that a neutral term would be preferable and, for lack of a better word, the term "saline deposits" will be chosen.

Conclusion. — This presentation of the ensemble of GOLDSCHMIDT's geochemical series, now reduced to five terms, summarizes the successive stages of general sedimentation. None of the interruptions, repetitions, oscillating reiterations of sedimentary series, nor the frequent facies associations have been taken into account. We are concerned here with a general view in geochemical terms, according to the model of GOLDSCHMIDT. Many great sedimentary series reproduce this scheme:

1° *Insoluble residues.*
2° *Hydrolysates.*
3° *Oxidates.*
4° *Carbonates.*
5° *Saline deposits.*

II. — GEODYNAMIC INTERPRETATION OF THE NORMAL GEOCHEMICAL SERIES OF SEDIMENTS

I have already had the occasion to present (MILLOT, 1949) the general geodynamic interpretation of GOLDSCHMIDT's geochemical series. The method consists of the attempt to reconstruct the evolution of the continent at the margins of sedimentary basins.

1° After an orogenic phase, relief is great and waters run rapidly. The material of which the uplands are composed is made mobile; pebbles, gravels and sands are carried to the margins of the basin and are deposited. This does not mean that fines and dissolved products are not also transported, but they are masked by the mass of the coarse products, and they appear only farther offshore. *Insoluble residues* are deposited.

2° Relief then decreases. The energy of running water diminishes and is sufficient for the transport of fine-grained particles only. The latter settle out, at first alternating with sands, and then alone, and, under appropriate conditions, with an increasing amount of carbonate. The majority of the dissolved products however is evacuated to the open sea. Argillaceous series accumulate in this manner; they are the *hydrolysates.*

3° Here in some cases the *sedimentary oolitic iron ores* occur, which represent an extraordinary concentration of an outstanding and very useful element. This iron is accompanied by many minor or trace elements that complicate its utilization. To explain such accumulations, it is necessary to look for a common natural mechanism capable of bringing about a great concentration of this element. In 1949 I thought of lateritization, which is the major process of liberating iron on the surface of the globe, and this was supported by a favorable coincidence. At the end of the period of argillaceous sedimentation topographic relief has been greatly lowered. Running water no longer has the energy to transport particulate matter; dissolved products only can be moved. In geological terms we are dealing with a *peneplain*, and it is well known that peneplanation is a result of continental weathering. Two cases are possible here. Either this weathering is of the temperate or cold type, and the cycle leads to carbonates, or the weathering is tropical and lateritization takes place, releasing great amounts of iron. By his works on the role of pedogenesis in the formation of sedimentary rocks, ERHART (1956, 1961) succeeded in reconstituting many phases in the genesis of sedimentary iron ores as a function of seasonal exudations, vegetation, and leaching of pedogenetic origin leading to the release and concretionment of iron. These works allow one to explain the place of iron ores in GOLDSCHMIDT's cycle, i.e., at the limit between clays and limestones, a position that does not prevent them from alternating with such sediments.

4° With or without this episode, let us emphasize that we have gone beyond the epoch of mechanical transportation to enter that of transportation in solution. In actuality, the peneplaned continent is traversed by running water, which can be abundant, but whose energy is weak. Such streams can still be responsible for considerable transportation, but in an ionic state, i.e., in solution. Thus the deposition of salts soluble in water occurs, and that of the least soluble ones first. Calcium and magnesium carbonates yield limestones and dolomites, with or without the intermediary of living creatures who construct their skeletons or tests of these elements. This is the stage of *carbonates*.

5° Finally, one reaches the most subdued paleogeographic state. The completely degraded topography is traversed by only weakly circulating streams, transporting only the most soluble salts. At this point one arrives at the epoch in which the effects of the initial tectonics are cancelled out. In fact, *differences of potential of every kind are reduced* in the realm of running water, as well as in that of climate and of life itself. It is at this time that the unrenewed waters of sedimentary basins become overloaded with salts, sulfates or chlorides, and give rise to saline deposits. The example of the Permian and Triassic on a world-wide scale and that of some Nummulitic basins in Europe demonstrate this point. We have arrived at the stage of *saline deposits*.

It is in this way that the succession of geological terms of GOLDSCHMIDT can be presented from the point of view of dynamic geology. It is immediately necessary to note that this vertical succession should theoretically be found also horizontally. At the shore line gravels and sands are deposited, farther offshore clays are deposited, and then carbonates. This is exact in some situations, not only when the oceans are observed as a whole, but also in certain favorable circumstances. However, the question is complicated by the intervention of currents, by oceanic climates, by the geography of the sea bottom and by biology. The simple scheme is quickly troubled. This brings us to the realm of oceanography, where patient analyses have shown that the driving role of sediments supplied from the continents is modified by a thousand nuances.

But, even if the horizontal distribution of sedimentary segregations is complicated, simple views of the vertical succession are recorded by geological sections at the margins of continents or of emerging mountain chains.

All this obviously shows that sedimentation brings about a gigantic spatial sorting of the bulk material eroded from the continents. The various constituents of continental rocks are separated from each other; this is one of the most fertile fields in geochemistry. If one has the impression that *metamorphism and depth within the crust produce mixtures and averages, the surface of the Earth is a powerful analyzer of elements.*

III. — POSITIVE AND NEGATIVE SEDIMENTARY SEQUENCES

LOMBARD (1953-1956)

Detailed analysis of the succession of sedimentary facies has received a great emphasis during the last twenty years. LOMBARD has given us numerous examples and made the use of vocabulary more precise (LOMBARD, 1953, 1956). It is necessary to define rigorously the terms series, cycle and sequence.

The **general,** or **fundamental, virtual series** which serves as a basis for every analysis is the "cycle", or normal series, of GOLDSCHMIDT, i.e., the common suite of deposits from coarse- to fine-grained sediments and from fine-grained to chemical ones. "Series" is the most modest and the most suggestive term to designate this succession. The word "virtual" means that it is unusual to find this series complete in any one locality and, *a fortiori*, in a single quarry or cliff outcrop, but fragments of it are continually encountered. The term "general" or "fundamental", evokes the characteristics of total generality which are well known to every geologist and which were expressed in geochemical and geodynamic terms in the preceding paragraphs.

The **lithologic sequences,** or simply *sequences*, are precisely those fragments of the virtual series which are so frequently encountered. The best examples are the following: clays - marnes - limestones, or coarse-grained sandstone - fine-grained sandstone - argillaceous sandstone, or anhydrite - halite - sylvite. Some sequences recur frequently in sedimentary series, giving them, in the field, a striking periodic and alternating aspect. In such a case we speak of a *rhythmic series*. On the contrary, in some deposits such superposition of sequences is lacking; these are *arhythmic series*. At this point we must clarify that which is a true **cyclic series** ; this latter would be characterized by an *a-b-c* sequence followed by a *c-b-a* sequence; this is very rare and therefore the word cycle in sedimentation, although traditional, is inappropriate in the general case.

Sequences can be studied on all scales of observation: on the microscopic scale, in varves, for instance; on the macroscopic scale in banded rocks, and in quarries, cliffs and mine shafts; on the megascopic scale, by means of large geological sections that link together the major stratigraphic units, those of stage or system.

It is necessary to refer to the works of LOMBARD and of other specialists of these sequential analyses in order to appreciate the great interest of such analyses and the service they have rendered in the study of the history of sedimentary deposits. Widely varying interpretations have been offered to explain these successions of sequences within sedimentary series: tectonic pulsations, prolonged or seasonal climatic variations,

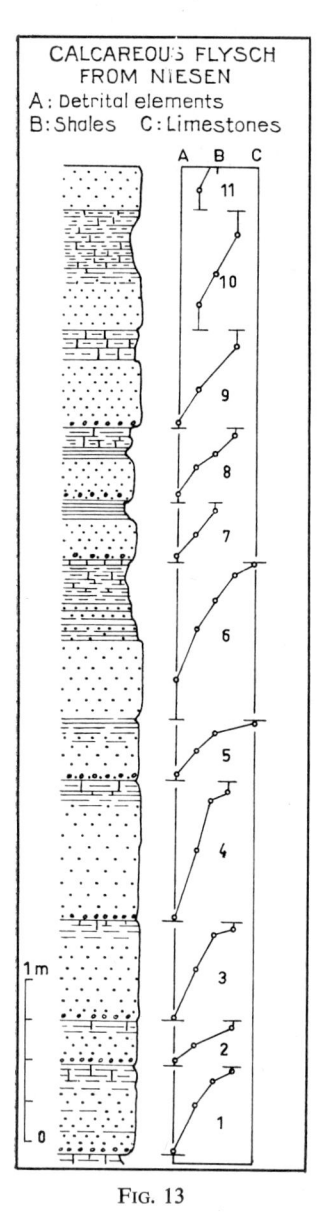

FIG. 13. — *Positive sequences in the detrital and calcareous flysch from the Niesen sheet (after LOMBARD, 1953).*

FIG. 14. — *Positive sequences in a typical section of the potash-bearing beds from the Basin of Mulhouse (LOMBARD, 1953, after MAIKOVSKI, 1941).*

FIG. 13

FIG. 14

shifting of streams bringing sediment from the continental platform, turbidity currents, spasmodic subsidence, and control by the bottom of the basin. Let us say that all this is related to the interference of crustal deformation, climatic variations, and circulation within the aquatic environment. In each case, the proportions of these factors are variable. But what interests us here, in studying the problem of sedimentary series geochemically, is the interpretation of *the sign of these sequences* (LOMBARD, 1953, 1956) (Fig. 13, 14 and 15).

Positive sequences are numerous and well known. They are those which evolve in the direction of the general virtual series, i.e., from detrital toward chemical sedi-

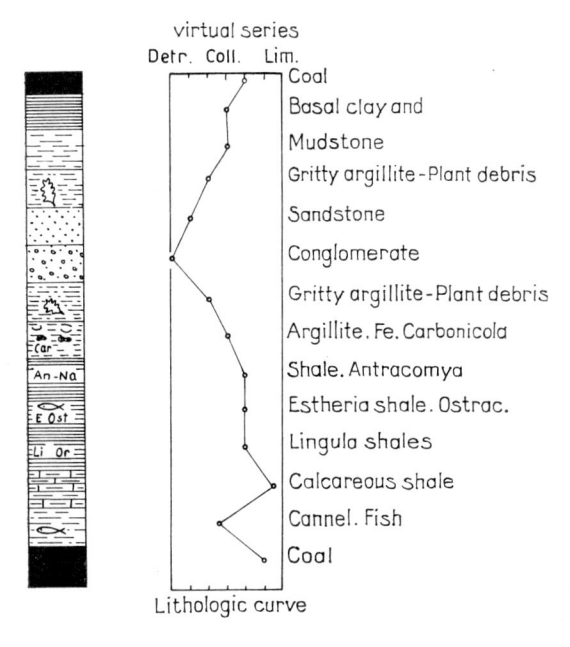

FIG. 15. — *Positive and negative sequences in the productive Carboniferous of Great Britain. The sequences are measured from coal to coal* (after LOMBARD, 1953).

FIG. 16. — *Negative oscillating sequences in the section of the principal borehole of Boko Songo, in the Niari Basin, Congo* (after BIGOTTE, 1956).
1. Argillaceous shale; 2. Calcareous shale; 3. Limestone; 4. Dolomitic limestone; 5. Dolomite; 6. Sulfatic dolomie; 7. Sulfates.

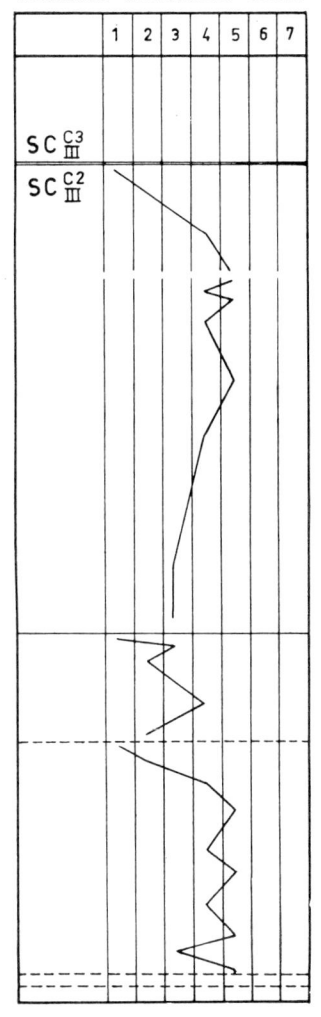

ments. These positive sequences correspond to progressive conditions of sedimentation: installation of a marine regime, subsidence of the basin floor, or transgression on shore line areas.

Negative sequences evolve in a direction contrary to the general virtual series, i.e., from chemical toward detrital sediments or, in detail, from clays to sandstones, from carbonate deposits to clays, from saline to carbonate sediments, etc. Negative sequences indicate regressive conditions of sedimentation: filling up of the basin, rising of the basin floor, return of terrigenous elements by closing of the basin or regression of its shores. An excellent example of a rhythmic series of negative character was studied by BIGOTTE (1956) in the Infracambrian of Niari, Congo (fig. 16).

Sequential analysis provides the fine details within the context of a concrete example of GOLDSCHMIDT's series. It allows one to grasp, within the overall view, its concrete, but hesitating and reiterated reality. It allows one, in addition, to appreciate by the sign of the sequences the order of the deposits, which can become inverted and negative. This is very precious information with which to approach the sense of their history.

IV. — THE EVOLUTIONARY SERIES OF PEDOLOGICAL ORIGIN

ERHART (1955, 1956)

ERHART has recently presented an interpretation of sedimentary succession based on a different sort of reflection, that of a pedologist. If the products involved in sedimentation originate on the continent, the role of weathering and pedogenesis must be of primary importance. Simply from this aspect alone his contribution is important.

The role of the forest. — In reviewing the present-day landscapes on the surface of the globe, ERHART notices that the reign of the forest was severely compromised by that of man. Formerly the forest had a considerably greater extent and alternated in space and time with deserts or with landscapes devoid of forest vegetation. Now, the essential point resides in the role of these immense forested domains during those periods when climates were favorable for their extension. It is under cover of this forest that weathering takes place; the forest acts as a separative filter.

In reality, during the time of its development the forest restricts erosion and retains in its roots and protective mantle the residual and neoformed minerals that constitute the weathering zone and the soil. On the contrary, by means of water percolating through these weathering products, the forest allows the lessivage of soluble products released by pedogenesis. *The forest acts, thus, as a kind of filter, that allows soluble products to reach the ground water reservoir, springs, and sedimentary basins, and keeps under its feet the structural material that constitutes the particulate residue of weathering.*

Biostasy and rhexistasy. — At this point two main periods are distinguished. First, the biostatic period, that of a state of biological equilibrium in which the phenomena discussed above take place: the *residual fraction* is stored up, the *migrating fraction* is removed. Then the rhexistatic period ($\rho\acute{\eta}\sigma\sigma\omega$ = I break) corresponds to the disturbance of this equilibrium for climatic or tectonic reasons. The forest is destroyed, and structural material which accumulated is released; the separative filter is broken. During the first period, the sedimentary basins peripheral to the continent receive mainly products in solution, and during the second, detrital elements.

The example of the tropical rain forest. — This succession of biostatic and rhexistatic periods concerns every climatic region, cold, temperate or tropical. Under cold and rainy climates the forest gives rise to podzolic soils, and in temperate countries, to brown soils. But the phenomenon manifests itself most fully in the case of shady tropical forests endowed with strong lateritic weathering activity. Under these conditions, which characterized widespread surfaces in the past, the elements produced by lateritization, mainly Na, K, Ca, Mg and Si (*pro parte*), were removed in solution during the biostatic period. On the other hand, during the period of rupture the products accumulated in the continental mantle are delivered: forest debris, lateritic crusts and clays. The main elements, accumulated as described previously, are Fe, Al and Si (*pro parte*) as well as vegetal material.

Thus ERHART, in his turn, arrives at the definition of a "geochemical cycle" or "evolutionary series", whose stages are the following:

1° *Carbonates.* Limestones and dolomites, and flint, are produced during the migratory biostatic phase.

2° *Coal.* Under certain favorable conditions only, coal is produced by the destruction of the forest layer.

3° *Iron and aluminum hydroxides and kaolinitic clays* are produced by the reworking of lateritic profiles.

4° *Sands and sandstones*, then *arkoses and conglomerates*, are produced by erosion of the lower part of soil profiles and then by the continental materials themselves.

We shall see in Chapter VIII that RADIER (1957) illustrated almost perfectly this type of series in the basin of Gao, in Soudan, now called Mali. Carbonates, coals, oxidates, kaolinitic clays and detrital products are found in succession.

Here we are, then, in the presence of a proposed cycle or series the elements of which succeed themselves in an order that is rigorously the opposite of the general geochemical series. A confrontation of the opposing schemes is necessary.

V. — CONFRONTATION AND GENERAL VIEW ON SEDIMENTARY SERIES

We find in front of us two propositions of opposite allure.

First, the geochemical series of GOLDSCHMIDT. In combination with the general virtual series of LOMBARD, it evolves from detrital to chemical sediments.

Then there is the evolutionary series of ERHART, which draws attention to sedimentary series of pedogenetic origin. It evolves from chemical to detrital sediments.

The stages of these two propositions can be represented by symbolic letters:

S = sands; A = argillaceous deposits; O = oxidates; C = carbonates.

Thus the two series oppose each other in the following way:

General series (GOLDSCHMIDT-LOMBARD): S.A.O.C.; S.A.O.C.; S.A.O.C.
Pedological series (ERHART): C.O.A.S.; C.O.A.S.; C.O.A.S.

The first interpretation which comes to mind is to see in the second series a negative form of the first one, according to the propositions of LOMBARD.

Examining this question in 1957, I committed the error of answering it in another way. I was impressed by the fact that the same symbolic letters represented different facies. For example, oolitic sedimentary iron ores are unlike siderolitic iron, and clays of major sedimentary series are unlike clays inherited from soils, etc. Therefore, I vizualized that, in spite of the apparent symmetry of the symbols, these two series were independent and not comparable the one to the other. This initial answer was in error. There is a confusion between the succession of deposits, on one hand, and their facies and origin, on the other. It is normal that continental facies differ from marine ones. It is normal that products reworked from soils are unlike products reworked from rocks. But what is important is the order of the succession of events. *The evolutionary series of* ERHART, *with its various facies, appears to be a negative series.*

The apparent contradiction is removed if one agrees to distinguish, two by two, some distinct and independent notions:

— On one hand, to distinguish between the *order* of the deposits and their *origin*;

— On the other hand, to distinguish, among the driving forces of sedimentation, between the role of *tectonics* and that of *pedogenesis*.

These *four independent notions* will be studied in turn.

The order of deposits. — LOMBARD examines first the order of the succession of deposits. When the sense of the evolution is from detrital to chemical sediments, the *series* is *positive* and corresponds, on the whole, to *transgressive conditions*. When the evolution goes from chemical to fine-grained, then to coarse-grained detrital sediments, the *series* is *negative*, corresponding to *regressive conditions*. The origin of the accumulated materials is another question, whether they are inherited from the continental rocks themselves or inherited from the pedogenetic cover of these rocks. Much has been said about the direction taken by the sedimentary process; it is oriented sometimes by decreasing transport energies, sometimes by increasing transport energies. It is easy to see that the decline of the detrital character of sediments brings out their chemical character and reciprocally.

In this way, ERHART's evolutionary series appears very simply to be a negative series characterized by an evolution from chemical to detrital sediments. In this case, following a quiet biostatic period in which the soluble salts are removed and carbonates are deposited, a rhexistatic period takes place. The components of the lateritic profile (arenaceous products, kaolinitic clays, ironstone crust, forest cover) are deposited "upside down" (coal, siderolitic iron, kaolinitic clays, sands). This invasion of sedimentation by detrital elements is typical of the regressive conditions of negative sequences.

It happens that this particular evolutionary series is negative. Other series of pedogenetic origin can be positive.

Every sedimentary series must undergo sequential analysis. This will indicate the order of the succession of the deposits. Whether positive or negative, this order is independent of the origin of the material deposited.

The origin of the deposits. — The origin of the deposited material is quite a different problem. It can be of two major categories: rock or soil. In reality, it can correspond, on one hand, to the *direct mechanical erosion of* continental *rocks* or of emerging massifs. On the other hand, it can correspond to the *reworking of the soils* which cover the continental surfaces.

The great glyptogenic series which follow the emergence of each mountain range are simple examples of evolutionary series originating from the continental rocks themselves. In the case of pebbles this is obvious; but even in the case of sands, often arkosic, and of argillaceous series, one usually finds himself dealing with the bulk material from the continent.

In the opposite case, siderolitic series provide the best example of deposits whose constituents originate on a continent by the processes of weathering and pedogenesis. The examples given by ERHART (1955) and RADIER (1957) show the predominent role of pedogenesis. Other illustrations are found in LOMBARD (1956), in this case with regard to coal-bearing series, the "pedological" origin of which, even if only partial, is obvious. These coal-bearing series provide examples of negative and positive sequences, where the coal appears sometimes in the positive sequence, sometimes in the negative one (LOMBARD, 1956).

Every sedimentary sequence must be submitted to mineralogical and petrographic analysis. Confronted with all the geological and paleontological informations on a given facies, these analyses allow one to reconstruct the origin of the sediments. This origin is a variable independent of the order of the deposits.

The role of tectonics. — Tectonics is one of the principal factors capable of regulating these variations. Continents that are levelled by erosion, that are alienated, or that become dormant produce positive sedimentary series that are less and less rich in solid elements and more and more nourished by chemical products brought in solution. In the opposite case, continents that are uplifted, reactivated, brought closer to the basin, or that emerge from the sea, supply the basins once again with a progression of detrital products.

Thus tectonics influence both the nature and the order of deposits at the same time. Active, it gives rise to detrital sediments; inactive, it leaves the field to chemical deposits; increasing tectonic activity favors negative series; decreasing, it orients the series in a positive sense. But tectonics does not choose, or only slightly, the origin of deposited products; the latter depend on the nature of the continental mass and on the type of weathering.

The role of climate. — There are climates that do not allow the development of vegetation and that, consequently, are prohibitive to soil formation. Such are the periglacial climates of cold regions or of elevated mountain ranges; such are also the arid climates. When erosion occurs, it is carried out by processes mainly of a physical character. Therefore, the products supplied by such weathering are fragments of bedrock. The sediments capable of being deposited issue directly from the continental mass. The same result is obtained when topographic relief is great and is maintained by orogenic forces long enough so that soils cannot be established nor retained. In every case, pedogenesis intervenes only slightly or not at all. It is the very substance of the continent itself which is delivered to the sedimentary process.

On the other hand, there are climates that can give rise to deep soils in which chemical and biochemical activity brings about an important reworking of the parent rocks: lessivage and neoformations. Such soils occur under various humid climates, but the most spectacular are the laterites, in the broad meaning of that term, which are at the present time typical of wet tropical or equatorial regions. These soils will supply to sedimentation an elaborated material of pedogenetic origin.

Moreover, climates change. If an arid climate succeeds a climate conducive to lateritization, the forest dies. The residual phase is released and succeeds the previous chemical sediments. Thus develops ERHART's evolutionary series, whose negative character we have already seen. But if the climate reverts to its original characteristics, and if the forest grows again, the continent is again colonized and detrital sediments decrease in size and amount and then pass over to chemical ones. We end up with a positive cycle, the material of which is still mainly pedogenetic.

Thus, climates influence, at the same time, both the nature and the order of the deposits. They regulate the nature of the sedimentary products. When climates are unable to form soils, they supply continental rock fragments; when they generate soils, they are at the origin of the reworking and the variety of these soils. They regulate the succession of deposits by the order of their own succession.

Climate and tectonics are two new and independent variables that are able to influence sedimentation. Let us examine their interference.

The interference of tectonics and pedogenesis. — All continents and all topographic entities have their own tectonic and their own climatic history. These two phenomena will thus interfere with each other. Interference is obvious, for example, when orogenic transformations disturb climates or pedogenesis directly; such is the case when tectonics gives rise to the destruction of forests by erosion.

But orogeny and pedogenesis interfere in a more general and more subtle manner. Mountain ranges exist, and mountain ranges have been formed under all sorts of climates, from periglacial to equatorial. Old stabilized shield areas, also, have been submitted to cold, temperate and tropical climates. Therefore, the kind of weathering and the intensity of erosion are the results of an equilibrium, moreover a changing equilibrium, between these two factors. *Continental dynamics and climatic evolution will interfere* in each case to furnish the materials transported and then deposited with an infinity of variations possible.

At the extremes we can see clearly. Sediments that are purely chemical or biochemical sediments, such as chalk, siliceous beds, attapulgite clays, phosphates, are in debt to tectonics solely for tranquillity and for the lack of detrital elements. They originate from products dissolved in water which, in the final analysis, result from weathering and pedogenesis.

At the other extreme, a detrital series some 2 000 meters thick accumulated in an alpine fore-trough, comprising conglomerates, gravels, sandstones and sands, and even clays, is the result of predominantly tectonic activity. This does not mean that all detrital series are due to orogeny alone; we shall see the example of important detrital red bed series in which weathering intervenes, and that of siderolitic series where it becomes important.

It is reasonable, however, to say that the detrital pole of the virtual series generally owes much to tectonics, whereas the chemical pole has a pedogenetic character. It is also proper to note that the normal sedimentary series of GOLDSCHMIDT presents a scope, a power and an extension that must be related to the orogenic cycles that govern it. On the contrary, series of pedogenetic origin, positive or negative, have a more discreet, thinner, more ephemeral character because they correspond more readily to the evolution of stabilized continents, which oscillate hardly at all in response to general tectonic movements (H. and G. TERMIER, 1956; MILLOT, 1957).

But these general views are confirmed only at the extremes of the series and must not hide the complexity of the possible interferences because every orogenic or epeirogenic evolution can coincide with any one type of pedogenesis, intense or nil.

The first and indispensable step in the study of a sedimentary series is the analysis of *the nature and the order of the sediments*. This must be distinguished from the study of *the primary origin of the materials* deposited. This origin, moreover, is the result of two distinct, but always intermingled factors: *the tectonic evolution and the climatic evolution* of the landscape.

Analysis of sedimentary series and research on the origin of their raw materials contribute sensitively to paleotectonics and paleoclimatology, in a single word, to paleogeography.

VI. — THE PLACE OF CLAYS IN THE SURFICIAL PART OF THE GEOCHEMICAL CYCLE

Now we have arrived at the point where we can examine the place of clays in the geochemical cycle, which we began arbitrarily at the stage of weathering followed by the stages of transportation, sedimentation, and diagenesis.

Clays of weathering and of soils. — The zone of weathering and of soils is one of the birthplaces of clays:

— Birth by fragmentation where weathering, even intense, has primarily a mechanical or physical character. This is the direct weathering of the rocks that form the continents, giving rise to *inherited clays*.

— Birth by pedogenesis, in those places where true soils occur; we shall see that, here, *clays inherited* from bedrock, or *transformed*, or *neoformed* within the soil itself are mixed together.

Detrital sedimentary clays. — After being transported, the fine fraction is carried off to basins of sedimentation. It is deposited in a sedimentary series, intermediate to, or alternating with sands and chemical sediments. These detrital clays issue from either the direct erosion of continental rocks or from the reworking of soils, with all the possible intermediate stages and mixtures. *Inheritance dominates* here.

Transformed sedimentary clays. — The clays transported to sedimentary basins, as in the two previous cases, obviously can be subjected to sorting of various kinds, and also to physico-chemical modifications. The latter are more difficult to demonstrate. These modifications, often tenuous but important from a geochemical point of view, will be called here "transformations". They will be discussed in the following chapters. These are physico-chemical transformations of inherited material; we shall speak of *clays transformed during sedimentation*.

Neoformed sedimentary clays. — In the strictly chemical stages of sedimentary series, chemical or biochemical neoformations are the rule: carbonates, flint and chert, phosphates, etc. Neoformations of silicates, and especially of clays, occur also. Clay minerals, in this case, are formed entirely from products dissolved in water, as are other minerals of chemical origin. Here inheritance is nil or weak, and neoformation dominates. We shall speak of *neoformed clays*.

Diagenetic clays. — After their deposition, sediments do not cease their evolution. Diagenesis occurs, both early and late. Clays are not insensitive to such changes, and we are able to grasp some examples of such evolutions. Here again we find transformation and neoformation. We shall speak of *diagenetic clays*.

Conclusions. — Such are the main positions occupied by clays in the first four stages of the geochemical cycle, before they vanish during metamorphism.

There are clays originating directly from the rocks constituting the continents. There are clays of pedogenetic origin whose complex history will be studied. Both these types of materials can be transported to basins of sedimentation and settle there without modification or with transformation. A new generation can form within the deposits by neoformation. Moreover, the whole can evolve in different ways during diagenesis.

This outline is fundamental and necessary as a backdrop for the inventory of the natural argillaceous products on the surface of the earth's crust.

This inventory will be conducted throughout the following three chapters:

— Clays of weathering and of soils.

— Continental clays.

— Marine clays.

CHAPTER V

WEATHERING AND SOIL CLAYS

Continental areas are subject to weathering and pedogenesis. Three distinct phenomena are to be considered (DUCHAUFOUR, 1960):

1° The disintegration and progressive alteration of the rocks leading to what pedologists call the "weathering complex" which, in the case of crystalline rocks, is represented most frequently by an *arène*. A part of the material is removed in solution;

2° The biological colonization by plants, microbes and animals, which brings organic matter into play;

3° The displacement or migration of the soluble or very fine-grained elements of the weathering complex under the influence of percolating solutions. These migrations lead to the development of impoverished horizons (A horizons) or to accumulative horizons (B horizons).

These three phenomena are not exactly successive, but can overlap. Biological colonization occurs very quickly, as soon as parent rocks are disintegrated, and the role of organic matter, and especially of humic material, begins early. Nevertheless, in many cases, and especially during arenization and lateritization, the pedological profile rests upon a weathered zone that can be several meters or tens of meters in thickness.

Usually one studies separately the physico-chemical mechanisms of true weathering and the complex processes of pedogenesis. Here we shall study:

I. *Weathering mechanisms;*

II. *Arenization;*

III. *Soil clays.*

I. — WEATHERING MECHANISMS

Here it is a question of meteoric alteration in the French meaning of the term and not in the broad sense used by Anglo-Saxons according to the Latin etymology of the word. *Alter* means "other" and alteration, for Anglo-Saxons, is the transformation of one product into another, whatever the mechanism: diagenesis, metasomatism, metamorphism, and so on. Here alteration is equivalent to the English word "weathering" and to the German *Verwitterung*.

It is traditional to distinguish, among the weathering factors, the physical mechanisms (freeze-thaw action, temperature, hydration, effects of salt crystallization) and the chemical ones (dissolution, oxidation-reduction, hydrolysis). However, it is very difficult today to distinguish which are uniquely physical from those uniquely chemical; several among them are physico-chemical. Moreover, these factors act together, assist each other and add their effects together. Lastly, organic matter also plays a role, adding biological effects to the physico-chemical ones.

For these reasons the weathering mechanisms will be described here, one after the other, without any distinction between categories, but according to a classification that moves in the direction of decreasing physical and increasing chemical character. Important documentation will be found in the works of CORRENS (1939), SAKAMOTO (1954), KELLER (1958) and BIROT, HÉNIN et al. (1962).

1° *Freeze and thaw. Cryoplasticism*

The effect of freeze and thaw on the rock fragmentation and disintegration is very important. It is the main weathering factor in cold regions and high mountains. It manifests itself in our regions by frost-shattered stones. Water enters the cracks and pores of rocks. Its increasing volume upon freezing makes joints widen and separates crystals from each other. This phenomenon has been studied experimentally, showing an optimum of porosity and increasing disintegration with the rapidity of the freeze-thaw alternations. Different quantitative measurements have been carried out (TRICART, 1953, 1956; MASSEPORT, 1958). A review of this problem was presented by BIROT (1962). CAILLEUX and TAYLOR (1954) have written a treatise on cryopedology wherein the role of frost in the shaping and the evolution of continental weathering is presented along with a considerable bibliography.

2° *Variations of temperature*

The differences of temperature in deserts can be considerable between day and night and between the exposure to the sun and rainfall. The weak thermal conductivity can give rise to stresses between the core and the exterior of boulders, or between sheltered and exposed parts. Those familiar with the desert know very well the sharp crack of these bursts; the visitor sees their effects. It is sometimes spoken of "thermoclasticism".

It is more difficult to know whether these variations of temperature can intervene in the granular disintegration of rocks. Some call upon the difference in the heat absorption by minerals of different colors (RUELLAN, 1931; MOHR, 1944). Another approach consists of an appeal to different coefficients of mineral thermal expansion; these coefficients are variable, according to the orientation of the crystallographic axis (anisotropy). Laboratory tests made according to this approach by GRIGGS (1936) were disappointing. BIROT (1962) showed, however, that samples submitted to daily temperature alternations of 50 °C during four months increased two or three times in porosity. The preexisting cracks were widened and could provide pathways for solutions and salts capable of disintegrating the rock.

But the main role of temperature is elsewhere: it is an indirect role that regulates in diverse ways all weathering phenomena. Variations of temperature govern the moisture in rock pores and also the solubility of gases or salts. Moreover, the rate of chemical reactions, especially that of hydrolysis, approximately doubles for a temperature increase of 10 °C. Thus, all the mechanisms are subject to temperature variations, and the different types of weathering, throughout the world, are largely regulated by this factor.

3° Humidity. Hydration and effect of salt crystallization

Humidity. Desiccation. — BIROT (1962) presents the current state of this question. Variations of water temperature within rock pores provoke a certain amount of disintegration because the dilatation coefficient of water is much higher than that of rocks. Moreover, humidity variations have the ability by themselves to disjoint crystals by the transformation of the intergranular water film into tiny drops, modifying the surface tension. But the maximum destructive effect on rocks is obtained by the union of the two effects of varying temperature and varying humidity.

These mechanisms have been tested in the laboratory. BIROT (1947) placed samples alternatively 12 hours in water at room temperature, then 12 hours in a drying oven at 70 °C. After 12 months he obtained a disintegration of 0.5 to 2 % for various samples.

Hydration. Salt crystallization. — CORRENS (1939) presented the possible cases.

— Formation of hydrated crystals. The best example is that of the crystallization of gypsum from anhydrite. This transformation requires 2 water molecules and progresses with an increase of volume, yielding a pressure of 20 atm. This example is well known in mines, tectonics and construction projects; it is often disastrous.

— Crystallization of a salt from its saturated solution. It is necessary that the volume of the crystal formed and that of the residual saturated solution are greater than that of the original supersaturated solution. It is also necessary that the intergranular space where crystallization occurs is closed, and this can be brought about by the initial stages of crystallization. One speaks of the explosive effect of crystallization.

— The growth of crystals can develop high pressures, but it is necessary that the surface tension of solution allows the solution to penetrate between the crystal and the wall.

It is immediately evident that such mechanisms will necessitate desiccation in order to saturate solution, or moisture to induce hydration, or a temperature variation to lower solubility. Here once more the action of temperature intervenes by means of its role on internal solutions in rocks in the process of disintegration.

Examples of the consequences of these crystallization effects are numerous. The corrosion of public monuments in industrial towns is the result of gypsum crystallization in the pores of limy rocks; this gypsum is the result of the reaction of atmospheric

SO_3, issuing from the burning of pyrite in coal furnaces, on the $CaCO_3$ of building stone. The choice of construction stone today depends on this illness. Another example has been furnished by the oxidation of pyrite in the Toarcian paper shales after drainage of the Saint-Jean marsh, Nancy (France). Pyrite was oxidized to iron sulfate which, by means of double decomposition with the calcium carbonate of marnes, resulted in gypsum. Gypsum crystallized between the shale layers provoking a volume increase that resulted in the heaving of floors in stores and banks.

Such effects are common in nature; they are the most readily observable in desertic and Mediterranean-type regions. Salts of eolian origin crystallize in the pores and disjoint the rocks. It is one of the mechanisms which explain the formation of *taffoni*. Moreover, the salts resulting from the alteration of the crystals themselves nourish the intergranular solutions and add their effect to that of exotic salts.

Tests on the crystallization effect of salts were carried out in the laboratory by BIROT (1954), HÉNIN, ROBICHET and DUROUCHET (1953) and CAILLÈRE, BIROT and HÉNIN (1954). Disintegration is 10 to 1 000 times greater than that obtained by the wetting-drying tests. Rocks are placed in drying oven at 70 °C, then immersed in a saturated saline solution until they become covered with efflorescences, and then only moistened morning and evening. After three months, initial granite fragments of about 100 g, treated with sodium hyposulfite, were completely arenized. With sodium carbonate, the loss reached only 5 g. There is, at the same time, physical disintegration with a well defined limit between the weathered zone and the sound rock, and chemical weathering. Proof of the latter is the increasing sericitization of plagioclases and the appearance of montmorillonite. After the action of magnesium chloride, blocks of Autunian schists yielded 15 % of their weight in sizes smaller than $2\,\mu$. Other tests show a variety of results differing according to the nature of the rock: granites or gabbros. This is confirmed by PÉDRO (1957). By means of comparative tests, he pointed out that better disintegration was obtained with NaCl (easy crystallization) than with $MgCl_2$ (difficult crystallization, hygroscopic salt). One can also see that the rock texture is very important: Volvic andesite is more sensitive than granite. The result is the inverse with H_2O_2 (HÉNIN, 1957; HÉNIN and PÉDRO, 1957). With H_2O_2, iron in biotites is oxidized and they are exfoliated and become golden brown, leading to arenization (PÉDRO, 1957). TRICART (1960) made systematic tests yielding the weight variation of disintegration as a function of time. Granites of Dahomey were submitted periodically to immersion in saline water. Some granites show accelerated weathering as a function of time, others have linear curves, and still other curves reach plateaux, indicating that the phenomenon is dying out. The last group was considered the best with which to build Cotonou harbor (Dahomey), and petrographic investigation accounted for the differences. The richness in biotite, previous weathering, even though mild, and the cataclasis of crystals, are factors of vulnerability.

4° *Oxidation and reduction*

The role of the oxidation and reduction phenomena in geology has been pointed out very often and especially by MASON (1949). In weathering, the most important reaction is that of iron. The latter frequently occurs in silicates (and carbonates) in the reduced

Fe^{2+} form. Reaching the oxygenated hydrosphere, it takes on its oxidized Fe^{3+} form. The equilibrium of crystalline lattice now breaks down, and observation shows that ferro-magnesian silicates are the most vulnerable to weathering.

The iron must then be transported. This can occur in a very acid environment, but such an environment only rarely is found under natural conditions. In contrast, waters rich in organic matter are reducing and transport iron in the ferrous state. Upon reaching an oxidizing environment, the organic matter is destroyed and the iron precipitates in the ferric state. Moreover, iron enters readily into complexes, either with organic anions, primarily humic matter, or with silica. These ferrohumic and ferro- or ferrisilicic complexes play a very important role in the geochemistry of iron in hydrosphere. They have been studied by BÉTREMIEUX (1951), BLOOMFIELD (1952, 1955) and LOSSAINT (1959).

5° *Hydrolysis and its variations*

Hydrolysis, i.e., decomposition of mineral salts by water, is the fundamental chemical mechanism of silicate weathering. Silicates can be considered to be the salts of weak acids and strong bases; the hydrolysis reaction is thus alkaline. The way in which this mechanism can be imagined at the present time was described in Chapter III. In spite of its still incomplete character, such an image is an useful guide in reasoning.

Intensity of hydrolysis. — A better understanding of hydrolysis allows one to grasp its dynamics. Thus one can account for the favorable effects of:

— temperature;

— leaching and cations present;

— low pH and the presence of acid anions.

a) Temperature increases thermal agitation and accelerates migration in the both directions. In this respect, there is a very important point about which errors are frequently made. Everybody knows that weathering is much stronger in warm regions, and especially in intertropical ones, than in cold regions. It is commonly attributed to the amount of CO_2 in rain water, and this is erroneous. The solubility of CO_2 in water decreases with temperature and, in fact, tropical waters are less rich in CO_2 than temperate waters. The factor which accelerates weathering in warm countries is temperature. For an increase of 10 °C, the dissociation of water and the rate of reaction is approximatively doubled. Assuming that this rule is exact, one can calculate that, if the rate of reaction is V at 10 °C, it will be $2V$ at 20 °C, $3V$ at 25.85 °C and $4V$ at 30 °C. Thus, there is as much difference between 26 and 30° as between 10 and 20°. Knowing that tropical rains fall during the warm season and that mean temperatures can reach 30 °C, one can understand that hydrolyses are accelerated.

b) Lessivage not only renews the free water which surrounds crystals, but also that of the most external sheets which are oriented on the crystal. Thus, the removal of released cations is maintained. On the contrary, if lessivage stops, the reaction is blocked and the silicate mineral subsists indefinitely, covered with a thin H^+ pellicle.

In the same way the presence of cations in the surrounding solution hinders the removal of released ions and, at a certain concentration, stops it. It is for these reasons that in environments poorly leached or rich in cations feldspars subsist, whereas they are destroyed under good conditions of natural lessivage.

c) Lastly, the removal of cations by exchange with the H^+ ions of the environment is facilitated by the *acidity of the solution*. The *presence of acid cations* increases the efficiency of hydrolysis. It is at this point that CO_2 intervenes, and also mineral anions, organic anions and even acid clays (GRAHAM, 1941). An excellent study on the role of biological acids and acid colloids has been presented by KELLER and FREDERICKSON (1952). One of the most striking conclusions is that the H^+ ions, which are released at the level of plant roots, are exchanged for nutritive mineral cations and that, in this way, a part of the biological energy is applied to weathering; thus one comes to one of the important phenomena of pedogenesis.

Quantitative estimations can be attempted. In a warm country where the temperature approaches 30 °C, the rate of hydrolysis is multiplied by four, relative to a cool country where it is only about 10 °C. Lessivage during the humid season can easily provoke percolation five times more intense than in temperate lands. Lastly, the decrease of one unit of pH, owing to the nitric acid in rain or to acid humic matter, multiplies the amount of hydrogen by ten. Assuming simple proportionalities, one can attain rates which are increased two hundred times. Therefore, one understands the nature of coincidences that make tropical countries regions of chemical weathering.

Differential hydrolysis of minerals. — The rate of weathering is not the same for all the minerals. This has been observed for a long time and the use of polarizing microscope confirms it every day.

A double sequence of common minerals was established by GOLDICH (1938) in the order of their decreasing vulnerabilities downwards:

Olivine	Ca plagioclase
Augite	Ca, Na plagioclase
Hornblende	Na, Ca plagioclase
Biotite	Na plagioclase
Potash feldspars	
Muscovite	
Quartz	

This typical scheme is, of course, a generalization and suffers many variations according to weathering conditions. Only two points will be emphasized, relating to micas and feldspars.

It is well known that there is a difference of vulnerability between biotite and muscovite. BASSETT (1960) accounted for this fact by a difference in the orientation of hydroxyl ions with respect to the plane of the mica : in trioctahedral micas, such as biotite and phlogopite, hydroxyl ions are perpendicular to (001), whereas they are inclined in dioctahedral micas such as muscovite. The large K^+ ions, situated precisely between the planes, are more tightly bound in muscovite because of the distance of separation from the proton H^+. The disjunction of micaceous layers, which is the first pathway of weathering, is made easier in the case of trioctahedral micas. This agrees with the observations of CAILLÈRE and HÉNIN (1951) on the weathering of the Malagasy phlogopite in quarry and mine exposures. In one case, they observed the following evolution: phlogopite —→ weathered mica —→ montmorillonite, tending to a dioctahedral arrangement and to an impoverishment in alumina in the tetrahedra. In the other case, they observed the evolution phlogopite —→ vermiculite.

As for feldspars, temperate climates have for a long time accustomed us to note that the weathering of plagioclase is much quicker and stronger than that of alkaline feldspars. The same rules were also found by LENEUF (1959) in the Ivory Coast, where climate is tropical and humid. However, RONDEAU (1958) described the contrary for arid lands. Some samples from the Sahara show strong weathering of alkaline feldspars and the freshness of plagioclases. This difference would be attributed to the composition of intergranular solutions, which is not the same in arid countries and which inverts the rate of hydrolysis. This is very important from a geomorphological point of view, and it is well known that the various granular rocks do not react to different climates in the same way. Under a humid climate alkaline rocks are vulnerable, and acid rocks dominate the topography. Under an arid climate it is the contrary. A second consequence would be important in sedimentology; statistical analysis of the two types of feldspars in arkosic sandstones should yield information about the climate under which arenization took place.

Experimental study of weathering of isolated minerals. — Studies on weathering of minerals are quite ancient. The first experiments are probably those of DAUBRÉE (1857). The kneading of orthoclase fragments together with water shows the fractionation of the mineral into silt and the dissolution of potash (2-3 %). Likewise, TAMM (1924) brought about experimental weathering with the release of alkalis, that which led him to obtain fine-grained products which he compared to Scandinavian glacial clays.

The most important works are those of CORRENS and his school, pursued from 1930 to the present day, first at Rostock, then at Göttingen (CORRENS and VON ENGELHARDT, 1938; CORRENS, 1939, 1940, 1961). After experimentation in which water was not renewed, CORRENS and his collaborators noted that in nature the system was open and continuously renewed. They constructed experimental apparatus that would insure a study with continuous percolation as close as possible to natural conditions. The measurements on orthoclase hydrolysis are the best known. At the beginning of leaching, the ground-up orthoclase releases potash principally; then, after 15 days of experimentation, i.e., a leaching of 5 liters per gram, silica and alumina appear. In an alkaline medium the dissolution is rapid and silica and alumina occur in equal amounts. In a neutral medium the attack is slow and silica is dissolved twice as fast as alumina. In an acid medium the attack is again rapid, but alumina predominates in the solution.

Many other works on the weathering of isolated minerals have been carried out under conditions similar to those in nature. Thus, GRAHAM (1950) also found experimentally the order of increasing vulnerability from alkaline feldspars to plagioclase, the order which had been observed in nature by GOLDICH (1938). The dynamics of weathering as a function of the ions of solution was studied by NASH and MARSHALL (1956). Other silicate minerals were submitted to many tests by CORRENS and his collaborators (1961). Electrodialysis allowed THA HLA (1945) to follow the destruction of lattices.

The attack on micas was conducted by DEKEYSER, VAN KEYMEULEN, HOEBEKE and VAN RYSSEN (1955) by means of various reagents: boiling water, CO_2 and H_2S.

But the fundamental results which interest us here were established by CORRENS. They modified profoundly all our conceptions on weathering, and there are two reasons for this:

— It was demonstrated that weathering releases ions into solution and not colloids as was commonly thought at that time.

This does not exclude the possibility that amorphous or colloidal products occur in some soils. They are called allophanes. They have been identified in the weathering products of some glassy volcanic ashes (SUDO, 1954; FIELDES, 1955; GASTUCHE and DE KIMPE, 1961), and in some soil horizons. But this is not a case of products of polymerization in soil solutions, which are always too poor in silica and alumina to allow polymerization. These allophanes are dissolution residues of silicates from which the monovalent and divalent cations, and little by little the silica, have been removed. These gels can only be dissolved in their turn, either to disappear or to provide solutions for neoformation. These questions will be examined in Chapter XI.

— Therefore, neoformed clays that originate during weathering are not the products of colloidal flocculation, as previously thought (MATTSON, 1935), but, as proposed by CORRENS, products neoformed from ions in solution.

6° Biochemical activity

Physico-chemical weathering is rarely independent of biological activity because of the rapid colonization of weathering zones by soils. Here we enter into the domain of pedogenesis and one can refer to the treatise of DUCHAUFOUR (1960). However, it must be realized that besides the biological phenomena which belong to the soil, the soil allows the biochemical products to percolate into the weathering zone.

If organic matter is mineralized, it gives rise to active soluble compounds, principal among which is carbonic acid, and also ammonia nitric acid, PO_4 and SO_4 ions.

If the organic matter is only slightly or very slowly mineralized, the results are humic compounds and products of organic decomposition, soluble or insoluble in soil solutions. The insoluble products combine with and fix clays. The soluble ones, on the contrary, disperse clays, make complexes with and remove iron, and probably play a role in neosynthesis, as we shall see.

Here is the frontier between organic geochemistry and the geochemistry of silicates, which is a very poorly known domain.

Conclusion

Like every enumeration, that of the weathering mechanisms cuts into several pieces a reality that at its earliest opportunity presents itself with all its factors united. If one wishes to raise a discussion, one sees that weathering uses at the same time mechanisms primarily of a physical character which have the effect of disjointing, separating and disintegrating rock crystals, and mechanisms primarily of a chemical character, the most important of which is hydrolysis, acting on the crystals themselves. Mutually assisting each other, the former offer new surfaces to attack, the latter provoke new weaknesses and ruptures.

The interplay of these associated mechanisms will be illustrated by the most common and important weathering process in crystalline terrains under a temperate climate: arenization.

II. — ARENIZATION

Many granitic or gneissic terrains are weathered to arenaceous products. The crystalline rock is still recognizable by its structure, by the veins that cross it and by traces of fractures, but it is decomposed or "rotten". The thickness of this arenization is quite variable, but in Europe it may reach more than ten meters and, under other climates, several tens of meters. This arenization is for a great part the result of meteoric alteration.

1° *History prior to weathering*

It is necessary, however, to evoke transformations that affect crystalline rocks before they are subject to weathering. Petrographers know very well that rocks called "sound" when taken out of quarries, boreholes and mine shafts already present modifications. Plagioclases contain scattered damourite crystallites; this is the *damouritization* or *sericitization* of feldspars. As for ferromagnesian minerals, and especially biotite, they are often chloritized. Lastly, alkaline feldspars are cloudy owing to many inclusions. KORZHINSKY (1940) studied these inclusions; he showed that they were liquid and did not seem to be connected with weathering. In fact, all these modifications result from what is called "retrograde metamorphism" or "retromorphosis". It is a question of new thermodynamic equilibria of the silicates when crystalline massifs are raised into a higher zone of the earth's crust.

2° *Petrographic data on the weathering of feldspathic rocks*

It is this material, already endowed with a long history, that weathering attacks. The first detailed petrographic analysis of feldspathic rock weathering is attributed to Jacques DE LAPPARENT (1909). Plagioclases are abundantly weathered to damourite or sericite crystallites which are scattered across thin sections and which, in some cases, completely replace the mineral. On the contrary, alkaline feldspars are more resistant; they are marked by inclusions and very fine granulations, making them cloudy. When weathering becomes stronger, it can reach the point where sericite is distributed in channels through the orthoclase or microcline, but this affects all the minerals of the rock, including quartz, and must be considered as a neoformation along fissures. During this same time biotites take on a golden brown aspect. Jacques DE LAPPARENT was able to show that this was due to interference effects of light between the planes of biotites and fine goethite crystals arranged on the biotite surfaces by exudation of iron. Muscovite and quartz remain intact. Moreover, Jacques DE LAPPARENT demonstrated by means of chemical analyses that the sericite of plagioclase was a potassic and not a sodic mica. This requires a potash supply, which is provided by neighboring minerals and from potassic feldspars which are being hydrolyzed. In the absence of the latter, biotite takes up the relay and releases potash as it is being transformed into chlorite.

The study of zoned plagioclases (MILLOT, 1949) defines rather delicate threshold mechanisms in the weathering of feldspars. In feldspars with alternating zones only the calcic zones are sericitized, whereas the zones rich in albite remain intact. Thus, in the composition of plagioclases there is a limit at which the sericitization sets in. The sericite crystallites are, moreover, mixed with small calcite rhombohedra, showing that the solutions are calcic. It can be admitted that solutions that circulate through the microfissures in a single zoned crystal are not very different from one point to another. However, within the feldspar lattice the presence of calcium in a determined amount results in the neoformation of the micaceous lattice, in which calcium does not intervene. Here is one of the pieces of the reasoning I used to demonstrate the eminent role played by calcium ions in the synthesis of minerals structured according to the mica type.

3° *Arenization and geochemistry of arenization*

Two weathering zones must be distinguished (J. DE LAPPARENT, 1941 and MILLOT, 1949). In the first one, the *cementation zone*, the rock maintains its appearance, mineralogical evolution begins, especially that of sericitization of feldspars, hydrolysis of alkaline feldspars and alteration of biotites, if present. This zone is characterized by extremely slow percolation of the solutions, producing subtle changes in an alkaline environment, but the rock remains massive. In the second or surficial zone, which can be called the *arenization zone properly speaking*, the microdivision which is carried on and the hydrolysis which intensifies hasten the disintegration of the crystals, promoting a more rapid percolation of solutions and leading to an *arène*.

Quantitative study of such arenization was made by COLLIER (1951, 1951) on granites from the Auvergne (France) and by DEJOU (1959, 1961) on granites and gneisses from the Morvan (France). A remarkable restatement on the arenization of granites under temperate climates has recently been presented by COLLIER (1961). Among the numerous measurements made by these authors, the following main features can be sorted out:

— The grain-size distribution allows one to estimate the intensity of disintegration. In the *arène* of the porphyroid biotite granite from Romagnat (Puy-de-Dôme), the fraction finer than $2\,\mu$ is about 2.3 %, whereas the fraction finer than 2 mm is about 54.3 %. In other profiles developed from other parent rocks, the fine fraction is only a few per cent, but the fraction finer than 2 mm can reach 70 %, a considerable value.

— Geochemical study allows one to estimate the order of removal of elements and the degree of weathering, i.e., the percentage of material removed. To perform these calculations, a standard of comparison must be chosen. This is the amount of quartz; all evidence, indeed, shows that quartz is stable. The order of removal of elements is the following:

$$Ca > Na > Mg > K\text{—}Si\text{—}Al.$$

Iron is situated among the mobile elements, but its place varies according to the specific case. On the right are found the constituents of alkaline feldspars, muscovite and quartz, and on the left the elements typical of plagioclase and the ferromagnesian minerals.

The degree of weathering measured by the loss of material does not exceed a few percent, between 1.6 and 4.7 in the cases studied by COLLIER. In the northern Morvan (France) DEJOU (1959) found the amount of weathering much greater: 18.18 % for granite from Lormes, 24.73 % for anatexites from Chastellux, and 50 % for the anatexites from Saint-Braucher. The order of removal of elements is similar to that given above, with this nuance: the removable elements Ca, Na, Mg and Fe can be interchanged according to the specific case. The quantitative results thus accumulated throw light on the intensity of arenization and its two specific features: disintegration of crystalline rocks and lessivage, which can reach very high values.

4° *Variations depending on climates*

The problem of the age of arenaceous mantles that cover crystalline basements is always difficult to solve. And it is this age alone that allows one to know the climate or the successive climates which prevailed during their formation.

In the Massif Central (France) arenization was earlier than the Quaternary, as testified by various reworked Quaternary deposits covering the *arène*. COLLIER (1961) rightly considered that these *arènes* post-date the Nummulitic, the tropical climates of which are well known. Specifically, the *arènes* which were studied show no tropical features, neither strong alteration to kaolinite, nor release of alumina, nor corrosion of the quartz, whereas ferruginous horizons corresponding to rubefactions or crusts can be formed under many humid climates and are not unique to the tropics. Thus, one is led to place the phenomenon in the Neogene, which is still rather vague and includes a variety of successive climates that can be considered on the whole warm and humid, with progressive cooling into the Quaternary.

The same difficulties arose in the interpretation of the age of arenization in northern Europe (Scandinavia, Scotland, Ireland). Arenization of more than 10 meters in thickness is well known there. We shall come back to the evidence given by the study of the clay fraction (GODARD, PAQUET and MILLOT, 1961). Since they pre-date glaciation, these *arènes* must be dated as late Tertiary, at which time warm, temperate climates prevailed.

On the other hand, LENEUF (1959) studied the weathering of calco-alkaline granites and granodiorites from the Ivory Coast, now under a tropical climate. It was discovered that at the beginning of this weathering there were, once again, many characteristics of arenization of temperate regions, particularly on well drained sites. One sees a microdivision of the minerals resulting in a chalky whitening of the feldspars; but, whereas plagioclases are sericitized, the alkaline feldspars are already giving rise to kaolinite and gibbsite. Very quickly the phenomenon accelerates, and we come to a lateritic profile, rich in kaolinite, gibbsite and goethite, with a loss of material which can reach 80 %.

The general view on the variations of arenization across climatic zones is the following. The two factors necessary for strong weathering are water and high temperature. Under periglacial and desertic climates, these two factors are not combined. On the contrary, under temperate and tropical climates, they are present. Thus it is in these two cases that arenization will develop fully; and we know that in the tropical zone it proceeds much further into the lateritic phenomenon that we shall examine later on.

Arkoses are *arènes* that are recemented. MAUREL and BROUSSE (1959) found kaolinite in the arkoses from Royat (Puy-de-Dôme, France), which are probably Eocene, thus tropical, but which seem to have been affected by hydrothermal activity. On the other hand, in the Permian arkoses in the Auvergne and in Provence they found montmorillonite, which eliminates tropical weathering as an explanation for their formation. In the arkosic sandstones of the Permo-Triassic from the Vosges (France) MILLOT, PERRIAUX and LUCAS (1961) identified illite as the predominent clay mineral. This corresponds to mild weathering, as witnessed by the presence of alkaline feldspars, and allows one to define the tectonic and climatic conditions during the genesis of these red sandstones; this will be analysed in Chapter VI.

5° *Experimental data on arenization and rock disintegration*

Exposure to weathering in observation boxes. — This method was employed by DEMOLON and BASTISSE beginning in 1930. Eight hundred kg of fresh granite reduced first to fragments 2 to 4 mm in size, were exposed to atmospheric agents in the absence of vegetation. The results of this experiment were given after five years and after fifteen years by DEMOLON and BASTISSE (1936, 1946). PÉDRO (1961) has recently presented summary of this experiment after thirty years. The total rainfall on the lysimetric box was about 17.648 m, 45 % of which, or 7.991 m, percolated through the granite fragments. The evolution of the granulometry was as follows:

Year	Duration	Grain-size distribution					pH
		Fraction between 2 and 4 mm %	Fraction finer than 2 mm				
			0.2 to 2 mm %	0.02 to 0.2 mm %	2 to 20 μ %	2 μ %	
1930	0	100	0	0	0	0	6.85
1935	5 years	—	71	25.25	2.95	0.80	—
1945	15 years	—	64.75	29.50	3.85	1.45	7.80
1960	30 years	8.7	31.00	62	4.60	2.40	7.75

The subdivision of the mineral and rock fragments was thus very great, since only 8.7 % of the original fragments still exceed 2 mm. Furthermore, the fraction finer than 2 μ, the clay-size material, passes from 0 at the beginning to 6.5 kg (in 800 kg of rock) after five years, 12 kg after fifteen years and 17.5 kg after thirty years. This mechanical or "physical" effect is considerable.

The geochemical evolution, on the contrary, is very weak. The amount of material exported during thirty years is about 542 g, i.e., less than 1 %. It is impossible to establish variations in the chemical composition in the case of such a small percentage. Lastly, there is no trace of clay minerals in the fraction finer than 2 μ.

Here, we have evidence of physical disintegration and not geochemical evolution.

Experimental weathering by continuous leaching. — Weathering experiments by continuous leaching on rocks have now been undertaken in the way that CORRENS experimented with isolated crystals. OBERLIN, HÉNIN and PÉDRO (1958) leached a granite in a Soxhlet extractor with 350 liters of distilled water. After two months, 0.54 % of the original weight was reduced to fragments finer than 2 mm and 82 % of these fragments were smaller than 2 μ. The examination of this fine fraction by electron microscope shows that particles are of two types. First, *very small monocrystals of the original minerals species*, then *amorphous materials* of globular form, either large spheres between 400 Å and 2 μ, or very small spheres between 30 and 40 Å. The weathering that produces this mass of fine particles was accompanied by a strong release of alkalines into the liquid. There is thus hydrolysis in the points and cleavages of the crystals, a hydrolysis resulting in their microdivision. PÉDRO (1958) described two successive weathering stages. The first stage is a peripheral chemical weathering which leads to the microdivision of the crystals (small crystals of feldspars and sheets of mica). The second stage is overall chemical weathering which produces, with dissolved ions, spherical globules, the analysis of which reveals 70 to 80 % SiO_2 and 10 to 15 % alumina (PÉDRO, 1960). They are the disordered debris of silicate lattices. These spherical globules have been identified in the weathering of volcanic lavas by SUDO (1954), AOMINE and YOSHINAGA (1955) and FIELDES (1955). In this case, the work of weathering is much more advanced than in the boxes of DEMOLON and BASTISSE. As a matter of fact, PÉDRO (1961) pointed out that the thirty years of natural drainage in the boxes correspond only to eight days of leaching in his apparatus. But still no clay mineral is obtained.

Experimental weathering with the appearance of clay minerals. — Caillère, Hénin and Birot (1957) and Birot, Caillère and Hénin (1959) have made weathering experiments with saline solutions on samples of gabbro from the Palatinate, gabbro from Morocco and granite from Louvigny, France. They observed the appearance of clay mineral of montmorillonite type in the weathered rock.

Pédro (1960), carrying on the experimentation by continuous leaching, introduces carbon dioxide. During the attack, the liquid which bathes the rock fragments becomes greenish. It turns out to be charged with calcium carbonate and montmorillonite. Thus an experimental neoformation of montmorillonite from the weathering products of crystalline rocks was obtained.

6° *The clay minerals of* arènes *and of the beginnings of weathering*

The nature of the clay minerals contained in the fine fraction of *arènes* is beginning to be well known. The difficulty is rather great because the clay fraction is hardly abundant. On the other hand, several sources of confusion are possible, if one does not pay close attention to the reflections greater than 10 Å. Indeed, the 7.15 and 3.57 Å reflections can be incorrectly attributed to kaolinite, even if they belong to chlorite or vermiculite. Tests with polyalcohols or by heating must be made in order to determine as rigorously as possible the minerals of weathering, such as vermiculite, montmorillonite and mixed layers.

In the present state of our knowledge (Millot, 1949; Godard, Paquet, Millot, 1961) we have arrived at the following outline:

1° The predominant clay mineral in *arènes* is illite, which corresponds to the fine fraction of sericites coming from feldspars and alumino-silicate minerals;

2° Its possible companion is chlorite from ferromagnesian minerals;

3° The weathering products of the illite and the chlorite are illite-vermiculite and chlorite-vermiculite mixed layers, and vermiculite. Even montmorillonite may be produced in the finest sizes;

4° If drainage is impeded, either in the past or recently, a neoformed montmorillonite can be found;

5° On the contrary, if the drainage is excellent, and if plagioclase residues and bivalent ions have been removed from the *arène* as a result of a long history, kaolinite will appear. The latter will continue to increase, then become massive, as the climate, always remaining humid, becomes warm, then tropical.

Clay minerals present in the meager amount of material finer than $2\,\mu$ that can be extracted from *arènes* can be classified genetically in three categories:

a) Clays neoformed from the silicates of the parent rock: illite and chlorite;

b) Clays weathered or transformed from the previous ones: vermiculite, mixed layers and possibly montmorillonite;

c) Clays neoformed from hydrolysis products under special conditions: montmorillonite or kaolinite, as the case may be.

Conclusion: the myth of the kaolinization of feldspars

A century-old patrimony teaches that the weathering of crystalline rocks is ruled by the kaolinization of feldspars: "If feldspars are kaolinized, the rock structure is destroyed and arenization occurs."

However, the presentation of the weathering mechanism which has just been given and that of arenization, which is the best example of weathering in temperate regions, show that kaolinization is not the prime mover.

It was Jacques DE LAPPARENT (1909) who pointed out that weathering resulted from the parallel effects of *fragmentation* and *hydrolysis* of rock minerals, each of these mechanisms leading to the other. Plagioclases are sericitized; alkaline feldspars are hydrolysed. All this precedes a possible kaolinization.

CORRENS and VON ENGELHARDT (1938) showed by their experiments that the hydrolysis of feldspars supplied neither clay minerals according to the model of the kaolinitic nucleus of VERNADSKY (1924), nor colloids capable of evolving directly to clays, as was thought at that time.

Reconsidering these questions of weathering fourteen years ago (MILLOT, 1949) by means of the microscope and X-ray diffraction, I concluded that the main mechanism of feldspathic weathering was hydrolysis and that the fate of the liberated ions could differ according to environmental conditions. They could be removed and abandon the arenized rock, which would remain poor in clay; this is commonly the case in our regions. They could be retained in part and in a favorable alkaline environment, become organized into one of the varieties of the micaceous lattice. They could also, under special conditions, such as in an acid and well drained environment, become organized into kaolinite. I concluded:

"Thus, the role played by kaolinization in teaching and common knowledge seems more and more abusive and is in fact erroneous."

I maintain my position, specifically that the hierarchy and chronology of these fundamental phenomena must be corrected in our text books. One can seek a formulation in the following direction:

1° The initial factor of feldspathic weathering is hydrolysis, which releases elements of the crystal into solution;

2° Under our climates these elements released by hydrolysis are, for the most part, removed. They abandon the weathered rock, which is called an arène.

3° Kaolinization is a later and facultative phenomenon. It is a further neoformation from the ions released by hydrolysis. Although rare in our regions, it becomes very important under tropical climates.

III. — THE SOIL CLAYS

The "clay" of soil is defined by a long tradition and codified by the decisions of the International Soil Science Association in 1926 as the granulometric fraction finer than $2\,\mu$. There is thus a size limit, and it is obvious that this fine-grained fraction, sometimes called "mineral colloids", can contain all kinds of products, crystallized or not, which have nothing to do with clay minerals. One can mention the extreme case of the "clay fraction" of a soil taken on a chalk of Champagne (France) which contains pulverized $CaCO_3$ almost exclusively. However, it is indeed in this fraction finer than $2\,\mu$ that the greatest part of the clay minerals are found whenever they occur. What we are interested in here is this "true clay" of the clay-size fraction, together with oxides and hydroxides. Carbonates, quartz and colloidal products, organic or not, which can occur in clay-size fraction will not be considered.

The purpose of this chapter is not to make a pedological study. There are excellent books on this subject, and this is not our field. But pedologists of France, Africa and other countries have often the kindness to ask our laboratory at Strasbourg to identify the clays of the soils they study. Thus, we have accumulated several thousand analyses. Comparing this knowledge with the information found in the literature, I shall present here the perspective of a geologist and geochemist on soil clays.

In order to establish this general view, I have followed the classifications and conceptions of French pedologists with whom I have had the honor to work. The main works are those of AUBERT (1954, 1956), DUCHAUFOUR (1956, 1960) and MAIGNIEN (1958, 1959). I have also used the works of SÉGALEN (1957), LENEUF (1959), SIEFFERMANN (1959), and the theses of BONIFAS (1958) and CAMEZ (1962), done at Strasbourg, which treat the geochemistry of lateritization and the evolution of soil clays.

1° History

The first inventories of soil clays. — The inventory of clay minerals in soils, as such, is not very ancient; they are hardly more than thirty years old. Whereas mineralogists had already defined the modern techniques of clay mineral determination, and especially X-ray diffraction, several years before, pedologists and geologists as well were still not making use of them. ROSS (1927) was the first to point out that crystallized clay minerals occurred commonly in soils. But for a long time yet, soil clay was commonly considered as an amorphous substance composed mainly of iron hydroxyde, alumina and silica. Furthermore, this tradition was maintained by the use of terms such as "colloids" or "colloidal fraction". Since such terms were imposed simultaneously by the size and the behavior of the products, they could only lead us astray from the true notion of clay minerals.

In 1929 the inventory began in a vigorous manner with the works of HENDRICKS and FRY (1929) and KELLEY (1929). They pointed out not only that many soils include a crystallized clay fraction but also that mineralogical methods allow one to identify the composition of this clay fraction: kaolinite, montmorillonite, and micaceous minerals quickly thereafter defined as illite by GRIM, BRAY and BRADLEY (1937).

Thus was organized the first effort necessary in the Science of nature, that of a systematic inventory. Great activity sprang up in all the countries under the direction of the early leaders in this field; in U.S.A., KELLEY and his collaborators (1931, 1939, 1941, 1942), ALEXANDER and HENDRICKS (1939); in the Netherlands and in the Dutch East Indies, EDELMAN, VAN BAREN and FAVEJEE (1939) and HARDON and FAVEJEE (1939); in Germany MÄGDEFRAU and HOFMANN (1937); in Great Britain and the Commonwealth MARSHALL (1935), HOSKING (1940) and NAGELSCHMIDT, DESAI and MUIR (1940); in Russia, where pedology was born, the team of SEDLETSKY (1939); in France, by the works of BRAJNIKOV (1937), DEMOLON (1939) and ERHART (1940).

This recitation of works performed on the eve of the Second World War could only be incomplete; it shows that the inventory has begun in all countries, and today this inventory includes thousands of works. The different clay minerals are identified in soils of all parts of the world, under the most diverse climates, and on various kinds of parent rocks.

The first research on causalities: inheritance and neoformation. — While the inventory was going on, the second problem for naturalists was developing, that of research on correlations and, in appropriate cases, of causalities.

In 1942, Kelley up-dated the notions about clays established in the field of agriculture. As always happens at the beginning of an inventory, he pointed out that the accumulation of facts led to contradictory interpretations. In fact, some investigators studying a single region observed that clays differed according to the nature of the parent rocks. Others, making a comparison of the frequent occurrence of kaolinite under tropical climates and its rarity under temperate regions, called upon climate. Finally, still others, considering a single parent rock type under an uniform climate, showed that relief resulted in variations in leaching and pH of the soil. They saw in that the explanation for the presence of montmorillonite here and kaolinite there.

In fact, we have succeeded in showing that all these factors play a role. The first analyses that allowed us to understand these factors are presented here.

VAN BAREN (1928) was the first to distinguish in the calcareous soils of Java what was inherited from parent rock and what was not. This inherited material could be argillaceous or non-argillaceous. As for the clays, many works have shown that in soils whose geochemical evolution is not very advanced a great part of clay minerals have in fact been inherited from the parent rocks. EDELMAN (1947) made a generalization on this notion of clay mineral inheritance in soils, and I have transposed it into the sedimentary domain (MILLOT, 1953).

Other works initiated our present knowledge about neoformation. In the Dutch East Indies, HARDON and FAVEJEE (1939) described the soils of Java; on the slopes of basaltic volcanoes soils were acid, and the soil clay was kaolinite; in the plains, alkaline and calcareous soils contained montmorillonite. In Australia, HOSKING (1940) showed that, from the same basic crystalline rock, a kaolinitic soil occurred under the influence of abundant rains and efficient drainage, whereas a montmorillonitic soil developed with less rain and impeded drainage. In India, NAGELSCHMIDT, DESAI and MUIR (1940) found in the same region, under the same climate, and from the same rocks, kaolinitic soils with a pH between 6.1 and 7.4 on slopes where drainage is good and a montmorillonitic soil with a pH between 8 and 8.4 in the plains where drainage is poor.

Drawing together such observations, EDELMAN (1947) studied the physico-chemical conditions of kaolinite and montmorillonite stability and inferred therefrom the most probable conditions for their genesis. HÉNIN (1947) made a more important step forward. Using the works of CORRENS that showed the release of ions from silicates into solution, he related the genesis of clay minerals, rich or poor in silica, to the amount of soluble silica in solutions. This was a rejection of the explanation of genesis of clays from colloids and opened the way for the direct genesis of clay minerals from solutions. This approach was followed in order to explain the genesis of sedimentary clays (MILLOT, 1949), and it guided also the magnificent works on the synthesis of clay minerals, works which will be presented in Chapter XI and for which the schools of CAILLÈRE and HÉNIN, FRIPIAT and GASTUCHE, and WEY and SIFFERT have become famous.

Thus, between 1939 and 1947, from the flood of the descriptive works, the outlines were being drawn of the mechanisms of the clay mineral genesis in soils: inheritance from parent rock and authigenesis according to the environment.

Recent works: the transformations of clay minerals. — At the same time that research was carried on, methods of identification were being perfected, and especially the possibilities of the identification of weathered clay minerals. In fact, muscovite, biotite and chlorite become vulnerable under certain conditions. But all the layers do not weather together, and here we have the mixed layers, mixed minerals or structures that are intermediate between the original clay minerals and vermiculites and montmorillonites. An entire dynamics of clay mineral evolution in soils is undertaken.

The precursor is incontestably BRAJNIKOV (1937, 1938, 1942) with his studies on the clay with flints (*argiles à silex*) and the silts of the Paris Basin. He identified there, with the means at his disposal at that time, a mineral E found in the environment of decalcification of the chalk or the silts. When all the carbonate is dissolved, when Mg and Ca ions are leached, when environment is no longer alkaline, the mineral E breaks down and disappears. Probably, the mineral E would be considered today as an evolving mixed layer. What is important in the work of BRAJNIKOV is the notion of clay mineral evolution or transformation as a function of the environment in which they are found.

JACKSON and his collaborators (1948, 1952), by means of modern techniques, brought these phenomena to light by their works on the "weathering sequences of clay minerals". The primary silicates are weathered, and one can define weathering sequences during which clay minerals are transformed successively under the influence of increasing weathering.

The study of the evolution of clay minerals within pedological profiles was begun by MACEWAN (1948) on Scottish soils. The English and Scottish schools followed up this initiative with the works of WALKER (1949, 1950), BROWN (1950, 1953), MITCHELL (1955, 1956, 1961), etc. In the Scottish soils one can observe the alteration of biotite into 12 and 14 Å minerals related to vermiculites and hydrobiotites. The weathering of dioctahedral micas leads to clay minerals of the vermiculite type (BROWN, 1953). MACEWAN applied to soil clays the possibilities of analysis of mixed layers (BROWN and MACEWAN, 1950).

In parallel fashion in the U.S.A., the school of JACKSON obtained similar results and their works have come to us regularly since 1948. A "hydrated mica" (vermiculite) was observed in many soils (JACKSON et al., 1952; ROLFE and JEFFRIES, 1953). Sequences of weathering were verified in numerous profiles (JACKSON et al., 1954). Binary, ternary and quaternary interstratifications were decoded (WHITTIG and JACKSON, 1955, 1956). In 1959 JACKSON summarized the numerous works available and described the possible pathways of transformations of clay minerals in soils. He pointed out the progressive stages of the weathering sequences. The illite-vermiculite stage was described by JACKSON et al. (1952), WHITTIG and JACKSON (1955, 1956), HATHAWAY (1955) and MORTLAND et al. (1956). The less frequently noted chlorite-montmorillonite stage has been reported by STEPHEN (1952), DROSTE (1956, 1958), BROWN and JACKSON (1958), RICH (1958), BROPHY (1959) and GJEMS (1960). The final stage of the transformation of the fine fraction into montmorillonite was pointed out by JACKSON et al. (1954), MURRAY and LEININGER (1956), WHITTIG and JACKSON (1956), BROWN and JACKSON (1958) and GJEMS (1960) in Scandinavia.

In France, the study of the evolution of the clay fraction of soils was undertaken by DUCHAUFOUR, MICHAUD and MILLOT in 1952. We observed an evolution, which is particularly strong in podzols, wherein the appearance of kaolinite was described. This must be modified today, thanks to modern techniques, to the recognition of vermiculite instead. New studies were undertaken on the loess and lehms of Alsace (France) by MILLOT, CAMEZ, WERNERT (1957) and CAMEZ and ROTH (1957), on the soil mixed layers (CAMEZ, LUCAS, MILLOT, 1959), on the vermiculite of podzols (CAMEZ, FRANC DE FERRIÈRE, LUCAS and MILLOT, 1960) and on the weathering of clay minerals in *arènes* (GODARD, PAQUET, MILLOT, 1961). Moreover, the recent thesis of CAMEZ (1962) gives the result of eight years' work on the transformations of clay minerals in soils. The reader is referred to this work for a more complete bibliography on the evolution of clay minerals.

In the Netherlands, parallel works were developed by VAN DER MAREL (1954, 1959 and 1960), who studied the weathered micaceous minerals by an investigation of the K fixation in soil.

Thus, the last fifteen years have shown unexpected progress in the study of soil clays. Taking everything into account, these clays have three possible origins:

— They can be collected by inheritance from clays of parent rock;

— They can be the product of evolution through a more or less strong transformation of those inherited clays;

— They can be synthetized by neoformation from products of hydrolysis.

Such are the different mechanisms that we shall see compete with each other in the study of some of the great families of soils.

2° *Non-evolved soils: cold countries and deserts*

These soils, in truth, are non-evolved weathering complexes. In fact, their characteristic is the almost complete lack of humus. Moreover, chemical weathering is practically zero, a fact easily explainable by what we know about hydrolysis. Either temperature is so low that, in spite of at least seasonable abundance of rain, hydrolysis is almost lacking; this characterizes the cold, arctic and mountainous climates. Or temperature is very high, but water is lacking; this is the case in desert weathering.

To these two categories of non-evolved soils, which are climatic, we can add, following DUCHAUFOUR, lithosols, regosols, and recent alluvial soils; also mountain rankers can be included, although they involve a humic A horizon, because, in fact, chemical weathering only occurs in this horizon and remains modest.

In all these cases the problem of the clay fraction, which is quite small, is simple. The modalities of weathering of the parent rocks are purely physical; only inheritance plays a role; pedogenesis is reduced, and the clay minerals are those of the parent rock. This inheritance can occur in two ways. In the case of crystalline rocks, clay minerals issue from the weathering of silicates: sericitization of feldspars, exfoliation of micas, chloritization of ferromagnesians. In the case of the argillaceous parent rocks, such as tills, marnes, argillites and clay shales, the weathered layer contains the same mixture as in the original rock.

Thus the clay minerals of these soils do not have a great genetic significance. They imitate those of the parent rock and consist mostly of illite and chlorite, which are the ubiquitous minerals and which are quickly gathered from the crystalline and metamorphic rocks and from the majority of sedimentary rocks. Here we are dealing with clay genesis only by inheritance.

3° Soils under an Atlantic climate

The soils of our countries are quite varied and have been studied in detail. Neither the mean temperature nor the rainfall are strong enough to induce massive hydrolysis, as occurs in the humic tropical zone. But already clayey silicates begin to evolve in soils. To describe these transformations, I shall follow the studies carried out in the laboratory at Strasbourg by CAMEZ (1962). Afterwards a general overview will be attempted.

Rendzinas and calcareous brown soils. — The argillaceous fraction of calcimorphic soils developed on limy parent rock shows little evolution in pedological profiles, many of which present a mineralogical composition which does not vary from the bottom to the top of profile. This is illustrated by the following example:

EXAMPLE 1. — *Weakly decalcified soil developed on the Senonian limestone of Le Rafidin-Poquancy (Marne, France).* P.-J.-J. FRANC DE FERRIÈRE collection.

	Illite	Chlorite	Kaolinite	Illite-montmorillonite mixed layer	pH	$CaCO_3$ %
A_1	50	40	10	Trace	8.1	69
C	50	40	10	Trace	8.7	81

However, differences can be found within the profile, if the soil is more highly evolved and if vulnerable minerals like montmorillonite and chlorite occur. In such a case, one can note the decrease of these clay minerals from bottom to top in the profile; this is shown by montmorillonite in the example 2 and by chlorite in the example 3.

EXAMPLE 2. — *Rendzina developed on the slope of Bischenberg (Bas-Rhin, France) on the "grande Oolite".* H. PAQUET collection.

Decrease of montmorillonite

	Illite	Montmorillonite	Illite-montmorillonite mixed layer	Kaolinite
A_1	80		10	10
C	60	20	10	10

EXAMPLE 3. — *Calcareous brown soil, weakly lessivé, developed on the molasse of Castelnaudary (Aude, France).* P.-J.-J. FRANC DE FERRIÈRE collection.

Decrease of chlorite

	Illite	Chlorite	pH	$CaCO_3$ (%)
A	50	40	8.0	26
B	60	50	8.0	28

Brown soils (sols bruns) **and leached soils** (sols lessivés). — The inventory of clay minerals and of their proportions in the series of soils evolved with mull, passing from brown soils to leached soils, gives the following results:

— Illite is uniformly distributed in brown soils, but as leaching becomes stronger from one example of soil to another, illite becomes more abundant near the surface than at depth.

— Chlorite shows an opposite behavior; in the majority of cases it decreases from bottom to top in the profile. Montmorillonite is still more fragile.

— Kaolinite, generally present in minor amounts, is constant.

— Vermiculite, when it occurs, decreases from bottom to top in the profile.

— Mixed layers are found in the upper horizons of profile.

EXAMPLE 1. — *Leached brown soil* (sol brun lessivé) *developed on lehm at Aspach-le-Bas (Haut-Rhin, France).* P.-J.-J. FRANC DE FERRIÈRE *collection.*

Increase of illite. Decrease of chlorite.

Appearance of mixed layers

	Illite	Chlorite	Illite-montmorillonite mixed layer	Chlorite-vermiculite mixed layer	Clay fraction %	pH
A	40	50	Trace	10	18.0	6.0
B	30	60	Trace	10	20.3	6.3
C	30	70		Trace	19.0	6.7

EXAMPLE 2. — *Leached brown soil* (sol brun lessivé) *developed on calcareous, sandy silts in the northern part of the Harth forest, plot 188 (Haut-Rhin, France).* P.-J.-J. FRANC DE FERRIÈRE *collection.*

Increase of illite. Decrease of montmorillonite.

Appearance of mixed layers

	Illite	Chlorite	Montmorillonite	Chlorite-illite mixed layer	Chlorite-montmorillonite mixed layer	Clay fraction %	pH
A	80	20		Trace		12.9	5.1
B	60	20	20	Trace	Trace	28.2	5.5
C	40	20	40			12.2	8.1

Podzolic soils and podzols. — Podzolic soils and podzols constitute the soils evolved with raw humus or mor, very acidic and with rapid decomposition. The variations in the mineralogical composition of the clay fraction throughout the profile increase. Not only does chlorite decrease from bottom to top of the profile, but illite does also. The intensity of weathering increases. Correlatively, illite-vermiculite and chlorite-vermiculite mixed layers appear, and then vermiculite itself.

EXAMPLE 1. — *Ochreous podzolic soil, highly evolved, in the forest of the Hospices de Nancy (Meurthe-et-Moselle, France).* DUCHAUFOUR collection.

Weathering of illite and chlorite.

Appearance of the illite-vermiculite mixed layer

	Illite	Chlorite	Illite-vermiculite mixed layer
A_1	30	20	50
B	40	20	40
C	60	40	
Bonhomme Granite	100 (mica)		

EXAMPLE 2. — *Podzolic soil developed on Vosges sandstone, in the forest of Salm (Bas-Rhin, France).* Institute of Botany, Strasbourg, collection.

Weathering of illite and chlorite.

Appearance of illite-vermiculite mixed layer and vermiculite

	Illite	Chlorite	Vermiculite	Illite-vermiculite mixed layer
A_0	40		40	20
A_1	70	10	10	10
A_2	70		20	10
B	90	10		
Niederbronn Sandstone	100 (mica)			

EXAMPLE 3. — *Ferruginous podzol on siliceous sands. Herrenwald (Bas-Rhin, France).*
Institute of Botany, Strasbourg, collection.

Decrease of illite and chlorite and appearance of chlorite-vermiculite mixed layer and vermiculite

	Illite	Chlorite	Vermiculite	Chlorite-vermiculite mixed layer
A_1	trace		50	50
A_2	10	10	40	40
A_3	40	40	20	Trace
C	100		Trace	

EXAMPLE 4. — *Humic ferruginous podzol on granite. Barembach (Bas-Rhin, France).*
Institute of Botany, Strasbourg, collection.

Strong evolution to vermiculite

	Illite	Chlorite	Vermiculite	Illite-vermiculite mixed layer
A_0	20		80	
A_1	20		80	
A_2	20		80	
B_1	20		80	
B_2	20	10	70	
C	60	20	20	Trace
Barembach Granite	80 (mica)	20		

These examples, chosen from the numerous profiles studied, allowed CAMEZ (1962) to announce a general view on the evolution of clays in soils under our climates. The starting point is inheritance from the parent rock or weathered parent rock, which can be identified at the base of the lower horizons of the profile. We must distinguish a double mechanism acting on this original stock: lessivage and transformation of clay minerals, which act together.

Dispersion. Physico-chemical fractionation. Lessivage. — Clay minerals are often grouped into aggregates, but some soil solutions are able to disperse them. The organic cements of the aggregates can be transformed to soluble products or simply oxidized. If the cements are ferruginous, the iron oxides can be mobilized in the form of complexes. The degree of flocculation of the clay micelles can also vary. Once dispersed, clay minerals can migrate, and the small size of the minerals facilitates this migration, which is especially easy for montmorillonite and the weathered minerals.

Clay minerals are not simply dispersed, they are also fractionated. This power of strong fractionation, which acts even on quartz, is well known in soils; it has been observed in hundreds of profiles where quartz finer than $2\,\mu$ is much more abundant in the upper horizons than in the parent rock. This corresponds to fractionation shown experimentally by PÉDRO (1958). Clay minerals coming from parent rocks are submitted to this exfoliation or fractionation which, moreover, is not solely physical. Thus, chlorite, because of its interlayer Mg ions, is more vulnerable than illite, which is itself more fragile than kaolinite.

Accordingly, lessivage, a well known phenomenon of pedogenesis, which nourishes the alluvial horizons, acts on the dispersed and fractionated minerals that have attained suitable sizes. Consequently, by the triple play of dispersion, physico-chemical fractionation and lessivage, clay minerals leave the upper horizons of profiles in the following order: montmorillonite, vermiculite, mixed layers, chlorite, illite and kaolinite.

Transformation of clay minerals in soils. — The triple mechanism mentioned above does not affect strongly the nature of the clay minerals. However, in the most evolved soils, and especially in podzolic soils and in podzols, clays are transformed. We shall limit ourselves here to a general outline, reserving a more exhaustive analysis of the phenomena for Chapter X on the genesis of clays.

By an attack on the connecting links between layers, chlorite becomes a chlorite-vermiculite mixed layer, then vermiculite itself. Likewise, but later, illite transforms to an illite-vermiculite mixed layer, then to vermiculite. It is important to note that the weathering of the two most commonly inherited minerals leads, by convergence, to the same destiny. And this is still more striking when, in the extreme, the transformation is observed of fine-grained vermiculite to a vermiculite-montmorillonite mixed layer, then to montmorillonite itself, while at the same time allophanes appear which correspond to complete disintegration.

Then the following outline is reached:

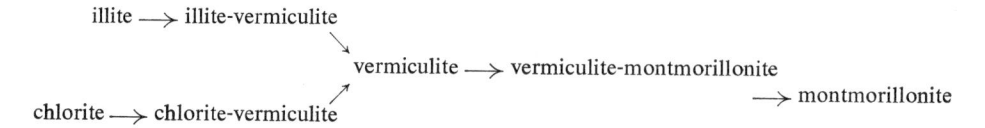

```
illite ——→ illite-vermiculite
                              ↘
                                vermiculite ——→ vermiculite-montmorillonite
                              ↗                                            ——→ montmorillonite
chlorite ——→ chlorite-vermiculite
```

It is necessary to specify that the term vermiculite has here the meaning that was defined in Chapter I with respect to its behavior in X-ray diffraction.

Likewise, this ultimate montmorillonite has a behavior in X-ray diffraction similar to that of true montmorillonite, i.e., its basal spacings expand under the action of polyalcohols. But we still do not know how these structures are organized at the end of their evolution. Their small size certainly facilitated the loosening of the layers with respect to each other, but their crystallochemistry is still poorly known. Here, as has been said already, we are in the presence of phenomena that are measurable but still have not been analysed.

The study of the evolution of soil clays under humid and temperate climates must be pursued by new methods on series of comparable soils and also on typical profiles analysed in detail inch by inch. Here and now, this evolution appears important from pedological and agronomical points of view, as well as from a geochemical viewpoint. It is one of the pieces in the study of the silicate cycle.

4° *The ferruginous soils of warm climate*

Introduction. — Leaving the temperate climates, one encounters a class of zonal soils designated in AUBERT and DUCHAUFOUR's classification under the name of ferruginous soils of warm climate (*sols ferrugineux de climat chaud*). The two principal examples of the latter are the red Mediterranean soils and the tropical ferruginous soils of the Sudanese zone.

These soils are commonly thicker than those of the temperate zone, but they are less deeply developed than those of the humid tropical zone. Their color is red, orange or ochre. They can be bleached, turning to beige, if iron is leached.

Ferruginous tropical soils (**sols ferrugineux tropicaux**). — These soils are found in the northern part of the humid tropical zone which is characterized by the lateritic or ferrallitic soils. They have been described by AUBERT (1951) in the zone of the Sudanese savanna, by MAIGNIEN (1959) in northern Guinea and in Senegal, and by SIEFFERMANN (1959) in North Cameroon. Rainfall is close to one meter per year and the humid season is spread over five or six months. These soils contain unweathered minerals of the parent rock, especially potash or lime soda feldspars, constituting a reserve of alkalis and alkaline-earth elements. The clay fraction often shows kaolinite associated with other clay minerals and iron hydroxides without free alumina. These soils can be lessivés in which case they show colors ranging from red to beige.

MAIGNIEN points out that red soils are developing today in Madagascar under an annual rainfall of about 1 000 to 1 500 mm and temperatures of about 21 °C to 23 °C and a climate with two well-marked seasons. These soils present many resemblances to the *sols rouges méditerranéens*.

Mediterranean red soils (**sols rouges méditerranéens**). — Many Mediterranean countries are characterized by red soils and their reworked products. The greater part of these formations correspond to climatic conditions that no longer exist, as can be proved in many places by the development at their expense of a brown or leached soil at present. Rubefaction occurs today only in those special sites where rainfall is more abundant and microclimate contrasty; it has been observed in Morocco, Oran, Portugal (AUBERT and MONTJAUZE, 1946; CHARLES, 1949; MENSCHING, 1956). The *sols rouges méditerranéens* developed during the great pluvials of Quaternary and Pliocene time (CHOUBERT, 1950; CHOUBERT *et al.*, 1956). One does not know if, during these pluvials, temperature was higher than that of present-day climates, but it is sure that rainfall was much more intense during the humid season, as testified by geomorphologic and geological observations. The pedological study of the red soils of Algeria was carried out quantitatively by DURAND (1959) who invoked lateritic action in their formation.

The clay fraction of ferruginous soils of warm climates. — The clay fraction of the ferruginous soils is still poorly known for the main reason that has already been emphasized, namely, the difficulty in distinguishing the 7 and 3.5 Å reflections of kaolinite, chlorites and vermiculites.

— The measurements made on *sols ferrugineux tropicaux* (SIEFFERMANN, 1959; PAQUET, 1961; PAQUET, MAIGNIEN and MILLOT, 1961) show several possibilities:
In the first place illite, montmorillonite and kaolinite occur in varied amounts in different cases. Thus, one cannot talk of the exclusive presence of kaolinite.
Illite, originating in the parent rock, subsists when soils are little evolved.
Kaolinite, on the contrary, characterizes a stronger weathering in a well drained and acid environment.
Montmorillonite becomes predominant wherever drainage is poor. However, the clay fraction of these soils commonly contains intact silicates: feldspars, micas, and chlorites. One is far from lateritization, the process resulting in a destruction of the silicate lattices.

— In the *sols rouges méditerranéens*, illite is the common clay mineral together with the weathering minerals of micas, mixed layers and even vermiculite. The latter, rather common in our measurements, renders the frequently reported presence of kaolinite doubtful. Kaolinite can occur in the most evolved soils, but naturally it is necessary to determine, by comparison with the composition of the parent rock, if this kaolinite is neoformed or only inherited. However, kaolinite has been identified with certainty in some *terra rossa* of Central Italy (LIPPI-BONCAMBI, MACKENZIE, MITCHELL, 1955).

All things considered, we find ourselves in a hesitant position. We begin with clays inherited either directly from parent rocks or indirectly from the weathered silicates of those rocks. Then, weathering proceeds with considerable variations in intensity. Some *terra rossa* show only illite and its weathering products; on the other hand, kaolinite can occur alone in the most evolved soils.

Everything indicates that we are in a position intermediate between temperate soils and the soils of the humid tropical zone. As in every humid climate, hydrolyses begin. They are intense enough to weather plagioclases and to release iron sesquioxides from ferromagnesian minerals. The iron sesquioxides are in the presence of too little organic matter to undergo lessivage and thus they impregnate the profile; REIFENBERG (1947) has described the formation of ferro- and ferrisilicic complexes. During the dry season, which is well marked, these complexes are destroyed, the iron oxides crystallize and the rubefaction becomes stable.

One can grasp the ensemble of mechanisms prevailing in rubifying climates in the following way:

— The humid season guarantees hydrolysis and the release of iron.

— The dry season mineralizes the organic matter and fixes the iron.

— The result depends on the alternative action of these two seasons.

— The intensity of vegetation, of hydrolysis and of the release of iron will depend on the duration and temperature of the humid season.

— The destruction of organic matter and rubefaction will depend on the vigor of the dry season.

It is well understood that these rubifying soils develop under rather diverse latitudes and conditions. They constitute an entire ensemble, the study of which needs to be carried on. The mean temperatures required include the whole gamut of possibilities between temperate and tropical climates; 20 °C can be considered as a median value. Rainfall is greater than the majority of the present-day Mediterranean precipitation and lower than that in the tropical zones. It can perhaps be said that it falls between 750 and 1 300 mm, with an average of about one meter. The variety of these possibilities accounts for the variety in the clay fraction of these soils.

5° *Calcimorphic soils*

The term "calcimorphic soils" is used in the sense of DUCHAUFOUR (1960).

Calcimorphic soils developed on limy parent rock, and chiefly represented by rendzinas, were studied above. The evolution of clays has been seen to be very weak within these soils.

The calcimorphic soils which are not connected with limestone outcrops have in common an abundance in Ca and Mg ions saturating the absorptive complex of soils devoid of carbonates. The most famous examples are the black earths, or *chernozems*, of temperate steppes, such as those in the Ukraine and the Dakotas, calcimorphic soils of semi-arid Mediterranean and sub-tropical American steppes, the *tirs* of Morocco and the dark clays of the tropics, which are also called *regurs* or "margallitic soils" (MOHR and VAN BAREN, 1959).

All these soils develop from parent rocks which contain bivalent Ca and Mg cations, whether they be volcanic or crystalline rocks with alkaline characteristics, marnes, loess, or limy alluvium. The upper horizons of the profile are leached of carbonates, but no lessivage of iron or clay occurs, and the absorptive complex is saturated with Ca and Mg ions. All the conditions are brought together for the neoformation of montmorillonite.

A detailed study of the clay fraction of dark tropical clays (*argiles foncées tropicales*) (PAQUET, 1961; PAQUET, MAIGNIEN and MILLOT, 1961) showed in fact the predominance of montmorillonite. In soils developed on attapulgite-bearing marnes, the great fragility of this mineral is confirmed; its degradation takes place at the same time as montmorillonite is appearing. In many profiles developed on crystalline rocks, montmorillonite is very abundant in the lower horizons where drainage is impeded and cations are abundant, while kaolinite relieves montmorillonite near the surface. In a general way, montmorillonite requires, under climates of limited rainfall, at the same time a parent rock rich in divalent cations and impeded drainage. The eutrophic brown soils (*sols bruns eutrophes*) of tropical areas constitute once again a series intermediate between the calcimorphic soils and the kaolinitic ferrallitic soils of the humid tropical zone.

6° *Lateritic soils*

Here I shall retain the term "laterite" as a name for the ensemble of weathering products of the intertropical zone. It is true that, for the last hundred and fifty years since this term was proposed, it has been used with extremely varied meanings by geographers, geologists, mineralogists, agronomists, travellers, civil engineers and, of course, by pedologists. Therefore, many pedologists, and especially the French school, have the tendency to abandon this term, and MAIGNIEN wrote recently (1958): "At present, this term is so widely used that it no longer has any pedological significance." It is preferable to use the locution "ferrallitic soils", instead of the old term "laterite".

Possibly I retain this word precisely because it no longer has a definite meaning. I think it is convenient to have a name for all weathering products whose predominant color is red and which cover great areas between the tropics. If I maintain that my friends, Messrs. AUBERT and MAIGNIEN, are specialists on laterites, everyone understands even if he is not pedologist, and this is what I am hoping for. The various new terms proposed are surrounded by a wealth of detailed definitions, giving rise to discussions and revisions whenever progress is made in our knowledge of the genetic phenomenon. As long as all the mechanisms have not been discovered, a pictorial term involving no causal explanation is preferable. I beg to be excused for this simplicity of a geologist, but it must be agreed upon that the words granite and gneiss also cover many things. The twenty last years have shown us this by hard experience; it is in the interest of all geologists to retain these words, and this is what they do in the difficult work of the explanation of their genesis.

However, if the word laterite is maintained for convenience, it will be necessary to make fundamental distinctions in order to analyse the vast ensemble covered by this term.

I must also beg to be excused for the small number of references I shall mention on laterites. Although I am well acquainted with bibliography and I have at my disposal several hundred books, I shall limit myself to the barest possible statement. As a geologist, I must present here a pedagogical view of the genesis of clay minerals and hydroxides in laterites. I shall do this in schematic terms, as I understand this genesis today, after ten years of my own work and the work and teaching of my friends and students. The essential part of my statement comes from the studies of AUBERT (1954, 1958), MAIGNIEN (1954, 1958), BONIFAS (1959), SIEFFERMANN (1959), LAJOINIE and BONIFAS (1961). I refer the reader to these works and especially to the more recent ones.

Fossil laterites and truncated lateritic profiles. — It is necessary to explain a primary point because of the errors that are still current in elementary documentation. Many profiles which have been described and analysed are at the same time fossil and truncated. There are numerous sections capped with iron crusts (cuirasses) giving the typical appearance to many African landscapes of the Sudanese or Sahelian zone. The explanation still persists that the tabular crusts at the top of the profiles were formed by the vertical rise of the iron sesquioxides and by hardening under the rays of sun. In reality, the iron oxides do not migrate vertically from the bottom to the top of the profiles, and the crust is not formed at the soil-atmosphere interface. The latter was exposed by erosion after it has developed within a pedological profile under a forest landscape.

Figure 17 shows schematically a profile truncated by erosion and a complete profile under forest.

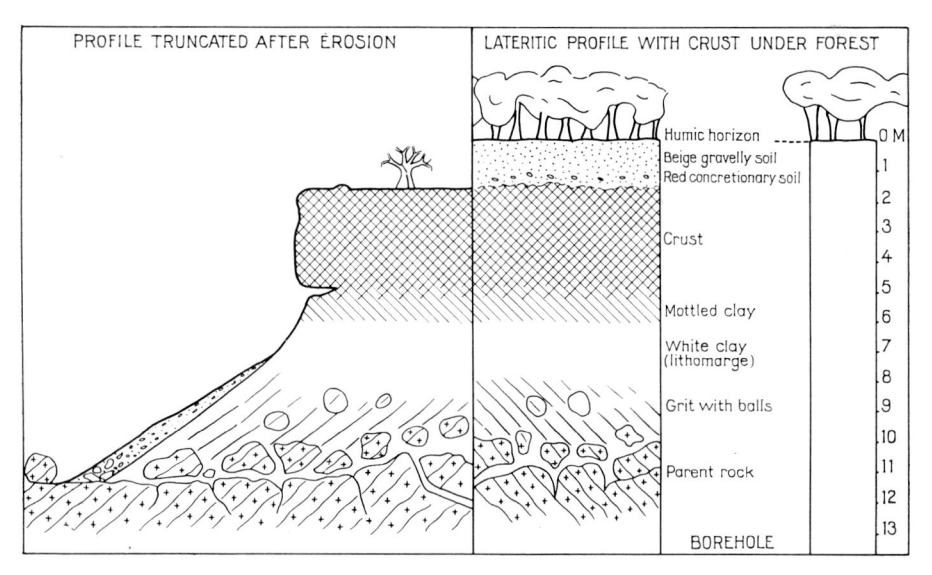

FIG. 17. — *Lateritic profile with ironpan under shade forest and its equivalent truncated by erosion in the sudanese region.*

Typical lateritic profile. — I shall choose the example described by AUBERT (1954) at the Fifth International Congress of Soil Science.

The profile is found in lowland Ivory Coast at Dakpadou, 50 km north of Sassandra, under a shade forest that is still rather beautiful, although somewhat degraded. The climate is sub-equatorial with minor dry seasons, rainfall of about 1 700 mm and a temperature of about 27 °C, varying from 23 to 32 °C during the year. The region is a plateau of limited extent, weakly undulating, underlain by a gneissic rock rich in ferromagnesian elements.

The section is as follows (fig. 17, right profile):

— At the soil surface, a bed of leaves, twigs and branches in the process of decomposition, which seems to be resting on the soil.

— From 0 to 110 cm: gray-brown horizon, slightly humic down to 35 cm; then beige gravelly clay, rich, especially in the first 40 to 50 cm, in very hard, round and very dark ferruginous concretions; from about 80 cm, this horizon becomes more compact, the concretions less hard, and the color brick red.

— From 110 to 175 cm: at 110 cm, rather sharp transition to a hardened horizon, but one which can still be broken by hand. Down to a depth of 175 cm, this horizon is formed of brown to red bands, more or less dark, which anastomose and delimit cavities occupied by an earthy material, whose color varies from ochre to beige.

— From 175 to 650 cm: the hardened horizon passes progressively to a non-hardened horizon, more compact near its base, brick red, with rather clearly delimited spots, beige, ochre or gray. It still contains some hardened nuclei, especially in its upper part, and the entire mass is crossed by numerous small channels. The quartz grains are much more perceptible here than in the upper horizons where they seem to have been pulverised.

— From 650 to 840 cm: the horizon described above passes progressively to one of mottled clay which is very similar but in which the spots are less clearly delimited and there is a greater number of light-colored spots, beige or gray, and more quartz grains. Already at the bottom of this horizon some white elements appear, having the form of feldspar crystals, but made up of a powdery mass.

— Below 840 cm and down to more than 1 100 cm, first there is an ochreous brown horizon, rich in quartz and powdery white elements of crystalline shape. Some less decomposed rock fragments show pyroxenes, which are recognizable but weathered. Then, below 9 m, a mass of gneissic *arène* with many friable white elements of feldspathic shape and quartz grains and highly weathered colored elements.

— Deeper, around 12 or 13 m, would be the parent rock, a gneiss rich in ferromagnesian minerals.

This is a complete lateritic or ferrallitic soil profile. From top to bottom below the forest it includes:

— An upper horizon, somewhat *lessivé*, weakly humic and rich in concretions (1 m).

— An eluvial horizon with crust and in the process of encrustation (6 m).

— A horizon of mottled clay or lithomarge (2 m).

— A horizon of initial decomposition.

— The parent rock.

FIG. 18. — *Isovolumetric weathering of dunite from Conakry, Guinea* (after MILLOT and BONIFAS, 1955).
1. Conakry (Mine). Sound dunite. Peridot surrounded with ribbons of serpentine (Polarized light; 110×).
2. Conakry (Mine). Slightly weathered dunite. Cleaved peridots and yellow antigorite (Natural light; 130×).
3. Conakry (Mine). Weathered dunite. Iron oxide penetrates into the cleavages of the olivine (Natural light; 110×).
4. Conakry (Mine). Soft bed of *Pain d'épice* (gingerbread). Complete transformation of the structure preserved by iron sesquioxides (Natural light; 110×).

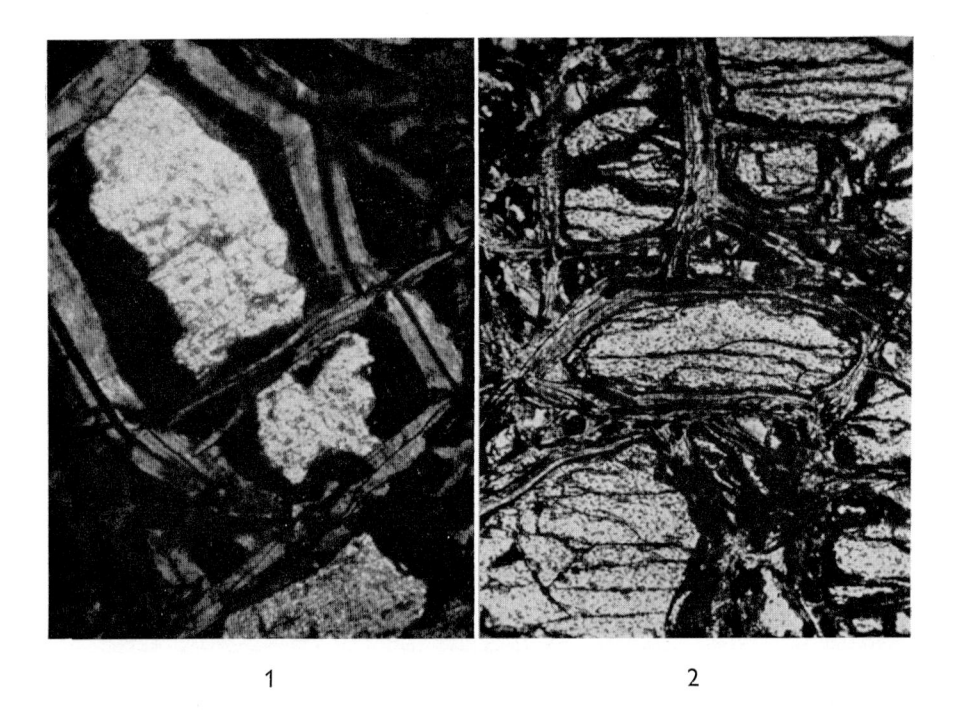

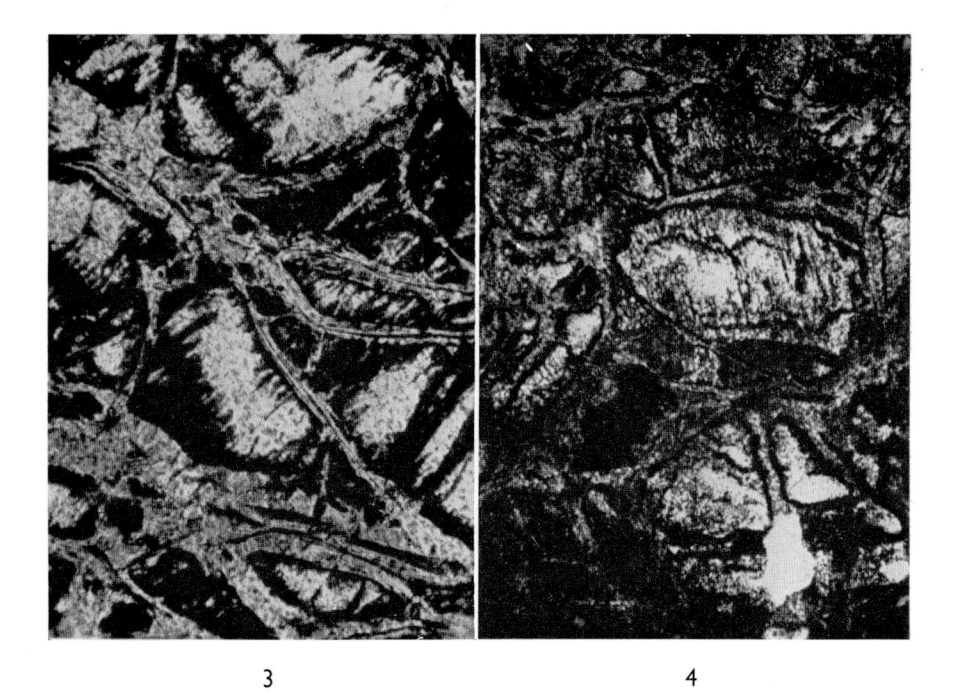

1 2

3 4

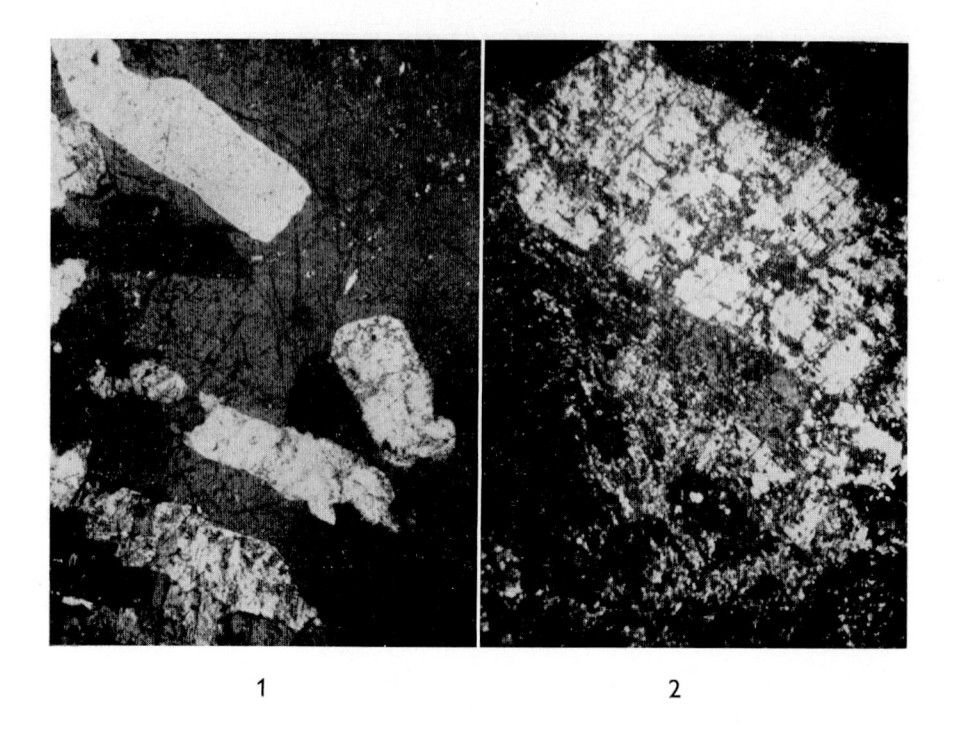

116   WEATHERING AND SOIL CLAYS

1

2

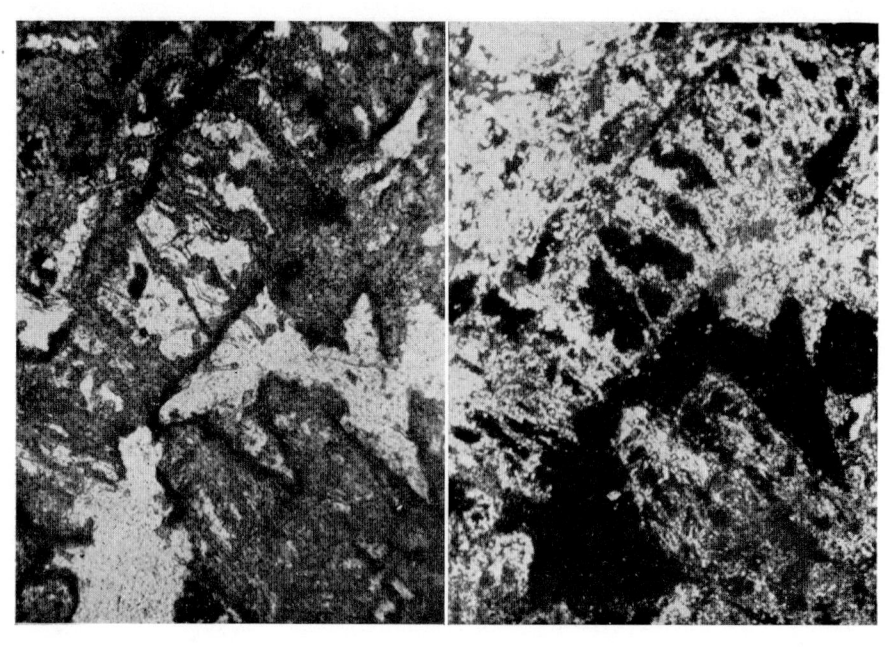

3

4

Fundamental distinction: lateritic weathering and encrustation (cuirassement).

— Many lateritic profiles, like the previous typical example, involve a horizon with ironcrust or in the process of iron encrustation, and rich in iron oxides. The usual manner of looking at profiles in wells or vertical sections has for a long time led our minds to conceive mechanisms of iron accumulation in the vertical sense, and, most often, by means of a migration from bottom to top under the influence of evaporation. We know today that to attribute the essential of the iron encrustation to migrations from bottom to top is an error (MAIGNIEN, 1958). In the first place, this impression was due to the fact that the study of truncated fossil profiles with their iron crust did not allow any pathway for iron migration other than an ascending one. But even when complete profiles developing under a forest were studied, the same vertical migration from bottom to top was evoked, especially because the ironcrust horizon overlies the much lighter colored mottled clays or lithomarge. In reality, evaporation is weak under forest cover, and there is no sufficient reserve of iron in the lower part of the profile to nourish the formation of the crust.

In fact, we know today that encrustation is a secondary and facultative phenomenon within profiles of lateritic weathering. It is the fruit of the immobilization of iron carried in part by vertical lessivage, but above all by lateral movement of ground water circulating in the weathering mantles.

Lateritic weathering and encrustation must be categorically *distinguished.*

The first category is that of *lateritic* or *ferrallitic weathering*. In tropical regions the annual rainfall is considerable and can amount to several meters per year. The mean temperature is very high, from 20 to 30 °C. The physical and chemical mechanisms of disintegration and hydrolysis, described in the beginning of this chapter, act in a very rigorous way. However, weathering goes very much beyond that observed in the arenization of temperate lands. Silicates are completely hydrolysed and even quartz is finally dissolved. The main ions, Si, Al, Mg, Ca, K, and Na which make up the silicate lattices, are released. The majority is removed by solutions percolating through the profile and reaches the ground water reservoir, rivers and the sea. But some elements are retained in part; these are mainly iron, aluminum and silicon, and they join together in the profiles to form the three main constituents of laterites:

— iron oxide, generally hydrated and crystallized into goethite, sometimes anhydrous and crystallized as hematite;

FIG. 19. — *Isovolumetric weathering of syenite from the Iles de Los, Guinea* (after MILLOT and BONIFAS, 1955).
1. Kassa. Iles de Los (Quarry). Sound syenite. Plagioclases within a mass of nepheline (Polarized light; 38×).
2. Kassa. Iles de Los (Quarry). Slightly weathered syenite, gibbsite in the cleavages and cracks of the feldspars (Polarized light, 38×).
3. Kassa. Iles de Los (Quarry). Syenite transformed into pumice. Gibbsite aggregates in the ghosts of the feldspars. Holes occur in sites formerly occupied by nepheline (Natural light; 38×).
4. Idem (Polarized light; 38×).

— hydrated aluminum oxide, generally crystallized as gibbsite, and sometimes as boehmite;

— aluminum silicate, crystallized in the form of kaolinite.

The second distinctive category is *encrustation* (cuirassement). This consists mainly of an accumulation of iron sesquioxides, and in some cases aluminum sesquioxides, in one of the profile horizons. It is sure that the vertical lessivage by means of the percolation of solutions will drain iron into the accumulation horizon, but this cannot account for the thickness and considerable extension of lateritic crusts. On the contrary, the principal phenomenon is the migration of ground water within the profiles. This ground water circulates in the same way as all ground water in the world situated in a permeable surficial material; it is charged with soluble iron which oxidizes and accumulates in the form of ferric hydroxide in the oscillating oxygenated ground water zone (fig. 20).

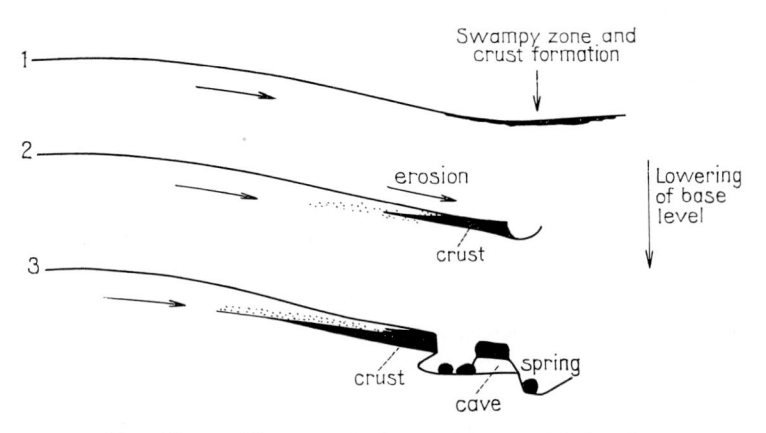

Fig. 20. — *Diagram of the emplacement of lateritic crust and of the formation of relief by lowering of base level* (after MAIGNIEN, 1958).

Iron encrustation (*cuirassement*) is common in lateritic profiles, but not at all essential, as we shall see; all depends on ground water hydrodynamics. Moreover, this phenomenon can occur in every kind of permeable formation traversed by waters charged with dissolved iron: alluvial terraces, sandy formations, slope breccias, etc. Finally, iron crust formation is found under climates other than lateralizing ones; alios is a ferriginous encrustation that develops under quite different latitudes.

Thus, encrustation is a phenomenon distinct from lateritic weathering. It consists of concretionary accumulation of iron and aluminum hydroxides in no matter what kind of permeable horizon. Lateritic profiles are frequently affected by iron encrustation, but this is a pedological phenomenon distinct from and secondary to lateritization.

Geochemistry of lateritization by the isovolumetric method. — During the beginning phases of lateritic weathering the texture of parent rocks is preserved. This has been confirmed by microscopic study, as well as it is shown by the photographs of figures 18 and 19. If the textures are preserved, volumes are not modified; this is the foundation of the isovolumetric method proposed by MILLOT and BONIFAS (1955) in order to study the geochemistry of weathering.

The principle is the following: if volumes are preserved, it is in the end possible to study the progress of rock weathering by comparison to a constant volume. This has been applied to many profiles (MILLOT and BONIFAS, 1955; BONIFAS, 1959).

1° WEATHERING OF DUNITES FROM CONAKRY (GUINEA) INTO IRON ORE. — Microscopic examination shows the replacement of the peridot and serpentine of the parent rock by iron oxide with respect for the textures, and thus for the volume.

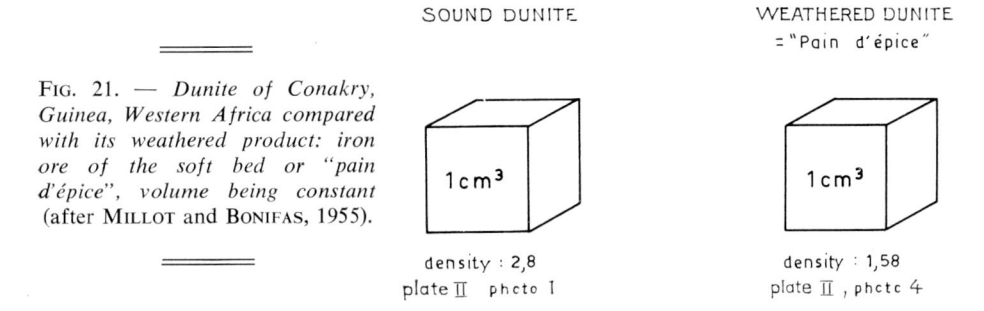

FIG. 21. — *Dunite of Conakry, Guinea, Western Africa compared with its weathered product: iron ore of the soft bed or "pain d'épice", volume being constant (after MILLOT and BONIFAS, 1955).*

SOUND DUNITE — 1 cm³ — density : 2,8 — plate II photo 1

WEATHERED DUNITE = "Pain d'épice" — 1 cm³ — density : 1,58 — plate II, photo 4

Let us compare 1 cm³ of fresh dunite to 1 cm³ of the *pain d'épice*, i.e., the weathered rock as presented in figure 21.

Geochemical calculations give the results shown in Table X.

TABLE X. — GEOCHEMICAL BALANCE BY THE ISOVOLUMETRIC METHOD. WEATHERING OF THE DUNITE FROM CONAKRY (after BONIFAS, 1959)

	Sound dunite	Weathered dunite	Differences (in absolute values)	Differences (in %) of the components of the dunite
Weight of 1 cm³ in cg	280	158		
SiO_2	95	2.5	− 92.5	− 97.5
Al_2O_3	4.5	2.5	− 2	− 44.5
Fe	32.8	93.5	+ 60.7	+185
CaO	1.4	0	− 1.4	−100
MgO	103	0.57	−102.5	− 99
Cr_2O_3	0.25	0.24	− 0.01	− 4
TiO_2	0.22	0.25	+ 0.03	+ 13.5
H_2O	29.7	17.7	− 12	− 40

The geochemical balance of weathering shows:

— a removal of silica, lime and magnesia;
— an important decrease of alumina and water;
— a slight increase of titanium;
— an important increase of iron (about 200 %);
— no change in chromium.

2° WEATHERING OF SYENITES FROM THE LOS ISLANDS (GUINEA) INTO BAUXITE. — Microscopic examination, shown in figure 19, shows the maintenance of the former texture and the development of gibbsite in the place of and instead of crystals, and especially of the feldspar crystals.

The geochemical calculations lead to the results presented in Table XI.

TABLE XI. — GEOCHEMICAL BALANCE BY ISOVOLUMETRIC METHOD. WEATHERING OF SYENITE FROM THE LOS ISLANDS (after BONIFAS, 1959)

	Sound syenite I	Weathered syenite II	Pumice III	Differences (in absolute values) I to III	Differences (in % of the components of the syenite)
Weight of 1 cm³ in cg	258	219	154		
SiO_2	148.0	127	4.8	−143	− 96
Al_2O_3	45.0	47.5	86.2	+ 41.2	+ 91
Fe	11.9	5.3	8.1	− 3.8	− 32
CaO	6.7	0.8	0.3	− 6.4	− 96
MgO	2.8	0.8	—	− 2.8	− 100
Na_2O	15.5	8.8	nd	nd	
K_2O	18.1	17.7	nd	nd	
TiO_2	1.8	1.3	2.6	+ 0.8	+ 44
MnO_2	1.0	0.2	0.1	− 0.9	− 90
H_2O	3.3	6.8	45.7	+ 42.4	+1 280

The geochemical balance of weathering shows:

— a removal of silica, lime, magnesia, to which must be added the alkaline elements, which remain only as trace elements;
— an important decrease of iron and manganese;
— a very important increase of alumina, which almost doubles, of titanium, and of water which increases by more than tenfold.

3° WEATHERING OF HORNFELS (CORNÉENNE) OF KOULOUBA, NEAR BAMAKO (SUDAN). — The hornfels of Koulouba was sampled at the big bend in the Bamako-Koulouba road. It is weathered to kaolinite. The geochemical comparison of these two examples is presented in Table XII.

TABLE XII. — GEOCHEMICAL BALANCE BY ISOVOLUMETRIC METHOD. WEATHERING OF HORNFELS FROM KOULOUBA (after BONIFAS, 1959)

	Sound hornfels	Kaolinized hornfels	Differences (in absolute values)	Differences (in % of the components of the hornfels)
Weight of 1 cm³ in cg	261	140		
SiO_2	172	66.1	−106	−61.5
Al_2O_3	37.4	46.5	+9.1	+24.3
Fe	13.8	6.3	−7.5	−54.4
CaO	1	0.4	−0.6	−60
MgO	4.2	0.1	−4.1	−97.5
Na_2O	1.3	—	−1.3	−100
K_2O	20.9	—	−20.9	−100
TiO_2	1	1.2	+0.2	+20
H_2O	3.9	16.7	+12.8	+328

The geochemical balance of this kaolinitic weathering shows:

— a complete removal of magnesia and alkaline elements;
— a release of about half of the silica, iron and lime;
— an increase of about a quarter of alumina and titanium.

4° GENERAL VIEW. — These calculations can be carried out for many profiles. They allow one to appreciate the geochemistry of lateritic weathering in a quantitative way. The main characteristics are the following:

a) The main point of lateritic weathering is *lessivage*. This lessivage affects chiefly the alkaline elements, lime and magnesia, and partially the silica, iron and alumina.

b) The transformations of parent rock are *transformations by removal*, and we shall look at these again. The least leached elements are organized into *neoformed products*, among which the principals are goethite, gibbsite and kaolinite.

c) *Silica is removed*, completely or partially, according to the individual case. When it remains in part, it gives rise, in combination with alumina, to kaolinite.

d) Iron and aluminum *decrease or increase*, as the case may be, in the volumetric unit. This gives evidence, in no uncertain manner, of two phenomena the appreciation of which is indispensable. Iron and aluminum can be leached, but they can also be imported. The numerical results give only the balance of this double mechanism: export and import. And this balance is quite variable from one case to another.

e) The removal of these elements along with the others is in the direction of the ground water reservoir and sedimentary basins. The importation can only occur *by gravity*, by vertical or oblique inflow of water from the upper weathering zones. The rock in the process of weathering is at the same time a place of lessivage and of accumulation.

f) It is obvious that at the moment when the texture of the parent rock, which is preserved during the initial stages of weathering, breaks down, the isovolumetric method becomes useless. It is convenient to use as a means of comparison an element which is suspected to be the least variable; thus we shall use methods of isochromium, isotitanium and isoalumina calculation. Every profile has its own geochemistry and necessitates the choice in each case of the best geochemical method of comparison.

Lateritic soils of the humid tropical zone. — The humid tropical zone, or Guinea forest zone, is characterized by annual rainfall of 1.3 to 3 m, with rainy seasons of four to five months and dry seasons lasting two to three months. The mean annual temperature is between 25 and 30 °C.

1° Acid parent rocks: granites, gneiss. — It is possible to follow two great itineraries.

The first itinerary concerns those cases where phreatic water does not stay in a prolonged or permanent manner in the profile.

Leneuf (1959) presented a systematic study of weathering in the forested Ivory Coast. His type I and II profiles correspond to this first itinerary.

— Type I occurs on bare granitic flagstones on the surface of large granitic domes subjected to ephemeral rainfall and strong variations of temperature. In truth, the alteration develops only through a few millimeters of thickness, and the microclimate on these arid walls is rather different from that which is appropriately called a tropical forest climate.

— Type II corresponds to the granitic walls of joints where water percolates easily. The thickness of weathering is a few centimeters.

In both types, one observes a bleaching of the granite which takes on a chalky aspect. The weathering is characterized by a microdivision of the minerals of the granite. Plagioclase and quartz are the most sensitive to weathering, alkaline feldspars and muscovite are the most resistant. Simultaneously, mineralogical measurements reveal the appearance of gibbsite and, as an accessory, goethite in rusty zones.

The example of the syenites of Los Islands is a typical but massive case of this kind of weathering. Excellent drainage excludes the permanence of ground water in the profile, and weathering leads directly to gibbsite (Bonifas, 1959).

— Type III of Leneuf includes a whole series intermediate between the first and second itineraries. It concerns massive underground alteration below the weathering mantle, but without permanent phreatic water because of good drainage. The bleached zone can attain several meters thickness and not only gibbsite develops there, but also kaolinite, according to the interplay of permeability and the water table in the upper horizons where the original rock texture has disappeared.

This second itinerary involves massive underground alteration with less intense drainage that maintains a permanent water table. This case is illustrated by type IV of Leneuf.

Lessivage at the base of the profile is so weak that montmorillonite and 12-14 Å mixed layers are found here along with kaolinite. But these minerals are transitory and decrease upward in the profile, whose thickness is between 1 and 4 m, giving way to kaolinite and goethite, but without gibbsite.

These examples show already the important role played by ground water within lateritic profiles.

2° Alkaline parent rocks: peridotites, dolerites, basalts. — These phenomena are even more marked on the basic rocks of petrographic classification, such as peridotites, gabbros, dolerites and basalts, and are dependent on hydrodynamics within the profile.

If drainage is excellent, the parent rock is transformed directly to goethite and gibbsite with permanence of structures and volumes; they are the *pains d'épice* (ginger bread) of Lacroix (1913). If the rock is aluminous, such as a dolerite, gibbsite is prevalent in the profile; if the rock contains few feldspars and is ferromagnesian, it is goethite that prevails. At the extremes, either a bauxite deposit or an iron ore will result.

If drainage is impeded, a humid horizon will be established at the bottom of the profile. The released alumina is then present in the form of kaolinite, and the rock weathering is argillaceous. The water table, however, oscillates with the seasons, and in the upper part of the weathering mantle seasonal desiccation can occur and gibbsite can develop. There is every reason to believe that, in this case, kaolinite, in its turn, acts as the parent rock in the upper zones and releases gibbsite. This case was studied by Ségalen (1957), Sieffermann (1959) and Gastuche and de Kimpe (1961).

If drainage is even more impeded and barely is able to evacuate all the rain that falls, the profile remains clogged with water. This does not mean that the profile is not leached since there is drainage, but the water level remains high. Thus, the conditions prevail for the genesis of kaolinite accompanied by goethite, but here these conditions occupy the whole profile. We obtain here that which French pedologists call weakly ferrallitic red soils.

If drainage is lacking, so is lessivage. The poorly evacuated basic ions accumulate with silica, and we have intrazonal soils such as the dark tropical clays (*argiles foncées tropicales*). They develop easily on poorly drained basalts, and they have been examined under the heading of calcimorphic soils.

Within profiles developed on basic parent rocks, the behavior of the water table regulates the outcome of lateritization. Gibbsite forms, either directly from the parent rock or indirectly from kaolinite, only in zones subject to intense drainage. On the other hand, kaolinite develops when circulating waters impregnate the profile.

Lateritic soils of the equatorial zone or the zone of dense forest. — In equatorial zones where the dense forest prevails, annual rainfall rises to or exceeds 2 and 3 meters to attain 4, 6 and sometimes 8 meters. The short dry season is hardly noticeable and does not reach the soil covered by the shade forest.

The weathering mantle, whose thickness can reach 10, 20 or 30 meters, always remains humid. Kaolinization prevails on acid parent rocks such as granites and gneiss, which are always clogged with water. Abundant lessivage removes the small amount of iron, and one arrives at tremendous areas of pale colored kaolinitic earths; these are the yellow equatorial clays (*argiles jaunes équatoriales*).

On basic parent rocks the saturation of the profile by ground water likewise allows only kaolinitic weathering. However, iron, which is more aboundant, persists; these are the red equatorial clays (*argiles rouges équatoriales*) which are, moreover, five to ten times thicker than the preceding ones.

In both cases, gibbsite is not very abundant, and these soils are classified with the weakly ferrallitic soils.

Moreover, the phenomena of encrustation (*cuirassement*) which require the circulation of an oscillating ground water supply within the profile, are absent or weak; this does not obviate the occurrence of fossil crusts corresponding to the former existence of other climates in these same latitudes.

Neoformation of clays and hydroxides in lateritic soils. — It is very difficult to encompass such extensive and so varied phenomena in a few lines. Simplifications involve great risks, but it seems that the ensemble of observations carried out in Africa, Guiana and Madagascar allows us to grasp the major rules of the mechanisms.

1° GOETHITE. GIBBSITE. KAOLINITE. — In an intertropical environment water and temperature favor the complete hydrolysis of silicates. From the released ions, the three minerals which are formed are goethite, gibbsite and kaolinite.

2° HYDRODYNAMIC EQUILIBRIUM. — The distribution of goethite is widespread. On the contrary, gibbsite and kaolinite are differently distributed. It seems that the fundamental rule is that approached by HARRISON in 1933, then codified by GOLDICH in 1948, and since that time illustrated by many authors. It is a question of the state of hydrodynamic equilibrium corresponding to the balance between the incoming water and its evacuation within the weathering mantle. It brings into play numerous factors: alimentation from upslope, annual rainfall, permeability of various horizons, fracturation of parent rock, etc.

In the whole, a balance is established which achieves a hydrodynamic equilibrium changing with the seasons and geological times.

In a very welldrained environment without permanent phreatic water, *gibbsite*, accompanied by goethite, is formed directly from solutions that result from hydrolysis of silicates.

In a well drained environment with permanently high water table, the amount of silica in solution increases and allows the neoformation of *kaolinite*.

In a poorly drained environment, cations are found, and montmorillonite can occur as a transitory mineral at the bottom of the profile, or definitively in calcimorphic soils.

In the average case, ground water circulates and nourishes, at the level of its oscillating surface, encrustations within the profile. This is a secondary, facultative, but important accumulation, whose pedological character is obvious.

3° CLIMATIC OR GEOLOGICAL VARIATIONS. — However, hydrodynamic equilibria change in response to climate variations and geological evolution (erosion, tectonism, etc.). Each profile is the result of a history in which several sets of controlling conditions could succeed one another. Horizons bearing gibbsite can be silicified into kaolinite. In the opposite case, horizons bearing kaolinite can be subjected to strong drainage or to pronounced periodic desiccation and give rise, in their turn, to gibbsite. One is obliged to study the reversible equilibrium between gibbsite and kaolinite, and the first works on this subject are encouraging (WOLLAST, 1961).

Likewise, lateritic crusts can follow an increasing development or, on the contrary, be leached or degraded (MAIGNIEN, 1958).

4° NOMENCLATURE. LATERITIZATION. FERRALLITIZATION. — Here we come to a discussion on nomenclature I have constantly avoided until now. Should we say "lateritization" or "ferrallitization"? The latter term has been preferred by the French school because it is much less vague than the former and indicates very well the simultaneous release of iron and alumina during weathering.

This term is especially convenient for lateritic products bearing goethite and gibbsite. However, as a matter of fact, analysis shows that numerous soils from the humid intertropical zone, and especially from the equatorial zone, contain little or no free alumina, but only kaolinite. In this case they are called weakly ferrallitic soils. It would be judicious to say "ferrisiallitic" soil, using the former term of PALLMANN (1947), to account for the combination of iron and kaolinite, but that term is already often used in an other meaning. In any case, this is disconcerting because we must choose between two points of view:

— Either our attention is called to the presence of free iron and alumina, and, in this case, the goethite- and kaolinite-bearing soils are not ferrallitic but ferrisiallitic, according to the etymological meaning of this term;

— Or our attention is called to the release of iron and alumina from the silicate framework and, in this case, the goethite- and kaolinite-bearing soils are pure ferrallites. In fact, the alumina of silicates was totally, and not weakly, released; it entered only into the kaolinite lattice. Although released, the alumina is no more free, but silicated.

The name that is to be chosen thus depends not only on the stage of the mechanism to which our attention is called, but also on improvements in our knowledge. All this is still changing and is difficult to codify. Therefore, I provisionally maintain the term "lateritization", whose purely descriptive character does not at all bias the progress of the works of specialists.

5° TRANSITORY NEOFORMATION IN THE ZONE OF INITIAL WEATHERING — A delicate problem remains, that of the detailed mechanisms of weathering in the zone of initial weathering. This concerns the environment where hydrolysis occurs in the parent rock that is being weathered, and where the pH is alkaline because of the massive release of

Ca and Mg ions from the parent rocks. Such are the genetic conditions of minerals structured after the mica type (vermiculite, montmorillonite), as I envisage them (MILLOT, 1949), and one could foresee their ephemeral appearance under these conditions. This has been verified by an increasing number of authors. CAPDECOMME and KULBICKI (1954) found montmorillonite in the zone of initial weathering on basalts from Thiès, Senegal; CAILLÈRE, HÉNIN and BIROT (1957) and BIROT, CAILLÈRE and HÉNIN (1959) in many, generally basic rocks in Guiana, Brazil and Madagascar; SÉGALEN (1957) in the initial weathering zone on basalts of Madagascar; GASTUCHE and DE KIMPE (1961) on basalts of the Congo. Elsewhere, BONIFAS (1959) described chlorite in the formation of *pain d'épice* on dolerites, and montmorillonite in the zone of initial weathering on peridotites of Conakry. Lastly, LENEUF (1959) showed the ephemerality of hydromicas and montmorillonite at the bottom of the profiles developed on poorly drained granites in the Ivory Coast.

All these facts are minor, but significant. I cite them only because they support the general mechanisms we are looking for. However, it must be admitted that these alterations have been rarely observed and that, in the greatest number of cases, they are fleeting. The microscopic examination of silicates, and especially of feldspars, reveal the main phenomena: epigenetic replacement of minerals by gibbsite and kaolinite accompanied by iron hydroxide. These are the fundamental neoformations during lateritization in the humid tropical climates.

Experimental "lateritization". — PÉDRO (1958, 1961) tried to reproduce conditions similar to those of lateritization with his continuous leaching apparatus. Samples of basalt and granite were immersed for the two thirds of their height, the upper third being out of the water. Leaching is guaranteed by warm, neutral water (70 °C).

Basalt, weathered in a humid atmosphere, is changed into a red material of the *pain d'épice* facies. Comparison of the intact and the weathered material can be effected by the isovolumetric argument (MILLOT and BONIFAS, 1955). 80 % of the basalt was decomposed and 60 % of that was removed by solutions. Silica, alkalis and alkaline-earth elements were removed, iron remained in place, and alumina went both ways. The weathering crust contains boehmite, gibbsite, hematite and goethite.

The weathering of granite displays the individualization of iron and aluminum hydroxides, and we approach mechanisms close to those of natural ferrallitization.

One finds in the last review of weathering by CORRENS (1961) the results of a test of continuous leaching at a pH of 7 of particles of kaolinite and montmorillonite smaller than 1 μ. Silica turns out to be ten times more soluble than alumina in the case of kaolinite and two hundred times more soluble in the case of montmorillonite. This is deduced from the curves of solubility and results in a model of the beginning of bauxitization.

7° **Lateritic crusts** (cuirasses)

We have seen that the phenomenon of encrustation is neither specific to certain climates nor to lateritic profiles. However, it occurs frequently and intensively in the latter.

Release of iron. — We have seen the ease with which ferromagnesian minerals release their iron by oxidation and hydrolysis under diverse climates. In an intertropical humid climate iron is released totally from the silicates which are completely hydrolysed.

Mobilization of iron. — The study of the migration of iron in soils has been the object of much research. An important documentation will be found in the works of REIFENBERG (1947), BÉTREMIEUX (1951), BLOOMFIELD (1955), MAIGNIEN (1958) and LOSSAINT (1959). Here are the principle results:

— The ferric ion is nearly insoluble in the pH conditions of tropical soils.

— The ferrous ion has an appreciable solubility which is maintained in a reducing medium, i.e., so long as reducing organic matter is present.

— Ferric and ferrous ions can give rise with silica to complexes soluble in water; these are ferro- and ferrisilicic complexes.

— Finally, biological phenomena, which participate in the decomposition of organic matter, play the greatest role in the migration of iron, either by the reduction phenomena connected with the life of the microorganisms, reducing the ferric forms, or by the residues of fermentation which give rise to pseudo-soluble complexes.

Migration of iron. — Once iron is in solution or in pseudo-solution in the soil water, its destiny coincides with that of the water. Now, this water is subject to gravity in two ways. In the first place by the percolation of rain water through the profile. This percolation is considerable; the water rapidly attains the level of the water table, which oscillates within the weathering mantle. Then, the phreatic water flows slowly away, as shown by piezometric maps. The soluble products migrate with it as long as the conditions of their stability in solution are maintained.

Immobilization of iron. — The principle mechanism provoking the immobilization of the iron carried by ground water is oxidation. This is obvious in the case of free ferrous ions. However, the ferro-organic complexes are also oxidized. The iron, no longer protected, is oxidized and precipitated. This mechanism of oxidation occurs in many cases: the approach to the soil surface where the ground water table crops out; the zone of oscillation of the water table; the crossing of coarse materials; the crossing of an older porous or cavernous crust; disappearance of the forest, cutting off the supplies of complexing organic matter; increase of natural drainage by accelerated aeration of the percolating water. These phenomena are seasonal; during the humid season there is mobilization and lessivage; during the dry season there is a violent immobilization in the zone of oscillation of the water table. In particular this accounts for the lack of crusts within profiles that are always humid, as in the equatorial zone; here lessivage prevails over immobilization. On the contrary, in the tropical zone with a well marked dry season, many seasonal circumstances give rise to the immobilization of the iron carried by ground water and, thus, to iron encrustation.

INDURATION OF THE IRONCRUSTS. — If an ironcrust is exposed at the surface, either by quarrying or by erosion, it tends to become indurated.

At the present time, induration is easily observable on bricks made of ironcrust when exposed to atmosphere, as well as in railway cuts. This induration is only few centimeters thick. The mechanism is poorly understood. There is desiccation that results in the formation of ferruginous films which cement the ensemble. Over the years, there is an alternation of moistening and desiccation, leading to some dissolution, and then to crystallization of the cement. Here the phenomena of capillary migration under the influence of evaporation can play a role.

The natural indurated ironcrusts are much thicker. They can attain several meters. The exposure of these crusts is a result of erosional phenomena; thus, it takes place slowly. Then, once exposed, the crust can have a history of thousands or millions of years. The grassy vegetation which succeeds to forest is able, by means of the organic products, to bring about once again a mobilization of iron in the upper part. This iron will fill the interstices of the lower levels by the crystallization of successive films which are observable under the microscope. When the ironcrust is in relief, its induration becomes very great, and it takes on its typical black, hard and scoriaceous aspect.

The chains of ironcrusts. — 1° INSTABILITY OF IRONCRUSTS. — It was believed that on the scale of the human life an ironcrust brought to surface by erosion was resistant and definitive. In reality, like all rocks, it turns out to be ephemeral. The conditions of the outcrop lead to its induration, granting it temporary solidity. However, its degradation starts slowly and will result in its disappearance. The encrusted mesas of the Sudanese landscapes are nothing but relicts of old ironcrusts.

2° WEATHERING. — Mechanical erosion attacks the crust along the valley sides. Overhanging crusts break off, cover slopes with talus and pediments with lateritic gravels. Vegetation establishes itself on the crust and climatic change can little by little restore some vigor to it. This savanna vegetation colonizes and rebuilds soils. The latter once more will permit percolation of these organic matters which have a redoutable power for the reduction and complexing of iron. In all the pores of scoriaceous or cavernous crusts the slow work of sapping begins which can lead to galleries and collapses. The mechanical actions are doubled by chemical ones.

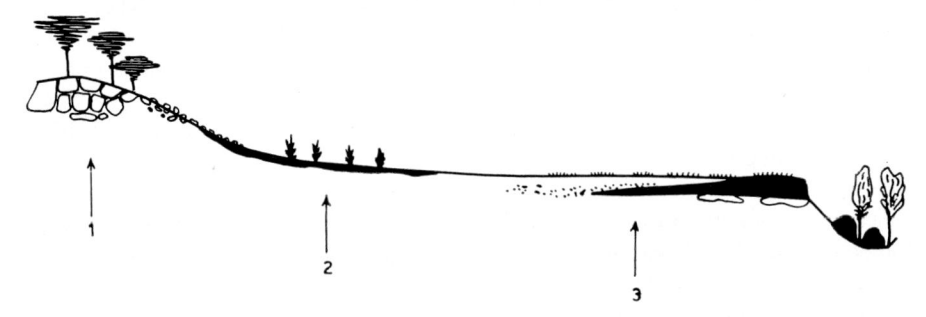

FIG. 22. — *Catena of soils showing a denuded ironpan* (1), *a colluvial ironpan* (2) *and ironpan formation in the terrace above the ravine* (3). Neighborhood of Labé, Guinea, Western Africa (after MAIGNIEN, 1958).

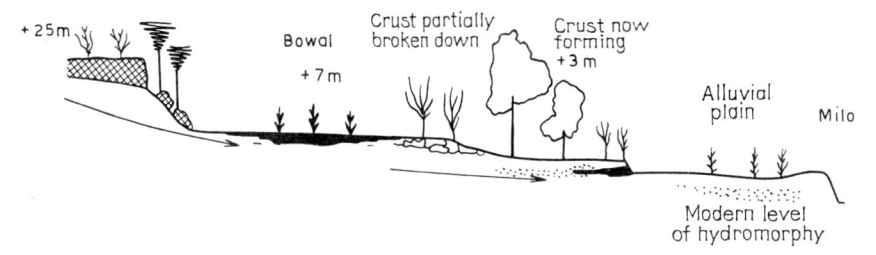

FIG. 23. — *Catena of soils in the neighborhood of Kankan, Milo river valley, Guinea, West Africa. Three levels of terraces with crusts and concretionnement of iron in the modern alluvial plain by hydromorphy (after MAIGNIEN, 1958).*

3° DIFFERENTIAL CHEMICAL LESSIVAGE. — Lateritic ironcrusts accumulate in a preferential manner iron, aluminum, and sometimes manganese oxides. When there is lessivage, the removal of each of these oxides does not occur with the same speed. Manganese is the most mobile, then iron, and finally aluminum, whose mobility is very limited. Thus, starting from a mixed crust, ferrallitic in the proper sense, the effect of lessivage carried on over geological spans of time impoverishes the crust in iron and enriches it in aluminum. The tendency, as we shall see, is toward a bauxite plateau.

4° THE FATE OF IRON OXIDES. — The fate of iron oxides is not indifferent. Obviously, a fraction can reach the rivers and be removed, but the greatest part circulates with ground-water in talus cones, in porous pediment gravels, under terraces and in alluvial ground-water. The rhythm of this circulation is controlled again by seasonal variations which are favorable to the phenomena of iron encrustation. Along its downstream trajectory, the iron will be engaged in new encrustations, only to be released later on, then immobilized little farther on, and so on. Thus, there will be formed as a kind of chain reaction slope crusts, glacis crusts, terrace crusts and plain crusts. This explains also that permanent supply within lateritic profiles of "fresh iron" brought in by lateral migration. This iron is only fresh relatively, however, since it travels from place to place along its slow gravitational itinerary. We are thus in the presence of the chains of ironcrusts and the cycles of encrustation so masterfully described by MAIGNIEN (1958), which have given rise to this modest summary. A few figures will illustrate better than words the different aspects of this chain of encrustation (figs. 22, 23 and 24).

5° IRON IN A LATERITIC LANDSCAPE. — One must read the work of MAIGNIEN to appreciate to what extent iron is "tenacious" throughout its history, even a prolonged

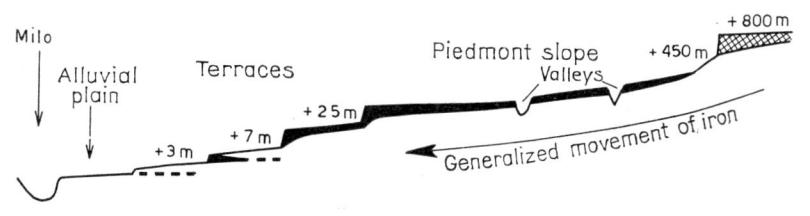

FIG. 24. — *Catena of soils in the neighborhood of Kankan, Milo river valley, Guinea, Western Africa. Encrusted alluvial terraces of the Milo river. The loose superficial horizons are partially eroded (after MAIGNIEN, 1958).*

history in a lateritic landscape. Many African ironcrusts belong to the beginning of the Tertiary, or even to the Upper Cretaceous. Part of the iron, released some tens of millions of years ago, continues to travel stepwise carrying with it all the dangerous consequences that this can have for vegetation and agriculture. A lateritic landscape must thus be understood as a kind of vast amphitheatre where the iron cascades from step to step, entering provisionally into ephemeral ironcrusts, followed by new releases which are just as provisional. Since ephemeral means "that which lasts only a day", we shall agree that it is a question of biblical days, but the sight of ironcrusts in the process of dissolution and dismantling does not leave any doubt concerning the mobilization of iron, once reducing waters attack it. The migrating iron oxides originate both from the uplands where they formerly accumulated by encrustation and from lateritization that occurs along their trajectory and that remove new quantities of these iron oxides. Juvenile iron of the continental hydrolyses is added to the old original stock. All this must lead to an impoverishment of uplands and to a general encrustation of depressions until a new history of erosion or tectonism takes over. By means of a lowering of base level, a new impulse is transmitted to this unceasing migration.

8° *Bauxites*

Bauxites are lateritic crusts rich enough in alumina to make industrial exploitation of them profitable.

In reality, this result can be obtained in nature in different ways. There is a considerable literature on this subject and the recent decades have thrown much light on possible mechanisms. If one confines himself to the domain of surficial geochemistry, one strays from the subject of clays; thus once again I shall present only a pedagogical model of this important subject about which geologists readily propagate an erroneous heritage.

The four principal itineraries that lead to a bauxite deposit are the following:

Direct genesis: parent rock $\rightarrow$ bauxite. — The direct itinerary requires a parent rock rich in alumina and direct weathering of silicates into free alumina. We know that this necessitates that lateritic weathering occur in a leached and well drained environment and above the phreatic water table.

The best example is that of the nepheline syenites of Los Islands (Guinea) and their transformation into bauxite, which was studied by LACROIX (1913), MILLOT and BONIFAS (1955) and BONIFAS (1959). The transformation of the nepheline syenite gives rise directly to a kind of "pumice". In the latter, one sees that feldspars are replaced by gibbsite with a preservation of the rock structures, that which guarantees the direct passage from the syenite to the aluminous rock (fig. 19). Higher in the profile this "pumice" is reworked, forms concretions, and becomes pisolitic and brecciated. This is a bauxitic encrustation. The isovolumetric argument (MILLOT and BONIFAS) has shown that, from the beginning, the lessivage of iron was adequate to ensure a high amount of alumina in the ore.

A quite comparable example is found in the deposits of Arkansas. This is the type I described by GORDON and TRACEY (1952), although later developments have kaolinized

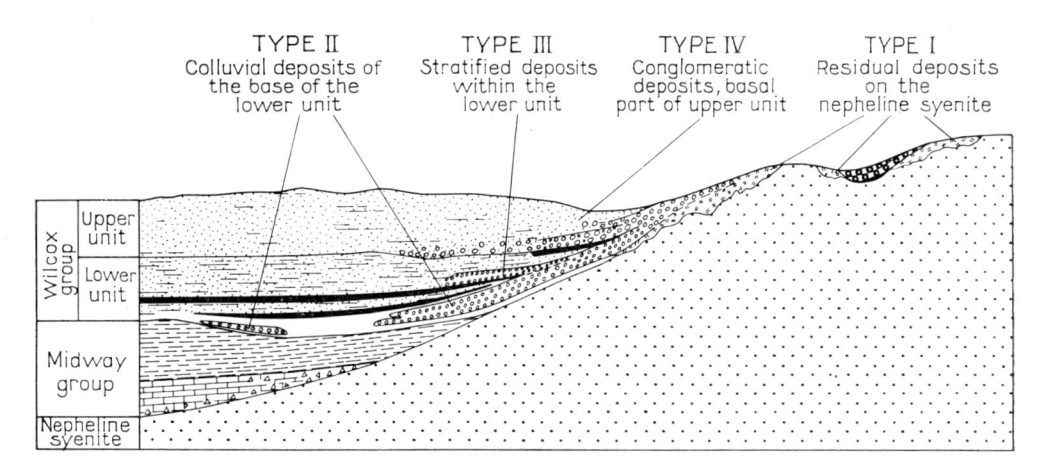

FIG. 25. — *Diagrammatic section of the principal types of deposits in the Arkansas bauxite region, U.S.A.* (after GORDON and TRACEY, 1952).

the zone of initial weathering so that the situation as presently seen is less demonstrative (fig. 25).

Indirect genesis: parent rock → weathered products → bauxite. — This second itinerary is indirect because it sets between the parent rock and the final bauxite a supplementary stage represented by weathered products. The latter are mainly composed of kaolinite which can naturally be associated with its common companions, gibbsite and goethite. Several examples can be given; however, it is a good idea to distinguish two kinds of indirect genesis, according to the condition that the weathered products are autochthonous or allochthonous.

— AUTOCHTHONOUS INDIRECT GENESIS. — A beautiful example is described in Jamaica by ZANS (1952) and ZANS, LEMOINE and ROCH (1961). The bedrock is that of a calcareous island with karstic relief. It is covered with andesitic pyroclastic products which are very thick. At the level of and below the ground water table weathering of the volcanic ash is of kaolinitic type. Above this water table the same pulverulent product has evolved into bauxite. Where the bedrock is not the karstic limestone the evolution to bauxite has been impeded owing to poor drainage, and the kaolinitic product remains intact.

Here in a model case one can appreciate the threshold mechanism: below the water table, kaolinization of volcanic ash; above the water table, bauxitization.

Likewise, in Cameroon, SIEFFERMANN (1959) compared the different degrees of weathering of basaltic surfaces. This weathering is kaolinitic, but when the profile is thick enough to allow pronounced periodic desiccation to take place in its upper part, gibbsite appears and develops. In the lower part of the profile the parent rock of the kaolinite is the basalt, and in the upper part, the parent rock of the gibbsite is the kaolinitic material.

— ALLOCHTHONOUS INDIRECT GENESIS. — The mechanism is the same, but the kaolinitic products are no longer in their original place. A good example is given by the

Arkansas deposits described as type II by Gordon and Tracey (1952). In this case, the weathering products of the nepheline syenite were transported to a piedmont zone; these are kaolinitic and bauxitic alluvium on which the bauxitization mechanism will act, resulting in a zone of concretions, thus of encrustation (*cuirassement*) (fig. 22).

— Mixed or parautochthonous indirect genesis. — The distinction between autochthonous and allochthonous is necessary for the discussion of European deposits of bauxite in France and Central Europe. Here we find once again the trace of a master, J. de Lapparent (1930), who pointed out the connection between the bauxites of southern France and the lateritization phenomenon. The works of Weisse (1948), Bonte (1958) and Bardossy (1958-1959, 1961) completed the interpretation of the genetic mechanism. The weathered products of an emergent zone have evolved on a well drained karst during the tropical periods of the Cretaceous. The material evolved in place, but with a tendency to migrate to the karstic depressions where it pursued its geochemical evolution.

So, in present-day deposits, autochthonous and parautochthonous products occur side-by-side and commonly are piled up one on the other. These products, dominantly argillaceous, were located in a very well drained environment of lessivage situated above the karstic water table; alumina was released, then later reworked, formed into concretions and took on the form of the present-day bauxites.

A great number of bauxite deposits have had an indirect genesis, i.e., they are the result of a double evolution: weathering of the parent rock into lateritic products, essentially kaolinitic, then weathering of kaolinitic products into bauxite. It is of little importance whether the second step occurs in place or somewhat removed; the main point is to see that the kaolinitic clays of a first stage of lateritization act as parent rock for the second stage (moreover, itself very complex) that merits the name of bauxitization.

Genesis by means of differential lessivage. Evolution of mixed crusts. — The important element in the preceding genetic mechanisms is that the lessivage of iron must be strong enough, compared to that of alumina, to make the resulting ore poor enough in iron to be exploitable. The richness in alumina of nepheline syenites, of kaolinitic weathering and of decalcified earths ensures from the beginning a predominance of alumina that has been maintained by lessivage.

In many cases, as we have seen, the products of weathering and of encrustation are mixed: goethite, gibbsite, kaolinite. In particular, this is the case on basic parent rocks, where the abundant and mobilized iron comes to supply lateritic products rich in this element. Although iron is retained less than alumina during migrations, it still dominates in the crusts.

It is the evolution of these crusts which can lead to bauxite ore deposits by means of relative enrichment or differential lessivage. We have seen how crusts that were eroded and exposed at the surface were impoverished preferentially in iron and manganese by lessivage. If the crust is old, if it occupies a high position in a lateritic landscape, it will become richer in alumina in the course of geological time. Thus in middle Guinea, the crusts of plateaus have an alumina content that increases with altitude, and the bauxite ores correspond to the oldest and highest surfaces.

The Guinean "bowé" are bauxitic plateaus whose iron has been leached through time. Some of this iron is found on the borders of large, cliffed crusts, but its greatest part has been spread throughout the landscape below.

Attention is called here to a belated mechanism. To know whether the release of alumina is direct or indirect is an interesting question in the case of each deposit, but it is secondary compared to the result. Likewise, the nature of the parent rock can be quite variable as long as it contains alumina and appropriate drainages have been ensured in the course of weathering. Thus the parent rocks may be basic rocks as well as more acid lavas, schists or sediments. Examples are numerous in Africa, in the Indies, and in South America. What is finally important is a prolonged history capable, by means of differential lessivage of mixed products, of giving rise to a high grade alumina ore.

Genesis by reworking. — Here, the geochemist yields to the geologist. Old crusts or bauxite deposits are attacked by erosion, transported as conglomerate, gravel or granules, and deposited again elsewhere.

Here again, the deposits of Arkansas give examples by means of types III and IV of GORDON and TRACEY (1952). Type III is an accumulation of cross-bedded gravels, pisolites and granules of bauxite. Type IV has a conglomeratic aspect (fig. 22). DARS (1960) described such a deposit in Sudan (Mali) at M'Pébougou. The rock is a recemented conglomerate of large bauxite spheres. It was used in part for the construction of the Markala dam on the Niger.

The reworked bauxites can attain considerable masses; they can lead back to the autochthonous deposits which nourished them, if the latter have not been completely destroyed.

IV. — CONCLUSIONS

At the end of this chapter on clays of weathering and of soils, it is a good idea to reassemble the essential points that have been presented. This will be carried out in three ways.

The two major types of weathering. — Two major types of weathering of continental masses can be set in opposition.

— The first is *weathering of a predominantly physical character*. It yields *detrital material* which is nothing more than the product of the disintegration of continental rocks. This is characteristic of cold climates and of *dry climates*, since in both cases hydrolysis is paralysed. It is likewise the case of *rising mountain chains* where erosion is much more rapid than the delicate mechanisms of weathering and pedogenesis. The very substance of the continents is fragmented and given up to sedimentation.

— The second is *weathering of a predominantly chemical character*. Still modest in temperate regions and in mediterranean or semi-tropical countries, it develops vigorously in the *humid intertropical zone*. Physical disintegration and chemical activity combine their effects to guarantee *dissolution, transformation and neoformation*. Prior to sedimentation, pedogenesis will have made its mark on the continental material.

The products of weathering. — Upon reaching its termination, weathering gives up a material to the agents of transportation. From a genetic point of view, these are the categories we can distinguish, while limiting ourselves to the silicates that are the subjects of our study:

a) *The residual silicates,* inert and unmodified fragments of original rocks. E.g., quartz, feldspar.

b) *The transformed silicates*, products of the weathering of modified original clay minerals. E.g., vermiculite, mixed-layered products.

c) *The colloidal products*, fragments of silicate structures, or colloidal hydroxides, which are still not well known and are likely to be ephemeral.

d) *The neoformed silicates* originating in weathering and in soils. E.g., montmorillonite, kaolinite.

e) *The dissolved elements:* Si, Al, Fe, Mg, Ca, K and Na capable of being dissolved in proportions that vary according to the climate.

The clays of weathering and of soils. — Weathering and soils deliver three kinds of clays to the sedimentation process:

— *Clays inherited* from parent rocks;

— *Clays transformed* by mild weathering;

— *Clays neoformed* in the soil.

The fate of these clays transported downstream and the fate of the ions removed by solutions remain to be studied in the sediments.

CHAPTER VI

CLAYS OF CONTINENTAL SEDIMENTS

Introduction

Continental sedimentation is subaerial, glacial or aquatic. The first two cases are easy to define, but the third one is complex. One can look for explanations either in geographical or in chemical terms.

In terms of geography, we must first consider rivers, streams and their estuaries. But not all rivers and streams empty into the sea by means of an estuary. Many of them now, and many in the past, leave behind traces of vast areas of sedimentation: alluvial cones, piedmont complexes and wide lacustrine plains. Here is the first series of cases to study.

In terms of chemistry, continental waters show the greatest variety: acid water, freshwater, alkaline water, hypersaline water and all the intermediate degrees. But the distribution of these "chemical facies" is not simple. Certainly, there are regions where the running water has similar properties all along its path to the sea. However, evolution will also occur along the way. Freshwater, which begins in a mountain torrent can reach a plain where evaporation will concentrate the dissolved salts, and its water will become saline or hypersaline. Chemical properties of waters vary in this way not only from one period to another in accord with climate, but also from one place in the landscape to another, according to alimentation and evaporation regimes.

It is through the midst of these varied cases that we must make our way, and our lack of knowledge is still so great that this can be done only by spot testing or sampling in those domains where studies have begun.

Thus we shall examine successively:

I. *Glacial deposits.*

II. *Eolian deposits.*

III. *Fluviatile and estuarine deposits.*

IV. *The siderolithic facies: a piedmont complex.*

V. *The great detrital red bed series: piedmont complexes.*

VI. *The coal measures, in some cases limnic, in others, paralic.*

VII. *The basic lakes of plains.*

VIII. *The still enigmatic analcimolites.*

I. — GLACIAL DEPOSITS

Glacial clays of Fennoscandia. — The glacial clays of the Scandinavian Peninsula have been studied with much care. Certainly they form considerable masses which are found principally in the Quaternary deposits of fjords, which are also population centers. They raise problems of civil engineering for the water supply, the corrosion of pipes, and the stability of buildings. It is a problem of masses of clay still saturated with water that occur as lenses in the fluvio-marino-glacial deposits. The ensemble of the studies carried out on these clays has been summarized in the critical study of COLLINI (1956).

These glacial clays are composed of a high proportion of illite and accessory chlorite (COLLINI, 1950; SOVERI, 1950). Three interpretations are given by the Scandinavians to account for the accumulation, in such great masses, of clays in which illite is very abundant.

The first interpretation is to see here simply the fine fraction coming directly from glacial weathering and glacial transport, with preferential settling of the fine clay particles in the fjords. But the tonnage is considerable, and ROSENQVIST (1955, 1961) emphasized that such quantities of material, evaluated at 10^{12} m³ for Scandinavia, must necessarily have released even greater volumes of sand and silt, of which similarly great tonnages are not found.

Two directions of research are proposed to remove this contradiction.

ROSENQVIST (1955) proposes the neoformation of illite from the products of the glacial weathering of crystalline rocks. These products, previously reported by TAMM (1924), are soluble substances and the debris of silicates that can result from intense crushing. Recently, ROSENQVIST (1961) has also insisted on preglacial weathering in which degraded micas and even montmorillonite were able to form from silicates. These degraded products, having reached the sea, would have formed mixed layers, and then illite, in order to arrive at the present state. Here we find locally the degradation-aggradation cycle to which we shall return later on.

Other authors, and COLLINI himself (1956), think that this great abundance is not explicable by the weathering of a continental area of mainly crystalline and metamorphic rocks, and they attribute the beds of Scandinavian Quaternary clays to the erosion of Mesozoic and Tertiary sedimentary mantles, now completely removed, which the glaciers easily reworked.

Immediately we see a double possibility in the case of Scandinavian glacial clays: *chemical evolution or mechanical reworking.* The problem is difficult because of the peculiar conditions of these clay deposits that are very rich in clay minerals in comparison to moraines, which are much coarser and heterogeneous.

Glacial moraines of Europe and U.S.A. — In Great Britain, PERRIN (1957) studied the tills of the Cambridge district. Clays are composed mainly of illite with traces of montmorillonite or kaolinite, depending on whether Jurassic or Cretaceous formations were crossed by the former glaciers. This allows one to reconstruct the glacier's route.

In France the varved clays of the former glacial lake of Eybens (SARROT-REYNAUD, 1953) show a predominance of illite. Likewise, LAFOND, RIVIÈRE and VERNHET (1961) studied the clays of glacial lakes and moraines of the Alps and the Pyrenees. In the three cases studied, illite dominates chlorite, and these minerals have been inherited directly from the drainage basins.

In Switzerland, VERNET (1959) showed that glacial deposits were composed of illite which dominates chlorite and its chlorite-montmorillonite mixed layer.

In Canada, ALLENS and JOHNS (1960) pointed out that tills and the sediments that were derived from them, such as varved clays, are composed of illite, chlorite and vermiculite. The latter suggest mild weathering, such as that in humid climates.

In Minnesota, ARNEMAN and WRIGHT (1959) showed that the major part of the fine fraction of tills is composed of quartz and feldspar. The clay fraction, properly speaking, would be derived from interglacial weathering and old preglacial sediments.

The origin of clays in glacial deposits. — Thus, illite first and then chlorite are the predominant minerals in glacial deposits.

In the case of the glacio-marine clays of Scandinavia the question remains concerning the *evolution of the weathered products* of silicate rocks and neoformation of illite in the sea.

For the ensemble of continental glacial deposits, one must consider at the same time:

— Former argillaceous sedimentary mantles, no longer present, that were able to take part in the supply of material to moraines;

— Former preglacial or interglacial weathered mantles that were able to develop illite (by sericitization of feldspars) and, accessorily, vermiculite and montmorillonite;

— Lastly, but of primary importance, *weathering of a predominantly physical character of the periglacial climates*, that gives rise to both illite and chlorite.

II. — EOLIAN DEPOSITS

Loess. — Many studies have been concerned with the clay fraction of loess. SCHROEDER (1955) studied the loess of Hannover in Germany; he identified illite and illite-montmorillonite mixed layers in the clay fraction. SWINEFORD and FRYE (1955) compared the loess of Kansas to those of western Europe. In Kansas, the clay fraction of the loess is montmorillonite, and this is related to the abundance of volcanic products in these countries. On the contrary, in Europe the predominant minerals are illite and chlorite, along with mixed layers; kaolinite and montmorillonite occur as subordinate minerals. CAILLÈRE and MALYCHEFF (1957) found that illite was predominant in the region around Paris, with a little kaolinite and montmorillonite. MILLOT, CAMEZ and WERNERT (1957), CAMEZ and ROTH (1957) and CAMEZ (1962) found in the loess formations of Alsace illite and chlorite in equal parts. One can see here the influence of the Alpine morainic terrains that supplied the eolian sediment. Chlorite is partly stable, partly interstratified with montmorillonitic layers.

Weathering of loess into lehm. — It is important to note that during the weathering of loess to lehm the clay composition can vary. In Alsace, chlorite and its mixed layers are degraded and tend to disappear in the upper parts of profiles, so that illite becomes predominant. This is one of the examples in which the fragility of chlorite relative to illite is well demonstrated (CAMEZ, 1960). The weathering of the loess of Wisconsin (U.S.A.) was studied by BEAVERS, JOHNS, GRIM and ODEL (1955), by the school of JACKSON (1959) and by GLENN, JACKSON et al. (1960). In these conditions, weathering by degradation transforms illite and chlorite into vermiculite mixed layers, then into vermiculite, and then into montmorillonite. Here one sees much stronger weathering of the loess than in Alsace.

Variations in the clay fraction of loess. — The clay fraction of loess is thus quite varied and there are at least three reasons for this:

1° Loess is inherited from the regions in which the wind picked up the dust. These regions are varied. If it is true, as the Abbé BREUIL thought, that the loess in Brittany was picked up from very wide beaches abandoned by regression of the sea during glacial times, its composition must be rather different from that of the loess peripheral to the contemporaneous, extensive morainic landscapes. Similar contrasts would result in the case of the Great Plains where important volcanic deposits supplied montmorillonitic dust.

2° Surficial weathering of loess to lehm is not indifferent with respect to the clay minerals. In leached soils, chlorite and montmorillonite are much more vulnerable than illite. The results of the measurements made on loess around the world thus depend on the degree of weathering of the loess.

3° The measurements which we have available today were made during the time when the methods of measurement of interstratified minerals were being established and improved. Since these methods are not always comparable, identifications must be compared only with discretion.

The "feich feich" of the Fezzan (eastern Sahara). — In the Sahara the name "feich feich" is applied to the dusty, eolian formations that are not gathered into dunes, but are distributed uniformly on the soil surface and especially in depressions. MULLER-FEUGA (1952) collected varied samples of "feich feich", and his measurements showed that the clay fraction of "feich feich" was similar to that of the formations of the surrounding landscape: kaolinite near Nubian outcrops, and illite in the neighborhood of Cretaceous formations. This confirms that which was observed in the case of loess, to which "feich feich" is very analogous, in spite of the differences in climate. CAILLEUX (1961) called our attention to the fact that the granulometry of the eolian dusts collected on Red Sea was identical to that of loess.

Conclusion. — The conclusion is obvious and expected. The *clay fraction* of modern and fossil eolian deposits has a *purely detrital origin* being inherited from the landscapes that gave rise to those deposits.

III. — DEPOSITS OF RIVERS AND ESTUARIES

Fluviatile clays. — HOLMS and HEARN (1942) studied the clay minerals carried by the Mississippi and its tributaries. The common clay minerals were present, but their proportions were different in the samples collected from the eastern tributaries, from the western tributaries and from the Mississippi itself. They were able to demonstrate that this was the reflection of the variations in composition of the soils of each drainage basin.

MILLOT (1953) has shown that the clays carried by rivers imitate either the bedrock or the soils of the drainage basin:

— Wadi Sébou, a Moroccan coastal river, flows through a landscape in which thick formations of gray Miocene marnes are found. Near its mouth it deposits modern alluvium that contains the same clay minerals as those of the Miocene formations, and in the same proportions.
— The middle Niger River in the Sudan (Mali) carries ochre-colored muds in which kaolinite is the predominant mineral; these clay minerals are inherited from the lateritized regions crossed by the river in which this mineral is typical.
— The upper Durance River in France shows in its muds an association of illite and chlorite that is a reflection of the schists of its upper drainage basin. The lower Durance River, arriving in the region of the *terres noires* of the Embrunais, contains a mixture of the former minerals along with kaolinite, which is found in the black Mesozoic marnes.

LAFOND (1961) studied the muds deposited along the course of the Vilaine, a coastal river of south Brittany. The clay minerals found in the river are illite, kaolinite, chlorite and also pyrophyllite. The ensemble has been reworked from soils and rocks of the surrounding region; in particular, the pyrophyllite comes from the Devonian schists. In the estuary the clay minerals are different, having been supplied by way of the sea.

PACKHAM, ROSAMAN and MIDGLEY (1961) have studied the material in suspension in nine rivers in England. Various clay minerals were identified along with quartz and calcite. The amount and nature of the clay minerals are related to the argillaceous geological formations crossed by the rivers.

The clays of estuaries. — TAGGART and KAISER (1960) studied the clay minerals carried by the Mississippi and the Red River. The two groups are different. The muds of the Mississippi are characterized by montmorillonite and those of Red River, which drains more eastern regions, contain kaolinite. The authors did not find any variation between the clay minerals deposited by the river and those of the delta. These results are contradictory to those of JOHNS and GRIM (1958) and MILNE and EARLY (1958), which will be examined with the subject of modern marine sedimentation; these authors gave evidence of an evolution of the clay minerals of the Mississippi and neighboring rivers when they reach saline waters.

POWERS (1954, 1957) also reported a modification of the clay fraction from fresh to marine water in Chesapeake Bay in Maryland. There was a transformation of illite into chlorite, as we shall see in the next chapter. But in 1959 POWERS reexamined the data collected in the estuary of the James River. This is the last river that empties into Chesapeake Bay, on the right bank, just before the ocean. One sees weathered illite change into chlorite by the intermediate illite-vermiculite-chlorite mixed-layer stage, while well crystallized illite is reconstituted by K fixation. The stability of chlorite upon heating increases from the river to the sea, the estuary occupying an intermediate position. Chemical analyses have confirmed this evolution; a relation exists between the chemistry of the interstitial water of the muds, that of the exchangeable ions of the clays, and that of the clay minerals. The mechanism of Mg fixation and of the mineral transformations will be studied in Chapter X (figs. 65 and 68).

GRIFFIN and INGRAM (1955) studied the clays brought into the estuary of Neuse River in North Carolina. These are kaolinite, chlorite, illite and montmorillonite, as well as amorphous or weakly crystalline material that can amount to 50 %. The estuary is not influenced by marine sediments. As one advances down the estuary, chlorite increases relative to kaolinite, and illite relative to kaolinite and chlorite. The author can explain this observation only by an evolution either of the minerals or of the amorphous material in transport. Here we find indications of the direction of transformations, or of neoformations, or of both.

The clays deposited on the bed and in the estuary of the Rappahannock River, Virginia, were studied by NELSON (1958) by means of the most accurate analytical methods. NELSON identified in mixtures kaolinite, illite, disorganized illite, dioctahedral vermiculite, 12 Å montmorillonite, 14 Å montmorillonite, chlorite, feldspar and quartz. If comparisons are made from stage to stage, along the course of the river to the termination of the estuary in the open sea, certain changes can be observed. Illite becomes better and better crystallized and more abundant. The disordered illite decreases. Vermiculite is stable. Montmorillonites decrease. Chlorite appears. The author discusses these results scrupulously. It is evident that it is tempting to explain these changes by mineralogical evolution of the clay minerals under the influence of the transition from freshwater to marine water. But the consequences are so important that verification is necessary. There is a risk of being deceived by supplemental, laterally supplied sediment, or by the conditions of transportation and dispersal of the sediments within the estuary.

Polders are fossil estuaries, and the clays of polders have been studied by DEKEYSER, HOEBEKE and VAN KEYMEULEN (1955). The clay fraction was composed of mica, but a long-term evolution has taken place, leading to illite, then to montmorillonite; here there is pedological weathering of the polder deposit, which introduces another aspect to the problem.

Piedmont deposits. — Many are the rivers that do not carry their solid load directly to sea, but leave an important part of it at the foot of the mountains. This means that when the rivers reach the plains, their energy decreases very much, and they immediately leave behind a great part of the products being transported. Thus, alluvial fans and piedmont complexes are formed here.

The clay particles that can be extracted from the piedmont complexes imitate those of the rocks or the soils of the drainage basin.

Whenever erosion is mainly physico-chemical, one finds illite and chlorite, which are the common products of the disintegration of highland areas.

When particular climates prevail, they are recognizable; this will be demonstrated for the siderolithic facies and the detrital red bed series.

There is another climate that leaves its mark on the reworked material in the piedmont region, it is the humid oceanic climate, for example, that gives rise at present to podzols under Atlantic landscapes.

We had the opportunity to study several examples of this phenomenon (FRANC DE FERRIÈRE, CAMEZ, MILLOT, 1959 ; CAMEZ, FRANC DE FERRIÈRE, LUCAS, MILLOT, 1960). This concerns detrital continental products of the end of the Pliocene and of the beginning of the Quaternary, for example, silts of the end of the Pliocene in the Forêt de Chaux and in the cover deposits of the Plaine de Bresse (La Chapelle Naude-La Saulcaie); the old alluviums of Aquitaine, sampled on a terrace of the Gers River at Masseube and at Pompignac (Gironde); and the old alluvium of the Sundgau in Alsace. In these cases, vermiculite is the principal clay mineral of the fine fraction of the detrital formations. This vermiculite belongs to the alluvial parent material and, if a soil develops on this, the vermiculite is degraded in the profile. We have here proof of former reworking of vermiculitic soils that formed during some cool, rainy seasons of the Early Quaternary.

The question is not closed, and it requires much more work. Whereas some great Quaternary pluvials seem to have given rise, in mountainous and continental regions, to weathering leading to vermiculite, other pluvials were conducive to a rubefaction that increases toward the SW. Here we grasp another way of characterizing the seasons of the Quaternary.

Conclusion. — *Clays transported by rivers and streams or deposited as alluvium are inherited* from parent rocks, weathering and soils of the drainage basin. This can lead to quite varied mixtures, but ones that are always of detrital origin.

When it is a question of fossil fluviatile deposits, reworked from soils with clear and easy-to-recognize features, one can begin to reconstruct paleoclimates.

A problem arises in the case of estuaries and in the case of minerals arriving in the marine domain. Some very careful studies have been made, taking into consideration, of course, that the estuaries are not provided with sediments from downstream by means of marine currents. These studies point to *progressive transformation of clays from the freshwater to the marine environment*. Is it a question of mineralogical transformations or of some differential sorting that could deceive us? The problem will be reexamined in relation to present-day marine sedimentation, but the preceding examples show that it is not possible to reject easily the hypothesis of mineralogical transformations.

IV. — THE SIDEROLITHIC FACIES

1° *Definition and history*

The word *sidérolitique*, also written *sidérolithique*, was proposed by the Swiss geologist THURMANN during the meeting of the *Société géologique de France* at Strasbourg in 1834.

This term was certainly chosen because iron is so very abundant in the form of nodules, crusts, pisolites, and concretions in certain formations in which it has been exploited. One of the most common aspects of these ferruginous concretions calls to mind green peas for French people, *pisolite*, and beans for German people, *Bohnerz*. However, many beds within siderolithic terrains are not iron bearing; layers of white sand, plastic clay and, in some cases, as a cover layer, lacustrine limestones or siliceous flagstones also occur. All these facies, easily recognizable by one who is well acquainted with them, also belong to the siderolithic facies, although this strange word may not come to mind. This term, however, has been retained to the present day, although subject to different interpretations as in the case of the word laterite, and I think that this traditional and odd term remains valuable as a name for the ensemble of the formations studied by the older authors, and which FLEURY (1909) defined in these terms:

"The word *sidérolitique* names an entire geological formation, very complex and very special, of quite variable behavior and aspect, which is commonly characterized by iron ores in grains or pisolites, better known by their German name *Bohnerz*, that represents only a small part of the deposits. In reality, they are always connected or mixed with, or subordinate to ferruginous clays (boles), refractory earths, siliceous sands and even limestones."

By careful consideration of this definition, one understands that this matter is not an easy one, but, in this domain as in many others, words are rather awkward for the description of a reality that can be readily recognized once you have seen it.

Use of siderolithic deposits. — Formerly, these deposits were actively exploited, primarily for iron, called *mine* or *minerai de fer fort*. It is in contradistinction to *mine* that the low-grade iron ore of Lorraine was called *minette* in the last century. The quarrying was effected by means of basket or dump cart. Many meadows of our landscape in the Haute-Marne, in Berry and in southwestern France show isolated groups of trees that have grown on the now-collapsed mouths of the diggings. Plastic clay or refractory earth was also exploited to line furnaces, as well as sand to make foundry molds. The forests provided the combustible material. In this manner an artisan equilibrium was maintained until its activity was suddenly interrupted by the introduction of the *minette* of Lorraine. The THOMAS process and the mining of coal proscribed open-pit mines and preserved our woods.

History. — The abundance of the mine workings and their interest allowed former geologists to make numerous observations and minute studies of the deposits, and to collect fossils. DE LAUNAY wrote that "science falls too quickly into anonymity". In the case of the *Sidérolitique*, it fell not only into anonymity, but also into oblivion, as we observe upon reading treatises more than half a century old wherein problems that we still raise too often are pratically solved.

However, an extraordinary divergence appeared very early among the old-time geologists concerning the origin of the siderolithic formations. Principally under the influence of BRONGNIART (1828) and DE GRESSLY (1828-1841), they considered the genesis of the *terrains pisiformes* to be catastrophic and attributed it to hydrothermal and semi-plutonic phenomena. Although the sedimentary origin was understood by some of the earliest workers, this hydrothermal hypothesis lasted a long time. One can note, with astonishment, that DOUVILLÉ who defended this thesis in 1872, still maintained it in 1936, when I was beginning my career in geology. This shows not only the tenacity of hypotheses that call upon unverifiable mechanisms, but also the tenacity of their supporters.

Catastrophism is confronted with uniformitarianism. We owe to the Belgian VAN DEN BROECK (1878) and to Professor DIEULAFAIT (1884-1885) from Marseille our understanding of the origin of the *Sidérolitique* as the result of the work of surficial waters. One can read in the treatise of VAN DEN BROECK (1878): "The ferruginous or plastic clays, the hydrated iron, and the granular iron ore are generally the residues of weathering, dissolution, concretionment and hydrochemical metamorphism of deposits subjected to accentuated weathering *in situ*." We are coming nearer to our present-day language, and, little by little, we come to the magnificent work of FLEURY, dating from 1909, where can be read, on page 235: "(The *Sidérolitique*) is the result of a long series of activities and developments that can be assigned to *two major phases*:

"1. *The lateritic phase*, during which, under special climatic conditions, the siderolithic products were formed by dissolution and disintegration of the rocks, following a process more or less comparable to that which at present forms laterites in tropical regions.

"2. *The siderolithic phase* properly speaking, characterized by slow working of infiltrating waters on the original laterite. The stratification of the deposits, the formation of pisolites, the concentration of certain elements are the main result of this *reactivation in situ*."

The Siderolithic is a facies. — As was the rule in the beginning of geological exploration, all the siderolithic formations were thought to be of the same age. One spoke of the "siderolithic formation". Then fossils were discovered and the formations could be dated. Then came a long period of discordant results. Several siderolithic formations turned out to have different ages and, in some cases, several of them were discovered in a single vertical section. At the special meeting in 1834 in Strasbourg, VOLTZ already offered the hypothesis that these formations constituted "a mineralogical accident which could recur at different epochs." This was confirmed by THURMANN at the meeting of Porrentruy in 1838.

Throughout the decades, siderolithic facies of different periods were recognized, and we shall find it mainly in the Carboniferous, Early Cretaceous and Eocene.

2° The siderolithic Eocene of northern Aquitaine

The siderolithic deposits of the Eocene of the northern part of the Aquitaine Basin will serve as a prime example. This siderolithic of Aquitaine has been studied for a century. A bibliographic list of the principal geological studies is given in the important work of KULBICKI (1953). From the point of view of conditions of clay genesis, the main works are the following: J. DE LAPPARENT (1930), SCHOELLER (1941, 1941), BERGOUGNIOUX (1957), VATAN (1948), MILLOT (1949) and KULBICKI (1953, 1953, 1954). The descriptions and works of KULBICKI will be followed in the presentation of the facts.

Distribution of the deposits. — At the end of the Cretaceous the sea receded from the flanks of the Massif Central. Northern Aquitaine appeared then as a limestone pediment resting against the Massif Central. The flora and fauna give proof of a tropical climate. The rivers that descended from the Massif Central spread over this pediment and deposited their sediments as far away as central Aquitaine; these

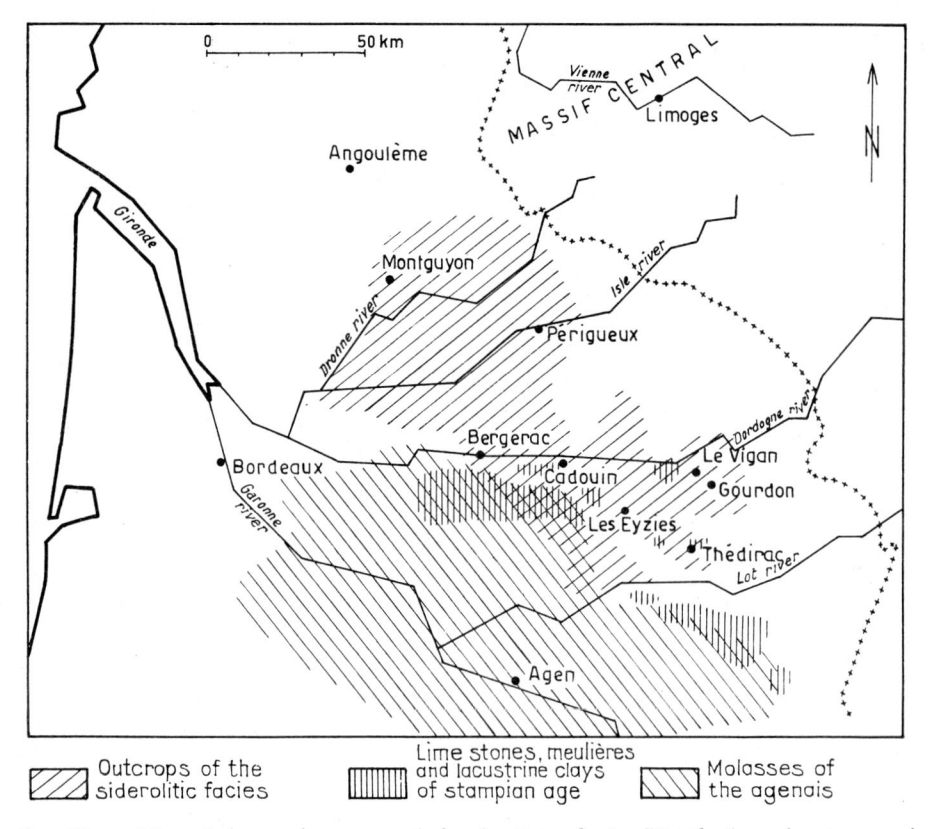

FIG. 26. — *Map of the northern part of the Aquitaine basin. Distribution of outcrops of the principal continental facies of the Nummulitic* (after KULBICKI, 1953).

THE SIDEROLITHIC FACIES

are continental sediments eroded from the crystalline Massif Central which was covered at that time with a thick weathered layer of lateritic type. The distribution of the siderolithic formations in the Aquitaine Basin is shown in figure 26.

History of the sedimentation. — *a)* COARSE SIDEROLITHIC. — Powerful streams issuing from the Massif Central travel across the wide piedmont zone situated southwest of this massif. Deposits are coarse, and stratification is irregular. In these deposits are found complete blocks of consolidated materials, pebbles, gravels, *arènes*, sands with kaolinitic cement, sometimes pieces of flint and ferruginous concretions, and even fragments of Cretaceous limestone.

This is the constitution of a coarse detrital mantle deposited under torrential fluvial conditions.

b) SANDY, CLAYEY SIDEROLITHIC. — The preceding coarse formations are covered by several tens of meters of finer sediments, sands and kaolinitic clays. Kaolinite is scattered through the sands, but it also forms enormous, very pure lenses, such as that at Montguyon that contains several hundred thousand tons of kaolinitic fire-clays.

The lack of quartz in the thick lenses of clay makes it difficult to explain their formation by means of a shifting torrent. It is necessary to call upon a separation of the clay from the mass of sandstone in the following way. Either in an abandoned pool, in a zone of compaction, in a zone of subsidence, or even in a zone subject to karstic underdrainage, a depression forms. The peripheral mass of kaolinitic sands loses its kaolinite through lessivage by rain and by springs that emerge round about. The basin is supplied with several meters of a fine-grained deposit containing no sand, except at the margins. Some beds are bituminous or pyritiferous, owing to the vegetation of the

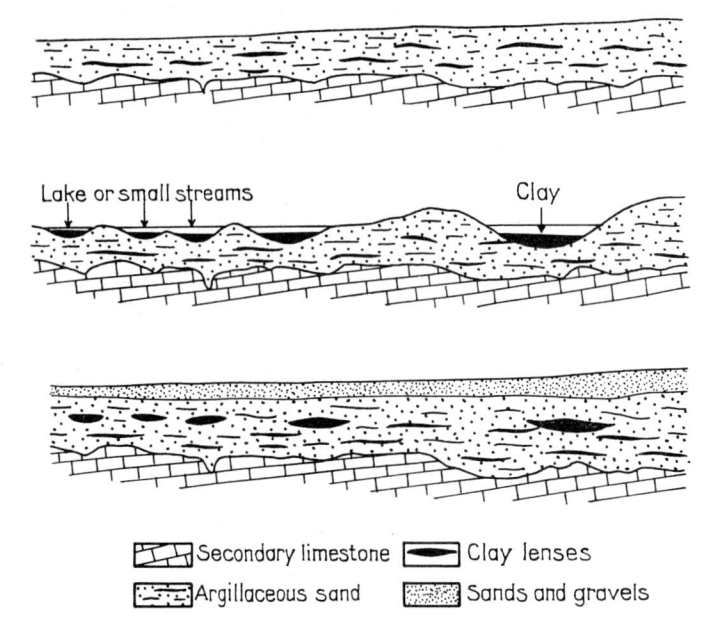

FIG. 27. — *Diagram showing the three stages of the formation of clay lenses* (after KULBICKI, 1953).

Secondary limestone Clay lenses
Argillaceous sand Sands and gravels

swamp. They show in some cases free gibbsite; then the sandy sedimentation spreads suddenly over the whole formation and the sandy series continues (KULBICKI, 1953) (fig. 27).

c) GREENISH SIDEROLITHIC WITH MICACEOUS CLAYS. — Throughout the region of the Double River, the preceding beds are covered by a mantle of greenish, clayey micaceous sands, about 1.5 m thick. The clay minerals are illite and montmorillonite. Therefore, conditions have changed without leaving any discordance or stratigraphic gap; the Massif Central is sending other kinds of clays downstream.

d) LEACHING OF THE DEPOSITS. ROLE OF KARST. — The deposits of the region of Cadouin and Les Eyzies show traces of leaching and intense deformation. Rain water and seasonal flood water percolated through the deposits, which were drained by their base through karst openings. An indication of this leaching can be seen in the neoformed halloysite found below pyritiferous horizons; the role of the sulfuric ion, produced by the oxidation of pyrite, is most likely. But karstic circulation creates a situation where the sediments are, little by little, drained basally and give rise to these siderolithic pockets filled with sands and clays of a great purity. The working at Cadouin and Les Eyzies show karstic walls, inclined stratification on the flanks, and a central chaotic zone.

e) LATERITIZATION OF THE DEPOSITS. — The seasonal nature of the surficial torrents allowed the deposit to be subjected to lateritic weathering under the influence of the same climate as that which gave rise to it. This effect is especially sensitive in the terminal deposits formed of micaceous clays. One sees that these deposits become kaolinitic and ferruginous, and that they show the beginnings of encrustation. Silica is carried to depth where it can react with aluminous minerals to give rise to montmorillonite or to form, as in the deposits of Les Eyzies, nodules of chalcedony. Herein is the origin of the very beautiful montmorillonites of Cadouin and Sauteloup.

Another indication of these excessive laterizing conditions can be found in megaluminous clays bearing gibbsite, described by LANQUINE and HALM (1951) and CAILLÈRE and JOURDAIN (1956).

f) SILICIFICATIONS. — This mobility of silica is typical of those climates that are conducive to hydrolysis. It leads to the formation of siliceous crusts or nodules a few meters from the surface. Quartzite flagstones also occur on the surface of the *Sidérolitique*; here we find a silicification in the form of plates or slabs, such as those we shall see again in Chapter IX, but we do not know if they belong to the end of this cycle or to later epochs.

g) STAMPIAN LAKES. — The region is submerged and covered with tranquil lakes with very different characteristics; their deposits consist of green, calcareous clays in many cases transformed to millstone grit (*meulière*). We shall study them in a later paragraph on basic lakes.

Genesis of clays. — The clays of the siderolithic deposits thus have multiple geneses.

However, the main mass of clays is kaolinitic, and there is no doubt that the origin of this kaolinite is in the reworking of the lateritic mantle of the Massif Central.

Within the deposit diagenetic evolution was able to take place, especially under the influence of acid solutions coming from the oxidation of pyrite. This gave rise to the neoformation of vermicular kaolinite, in many cases developed from a detrital mica, from dickite and hydrargillite, as well as from halloysite.

Lastly, pedogenetic activity at the surface was able to bring about the neoformation of montmorillonite and nodules of chalcedony at a few meters below the surface, adding a final surficial effect to an already complex history.

In this case, at least, three mechanisms of clay genesis can be added together:

— Detrital heritage for the great kaolinitic mass;

— Diagenesis within this mass;

— Pedological neoformation at its surface.

The first mechanism is the primary one; it can be summed up in one sentence: *"The Sidérolitique is reworked laterite."*

3° The siderolithic facies in Europe

Thus it is well agreed that *sidérolitique* is a facies name. It must be recognized that in France this term was readily used to designate the siderolithic Eocene of the central, southwestern and southeastern parts of the country. But this is an abuse of language. Throughout the world, one finds siderolithic formations of different ages. Let us begin with Europe.

The Siderolithic of the Carboniferous Period. — J. DE LAPPARENT (1934) studied kaolinitic clays and bauxitic clays that crop out in Ayrshire, Scotland. He described their mode of genesis and the association of kaolinite with aluminous pisolites. The examination of the latter shows very beautiful crystals of boehmite. These bauxitic and kaolinitic clays, formed from basalts, occur at the contact between millstone grits and coal measures. They are immediately covered by carbonaceous layers, from which J. DE LAPPARENT deduced the existence in the Scottish Carboniferous of physical conditions analogous to those of the Cretaceous of Provence.

KULBICKI and VETTER (1955) have described a lateritic profile on the eastern border of the Decazeville basin in Stephanian formations. They described a kaolinized parent rock with feldspathic crystals altered by epigenesis into gibbsite, above that a kaolinitic clay reddened by hematite and surmounted by a clay with oolites of boehmite. They interpret this to be a lateritic profile, typical of laterizing conditions found on the flanks of the Massif Central during Stephanian time. ERHART (1962), who visited this deposit, considered it, as well as that of Ayrshire, allochthonous. He thinks that the emplacement was later than the epoch of bauxitization, which would be perhaps dated Dinantian. In many cases it would be a question of a laterizing epoch within the Carboniferous in the broad sense, that which goes along with reports from Russia (PETROV, 1958) where siderolithic deposits of the Carboniferous are reported to be the result of a reworked lateritic mantle formed during Devono-Carboniferous time.

The Wealdian: Siderolithic of the lower Cretaceous. — The Wealdian has been defined in the southern part of the London Basin as a term including the Valanginian, Hauterivian, and Barremian stages, represented by a continental series whose thickness reaches 600 m (EDMUNDS, 1948). The lower half is composed of the fluviatile and lacustrine Hastings sands and the upper half of Weald clays, with a fauna of freshwater mollusks, fragments of plants, and horizons of refractory clays.

The Wealdian extends to Belgium where VAN DEN BROECK (1898) studied the deposit of Bernissart, famous for its fossil Iguanodons. At Bernissart the series of clays was drawn off by the subjacent karst, and it is in this accumulation that numerous giant skeletons were discovered. The Wealdian of Hainaut was studied by MARLIÈRE (1946-1947) who reconstructed a great fossil delta situated in a north-south channel scoured in the Paleozoic basement and forming a wedge under the Cretaceous mantle. MARLIÈRE was kind enough to send me samples which I studied (MILLOT, 1949). One of them contained siderolithic nodules whose main constituent was kaolinite.

The Wealdian facies is also found in the Boulonnais and the Pays-de-Bray, northwestern France, where it is exploited, and in the center of the Paris Basin. In the Haute-Marne, the iron ore of Vassy and the variegated clays of the Barremian stage represent an episode of this facies, as well as in Berry. The samples from the Pays-de-Bray and the Haute-Marne corroborated the predominance of kaolinite (MILLOT, 1949).

All these kaolinitic and ferruginous deposits of the lower Cretaceous must be seen as an extension of a siderolithic facies that corresponds to the dismantling of kaolinitic and lateritic mantles of emergent massifs, such as the Ardennes, the Massif Central and the Vosges. This same phenomenon occurs in Spain (RAT, 1960). It is a witness of the extension of tropical climates over Europe during this epoch, which is the time of bauxitization in the Ariège and the Pyrénées-Orientales (J. DE LAPPARENT, 1930).

The siderolithic Eocene. — Following the Cretaceous transgressions the siderolithic facies manifests itself with a still greater amplitude in the French sedimentary basins of the Eocene.

The Eocene of northern Aquitaine serves as a detailed example. The Sidérolitique is found again in the Charentes as deposits covering the Mesozoic karst or the crystalline basement. In the latter case, KRAUT and VATAN (1938) found lateritic weathering profiles still in place. Furthermore, KLEIN (1961) described kaolinitic sandstones, pisolitic clays and chert, gravels, and defined their age in this region as Middle and Upper Bartonian. From here we go to the Berry where AUFRÈRE (1930) showed the relations between the Sidérolitique and more or less silicified continental limestones, and where VATAN (1947) reconstructed the geology of these formations. The studies were carried on in the Sancerrois and the Allier by DESCHAMPS (1957, 1958, 1960), who reconstructed step by step the variety of these continental facies as far as the Département du Puy-de-Dôme (1962). Thus we arrive at the collapsed basins in the heart of the Massif Central. At the base of the Tertiary series which have accumulated in these basins kaolinitic clays are found once again; this is the situation in the basin of Salins described by JUNG (1954) and that of the Velay analysed by GABIS (1958, 1959).

In the Paris Basin the products of lessivage from the Massif Central reached far to the north across a wide piedmont. One passes through the deposits of La Puisaye (LANQUINE and CUVILLIER, 1941) and those of Breuillet (BERTRAND and LANQUINE, 1919), which were used to make siliceous bricks during the First World War, and one reaches the sands and plastic clays of Provins, which are kaolinitic (MILLOT, 1949) and refractory clays in their purest varieties. Then we come to the lignitic clays of the Soissonnais, which include brackish horizons.

Concerning the southern and southeastern flanks of the Armorican Massif, MILON showed as early as 1930 how the analysis of the Armorican Eocene and of the sediments of the border of the old shield areas could permit the reconstruction of former climates. The siderolithic sediments of the Armorican Eocene were related to the weathering of the old peneplain that had been attacked by what MILON called "Tertiary sickness". Since that time these works have been extended and completed by DURAND (1960) who showed the development of siliceous and kaolinitic sedimentation related to these former climates, that which permitted the interpretation of the whole of the Cretaceous-Tertiary sedimentation as a function of successive climates (DURAND and MILON, 1939). More recently BOILLOT and MILLOT (1962) have found a very probable siderolithic facies under Lutetian limestones of submarine rises in the English Channel off Roscoff. These phenomena can be followed into the Vendée where weathering of the basement rocks, siderolithic expansion and silicifications are to be seen (KLEIN, 1961). Then one comes to the Poitou (STEINBERG, 1961) where the same association, characterized by the predominance of kaolinite, fills the calcareous karst. This leads us again to the Berry and to the Charentes.

Such siderolithic facies prevail also in the Rhone valley where they have in many cases filled the karst of limestone plateaus and where they have been exploited for fireclays (GIOT, 1944). Former pockets, emptied of their fillings, can be seen in the Chartreuse, the Jura, and the Alsatian Jura, where the ferruginous pisolites are called *Bohnerz*. The siderolithic Eocene has also been found by the bore holes of southern Alsace and in the Sundgau where its thickness can reach 30 m. We have now arrived at the Swiss frontier.

In Switzerland the pisiform iron ore has been studied for more than a hundred and fifty years. At the special meeting of the *Société géologique* at Porrentruy in 1838, THURMANN who presided showed not only that *Bohnerz* was a facies, but also that *Bohnerz* properly speaking was situated between the Portlandian stage and the Molasse. Works were numerous prior to the appearance of the fundamental treatise of FLEURY (1909) on the Swiss *Sidérolitique*. He described the siderolithic types, the geographic distribution throughout Switzerland, the stratigraphy, paleontology and origin, drawing a parallel between "lateritization" and "siderolification". Every geochemical aspect of dissolution and concentration is considered as well as the possibilities of reworking and lessivage. More recently, mineralogical studies have been undertaken by HOFMANN (1959) and VERNET (1962).

The Hercynian basement in Germany and Czechoslovakia has been deeply altered by the Tertiary weathering. Below the Oligocene beds in the Hunsrück massif, Germany, ECKART (1960) described a progressive Tertiary weathering of Devonian shales to a depth of 20 m. Kaolinite is predominant at the surface and montmorillonite at depth.

The parent rock is composed of chlorite and muscovite, and the first of these two minerals is much more fragile than the second. One has here a profile developed on a shale, a slightly permeable and always poorly drained rock, that gives us a kind of developed, slow-motion picture of the mechanisms of kaolinitic weathering that we know.

The Moldanubian massif (WALDMANN, 1938) is deeply weathered to a depth of several tens of meters. As usual, coarse-grained rocks turn out to be more vulnerable to this type of weathering than shales. The kaolinitic clays, ferruginous sandstones and silicifications have been studied in detail in the prisoners' camp Oflag XVII A by a team led by ELLENBERGER (1948). The samples collected in the camp enclosure were cut into thin sections and minutely studied with a makeshift microscope under the conditions of camp life.

The great kaolinitic deposits of the Pilsen basin in Czechoslovakia were studied by SLANSKY (1956) and KONTA and POUBA (1961). They think that these are deposits of kaolinitic arkose, the kaolinization of feldspars being carried on during transport and deposition. The ferruginous levels give a siderolithic character to these deposits. In contrast, the great kaolinitic deposits worked at Karlov have been attributed by KONTA and POUBA (1961) to strong kaolinitic weathering of Early Tertiary age. The parent rock is a tectonized granite that is kaolinized to a depth of more than 50 m.

PETROV, in 1958, presented us the history of research on the large deposits of sedimentary kaolinite exploited in Russia, in the Urals and in the Ukraine. It was GINSBURG who demonstrated in 1912 and 1915 that these deposits must be understood as a reworking of a kaolinitic weathering mantle that resulted from different climates throughout several periods, principally Late Triassic, Early Jurassic and Early Tertiary.

Conclusion. — To summarize this tour of the siderolithic facies of Europe, it can be said that this facies gives evidence of its extension at three periods: Carboniferous, Early Cretaceous and Eocene. This evidence is, as usual, more abundant in the case of the more recent periods than for the older ones.

Does this mean that we must distinguish three periods of laterization corresponding to the three periods of reworking? Very probably not. In fact, the laterizing climates must have prevailed throughout prolonged periods of time, and especially prolonged in the past, relative to the major siderolithic periods. We can find some evidence for this by means of local observations in Mesozoic rocks. In Hannover, VALETON (1957) described a lateritic profile interstratified in the limestone series of Late Jurassic age. In the region of Fenouillet (Pyrénées-Orientales), bauxites are included in the Aptian stage, and in Provence there are many examples of bauxites overlain by the Cenomanian or Danian and underlain by the Albian (J. DE LAPPARENT, 1930). This means that if, during Mesozoic calcareous sedimentation, a region was by chance emerged and has been preserved to the present day, we have the possibility of finding there the trace of tropical and humid climates.

Thus, during the Mesozoic Era there was a long period during which laterizing climates prevailed over Europe; the sedimentation of chalk is an echo of this phenomenon, as we shall see. On the other hand, certain periods were privileged with respect to the reworking of the laterite, and their ages are Wealdian and Eocene. It is important that we *distinguish the generative climate from the epoch of reworking*. The former ensures the slow genesis of materials, the latter brings about their dispersal during periods of regression.

4° *The siderolithic facies in Africa*

Credit is due to KILIAN (1931) for the main outlines of the distribution in northern Africa of continental and marine series in the stratigraphic succession, as well as for other subjects. KILIAN distinguished three "Continental series" in the Sahara.

— The Basal Continental Series composed of the ensemble of Cambro-Ordovician sandstones. One cannot speak of siderolithic facies here, but we shall come back to them with regard to the great sedimentary series (Chapter VIII).

— The Intercalated Continental Series, situated between the Tassili formations and the Middle Cretaceous transgression. This is the equivalent of the Nubian sandstones.

— The Terminal Continental Series, which is tertiary and posterior to the Late Cretaceous of the hammadas.

The term Intercalated Continental Series is ambiguous. In reality, the top of the Paleozoic Tassili formations commonly includes continental facies, so much that KILIAN himself said that the Nubian sandstones of the northwestern Sahara were formed by the addition of Carboniferous, post-Tassilian continental sediments and those of the Intercalated Continental Series, which are principally Mesozoic. Thus, one should speak of an Intercalated Continental Series in the broad sense, that is "intercalated" between Paleozoic and Cretaceous marine series—these are the Nubian sandstones—and of an Intercalated Continental Series, in a narrow sense, that corresponds to the Mesozoic Era.

A. F. DE LAPPARENT (1952) showed in fact that the Nubian sandstones in the central Sahara were composed of two ensembles; the one from the Middle and Late Carboniferous, which is the post-Tassilian Continental Series, and the other, Wealdian in the broad sense, which is the Intercalated Continental Series properly speaking. In the Fezzan, continental facies tend to be numerous within the Carboniferous System (MULLER-FEUGA, 1952), whereas geological and paleobotanical studies have shown the extension of the Nubian sandstones throughout the whole of the Mesozoic with three principal series in the Triassic, Late Jurassic and Early Cretaceous (LEFRANC, 1958, 1959; BOURREAU and FREULON, 1959). In any case, it is this Lower Cretaceous of Wealdian facies that has the greatest extension throughout Africa. The siderolithic facies is not common in these continental series, but it does reveal itself with a great amplitude on several occasions. It is characterized by cross-bedded sandy sedimentation, concretionary or pisolitic ferruginous horizons, lenses of kaolinite, variegated coloration from white to violet, silicified wood, and its predominant clay mineral, kaolinite.

With MULLER-FEUGA (1952) we have identified kaolinite in the continental facies of the Nubian sandstones, that which has been confirmed later by OBERLIN, FREULON and LEFRANC (1958). The dominating kaolinite is also the typical mineral in the Intercalated Continental Series of the Nara basin in western Sudan (Mali), as we found with DARS (1957). Finally, by way of comparison, one can point out that BENTOR (1957) considered kaolinite to be typical of the Nubian sandstones in Israel.

All this has a profound geological and paleoclimatic significance. It is easy to deduce that during the Lower Cretaceous, in Africa as well as in Europe, countries that were emergent during the great regression were supplied with continental deposits reworked from laterites. The contemporaneous or anterior climates must have been laterizing.

The Terminal Continental Series reproduces these phenomena on a gigantic scale. During the Tertiary, earlier or later according to the location, regression and erosion set in again, and the sedimentary basins are invaded by the products of the demolition of the lateritic weathered mantle wherein kaolinite and goethite abound.

— In the basin of Gao, eastern Sudan, RADIER (1958) showed that the arrival of the *Sidérolitique* corresponded to the reworking of a lateritic mantle. Farther to the east, in Niger, ferruginous oolites, which are equivalent to pisiform iron and *Bohnerz*, are abundant; TESSIER (1954) studied them in detail. In the basin of the Niger, FAURE (1961) not only described the siderolithic facies of the Terminal Continental Series, but he studied also the distribution of weathered basement rock in the northern part of the country on the flank of the crystalline Aïr massif. With GREIGERT (1960) he showed that, in the central part of western Niger, the basement that is covered by the marine Eocene was not weathered. In contrast, the basement elsewhere is deeply weathered, kaolinized and shows concentrations of bauxites and the presence of alunite. This reconstruction is very important because it permits direct dating of the periods of intense weathering; the parts only of the landscape that were emergent during the Eocene were deeply weathered. These weathering mantles fed the siderolithic sedimentation in the basin. In Dahomey and Togo (SLANSKY, 1959), the Terminal Continental Series with its very widespread siderolithic facies, is also post-Lutetian; in Senegal and southern Mauritania (TESSIER, 1952; ELOUARD, 1959) it is still later. Thus, this facies of reworked continental weathering spreads out in the different basins earlier or later, according to the deformations of the African shield. As in Europe, this facies develops as a function of local tectonic evolution.

— In central Africa, likewise, formations with kaolinite and characterized by a siderolithic facies exist. The great sedimentary series of Gabon studied by HOURCQ and REYRE (1956) and DEVIGNE and REYRE (1957) shows a tremendous development of a continental series with a siderolithic facies in the Upper Cretaceous, i.e., in the Turonian and Senonian stages, including argillaceous sandstones, ferruginous crusts, white or purple argillites, and siliceous sandstones. Our clay identifications performed for M. REYRE show only kaolinite. One must consider that it is a question of reworking of the lateritic weathering of the Mayombe massif during the Late Cretaceous. A new, analogous expansion is formed by the series of cirques that follow a lateritic surface covering all the members of the Cretaceous System. But this new detrital sheet is much more recent; it is considered to be Pliocene.

Such developments with identical facies are well known in the Karroo and Kalahari series in Congo and South Africa, as well as in Madagascar. In South Africa BOSAZZA (1948) has shown the development of kaolinitic series during the Late Carboniferous and the Late Cretaceous.

5° The siderolithic facies in the U.S.A.

In the U.S.A. the siderolithic facies is found in several epochs that are not without analogy to those of Europe and Africa.

KELLER et al. (1953) described deposits of the flint clays, i.e., very kaolinitic, aluminous and refractory rocks of Missouri. This is a Carboniferous series, dated Early Pennsylvanian, that is represented by continental formations deposited on a primary karst. One can follow a succession starting from the regions that were the highest and the best drained of diaspore-bearing flint clays, flint clays, semi-flint clays, and marine shales. All these formations are characterized by kaolinite except the last one, which is marine and composed of illite. The successive facies are closely controlled by the conditions that prevailed at the place of deposition; the aluminum sesquioxides occurred in the highest and best drained points, and the flint clays with kaolinite on well leached slopes; then the semi-flint clays are mixed with illite, and the marine shales are illitic. This is a spectacular example of the control of clayey facies by the environment, but it is important to note that this control is subaerial, which is to say that the processes of continental weathering are determinant, not the conditions of sedimentation under water.

The kaolinizing conditions that prevailed during the Carboniferous Period in the U.S.A. are well known, and the quarried deposits are numerous in Pennsylvania (BOLGER and WEITZ, 1952), Kentucky, Maryland, etc.

In the Lower Cretaceous of the southeastern U.S.A. the continental facies appears again, with many characteristics of the European Wealdian. Studies show the importance of kaolinite in the sediments (DUNCAN and HERON, 1960).

But it was during the Late Cretaceous and Eocene that in the Atlantic portion of the U.S.A. the siderolithic facies reached its greatest amplitude with, once again, deposits of kaolin, refractory earth and bauxite. The deposits of Arkansas (GORDON and TRACEY, 1952) have served as examples in the study of autochthonous and allochthonous bauxite deposits. The kaolin deposits of Georgia (KELLER, 1952) and of South Carolina were deposited on the piedmont of the Precambrian crystalline basement.

Conclusion

The siderolithic facies is a continental facies characterized by a great extent and economic interest, by its kaolin, fire clay, bauxite and sands with various properties. Moreover, it has a definite climatic significance.

In brief, the siderolithic facies represents a residue, reworked or not, of the intense weathering of the tropical humid type, that is, lateritic in the broad sense. Today, such weathering is limited to the equatorial belt of the globe. It was not always the same in the past. On several occasions these climates invaded high latitude countries, such as Europe, South Africa, both Americas, not to mention Asia and Australia. The immediate effect consisted of subjecting the emergent lands to the strong lateritic weathering that we have studied. However, during periods of transgression the emergent lands were limited in extent and the residues of their weathered mantle often escape us. In contrast, during the periods of regression the abundance of these manifestations jumps before our eyes as the result of a double effect. In the first place, the surfaces subjected to weathering become very large, and they are preserved under later sedimentary mantles. Then, erosion distributes the products across piedmonts into sedimentary basins. It is possible in this way to date the reworkings and *to situate more or less accurately the great climatic pulsations of the past.*

V. — THE GREAT DETRITAL RED BED SERIES

Red sandstones occur frequently within the great series resulting from the demolition of old mountain ranges. Their color has been the object of various interpretations aimed at an understanding of the climates that prevailed during their formation. The study of the clay fraction is one argument that can be added to the others in pleading this case. I shall take up here again a recent work on this subject (MILLOT, PERRIAUX, LUCAS, 1961).

Deserts or tropics. — The study of the "Old Red Sandstones" and "New Red Sandstones" forms a part of introductory geology teaching. A long-standing tradition, still continued in some textbooks, has proposed long ago that this red color be attributed to a desert environment. It is commonly agreed that this tradition dates back to WALTHER (1900), but it is likely that traces of it could be found in earlier literature. It is a question of a period in which the science of geology was above all European and when geologists who travelled far were rare.

This tradition has been criticized by geologists of the New World, who are better informed about climatic variety through everyday experience. Red sands and sandstones are formations of reworked weathered products, and one asks himself what kind of climates brought about this weathering. From the beginning, BARRELL (1908), with an astonishing insight, attributed the origin of the red color to climates that are warm, but alternatively humid and dry owing to the succession of rainy and sub-arid seasons. TOMLINSON (1916) insisted on the rapidity of erosion relative to weathering and the relative aridity of the place of deposition. DORSEY (1926) and RAYMOND (1927) emphasized that red colors do not occur in deserts, but in the intertropical zone characterized by a warm and humid climate and a thick vegetal mantle. In Europe, the great travellers, such as BOURCART (1937, 1938) and J. DE LAPPARENT (1937), pointed out that desert weathering is not red, and if red colors do appear, they originated in the humid past. The origin of red deposits is to be found in the redistribution of weathered products furnished by warm, humid zones of tropical or subtropical character wherein red weathered products are widespread.

During more recent times consideration of these red colors has been pursued and has become more precise. GÈZE (1947) has shown the diversity in age of the red beds of Mediterranean Languedoc, which are considered to result from formations, reworked or not, of more or less lateritic type, that are found frequently under a humid tropical climate. GÈZE noted correctly that this characteristic can occur either between the dry tropical (desertic) zone and the equatorial zone, or between the dry tropical zone and the temperate zone. VAN HOUTEN (1948) has shown that the red beds of the lower Cenozoic in the Rocky Mountains were derived from red soils of the hinterland that originated in a warm, humid region with a warm temperate to subtropical climate. KRYNINE (1949) has reviewed the variety of possibilities, and they are numerous; the red color may be primary, postdepositional, secondary or chemical. As for primary red beds, which are our subject, he defined the climate of their land of origin as warm and humid.

G. CHOUBERT (1950) defined *climats rubéfiants d'accumulation* (climates of rubefaction and accumulation) in the history of Morocco. He distinguishes them categorically from humid tropical climates, and shows that some humidity is necessary because as one travels toward arid Morocco, the colors become paler. He also showed that these climates must have been fairly warm. In 1953, G. CHOUBERT was able to demonstrate that the climates of rubefaction correlate with the great pluvials of the Pliocene and Quaternary. One understands at the same time the role of rain and that of temperature. A climate that can include a warm season and a little marked winter cannot be considered "tropical".

DUNHAM (1952) attributed the red color of the Permo-Triassic to humid tropical or subtropical climates. In 1956, ERHART, in his fresco on the relations between Pedology and Geology, discussed the Old Red and New Red sandstones and said that, contrary to opinion received in textbooks, "only a long period of forest pedogenesis was able to disintegrate rocks to a very great depth, remove all the soluble substances and separate out the iron hydroxides that color the rock." ERHART was cautious to note that the period of weathering must be distinguished from that of reworking; they could be separated by a long time. RICOUR (1960) applied this information to the French Triassic; the red color is inherited from humid tropical climatic conditions. LIENHARDT (1961) took his bearings in the same way.

Thus the teaching that we received has been profondly modified today.

— All specialists deny the possibility of attributing the origin of the red color of the great detrital red bed series to a desert climate.

— In contrast, and in spite of the indications of BARRELL (1908), GÈZE (1947) and CHOUBERT (1950), we are at present oriented toward humid and warm tropical or intertropical climates that allow the development of a dense tropical forest.

— Now then, the products of reworking of the weathered intertropical mantle is the *Sidérolitique*. From the beginning it appeared that this hypothesis was impossible. All measurements contradicted it. The conditions of genesis of these red sandstones remain to be defined.

Information provided by petrographic study of the Permo-Triassic red sandstones of the Vosges. — Parallel to petrographic studies on the Permo-Triassic red sandstones made by one of our group (PERRIAUX, 1961), we have studied in common the nature of the fine fraction of these sandstones (MILLOT, PERRIAUX, LUCAS, 1961). The results of the ensemble of studies are the following:

— The Permian sandstones very commonly contain pebbles of granite, gneiss, schist and all the related crystalline rocks that occur around the borders of their basins of accumulation. The Buntsandstein conglomerates also contain polycrystalline pebbles in the region of Monthureux-sur-Saône and Darney, that is to say, close enough to the uplands so that transportation did not destroy them.

— Feldspars are so abundant in the Permo-Triassic sandstones that they are called either arkoses or arkosic sandstones, depending on the individual outcrop. Let us add that these feldspars are exclusively alkaline; plagioclases are rare in the Permian and very rare in the Buntsandstein. Feldspars are fresh in the Permian sandstones and slightly weathered in the Triassic sandstones. The following table shows the quantity of feldspars of the diverse sandstone formations:

Formation	Percentage of the number of feldspathic grains relative to the total number of grains
Permian sandstone	35 to 60 %
Annweiler sandstone	16 to 45 %
Vosgien sandstone	5 to 38 % (mean = 15 %)
Principal conglomerate	10 to 15 %
Lower intermediate beds	20 %
Upper intermediate beds	25 %
Voltzia sandstone	17 to 35 %

— The grains of quartz are neither cracked nor decayed (only 0.1 to 0.5 % are), and the ferruginous coating is external and very different from the ferruginous cavities observed on quartz grains of the tropical zone. In contrast, ferruginous aggregates of quartz grains are common and are reminiscent of the "pseudo-sands" of red soils of tropical countries. Let us recall there that hematite is the main constituent of the pigment of red sandstones, as has been confirmed in X-ray diffraction studies by many workers in all countries (ROBB, 1949; cf. documentation in STEINWEHR, 1954 and VAN HOUTEN, 1961).

— The clays that make up the fine fractions sampled from ten different formations or horizons in the Permo-Triassic sandstone series are almost exclusively illite. It is a question of the most common clay mineral that comes from crystalline massifs during mild chemical weathering. This type of weathering induces physico-chemical disintegration of the micas themselves and also chemical attack on the more vulnerable silicates: plagioclases, cordierite, etc.

Lastly, PERRIAUX (1961) was able to characterize some horizons of the Buntsandstein as soils. In them could be found an altered illite interstratified with layers of montmorillonite, and a strong micro-subdivision of quartz, which our long experience with the clay fraction of soils and sediments allows us to attribute to pedogenetic processes. These soils do not have the characteristics of humid tropical soils; their association with dolomitization and silicification points rather to a semi-arid climate. Climates with arid characteristics must have prevailed over the area of sedimentation during certain periods.

Thus the principal petrographic characteristics of these red sandstones are:

— The presence of polycrystalline pebbles;

— The presence of abundant alkaline feldspars;

— Quartz neither decayed, nor cracked, nor ferruginized in the solution cavities;

— A clay fraction composed chiefly of illite;

— The occasional occurrence of soils with arid characteristics.

These features can be compared with information supplied by pedology.

Information given by the study of soils. — Laterotization, in the broad sense, is the weathering process that prevails under humid tropical climates. The product resulting from the reworking of laterites is the *Sidérolitique* that we have just studied. Let us summarize in a few words the petrographic characteristics of a siderolithic product:

— Never any polycrystalline pebbles; hydrolysis has decomposed the silicates, and the rocks fall apart;

— Never any feldspars remain except in the zones of initial weathering where weathering is only partial;

— Quartz grains, which are decayed, cracked, and ferruginized in the cavities;

— A clay fraction represented exclusively by kaolinite;

— If soils are interstratified within siderolithic formations, they are lateritic soils with kaolinite.

These characteristics are in contrast to those of red sandstones. Even if "red sandstones" and the siderolithic facies were not distinguishable at first glance by an expert, this contrast would not allow us to group them together.

We then searched (MILLOT, PERRIAUX and LUCAS, 1961) in the direction of the ferruginous soils of warm climates, as defined and studied in Chapter V. We are not concerned here with a localized, limited or definite soil type but with the ensemble of reddened soils that form over wide areas south of the temperate zones and north of the humid tropical zones. Hydrolysis begins, rubefaction occurs, plagioclases break down, alkaline feldspars are preserved and, in the less evolved soils, illite is the predominant clay mineral; let us keep in mind that illite is the common weathering product of many silicates.

Thus one arrives at a facies of weathering in which:

— polycrystalline pebbles are possible;

— alkaline feldspars are preserved;

— quartz grains are ferruginized but not decayed;

— illite is predominant.

These characteristics are those of red sandstones; we are close to our goal.

The conditions of deposition of the Permian and Buntsandstein red sandstones of the Vosges have been minutely reconstructed by PERRIAUX (1961). During the Permian, depressions in the still young mountains are filled with red sandstones originating locally from the weathering of the surrounding uplands. During Buntsandstein time, sediments arrive from the southwest of this "Gallic continent", which has been reconstructed from its remains in the area of Burgundy (Bourgogne). Sedimentation was seasonal in the form of meandering streams, giving rise to cross-bedding and to many of the sedimentological characters of these sandstone formations. The red color of the deposits is certainly allochthonous, being brought in with the sedimentary material. Moreover, whenever this sedimentation was interrupted, soils with semi-arid characteristics were able to establish themselves, along with the possibility of dolomitization and silicification, as has been described by PERRIAUX. Humid zones could likewise have been established in the lowlands.

There are so many convergences between the petrographic characteristics of the Permo-Triassic sandstones and those of ferruginous soils we have studied, that it is possible to group them together. The Permo-Triassic sandstones are the product of the reworking of rubified *arènes* originating on the Gallic continent. This Gallic continent was subjected to a climate neither desertic nor tropical, but intermediate. Essentially, one can speak of the weathering of crystalline basement rocks in a warm region characterized by alternatively rainy and dry seasons. The humidity gives rise to hydrolysis, especially of ferromagnesian silicates that release iron, whereas the well marked dry season fixes the iron sesquioxides.

The great detrital red bed series. — Great detrital red bed series are numerous throughout the stratigraphic column, and the interpretation of this phenomenon, which is at present still not well known, has always been difficult. The most penetrating reflection on this subject was made by CHOUBERT between 1945 and 1959. Finding this peculiar facies on many occasions in the Moroccan series, CHOUBERT attributed it to the *climat rubéfiant d'accumulation*; patiently he defined the characteristics that the contribution of modern pedology allows us to state with precision today.

CHOUBERT (1959) described the earliest detrital red accumulation in Morocco, which is anterior to the Cambrian and called "Precambrian III" or the upper part of the *série d'Ouarzazate*; he firmly emphasized its analogies with the Permo-Triassic. The same facies were found again in the Adoudounian stage, corresponding without doubt to the Infracambrian. There are also analogies with the purple series of Ahnet in the Central Sahara. Likewise, in Scandinavia great glyptogenic series of violet color occur in the Jotnian formations. In the Cambrian, the Moroccan series again shows this facies in the Upper Georgian (CHOUBERT, 1959). In the Sahara, in detrital formations formerly called *grès horizontaux* (flat-lying sandstones) this facies recurs as pink feldspathic sandstones of the Early Cambrian (KULBICKI and MILLOT, 1961). Some comparisons have been made with the Cambrian of Normandy.

Next we come to the famous Old Red Sandstones, known from Ireland, Scotland, Scandinavia and Russia. In this case once again, a very great detrital, continental series in which red colors are common followed the emergence of a mountain range.

We reach finally the Permo-Triassic New Red Sandstones of which the Vosgian ensemble is a part. Here it is a question of the consequences of Hercynian tectonics. Detrital red beds spread over the entire Eur-African area.

Following the Alpine movements, the process becomes more hesitant; this sedimentation no longer occurs widespread, but as more local facies. G. CHOUBERT and specialists from Morocco (1945, 1950, 1953, 1956, 1957) have reconstructed little by little the characteristics of these successive detrital red bed series and the *climats rubéfiants d'accumulation*. Considering only Tertiary and Quaternary time, such detrital pink and red deposits developed at the end of the Oligocene, during the Pontian, in the Late Villafranchian or Moulouyan, in the Amirian (Mindel) and in the Soltanian (Würm). In the older nomenclature, the Amirian was the second pluvial and the Soltanian the fourth. If the Moulouyan is called the first pluvial, everything must be displaced, and they become the third and fifth pluvials (CHOUBERT, JOLY, GIGOUT, MARÇAIS, MARGAT and RAYNAL, 1956). These divisions are well observed, especially in maritime regions because the synchronous continental series are much less rubified.

Strong rubefactions have been studied in Europe, either in reworked red detrital sediments or in secondary rubefactions *in situ*. Some examples of these studies can be cited: the detrital Miocene series of Vallés-Pennedés (MARTIN-VIVALDI, FONTBOTE et al., 1957); the Miocene of the Massif Central in France (GRANGEON, 1959, 1960; DERRUAU, 1960); the Pliocene and Quaternary of western and southwestern France (CAILLEUX, 1953; ALIMEN, 1954); the Quaternary of northern Italy with its *ferretto* facies (VENZO, 1952; GABERT, 1961) where, moreover, hydrolysis has been stronger and kaolinite appears.

Geodynamic and climatic interpretation of the great detrital red bed series. — The red beds or rubified deposits of the recent geological past are interesting and will little by little permit detailed reconstruction and lead to greater precision in interpretation.

But that which is most important for our subject here is the examination of the ancient great red bed series dated Infracambrian, Cambrian, Caledonian and Hercynian. In these cases there was detrital sedimentation over large areas and in great thicknesses during epochs of demolition in successive mountain chains.

There are two ways of reconstructing the past. The first way is that which we used in comparing the petrographic character of sandstones with the present-day weathering. By this path we must reject both desert and humid tropical climates because the former are not red and the latter give rise by reworking to the siderolithic facies, which is quite different. This leads us to an intermediate, seasonally humid, warm climate in which rubefaction can occur without allowing the strong hydrolysis of the lateritic regions.

The other path, more properly geological, consists of placing these great "glyptogenic" series into the perspective of geological history. This becomes a question of the demolition of mountain ranges that have just been uplifted. Erosion attacks the highlands, which are of considerable extent, if not high in altitude. The attack continues over a long time, progressively as these uplands are rejuvenated, giving rise to a special type of detrital sedimentation that covers the whole region and sometimes entire continents. It is this kind of erosion—of long term, of rubefaction, and of a special sort—

that interests us; it has taken place at several different epochs and on widespread areas, accumulating great thicknesses of sediments.

Therefore, it is not necessary for us to invoke a precisely and narrowly defined climate, painstakingly reconstructed by detailed observations of modern soils. A rather ordinary, common climate can account for a result that is itself also ordinary and common in occurrence and of great amplitude. Here then is the conclusion that we reach.

On highlands that are being eroded, climates in which long rainy periods alternate with strong dry periods give rise to this arenization and rubefaction that, upon being reworked, results in red sandstones. We are a long way from cold climates, from desert climates, and from climates in which a thick forest allows intense ferrallitic weathering under its roots.

Contrary to these conditions, this process takes place in warm countries characterized by showers and monsoons, heavy rainfall alternating with periods of desiccation, rubefaction, and even sometimes aridity. The reasons for the mild degree of weathering would lie much less in the moderate rainfall and temperature, such as we reconstruct them from modern soils, than in vigorous, long-term erosion continuously moving the rubified *arènes* from the uplands and accumulating them in sedimentary plains during each seasonal wet period.

Let us turn the perspective around; each time that a climate of alternating warm periods, tropical or not, and with humid and dry periods has acted upon the rising uplands of a growing mountain range, it has engendered this sort of hasty weathering, not very advanced from the point of view of hydrolysis, but nevertheless producing rubefaction. It is this kind of weathering that has resulted in large accumulations of red sandstones.

Let us remark here that that which we have said about the intermixing of facies in the Old Red and New Red Sandstones agrees with this reconstruction. Re-reading the descriptions in textbooks, while thinking about our landscape, lessens the number of contradictions in which we found ourselves caught up. Cross-bedding, pebble layers, *fine-grained layers of crushed debris, in some cases rolled,* dolomitic or chalcedonic horizons, the contrasting character of the flora and fauna, in some cases really humid, in others lagoonal or arid, all these whims of nature find their places on vast piedmont areas subject to such seasonal climates. The red color is but one effect, among others, of these alternating climates that we are invoking. FALKE (1961) arrived at a similar conclusion in his study of the faunas, floras, and sediments of the Permian of central and western Europe, namely, warm, humid climate interrupted by a few dry periods. As at the present day, relief must have had a great influence on climate.

What can one say about the forest cover on such landscapes? The question is quite difficult to approach today. However, it is necessary to note that intense and rapid erosion is not favorable to a heavy forest; in the language of ERHART (1956), it is much more a question of rhexistasy than of biostasy. Moreover, a heavy forest cover maintains humidity all year around, brings about total hydrolysis of silicates, and gives rise to kaolinite. It seems that the uplands, as well as the areas of sedimentation, must have been lightly populated, continuously reworked and rejuvenated, and covered by a meager vegetation except in the lowest areas where there was permanent humidity.

Thus one arrives at the idea that the red color is due precisely to the fact that the vegetation was scrubby. The quantity of organic matter is too small to complex the iron and to evacuate it across the landscape. As soon as the iron is released by the initial hydrolysis it is fixed on the *arène* itself by means of the desiccation that is permitted by the absence of forest. An entirely different path might be followed, according to the outline of ERHART (1956). This consists of the consideration that thick weathered layers of red color are stored up on the continent during a period of forest cover and then delivered to the sedimentation process after the destruction of that forest. We do not think that this mechanism can be invoked here. The forest does not have the capacity to store up thick *arènes*; on the contrary, it favorizes total hydrolysis beneath its feet. Moreover, the forest evacuates iron by means of *lessivage*, as is shown by the light color of the great ferrallitic profiles under the equatorial forest. Only the ironcrusts (*cuirasses*) are red, and they are emplaced within the profiles by ground water circulation. The role of the forest is not that of rubefaction, but that of hydrolysis and lessivage. Rubefaction is a phenomenon of oxidation and desiccation that is produced either in ironcrusts, which are secondary features and subordinate to ferrallitization, or at the surface in red soils, which are precisely not under the forest.

Conclusion. — Comparison of the petrographic study of red sandstones to the mineralogical characters of ferruginous warm climate soils allows us to understand the origin of the color of the great detrital red bed series and the climates that prevailed during the formation of that color.

These *climats rubéfiants d'accumulation*, in the descriptive terms of CHOUBERT, are equidistant between temperate climates and humid tropical climates. The possibilities are extremely varied, but their allure can be reconstructed: warm climates, alternately humid and dry.

An encouraging convergence is found here with the works of ERHART (1962) and with the results of discussions that he led at the Colloquium of the *Société de Biogéographie* in the winter 1961-1962. In fact, ERHART arrives, by a different route, at the following proposition: "We envisage for the Permian a warm climate with alternating seasons, a tropophilous forest vegetation at low altitudes and a xerophytic 'bush' higher up or in dry areas."

Tectonic considerations are to be added to this. They are important because the material originates on uplands and especially on *active uplands*. In this case, weathering is without doubt moderated more by the rapidity of erosion than by the controlled quantities of precipitation and the temperature the variations of which are evidently tolerant. This mild weathering produces rubified *arènes* that, with each major downpour or each rainy season, emigrate and form flat-lying or cross-bedded layers on the piedmont.

If this line of reasoning is confirmed, the *climats rubéfiants d'accumulation* of CHOUBERT will have to take their place in paleogeographic reconstructions along side the other well typified climates: glacial, desert, and tropical laterizing climates.

VI. — COAL MEASURES AND TONSTEIN

As in the case of the great detrital red bed series, petrographic studies and, in particular, studies of the clays are very important in the reconstruction of the conditions of sedimentation and of the climate that gave rise to coal measures. We shall distinguish the study of coal measures from that of peculiar beds called Tonstein.

1° *The Coal Measures*

The silicates of coal-bearing sediments. — The clay minerals of coal-bearing sediments are now well known. At the start we have available the works of GRIM (1935) and of GRIM, BRAY and BRADLEY (1937), and it is extremely significant that illite was defined principally on the basis of the study of the coal measures of Illinois, and that one of the two reference samples of illite is precisely a mineral taken from a coal-bearing sediment of Pennsylvanian age in Vermilion County, Illinois. NAGELSCHMIDT (1943) found in the coal-bearing shales of Wales 90 % illite and 10 % kaolinite, and MILLOT (1953) found an association of "open" illite along with a little kaolinite (about 20 %). ENDELL (1955) gave comparable results.

Since then, numerous studies have been made in America, and there is in particular the study of SCHULTZ (1958) who was able to benefit from recent methods of identification of mixed layers. The assemblages are varied, but the most frequent mineral is illite accompanied by its usual cortege of chlorite and various mixed layers. Kaolinite is subordinate, although more common in the coal than in the barren layers.

A very detailed study of coal-measure sedimentation has been presented by GLASS (1958). This concerns the coal basin of Illinois where sedimentation was rapid because of intense subsidence. The mixture of minerals are varied as is usually the case in detrital sedimentation. Illite, kaolinite, chlorite and mixed layers are present in the sediments in varied proportions. These variations were precisely what GLASS studied; he was able to show evidence of multiple influences. In the first place, there is the influence of the source of the argillaceous particles, which undergoes cyclic changes from times of fluvio-lacustrine sedimentation to times of marine sedimentation; next, the influence of the environment by means of the rapidity of deposition and the milieu established in the basin during the coal-generating and the detrital periods; then, pedogenetic alteration of the underclays under the influence of swamp waters; then, after lithification, diagenesis, the effect of which is to develop kaolinite in the porous sandstones; and, finally, exposure as an outcrop in which the differences between core samples and surface samples become apparent. Here we see a succession of heritage, transformation, pedogenesis, diagenesis, and weathering. Such a detailed study of some 1 800 feet of coal-bearing sediments averts rapid interpretations.

In addition to the argillaceous fraction, microscopic examination shows in the barren coal measures, the abundance of larger flaky minerals called "sericite" by DUPARQUE (1946) and, in certain basins and in certain horizons, the abundance of feldspars (GLASS, POTTER, SIEVER, 1956; LIENHARDT, 1961).

One can say right away that illite, chlorite, sericite, mixed layers and feldspars are incompatible with the existence of a humid, tropical climate in the regions supplying coal-measure sedimentation.

Underclays and coal-measure paleosols. — Americans call those argillaceous rocks that immediately underlie coal beds "underclays". We are grateful to SCHULTZ (1958) for an extremely detailed study of the underclays of the Carboniferous (Pennsylvanian) coal basins of the Appalachians, of the Mid-Continent region, and of Illinois. The comparison of the underclays with the coal-bearing shales with which they are associated was systematically made. It is evident that the mineralogy of the underclays is varied because of the variety of rocks from which they originate, but there are additional variations.

SCHULTZ shows the variations in composition of the clay minerals in the underclay profiles:

	Number of samples that		
	increase at the top of the profile	remain constant	decrease at the base of the profile
Kaolinite	31	38	33
Montmorillonite	29	60	6
Feldspars	10	13	8
Mica	4	9	39
14 Å (chlorite)	6	12	31
14 Å (vermiculite)	5	21	4

Kaolinite and the feldspars show varied behavior. Mica and chlorite have the tendency to be altered. Montmorillonite has the tendency to increase at the top of the profile, and vermiculite is stable.

This statistical presentation hides various possibilities, but the essence of these variations is well known to us. It is a question of weathering under humid climate that induces the opening of the micaceous minerals; this is what we have called the transformation of clay minerals in soils, and it has been verified in the text that the mixed layers increase in abundance from bottom to top in the soil profile. The increase of montmorillonite in certain profiles calls for hydromorphy. This is exactly what MACMILLAN (1956) showed in his very beautiful study of coal-measure soil. He reconstructed a gleyed soil showing the weathering of chlorite into interstratified illite-montmorillonite.

GRAFF-PETERSEN (1961) has carefully studied the Jurassic limnic and coal-bearing basin of Bornholm, which is one of the Danish islands. The constituents of these sediments are, in order of abundance, illite, mixed layers, kaolinite, vermiculite, chlorite, and montmorillonite. In the course of lessivage of the underclays by the acid waters of the lagoon, chlorite is the most fragile and is weathered into vermiculite, and illite is weathered to mixed layers. Vermiculite and mixed layers can even evolve into a small amount of kaolinite, or even be hydrolysed. Kaolinite is stable and, therefore, shows a relative increase. We have here another and an excellent example of pedological degradations beneath the coal beds.

Here we are very far from humid tropical soils in which there is a massive development of kaolinite. However, such development does occur in certain American underclays, which are commercially exploited for refractory clays. This has been described many times. Let's take the example cited by PATTERSON and HOSTERMAN (1960) from Kentucky; they describe an underclay with Stigmaria that has the characteristics of a soil. The varied minerals of the coal-bearing shales, formed of a mixture of kaolinite, illite, and mixed layers, are progressively transformed into kaolinite with the disappearance of feldspars and even partially of quartz. This weathering is autochthonous and is produced under the influence of lessivage by organic matter coming from the coal swamp. The same mechanism has been envisaged for the Carboniferous swamps of Missouri (KELLER et al., 1953). It is a question of kaolinization under the swamp by means of leaching by waters charged with humid acids.

The climate of coal-measures time. — If one tries to reconstruct the climate that prevailed during the formation of coal basins, the first point to be made is that petrographic information orients us towards conditions opposite to those of humid, tropical climates; this is contrary to our scientific heritage. In fact, humid tropical climates develop a powerful hydrolysis that ends, on the whole, with the invasive and gigantic neoformation of kaolinite. This cannot be the case in coal-measure landscapes where thousands of meters of deposits accumulate in which illite and micaceous minerals are largely dominant, feldspars are often present, and kaolinite is subordinate. We must look elsewhere, and mineralogy is rather impotent in this search because the argillaceous assemblages of coal measures are characteristic of a common type of weathering that is produced under many kinds of humid climates.

Other techniques are more penetrating; at the Colloquium of the *Société de Biogéographie* organized by ERHART (1961-1962), paleogeography (FABRE and FEYS, 1962), paleobotany (GREBER, 1962) and paleopedology (ERHART, 1962) added their arguments. In each coal basin the final part of the sedimentation passes from gray or black to red, with more and more frequent alternations, but this transition occurred at different times in different basins. The autochthonous and allochthonous vegetation is varied and corresponds not only to that of swamps but also to that more xerophytic vegetation of the uplands. Even the swamp vegetation gives the impression, by means of its anatomical characters, to have lived "with its feet in the water, but its foliage in the dryness." ERHART has noted that the barren horizons of the coal measures are depleted in hydroxides, in silica, in red products, and in carbonates, and that this does not correspond to the results of lessivage under a humid, tropical forest. He restudied entirely the paleosols and the clayey particles they contain. It is the concretions, which may be ferruginous, dolomitic or siliceous, that orient us once again, by comparison with modern phenomena, toward the idea of an alternately humid and dry climate.

The question of illite has been brought forth, and there is hesitation whether to call it a detrital or a pedogenetic product. Let us say first of all that the coal-bearing shales contain the suite of micaceous minerals: illite, chlorite, and the mixed layers of these minerals with vermiculite and montmorillonite; kaolinite is only subordinate. What is the origin of these clay minerals? It can be one of three kinds:

— either direct reworking of previous argillites, shales, schists or mica schists;

— or the products of weathering of silicates by the processes that we looked at in Chapter V; here it is a question of the common weathering by sericitization of the feldspars and pulverization of the micas;

— or a pedogenetic origin by neoformation within the soil.

Our knowledge is now advanced enough so that we can say that the third type of origin cannot be envisaged for illite itself. This mineral is stable in soils as long as lessivage and the role of organic matter are discreet, and beyond that it is weathered. Illite is a common mineral of neoformation during the physico-chemical alterations of diagenesis, it is either inert or fragile whenever organic matter comes in or hydrolysis is amplified.

On the other hand, mixed layers are characteristic of weathering and of soils. For this reason the suite of clay minerals of barren coal measures seems to us to owe much more to the first factor (reworking), and everyone knows that the cover rocks on the Hercynian massifs, before their emergence, were made up of shales and culm in enormous beds that have been eroded. The second factor, weathering, accompanied the first, of course, because erosion of the mountains must have been attended by all the mechanisms of weathering and disintegration of rocks. The third factor certainly existed also; it would have been responsible, in part, for the mixed layers, but there is no comparison in tonnage during a period of active erosion between the effects of erosion and those of pedogenetic neoformation.

In conclusion, the coal-producing landscape appears to be a tectonically active landscape and to be connected with Hercynian orogeny. In the plains near base level, the water table crops out and swamps form. On the heights and the hills, everything indicates a prevailing warm climate with alternating seasons. As ERHART (1962) has said, these conditions are not so different from those reconstructed for the Permian. One could then approach the following outlines:

— Coal-measure sedimentation is the result of ordinary weathering of Hercynian mountains which were rising and being eroded under a warm climate of alternating seasons. The forest proliferates in the well-watered lowlands. The heights are drier, and it is even not excluded that the weathered products may be rubified, but, in this case, they would be reduced by organic matter in the coal swamp.

— In fact, no matter at what epoch the coal-measure sedimentation died out, its place was taken by red sedimentation. Tongues of red products spread out across the lagoon that no longer had the capacity to reduce them. But here is an alternation of gray and red deposits over a long period, showing well the possible coexistence of these facies, one being subaerial and the other aquatic.

— Finally, the red sedimentation wins out, and little by little, between the Westphalian and the Autunian (Fabre and Feys, 1962), it spreads everywhere just as if it had been waiting only for the drying up of the coal swamps in order to invade the basins. More detailed studies will show us, perhaps, whether the appearance of the crystalline basement after the erosion of its sedimentary mantle will have an effect on sedimentation and whether under a whole series of climates, the modifications can be understood.

We arrive once more, by means of the concurrence of disciplines, at a considerable renewal of our knowledge. Mineralogical and pedological studies oblige us to look for the type of weathering hidden within the sediments. Weathering occurs prior to sedimentation and is much more dependent on past climatic conditions than the latter.

2° Tonstein

Tonstein was named from the Sarre basin. These are beds of little and, moreover, variable thickness, being a few decimeters thick. *Tonstein* is extremely widespread and is readily recognizable by its pale color, beige or brown. They have served as guide horizons for miners. In the Massif Central they have been known for a long time under the name of *gores* or *roc seda*. In the basins of the north of France also they are called *Tonstein*. They have considerable stratigraphic interest for geologists working in coal basins. We shall examine here their composition and their genesis.

Leverrierite. — Microscopic examination of *Tonstein* shows elongated, twisted and vermiform crystals, as much as 15 mm long, against a background that is generally cryptocrystalline. P. Termier (1890) called these crystals leverrierite, the gross formula of which is $2 SiO_2, Al_2O_3, (H_2K_2)O$. It was considered to be intermediate to kaolinite and mica.

In 1931, X-ray analysis permitted Ross and Kerr to announce that leverrierite was nothing other than kaolinite. However, J. de Lapparent (1934) was able to demonstrate that leverrierite presents itself as an association of kaolinite and mica. The epitaxic association can occur in variable proportions. Kulbicki and Vetter (1955) confirmed this alternation, and it is reasonable today to consider leverrierite no longer as a mineral, but as a vermicular facies of kaolinite with the possibility of an alternation with layers of mica.

These vermicules of kaolinite developed from mica are not, moreover, specific to *Tonstein*; they have been described by J. de Lapparent and Hocart (1939) from the lateritic formations of French West Africa, by Vatan (1939) from the sands with kaolinitic cement in the southern part of the Paris Basin, by Kulbicki (1953) from the *Sidérolitique* of Aquitaine, and by Kulbicki and Millot (1960) from the oil-bearing sands with argillaceous cement in the central Sahara.

The genesis of leverrierite. — J. de Lapparent (1934) interpreted leverrierite as the kaolinitic alteration of detrital micas. Kulbicki and Vetter (1955) reconstructed in great detail its conditions of genesis. The micas are expanded like accordions by the epitaxic growth of kaolinite between the layers. This phenomenon cannot take place before transportation, nor after consolidation. It is therefore a question of neoformation within the muddy sediments under the influence of acid conditions maintained by the percolation of organic matter. These arguments are important in the explanation of the genesis of *Tonstein*.

Petrography of **Tonstein**. — Leverrierite is not the only constituent of *Tonstein*; detrital minerals, micaceous clay minerals, siderite, a cryptocrystalline background of kaolinite, pseudomorphs of kaolinite, coal and vegetal debris have been described therein.

An important point emphasized by a number of authors is the remarkable petrographic constancy of *Tonstein* across distances of tens, and in some cases hundreds of kilometers, and this is the reason that they serve as guide horizons.

Hypotheses for the genesis of **Tonstein**. — INHERITANCE OF KAOLINITE. — The traditional hypothesis proposed some 30 years ago by PRUVOST (1934), and then taken up again by LIENHARDT (1960), is that of sedimentation in a coal basin of kaolinite inherited from the surrounding landscape. This is incompatible with that which we have said about the climate of coal-forming times.

— CHEMICAL OR BIOCHEMICAL NEOFORMATION WITHIN THE SEDIMENT (SCHULLER, 1951; SCHEERE, 1959). — It is certain that leverrierite is neoformed within the sediment, as we have seen, and it is possible that the neoformation of cryptocrystalline kaolinite occurs there. However, it is difficult to see why "kaolinizing periods" should set in for a brief interval of time over enormous distances within a coal basin, when the *Tonstein* horizons are interstratified in some cases below, in some cases above, and even within the coal itself.

— PEDOLOGICAL INHERITANCE FROM SWAMP SOILS. — ERHART (1962) suggests an interesting possibility based on the reworking of swamp soils. His proposition runs like this:

"Now then, the conditions of an arid climate with alternating seasons does not square itself at all with the conditions required for the genesis of kaolinite. On the other hand, the genesis of kaolinite can be perfectly understood if one admits that it formed under an acid swamp.... Then, an extremely weak epeirogenic movement suffices to bring the swamp into a position of complete emergence. The hydrophytic vegetation dies, and the kaolinitic soils are delivered to the erosional process. This is 'exundation rhexistasy'.

The kaolinitic products are then spread about because of their extremely small size, but this reworking does not last long, as shown by the thinness of the *Tonstein* layers. In the deposit the kaolinite will be mixed with intact minerals from zones of little weathering, that which explains the amount of alkalis shown by analyses of *Tonstein*."

This hypothesis is compatible with the neoformation of kaolinite within the leverrierite and the kaolinized feldspars within the sediment, which is necessary because the pseudomorphs would not have endured through transportation. The evolution can continue within the *Tonstein* before its consolidation. It is more difficult to understand why the soils reworked on the margins of the basin have not "disappeared" into the coal basin, becoming mixed with all the other products that had been reworked there permanently. The sediments of limnic or paralic basins rarely show any such clear cut, instantaneous deposits isolated within a series spread out in all directions.

— HYPOTHESIS OF VOLCANIC ORIGIN. — *Tonstein* beds are so curious because of their vast extent that hypotheses have been proposed that take that character especially into account. J. DE LAPPARENT (1934) had thought of the burning of the coal swamp forest and of the evolution of volcanic ash in the swamp. This is an old hypothesis and it had fallen into disuse. STACH (1950) took it up again, proposing that volcanic ash fell into the coal swamp and evolved there into kaolinite under the influence of organic matter. The arguments, besides that of great extent, are based on petrographic observation of crystals attributed to the ash. These arguments were fought against, but the geological team from the coal basin of the north of France revived them. CHALARD (1952) and BOUROZ, DOLLE and PUIRABAUD (1958) compared and correlated the Basilic *Tonstein* of the *Bassin du Nord* with the Baldur *Tonstein* of the Ruhr basin. The petrographic resemblances are so great across several hundreds of kilometers that they think that only volcanic ash could explain their distribution. Moreover, they find that this *Tonstein* occurs indifferently below, above, or within the coal seam, while always retaining its identity. Such an indifference to conditions of sedimentation would be extraordinary for a chemical or biochemical sediment. Since then, arguments for a pyroclastic origin have been put forward by FRANCIS (1961) who observed the transition of a volcanic tuff into a kaolinitic horizon.

Here we are confronted with a new hypothesis that satisfactorily explains the extraordinary spatial distribution of *Tonstein* beds. One might, then, be surprized to read that the volcanic silicates and glass are so sensitive to kaolinization in the lagoon, whereas the silicates of the coal-bearing shales are normally indifferent there. There is really nothing astonishing in this; the silicates of coal-bearing sediments are the remains of all those that could resist ordinary chemical weathering: quartz, alkaline feldspars, and illite and its suite. On the other hand, volcanic silicates and glass come directly through the air from the volcano to the lagoon. This is an extraordinary change of thermodynamic conditions for these minerals. It is not a question of an old, resistant detrital residue, but of silicates that are stable at high temperatures and that are abruptly subjected to lessivage by humic solutions. All the vulnerable minerals are transformed to pseudomorphs of kaolinite, and the others change more slowly or resist entirely so that analyses of the whole mass show the persistence of alkalis.

This is exactly the same phenomenon that transforms ash to montmorillonite in bentonite deposits and crystals of mica to illite and glauconite with vermicular facies, which has been described by NORIN (1953), MULLER (1961) and GRIM and VERNET (1961) from the Gulf of Naples (Italy) in the case of the illites and by GALLIHER (1935) and CAROZZI (1951) in the case of glauconites. The volcanic silicates are extremely sensitive in an aqueous environment, and their evolution varies according to the properties of that environment, which is understandable.

Conclusion. — The genesis of *Tonstein* is, therefore, in controversy.

— The hypothesis of inheritance of kaolinite from the landscape during a period of extreme calm comes up against the fact that the landscape does not seem capable of delivering this mineral because its weathering is hardly kaolinizing.
— The hypothesis of chemical or biochemical neoformation within the lagoon does not lead to an understanding of the reasons why a coal basin suddenly has the peculiar property of bringing its humic solutions to bear on these sediments independently of the coal-measure sedimentation.
— The hypothesis of reworking of the littoral soils of the swamp explains well the genesis of the fine-grained kaolinite, but less well the genesis of leverrierite, which forms *in situ*. It is also difficult to understand why the reworked littoral soils are not mixed up and lost in the mass of material swept across tens of kilometers into the lagoon.
— The volcanic hypothesis explains well the vast extent of a *Tonstein* bed. Moreover, it comes as no surprise that the material is hypersensitive to kaolinization. It is a question of silicates and glass of volcanic origin that are very easily attacked by natural waters. It would be helpful, however, to know that the necessary minerals are present and, if possible, to specify the source of the volcanism.

Can we suppose that not all *Tonstein* have the same origin? Future research will clarify this question. The point in common should be, in any case, the abrupt, instantaneous, and partial kaolinitic evolution that gives them their mineralogical character.

VII. — ALKALINE LAKES IN PLAINS

The siderolithic facies is a fluvio-lacustrine piedmont facies of acid character. The red sandstone series are fluvio-lacustrine piedmont complexes involving water that is still fresh but incapable of acting on the silicates being transported. Coal measures are lacustrine deposits or fluvio-lacustrine freshwater deposits of the continental margin, wherein the water becomes acid under the influence of swamp vegetation.

Let us look now at a series of fluvio-lacustrine deposits of alkaline reaction. From a geographic point of view we shall find all the intermediate stages between a small lake that is nothing more than a temporary enlargement of a stream and those great continental aquatic expanses. From a chemical point of view we shall deal with alkaline waters of zero or uncertain effect, then with more and more accentuated chemical composition, ending up with chemical lacustrine limestone deposition.

I envisaged (MILLOT, 1949) argillaceous sedimentation in calcareous lakes. If one considers only those that contain no attapulgite, analysis shows 100 % micaceous minerals, with illite at the head of the list. This was the case for samples collected from the lacustrine limestone series of the Paris Basin and of the Limagne graben. Only one sample, that from the Aquitanian lake of Donnery in the Loiret, showed 30 % kaolinite and 70 % montmorillonite.

In 1949 I arrived at the idea that the alkaline environment formed the particles and guaranteed the neoformation of three-layer minerals. I saw in the study of marny or calcareous lacustrine deposits a confirmation of that idea. Let us examine the present state of that question.

1° Clays of present-day lakes

Even today argillaceous sedimentation in present-day lakes is known in only a rather fragmentary manner.

CUTHBERT (1944) studied the clay minerals in Lake Erie in North America. Illite is the dominant mineral, reflecting the shales, principally Devonian, of the drainage basin, as well as the soils and the glacial deposits that cover the landscape.

The playas of the Mohave Desert (California, U.S.A.) have been studied by DROSTE (1959). He showed that there is a correspondence in the case of each playa between the clays deposited and those brought in from the surrounding landscape.

Great Salt Lake (Utah, U.S.A.) has been studied by GRIM, KULBICKI and CAROZZI (1960). The clays that it contains are formed principally of rather poorly organized montmorillonite, well crystallized illite which is certainly detrital, and a minor fraction of kaolinite, likewise certainly detrital. The clays of the terrace that immediately dominates the lake are different; they show a degraded chlorite, and the authors raise the question whether the montmorillonite of the Salt Lake is not a degraded chlorite. The argillaceous fraction of deposits of the Chott ech Chergui in Algeria has been studied (MILLOT, 1949). The spring water that feeds the chott is rather fresh (dry residue = 1.56 %), but evaporation concentrated the salts. The clay fraction is composed of 90 % illite. Nothing is known about its origin.

ROLFE (1957) studied Lake Mead at the confluence of the Virgin and the Colorado Rivers downstream from the Grand Canyon. The argillaceous fraction of the most recent deposits is different above and below the confluence, in the Virgin basin and the Boulder basin, in particular in the amount of montmorillonite and the size of the particles. This is due to the flocculating power and the velocity of flow. Everything indicates that the clays are inherited directly from the landscape upstream.

There is too little of this kind of information; it needs to be augmented. However, one can summarize that which is most probable:

— As long as a lake is nothing more than an enlargement of a stream, its waters, even if calcareous, remain fresh, and there is no reason for the clay particles to be different from those transported by the stream. The clay of these lakes will be, therefore, inherited from the rocks and soils of the drainage basin.

— Whenever a lake expands, it concentrates its solutions by evaporation; the initial stock of clay minerals is inherited, but the problem of transformations arises, just as it does in other analogous environments. This question remains to be studied and to be demonstrated.

2° Oligocene lacustrine clays of France

We have followed, with KULBICKI, the evolution of the *Sidérolitique* in the Aquitaine basin. This sedimentation, dominated by kaolinite, spread quite far to the south, and it was succeeded by that of molasse and of lakes. It was VATAN (1948) who showed that the material of the molasse came from the Pyrenees, whereas that of the *Sidérolitique* came from the Massif Central. Thus a considerable change was being established in Aquitaine during this sedimentation. The molasse coming from the south succeeded the *Sidérolitique* in the north. The map in figure 26 shows, for the Stampian, the distribution of molasse and its dependent calcareous lakes. In fact, north of the zone of molasse sedimentation, a large number of small lakes were established during the Stampian.

KULBICKI (1953) has studied the sedimentation and the clays of these lakes that rest upon the *Sidérolitique*. The sedimentation was much more calm and, away from the margins, consisted only of fine-grained products: clay, marne, limestone, and millstone grit (*meulière*).

The dominant clay mineral in these deposits is illite, and is essentially all detrital illite. This is shown by the fact that all the products of the destruction of the Pyrenees are made of illite and, in particular, that the grain size of the illite becomes finer as one moves away from those mountains. KULBICKI also gave some thought to phenomena of transformation. Not only did he observe large and well-formed crystals, which could only be neoformed, but he also thought that the illite which had been degraded by weathering and transportation had been able to reconstitute itself during slow sedimentation in a lake in which all the necessary ions were present. He thus envisaged the modification and perfecting of the illite by the lacustrine milieu.

Attapulgite has been identified at the contact with limestone beds in certain lakes. KULBICKI thought that it was the result of neoformation. In summary, KULBICKI came to the following conclusion: "The environmental conditions determined in the lake by a massive introduction of illite from the Pyrenees brought about partial solution of those minerals. The liberated ions were recombined, in most cases onto a detrital micaceous lattice, forming illite once again; locally they were able to form some attapulgite."

An echo of the phenomena seen in the northern part of the Aquitaine basin is presented by the lacustrine basin of Salins, previously examined by KULBICKI and then studied in detail by JUNG (1954). This basin is located in the Cantal between the Dordogne and the Cère rivers, thus in the southwestern part of the Massif Central. JUNG describes a dark green argillaceous formation with intercalations of magnesian limestones and gypsum which succeeds the siderolithic sedimentation that was, first, kaolinitic and later mixed, with both kaolinite and illite present. The clay fraction, which was formerly quarried, is composed of an extremely pure illite. However, chemical analysis showed that this mineral has a chemical composition intermediate to illite and glauconite. One could call it ferric illite or aluminous glauconite; this was the first discovery of this intermediate clay mineral that was reported again by KELLER (1958) and by NICOLAS (1961). JUNG interpreted this extremely fine-grained and very pure clay mineral to be a neoformation within an alkaline environment.

In the Limagne graben, the Stampian is represented by an alternation of limestones and green marnes. Two samples have been studied (MILLOT, 1949). The environment is lacustrine, as witnessed by shells of the mollusks *Cypris, Helix, Limnea* and *Planorbis* and the presence of nests of fossil turtles. Certain horizons are formed of sheets of chert. The clay minerals extracted from these sediments are pure illite, and this has also been found in the argillaceous fraction of concretionnary limestones of the Limagne by RAYNAL (1953).

GABIS (1958, 1959) studied the lacustrine sedimentary series of the Oligocene basin of Le Puy-en-Velay. Following a history comparable to that of the basin of Salins, the top of the series shows gray clays with limestone horizons wherein illite is the only identifiable mineral. GABIS considers the clays to be the result of reworking and flotation of weathering clays from the neighboring landscape.

In general, the Late Eocene and Oligocene sedimentation in the basins of subsidence in the Massif Central, as well as around its periphery, shows a kind of clay totally different from that of the Eocene. The siderolithic regime is followed by a regime of calcareous or marny lakes in which illite is the dominant mineral. These deposits are not all synchronous, but their facies and the nature of the clays are similar.

Genetic interpretation is difficult, and each author views it in a different way:

— In favor of neoformation there is the general character of their extent across all kinds of landscapes, the purity of the illite, and the peculiar composition of that of Salins, for there is nothing that leads us to believe today that such aluminous glauconite could form as a result of rock weathering.

— In favor of inheritance, there is no dearth of arguments. It would be a question of inheritance of weathering products of mountain ranges under a totally different climate. Moreover, the evolution of grain size geographically, according to the outline of KULBICKI, as well as sorting in time, evoked by GABIS, are interesting arguments.

— It is nevertheless surprising that detrital minerals would be so pure. We are not accustomed to such purity in the detrital sedimentation of present-day streams where it is always a question of a very complex mixture of mica, chlorite and mixed layers. For this reason, perhaps, it will be necessary to call upon a mixed mechanism, that is, inheritance of diverse micaceous clay minerals, followed by transformations that orient all of them toward the illite structure in the course of sedimentation in calcareous lakes. Let us say that the problem is interesting, but it is not resolved.

3° *Lakes with attapulgite and sepiolite*

In 1949 I summarized our knowledge on this subject, but I did not know at that time of marine attapulgite. Here is a list of occurrences of lacustrine attapulgite and sepiolite that I know of.

— *Mormoiron* (Vaucluse). — Early Ludian. The environment is continental. The series enclosing the horizons of bleaching clay is marny, but it contains lacustrine limestone beds with *Limnea* as well as chert. The clay fraction is composed of attapulgite and montmorillonite.

— *Cormeilles-en-Parisis* (Seine-et-Oise). — Ludian. Marny white limestone with spearhead gypsum. The clay fraction contains attapulgite. It was sampled between the first and second gypsum masses, and the presence of crystals within the chalky and marny limestone studied indicates hypersalinity.

— *Salinelles and Sommières* (Gard). — Stampian. The deposits had already been studied by LONGCHAMBON and MOURGUES (1927). The beds with sepiolite, or fuller's earth, alternate with lacustrine limestones with *Limnea* and chert. It is, therefore, a question of lacustrine deposits.

— *Herbeville* (Seine-et-Oise). — Late Lutetian. The environment is continental. Marne is interstratified between large siliceous blocks and a calcareous limestone, both characterized by freshwater fossils. The clay fraction is composed of attapulgite.

Some older and more recently known examples can be added.

BRONGNIART (1822) gave us an admirable little memoir on "magnesite" of the Paris Basin and elsewhere. "Magnesite" was the former name for sepiolite. BRONGNIART described for us the deposits at Coulomnier, at Saint-Ouen, at Montmartre, at Sommières (which had already been described by DE SERRE in 1818) and at Madrid. The sedimentary features in common with chert, chalcedony and opal were emphasized: "Such a remarkable constancy in the association of silica and magnesia, two substances that have no chemical analogy between them, is reported in order to attract the attention of geologists, and perhaps this will contribute to our search for the origin of these deposits."

LACROIX (1893-1913) described the "magnesite" deposits of the Paris Basin. He reported it to be common "in the Eocene and Oligocene lacustrine limestones of the Paris Basin, where it is often associated with impure opal (*ménilite*)."

BRADLEY (1929) described the lacustrine deposits of the Green River Formation in Utah and Colorado, U.S.A. This Eocene deposit contains sepiolite accompanied by glauberite and magnesite, and impregnated with organic matter.

J. DE LAPPARENT (1937) mentions, along with the sediments at Attapulgus, Georgia, U.S.A., of which I do not know the conditions of deposition, those of Mormoiron, cited above, and those of Lagny in the Seine-et-Marne, which are without doubt similar in age to the attapulgite earths of the lacustrine marny limestones in the vicinity of Sézanne (RIVIÈRE, 1946).

URBAIN (1951) identified attapulgite and sepiolite in lacustrine marnes of Eocene and Oligocene age in Paris and at Cormeilles-en-Parisis.

KULBICKI (1953) demonstrated the presence of beds of attapulgite at the contact with limestone beds of the Stampian lakes of Aquitaine.

KLINGEBIEL (1961) identified attapulgite in the dolomitic zones of the Late Eocene lacustrine molasse of Aquitaine at Issigeac and at Sainte-Foix-la-Grande.

GRIM (1953) reported the presence of the same minerals in the deposits of present-day arid lakes in Australia. The reports of PERELMAN (1950) on desertic relics of central Asia are perhaps to be interpreted in the same manner.

HEYSTECK and SCHMIDT (1954) described sedimentary masses within the lacustrine dolomites of the Transvaal.

ROGERS, MARTIN and NORRISH (1954) found very pure masses of attapulgite in dolomites of the Tertiary lakes of Ipswich (Queensland, Australia).

BONYTHON (1956) described the presence of palygorskite in the hypersaline sediments of Lake Eyre in Australia.

CAILLÈRE and ROUAIX (1958) identified a very aluminous attapulgite in a lacustrine zone within the Intercalated Continental Series of the northern Sudanese Sahara near Taguenout Hagueret. On the other hand, the milieu of formation of two samples of fibrous, yet sedimentary palygorskite brought by MONOD from Taodeni cannot be determined (CAILLÈRE, 1939).

KUBLER (1959) studied the freshwater formations of Le Locle, Switzerland, which are deposits of the Oeningian Stage of the Miocene Series. Attapulgite appears as intercalations of lacustrine magnesian chalks. There is an antagonism here between attapulgite and kaolinite in the sediments.

All these reports are in agreement; attapulgite and sepiolite occur in certain fossil and modern lakes that have the following characters in common: presence of carbonates, common occurrence of chert, and, facultatively, hypersalinity. In one word, it is a question of alkaline chemical sedimentation.

The origin of these minerals can be interpreted only as neoformation:

— Certainly attapulgite does not exist in soils.

— If present in bedrocks, they are rapidly weathered therein.

— If they cannot come from an inheritance, they must be neoformed.

Marine attapulgite and sepiolite had not yet been discovered in 1949; thus, I thought that these minerals were characteristic of lakes with alkaline chemical sedimentation. But this was an error. The problem of their genesis will, therefore, be taken up again after the study of marine sediments. Let us say, in brief, that it is a question of neoformation of silicates that are the products of lessivage of continental areas under laterizing or strongly hydrolyzing climates.

Conclusion

Thus, alkaline lakes have shown us, once again, a gradation.

— In the first place, *detrital sedimentation* in which the clays deposited are copies of the argillaceous stock of the rocks, the weathering, and the soils of the drainage basin.

— Next, *chemical sedimentation*. The effect of this chemical treatment is, in the early stages, uncertain and open to discussion, giving rise to *transformations* of the inherited clays that are difficult to grasp and to demonstrate. Then, in lakes with calcareous and siliceous alkaline chemical sedimentation, *indiscutable argillaceous neoformations* take place.

VIII. — ANALCIMOLITES

Analcimolites in sedimentary formations are rare rocks. Moreover, they pose a complicated problem because, even though certain deposits are manifestly the products of the evolution of pyroclastic material, others are, perhaps, of sedimentary origin and would hold a very great interest for us. We have had the good fortune to be able to study in great detail a formation of analcimolites discovered by JOULIA in the Niger. This case history will be described before presenting hypotheses on this subject and furnishing at the same time the principal documentation.

Analcimolites of the Intercalated Continental Series of the Niger (Central Sahara) (JOULIAN, BONIFAS, CAMEZ, MILLOT and WEIL, 1958, 1959).

— THE DEPOSIT. — JOULIA discovered a curious formation in the course of his surveys of the Intercalated Continental Series of the Niger basin. We shall present only the résumé of his study here.

This formation, dated Early Cretaceous, contains in its lower division (Agadès sandstone: 250 m) some sandstone with silicified wood (Tchirezrine beds) interstratified with massive, red, glomerular and enigmatic rocks. Field study shows that this rock is a stratified sedimentary deposit that alternates with, or occurs as lenses within continental sandstones.

The samples were first examined for their clay minerals. However, the X-ray spectra showed an improbable result: analcime. In fact, all the diagnostic methods confirmed this identification. This facies, belonging to the Intercalated Continental Series south of the Aïr massif, is formed essentially of analcime. We have proposed the name "glomerular analcimolite".

LABORATORY STUDIES. — To the naked eye this massive, dark red rock, spotted with white glomerules can appear sandy. We shall thus distinguish between analcimolites and sandy analcimolites.

Under the microscope, as shown in a detailed study by WEIL, the glomerules, which are neither radiating nor oolitic, are seen to be made of analcime; they are 1 to 2 mm in diameter. The matrix is made up of microglomerules, 0.01 to 0.1 mm in diameter, of the same mineral. The clastic minerals are angular; they include quartz, feldspar, and rare, rubified biotite, but no other colored minerals.

In X-ray analysis, the fraction finer than $2\,\mu$ shows only infinitesimal traces of clay minerals, either illite or montmorillonite or a mixture of the two, but, on the other hand, the characteristic reflections of analcime are present and always dominant.

Chemical analysis yields a composition that can be represented by the following formula:

$$4\,SiO_2,\ Al_2O_3,\ 0.8\,Na_2O,\ 2.2\,H_2O$$

which is very close to the classic formula for analcime:

$$4\,SiO_2,\ Al_2O_3,\ Na_2O,\ 2\,H_2O\,.$$

Density measurements, as well as thermal analysis, confirm the diagnosis.

Discussion of the origin of analcimolites. — The best method will be to follow the historical development.

Already a century ago formations with analcime had been described among sediments apparently having no connection whatsoever with volcanic material (VON SEEBACH, 1862, and KLOSS, 1899, cited by BRADLEY, 1929).

In the United States, ROSS (1928, 1941) described beds with analcimolites from the playas of Arizona. These beds are made up exclusively of automorphic analcime crystallized in the form of small trapezohedra surrounded by a thin film of glauconite. In his first study, ROSS proposed two possible explanations: either an evolution of volcanic products in contact with the waters of the playa or a reaction between the saline water and the clays of the playa muds.

BRADLEY (1929), next, described the analcimolite deposits in the Green River formation of Utah, Colorado and Wyoming. The analcime here forms automorphic crystals, more or less connected, that developed in a volcanic tuff; the volcanic origin is evident here as well as for the deposit discovered in Iceland (TYRRELL and PEACOCK, 1926) and the montmorillonite and analcime earths in California (KERR and CAMERON, 1936).

ROSS (1941), in his second study, demonstrated the association of analcimolite and volcanic ash and opted for the hypothesis of transformation of pyroclastic material.

ERMOLOVA (1955) studied the sandy Oligocene and Miocene formations of the Caucasus. Moreover, she lists the Russian literature. In the Russian formations with analcime and mordenite, one frequently finds the trace of volcanic material. In addition to zeolites, one finds many neoformed minerals: calcite, dolomite, chlorite, gypsum, montmorillonite, etc. The origin by means of evolution of volcanic products is evident.

KELLER (1952) studied in great detail a formation comparable to the analcimolites of the Niger. This is the Triassic Popo Agie beds in western Wyoming (U.S.A.). His description could apply word for word to the Saharan deposit, and the descriptions of the glomerules are identical. After a detailed discussion, KELLER, having discovered no trace at all of volcanic material, accepts without enthusiasm the hypothesis of sedimentary origin through the action of sodic saline solutions upon lacustrine clays.

NORIN (1953) and then MULLER (1961) were able to demonstrate the neoformation of analcime from volcanic glass within the sediments of the Bay of Naples; zeolites had already been collected there by *Challenger* expeditions (MURRAY and RENARD, 1891).

A second African deposit has just recently been discovered by VANDERSTAPPEN and VERBEEK (1959). The analcimolites, very commonly glomerular, are present in stratiform horizons, certain of which reach a thickness of several tens of meters. They are interstratified in an ensemble of continental facies of Mesozoic age, from Late Jurassic to Early Cretaceous. The total thickness can reach 700 m, and the outcrops recognized so far can be as much as 400 km apart. In the Congo, stratigraphic conditions allow the authors to exclude a volcanic origin and to attribute to the deposits a sedimentary origin. VERNET (1961) studied independently some of the samples from this Congolese series and, in consideration of the fact that the geology of these vast regions is not completely known and that montmorillonite is associated with these analcimolites, he thinks that a volcanic origin is not impossible.

Let us report that, within the Permian red sandstones of the Lodève basin, a young Dutch scientist, KRUSEMAN (1962), has discovered the common presence of analcime in isolated crystals or as nests within the sediment. This analcime has been identified likewise, incidently, by MAUREL (1962) who, himself, demonstrated the abundance of neoformed albite in the red sandstones of Saint-Affrique and of Lodève.

In the study by DEFFEYES (1959) one finds a very important bibliography on zeolites in sediments and the mechanisms of evolution of volcanic glass.

Here then is the balance sheet of these studies on analcimolites.

In a great number of cases the volcanic origin can be demonstrated either by the association with or the transition to volcanic tuffs or glass, or by the observation of residues of volcanic minerals in the deposit.

In other cases, such as those of the older German authors of the xixth century, that of Wyoming studied by KELLER, and that of our analcimolites in the Niger, no trace of volcanic minerals or glass has been able to be shown. Like KELLER, we are therefore led to imagine a direct sedimentary process that takes place in the large basins with playas of the Intercalated Continental Series of Africa. It is possible that the Congolese deposits are of the same kind. It is not without significance to note that the paleogeography and the age of formations in the Sahara and in the Congo are similar.

It is not possible to draw conclusions from negative arguments. We think that work now in progress will throw some light on the genesis of the Niger analcimolites.

The epigenesis of pyroclastic products in a sedimentary milieu gives us a new example of the great sensitivity of volcanic silicates in an aquatic environment.

This sedimentary neoformation would be of great interest, but it must be demonstrated.

CONCLUSION

Clays of lacustrine deposits show us already a great variety of genetic possibilities:

— *Mechanical inheritance* either from the substance of the rock itself, or from its weathered products, or from neoformed clay minerals in the soil.
— Possible *transformation* of clay minerals in certain kinds of estuaries or certain kinds of lakes.
— Sedimentary *neoformation* in water bodies dominated by chemical activity.
— *Pedogenetic or diagenetic evolution*, after deposition, very few cases of which have yet been recognized.

CHAPTER VII

CLAYS OF MARINE SEDIMENTS

The great inertia and the monotony of sea water could give the impression that marine clays show hardly any variety. This is certainly not the case for two principal reasons. In the first place, marine clay sedimentation is not only the reflection of that which takes place in the water, but also of that which takes place within the muddy sediment. It is well known that the muddy milieu is the locus of important chemical and biochemical reactions. Secondly, marine clay sedimentation has taken place throughout all of geologic time, and the relations of oceans to continents were not always the same as now. In particular, in the periods of great transgressions, epicontinental chemical sedimentation had a rather different character from the predominantly mechanical sedimentation that we observe today.

The study of marine sediments will, therefore, be made by means of studying successive problems. After an exposé on research on present-day sedimentation, we shall pass from the detrital to the chemical, from general cases to particular ones, throughout the following paragraphs:

I. *Present-day marine sediments.*

II. *The detrital series: from sandstones to shales and argillites.*

III. *The marno-carbonate series: from argillites to limestones.*

IV. *Alkaline chemical sedimentation.*

V. *Glauconitic sediments.*

VI. *Clay minerals of iron ores.*

VII. *Hypersaline facies.*

I. — PRESENT-DAY MARINE SEDIMENTS

Here we arrive in the marine domain. Marine sediment is supplied by streams and coastal rivers. We have abandoned in the preceding chapter the study of the particles transported by the rivers into the estuaries. We have seen that rivers transport and deposit clay minerals inherited directly from their drainage basin. In the estuaries, however, a problem arises. Modifications of the clay minerals are envisaged by certain authors, in particular, GRIFFIN and INGRAM (1955), NELSON (1958), and POWERS (1959), relative to the estuaries of the Neuse, Rappahannock and James rivers on the Atlantic coast of North Carolina, Virginia and Maryland. On the other hand, TAGGART and

KAISER (1960) do not see any modification in the estuaries of the Mississippi and the Red River.

We must now study the succession where sediments debouch into the marine domain. There are three questions:

— Is there or is there not a change in clay minerals when they pass from the fluvial to the marine environment?

— In addition, can an evolution be observed in the sediments after burial?

— And, if there is a change, what can be the reasons for it?

1° *The facts*

The early works. — We are indebted to CORRENS (1937, 1939) for the first documentation on clays deposited in the ocean. In the equatorial Atlantic Ocean kaolinite is the dominant clay mineral. In the North Atlantic it is mica, and in the vicinity of Cape Verde, where volcanic phenomena have been important, it is montmorillonite. Immediately it was possible to deduce from this information, as DIETZ (1941) did, that the relative abundance of kaolinite was due to the proximity of tropical regions and that of montmorillonite and of illite were related to temperate and polar regions. The marine clays have been derived from continental supplies. This point is evident today and is unanimously recognized.

Since 1941, however, and the oceanographic research of DIETZ, the question of transformations has arisen. DIETZ studied the clay minerals of 39 samples gathered by the *Challenger*. In 23 of them illite was dominant. In the others, it was kaolinite or montmorillonite, but these samples were taken rather close to coasts. DIETZ thought that the illite was derived from montmorillonite by *fixation* of potassium ions from sea water.

Here then the problem has arisen. The clays of marine sediments are of continental origin, but, from the time of the very first studies, there is talk of evolution of the clays during the passage from fresh to marine water.

Discovery of the modification of argillaceous composition in the marine environment. — The first series of studies concerning the Pacific off California, the Gulf of Mexico, and the Atlantic coast of the south-eastern United States described appreciable changes in various ways.

GRIM, DIETZ, and BRADLEY (1949) identified, using improved methods, mixtures of clay minerals in sediments of the Pacific and of the coast and the Gulf of California. They found, in general, illite dominant, then kaolinite, chlorite and a small amount of montmorillonite. Samples taken near the coast were richer in kaolinite and montmorillonite, and the illite was degraded. The authors supposed, therefore, that there was a transformation of the minerals into illite and chlorite under the influence of K and Mg ions in the saline ocean water.

GRIM and JOHNS (1954) studied the clayey sediments of the margin of the Gulf of Mexico in the region of Rockport, Texas. The continental detritus transported by the rivers is composed essentially of montmorillonite with traces of chlorite and illite. However, as the sediments carried by the Guadelupe River move toward the sea, the montmorillonite is replaced by illite and chlorite that become equal in proportion to it. These changes are shown in figure 28.

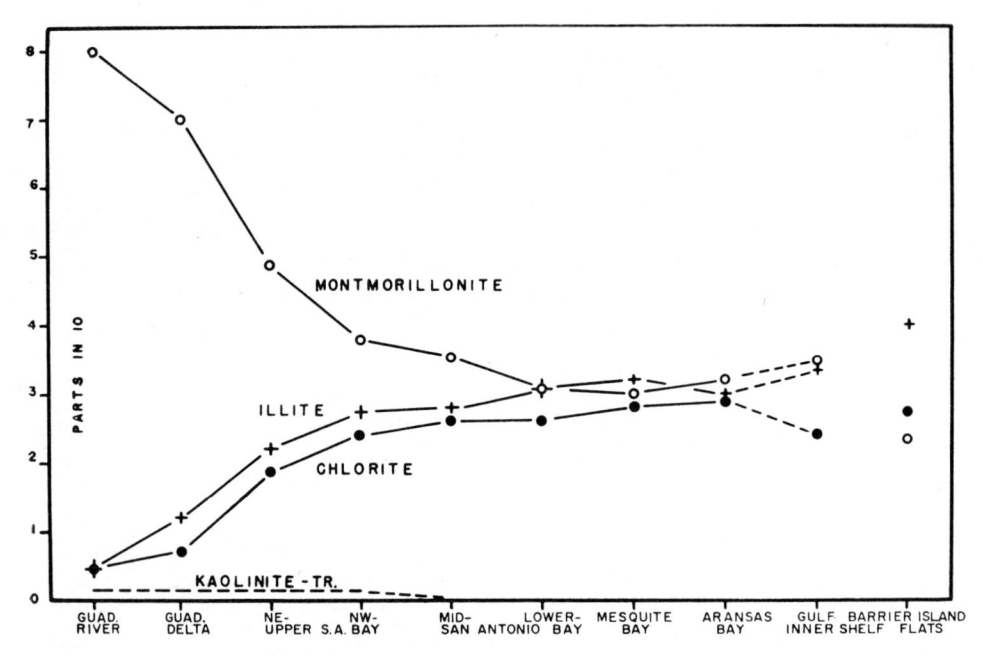

FIG. 28. — *Argillaceous sedimentation in the Rockport area, Texas, U.S.A., in the Gulf of Mexico.* Relation of clay mineral composition to geographical divisions, from the river toward the sea (after GRIM and JOHNS, 1954).

JOHNS and GRIM (1958) studied the sediments of the great Mississippi delta. Montmorillonite is the dominant material supplied, accompanied by degraded illite and chlorite. All this originates in the Mississippi drainage basin, which is rich in montmorillonite owing to the presence of volcanic or sedimentary rocks and loess, as well as semi-arid soils. Upon arrival in saline water, a fraction of the montmorillonite is transformed into illite; this is a smaller fraction than in California and in Texas.

MILNE and EARLY (1958) studied the clay minerals in that part of the Gulf of Mexico situated between the Mississippi delta and the Mobile River. They noted the modifications of minerals inherited from the Mississippi and other rivers upon arrival in the sea. It is, above all, a question of increase in illite, and that only where sedimentation is slow. They explain this by means of the time required for the transformation of montmorillonite.

PINSAK and MURRAY (1960) have studied the entire Gulf of Mexico by means of oceanographic soundings. They show that the dominant factor in the distribution of clay minerals is the continental supply. The most important mineral is montmorillonite, then illite, chlorite, kaolinite and the mixed layers. However, transformations are similarly notable. They are abundant in the center of the Gulf and in the abyssal zones. They are, moreover, complementary in quantity with montmorillonite, that which suggests that it is the latter which is transformed during its transportation. Likewise, a swelling chlorite appears, even in places where sedimentation is rapid.

POWERS (1957) studied Chesapeake Bay in Maryland, where the dominant mineral is illite, in some cases well formed, in others altered. Mixed layers and then vermiculite and chlorite are formed within the estuary. The stability of these minerals upon heating increases with the increasing salinity of the environment. There appears to be, therefore, an evolution of the disordered micaceous lattices inherited from the continent toward chlorite (Chap. X, figs. 65 and 66).

Modifications weak or absent in the marine environment. — Other studies note no mineralogical changes, or at the very most an increase in crystallinity.

MURRAY and SAYYAB (1955) have studied the recent sediments of the North Atlantic along the coast of North Carolina. The clay minerals present are principally illite, chlorite and mixed layers of these minerals with montmorillonite. The clay minerals do not change with distance off shore or with depth, but the crystallinity does improve. In cores, likewise, one sees the crystallinity improve with the depth of the sample. Transformations, therefore, do not affect the nature of the minerals, but they effect an increase in crystallinity, which corresponds to a decrease in the mixed layers and a regularization of the lattices.

FRIDMAN (1953) studied the coastal muds of the Vendée in France. The clay minerals are illite, kaolinite and chlorite; they show no differences from those which come from the continent, but one can observe phenomena of sorting. Likewise, VERNHET (1956) showed that along the coast of Monaco the marine clays, which are rich in illite, mimic the continental supplies. Such reports led RIVIÈRE and VISSE (1954) to think that marine clays are exclusively detrital.

Epigenesis of volcanic material. — Finally, oceanographic research has been able to demonstrate the neoformation of clay minerals from volcanic material, either minerals or ash. The epigenesis of volcanic glass into montmorillonite has been recognized for a long time, but other, more curious neoformations have been discovered. NORIN (1953) discovered, in the sediments of the Tyrrhenian Sea, an alteration of biotite by exfoliation, ending up with an accordion-like appearance. Altered in this way, the mineral gives the X-ray reflections of illite and chlorite. Similarly, MULLER (1961) showed, in the Bay of Naples, the formation of accordions not only of illite, but also of kaolinite, analcime, chalcedony, etc. Here it is, thus, a question of transformation, but transformation of unstable volcanic material, and this has an entirely different significance for us. Let us note, moreover, that GRIM and VERNHET (1961) have been able to demonstrate further evolution of montmorillonite of volcanic origin towards mica or chlorite by the intermediary of mixed layers. Here once again we find transformations.

Evolution after burial. — In 1938, CORRENS studied 86-cm-long cores taken from the equatorial Atlantic. Neither neoformation of mica, nor transformation of halloysite into kaolinite was visible. Similarly, LOCHER (1952) demonstrated that the kaolinite-illite-montmorillonite assemblage present in a core taken 1 000 km off Brazil did not vary at all through 9 and 14 cm thickness. Likewise, DEBYSER (1959) studied cores of 10 to 20 m taken in the Baltic Sea, in the Bay of Arcachon, and in the Abidjan lagoon. He showed that the clay minerals are the same throughout the cores.

However, these measurements were made through only a few meters thickness. If, by means of modern techniques, one studies the evolution of the sediments through several hundreds or several thousands of meters, the perspective changes.

MILNE and EARLY (1958) have studied a thickness of 1 000 feet of Mississippi River sediments in Scott Bay in Louisiana. Comparison with the present-day Mississippi muds shows that the only change was a slight increase in illite. This could be attributed to a neogenesis of mica by adsorption of K ions by montmorillonite, but the phenomenon is neither large scale nor certain. Oligocene and Miocene sediments of the Gulf Coast have been studied from samples taken from drill holes at Galveston, Texas. The clay mineral assemblages are identical to those that come from the east of the present drainage basin of the Mississippi, and the authors see no reason to think that modifications have occurred. On the contrary, Late Miocene sediments of Los Angeles, California, do show modifications. Samples have been studied from 9 000 feet of section, and the clay fraction of the sands, fine sands, and shales have been compared. The shales contain a mixture rich in montmorillonite, and the sands and fine sands show an assemblage rich in chlorite. MILNE and EARLY see in these observations the trace of diagenesis under the influence of solutions that migrate out of the porous sediments but are imprisoned in the clayey sediments. These results are shown in figure 29.

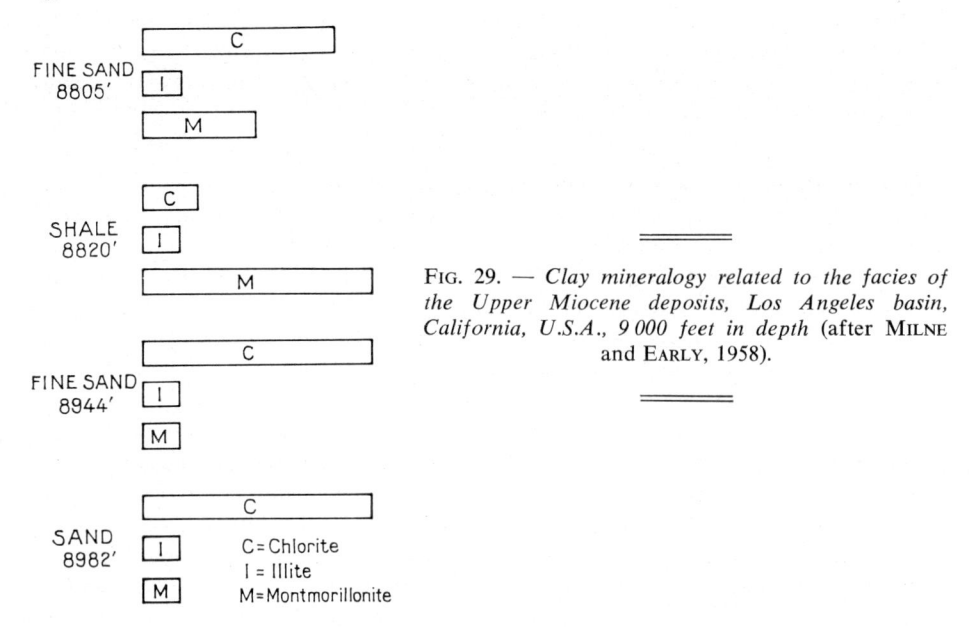

FIG. 29. — *Clay mineralogy related to the facies of the Upper Miocene deposits, Los Angeles basin, California, U.S.A., 9 000 feet in depth* (after MILNE and EARLY, 1958).

BURST (1959) and POWERS (1959) have studied the evolution of clay minerals in the Tertiary series of the Gulf Coast of the southeastern United States. Montmorillonite decreases and then disappears as a function of depth. Its place is taken by illite-montmorillonite mixed layers and by illite itself. This corresponds to a decrease in the capacity for swelling of those minerals with a variable basal spacing, as is also seen in the increasing stability of chlorite upon heating. Thus, the peak for montmorillonite treated with glycol occurs exactly at 17 Å at the 7 000-foot depth, but at 16-17 Å at 9 000 feet. Beyond that, the peak corresponds to the illite-montmorillonite mixed layers, whose capacity for swelling is increasing.

WEAVER (1959) reviewed the modifications as a function of depth observed in borings in recent, still unconsolidated deposits in California and on the Gulf Coast. He reports that montmorillonite decreases in favor of illite-montmorillonite mixed layers and then illite. He thinks that these variations can be explained by changes in the source of the sediment or in the distance from the shore. It is obviously not possible to know whether the supply of clayey sediments remained constant during such long periods as the Miocene, Pliocene and Quaternary. However, it is curious that the modifications of the clay minerals during burial always take place in the same direction, namely that of the construction of more and more stable minerals from open minerals. But here we have come little by little from the domain of modern sediments into that of the diagenesis of sediments; we shall take up this problem again later on.

2° *Interpretation*

Based on the very great number of measurements made by the most modern methods, the majority of authors thus show evidence of changes in the mineralogical composition of clays as they pass from the continental to the marine environment. The interpretation of the changes is not easy and has given rise to numerous works and to a great number of discussions among specialists.

We shall leave aside for the moment the neoformation or epigenesis of volcanic material, on which every one is in agreement.

We shall admit likewise, at the start, that there is unanimity on the continental origin of marine clays, which in each ocean or in each part of an ocean reflect, in general, the nature of the rocks, the weathering, and the soils of the neighboring continents. This has been understood since the beginning of observations by the first workers (CORRENS, 1937, 1939; DIETZ, 1941). All oceanographic studies have since confirmed it. Recent works emphasize it. HEEZEN, NESTEROFF, and SABATIER (1960) and NESTEROFF and SABATIER (1961) studied samples taken from three oceanographic domains and measured the compositions. In the North Atlantic, south of Newfoundland, illite made up 80 % of the clay fraction along with a little chlorite but no mixed layers. In the central Atlantic, east of Brazil, montmorillonite, illite and kaolinite occur in equal proportions. In the Pacific, SSE of Tahiti, montmorillonite is dominant along with a little illite and phillipsite. These results reflect the continental supplies which come, respectively, from the northern United States and Canada with cool climate weathering, from humid tropical lands with lateritic weathering, and from volcanic regions such as the Pacific.

If, therefore, there is no doubt that marine clays are the echo of clays inherited from rivers, the majority of authors show us that the echo is distorted. Why?

We have at our disposal at the present time two kinds of explanations: transformations and differential sedimentation.

Transformations. — Reading of the works that have been summarized in this chapter shows that this is the explanation given by the majority of authors, and in particular by that great specialist on clays, GRIM, as well as by his collaborators. For the sake of clarity we shall not consider the scheme of each author in detail; the essentials of their arguments can be presented in the following way. We have seen that clays from weathering and of soils commonly are open minerals. As a matter of fact, the fractionation, lessivage, and initial hydrolysis act first on the interlayer ions. And in the weathering process we see a great number of "yawning minerals" appear at the expense of the two initial, primary minerals. These are degraded illite, vermiculite, all the mixed layers of illite and chlorite with montmorillonite and vermiculite, and finally montmorillonite itself. The evacuated ions are carried away in ground water, rivers, and the sea.

These "yawning minerals" then find themselves in the sea in the presence of the ions that had been taken away from them. It is possible to understand, and we shall see that this has been done experimentally, that the ions in sea water, and in particular K and Mg, occupy once again their places between the layers, reactivate but in the opposite sense the mechanisms that had liberated them, and reconstruct in reverse order the various mixed layers from montmorillonite and chlorite and illite from the mixed layers.

We shall take up the subject of the mechanisms envisaged in Chapter X again in order to examine them as a group. The arguments and the experiments will be presented there, but, for the time being, we have at our disposal the results of measurements and oceanographic maps. The bundle of accumulated facts is rather impressive.

Differential sedimentation. — The second explanation is differential sedimentation. This is the term that seems to be the most general, and that is preferable to the terms "differential or preferential flocculation", that evoke a more precise mechanism. The phenomenon can be defined in this way: the stability of particles in suspension depends on the nature of the particle, its size, and the products contained in the suspension.

As early as 1949 KELLER and FOLEY sampled mud-laden water of the Missouri River. They studied the dispersion of particles in the river water, in a 100 N solution of sodium oxalate, and in sea water; the results were strinkingly different. The fraction smaller than 4 μ constituted 62.6 % of the material in suspension in river water, 71.4 % in the deflocculating oxalate solution, and 16.4 % in the flocculating saline water.

RIVIÈRE and VERNHET (1951) have studied the speed of flocculation of illite, montmorillonite and kaolinite in the presence of trace amounts of humic matter. Kaolinite acquired a great resistance to the flocculation provoked by bivalent ions and notably by those of sea water.

WHITEHOUSE, JEFFREY and DEBRECHT (1960) presented in 1958 the voluminous summary of their works on differential settling. They studied the settling velocity in marine water and in brackish water under controlled conditions of temperature and pH. The influence of organic matter was also examined. The behavior of illite, montmorillonite, kaolinite, mixed layers and vermiculite were quantified. Let us say only that in calm sea water at 18 $\%_0$ salinity and 26 °C the settling velocity of illite, kaolinite and montmorillonite are, respectively, 15.8, 11.8 and 1.3 m/day. Chlorite settles slightly faster than kaolinite, and vermiculite slightly faster than illite. These tests were made on 1 500 samples of clay from all kinds of sources, and they constitute a gold mine of information. The role of organic matter was studied; hydrocarbons increase the settling velocity of montmorillonite, but proteins decrease it. Humic matter and proteins stabilize kaolinite considerably. Illite is not affected.

WEAVER (1957) found much support in these studies for this attack on the transformation mechanism. He thought that the latter was of little importance. He pointed out, in the same vein, that currents, lateral supplies, and seasonal floods must be taken into account. Floods could act in the following way: material in transport during the normal season would be richer in products reworked from soils. During the flood, less weathered minerals coming from outcrops would be more numerous. Then, during the low-water period the material would be deposited near the coast, and later it would be licked up again and carried into the sea in time of flood. Comparison would become difficult, if the materials being compared did not come from the same supplies.

Conclusion

The balance sheet for research on present-day marine clays can be presented in this fashion today:

1° *The most important factor is inheritance* of material furnished by streams.

2° *Differential sedimentation* certainly plays a role. This is a physical, quantitative, and reproducible phenomenon. Certain minerals must settle out more rapidly than others under the same conditions. It must, however, be agreed that this phenomenon cannot explain all the observed facts. It is montmorillonite that has the slowest settling velocity, and, nevertheless, it is the mineral that we see disappear progressively away from the shore line. Moreover, if differential sedimentation plays a role, it would be necessary to find, by sampling along the coast, those preferred locations in which minerals that settle rapidly accumulate. Such locations are not found, and the changes are apparently continuous. Finally, POWERS (1959) pointed out that the clays are in a flocculated state immediately after their entry into the estuary and long before their arrival in the sea. The minerals are, therefore, flocculated as they work their way along the coast and reach the open sea. The role of differential flocculation is played prior to the changes seen in marine sedimentation.

3° *Transformation is the most probable mechanism* to explain the thousands of measurements now accumulated that demonstrate discreet, slow, but significant mineralogical changes between the fluviatile and marine environments. It is a question of transformations that take place in the opposite sense to those that we have studied in soils. Pedogenetic activity leads to the degradation of minerals. Here there are "aggradations" which begin. This is a modest beginning that we shall see pursued during diagenesis. The possible mechanisms will be studied in Chapter X, which is reserved for the problems of genesis.

II. — THE DETRITAL SERIES: FROM SANDSTONES TO SHALES AND ARGILLITES

Under this heading we want to discuss that which is often called the clay-sandstone series. Sands and sandstones are, by definition, detrital deposits, the material of which has come from the continents. The clays that accompany them can be found either in the cement or in silty or clayey beds that alternate with the sandstone.

It is completely normal that the clays accompanying the detrital material are, themselves also, inherited from the continents, be it by direct erosion of bedrock, or by weathering of the bedrock, or from soils. We shall examine several major clay-sandstone facies.

1° *Basal sandstone series*

At the base of the normal sedimentary series, in its most common positive development, are commonly conglomerate and sandstone. These detrital deposits still have rather frequently a continental character, such as siderolithic sandstones with kaolinite, red arkosic sandstones with a dominance of illite, and the various piedmont complexes. They have been studied in the preceding chapter. We have been able to demonstrate, however, that the clays of these sandstones had been inherited from direct erosion of continental rocks or from the reworking of weathered products and soils.

If one examines a great number of these series, it is frequently seen that the base level of the waters rises and the deposits take on little by little a marine facies. Examples are numerous; let us cite the Triassic sandstones that imperceptibly, although with alternations, end in the lower, marine Muschelkalk; let us also cite the deposits of "flat-lying sandstones" (*grès horizontaux*) of the Sahara that, between the Cambrian and the Gothlandian, attained little by little a fully marine facies. After the emergence of the uplands, transgression once again enveloped the continental surfaces. In general, a permanence of the clay supplies can be observed in these imperceptible transitions, the dominantly silty alternations near the top having received the same clays as the dominantly sandy alternations of the base of the section.

It also happens that the basal detrital series are marine from the start. Although the facies of the deposit has changed, the origin of the material is the same: continental. Specific examples that will be discussed later on bring into play pedological or diagenetic action, but they do not diminish the interest of this rule: detrital sediments contain an initial supply of detrital clays.

2° Graywackes

Graywackes in the classic sense of PETTIJOHN (1956) are detrital rocks composed of feldspars, quartz, and rock debris in a matrix that is itself detrital, but without cement. According to numerous authors, the clay minerals extracted from graywackes are illite and chlorite. However, their origin is difficult to reconstruct. In fact, graywackes have been described by KRYNINE (1945) as "autocannibalistic" sediments. This means that they feed off themselves by continuous reworking. For this reason the clay minerals can have undergone numerous stages of liberation-degradation and aggradation-crystallization (GRIFFITHS, BATES and SHADLE, 1955). According to GLASS, POTTER and SIEVER (1956) and GLASS (1958), the subgraywackes are particularly rich in illite and chlorite, kaolinite being subordinate but more abundant in the sandy horizons; they think that diagenetic action has been added to the original sedimentation.

3° Flysch and molasse

Around the periphery of rising mountain chains one finds the flysch facies within Mesozoic and Tertiary series. They correspond to the culm of the Hercynian series. Laboratory measurements confirm the presence of massive detrital sedimentation shown by illite and chlorite. These are the "run of the mill" minerals from the direct erosion of cordilleras that are being dismantled.

Molasse characterizes the erosion that nourishes peripheral basins in the final stages of an orogeny.

Petrography of the Alpine Foreland between Ulm and Munich has been carried out by LEMCKE, VON ENGELHARDT and FÜCHTBAUER (1953). Three-fourths of the clay fraction is composed of illite, which dominates montmorillonite and accessory chlorite and kaolinite.

The suite of minerals in the Swiss Tertiary molasse has been carefully reconstructed by VERNET (1959); it is composed of illite, chlorite and mixed-layer chlorite-montmorillonite. The mixed layers are lacking in the Rupelian, they are abundant in the Chattian and decrease in the Aquitanian before increasing again in the Burdigalian; they are the only minerals that permit subdivision of the molasse series because illite and chlorite are common minerals coming from the erosion of the Alps. Interpretation of the variations in the mixed layers is difficult and hypothetical, being attribuable either to changes in the supply or in the conditions of sedimentation, or to tectonic influence.

HOFMANN et al. (1949) demonstrated the presence within the molasse of a montmorillonite horizon that corresponded to the evolution of a volcanic ash; there are, therefore, bentonites interstratified in the molasse. Moreover, HOFMANN (1959) has been able to show the contrast that exists between the clay fraction of the molasse of the Nagelfluh, rich in calcite and illite, and the entirely different siderolithic with kaolinite.

The *macigno* is a molasse facies. WEZEL (1960) has described the monotony of the clay supplies in this detrital facies: illite, chlorite, and kaolinite.

4° *Alternating series*

The most valuable information is obtained by the study of formations whose complete history can be studied as a function of paleogeography, paleontology and succession; alternating series of sandstones and shales are among the most instructive.

GLASS, POTTER and SIEVER (1956) and GLASS (1958) made a remarkable study of Pennsylvanian sediments of southern Illinois. The series includes principally shales and sandstones with some coal beds. The sandstones are either quartzite, subgraywackes or feldspathic sandstones. It is within the latter facies that the coal beds are found in cyclic sedimentation. Study of this detrital series brings out four contrasts from the point of view of clay minerals.

1) Contrast between surficial outcrop and core sampling.
2) Contrast between sandstone and shale.
3) Contrast between shales below and above the coal.
4) Contrast between subgraywacke shales and orthoquartzite shales.

Let us study these phenomena successively.

1) *Contrast between surficial outcrop and core sampling.* — Samples taken from the outcrop are different from samples taken from cores. The former are richer in kaolinite, less rich in illite and chlorite, and richer in mixed layers. Atmospheric weathering has occurred near the surface, giving rise to mixed layers by degradation of mica and illite and by kaolinization. Moreover, these modifications are much more important in the sandstone horizons than in the shales because of the permeability that opens the way for weathering.

In order to eliminate this variable, we shall exclude samples taken from the outcrop and study the cores.

2) *Contrast between sandstone and shales.* — Samples of sandstone show a different composition of clay minerals than that of the shale. There is more kaolinite, less illite and chlorite, and fewer mixed layers. These variations are due to the circulation of water through the rocks during diagenesis. They are once more regulated by the permeability. The principal modification is kaolinization.

In order to eliminate this variable, we shall exclude samples of sandstone; they are the most modified. The low permeability of the shale allows it to approach as closely as possible the original mineral assemblages.

3) *Contrast between shales below and above the coal.* — This phenomenon has been already evoked in relation to coal-measure sedimentation. The shales of underclays are different from the covering shales. They contain more kaolinite, less illite and chlorite, and more mixed layers. This results from the conditions of sedimentation and pedogenetic action on coal-measure soils, whether or not these soils are reworked, in the coal swamp. There has been weathering, in the usual sense of degradation, of illite and of chlorite which is the most fragile. Likewise, there has been kaolinization.

In order to eliminate this variable, we shall exclude the shales situated below the coal. Let us now compare shales associated with subgraywackes and those associated with quartzite beds.

4) *Contrast between subgraywacke shales and shales associated with quartzites.* — The shales overlying the coal are interstratified with feldspathic sandstones or subgraywackes. The quartzites form sedimentary units within the Pennsylvanian series, but they are barren; they contain thin beds of shale. The shales of feldspathic sandstones contain more chlorite, more kaolinite, and less illite. This is related to continental supplies. The continent passes through diverse periods in which it delivers either little evolved products of weathering rich in feldspars and chlorite or more evolved products rich in quartz and illite. Here we are able to grasp the inheritance, assuming that diagenetic changes have been weak.

Here then is an analysis that shows us the successive effects of:

— continental supplies;

— environmental conditions with pedogenetic action;

— diagenesis;

— atmospheric weathering of the outcrop.

Such a careful study shows us to what extent the history of a sediment is complicated. It is possible to work it out only through the complete study of a basin by means of all the methods of geology.

The works of SMOOT (1960, 1960) and SMOOT and NARAIN (1960) follow the same path. They also studied sediments in Illinois, but of pre-Pennsylvanian age, principally Mississippian and Cambro-Ordovician sandstones and shales.

The sediments include sandstones and shales as well as intermediate clayey sandstones and sandy shales. Core samples were taken of rocks in which the rhythm of sedimentation had produced rapid alternations. One can be certain, in this case, that the alimentation and the environment were identical. Samples of the shales showed well crystallized illite and chlorite, few mixed layers, and kaolinite was practically absent. On the other hand, the sandstones produced much more varied clay minerals assemblages in which the dominant species was in some cases kaolinite, in others illite or much degraded chlorite, and in still others mixed layers. This variety results from transformations during diagenesis, which is important in the permeable facies, but not appreciable in the impermeable shales. Solutions migrating through the pre-coal-measures series have modified the permeable horizons with the passage of geologic time. In this series the dominant effect has been the degradation of three-layer minerals. In outcrops this effect is naturally exaggerated, and montmorillonite and its mixed layers appear in the permeable sandstones, whereas the shales resist much better and preserve their characteristics.

SMOOT and NARAIN (1960) then studied the clay composition of oil-bearing sandstones and barren sandstones in the same pre-coal-measures formations in Illinois. The differences jump before your eyes, as can be seen in figure 30, which shows also the composition of the shales for purposes of comparison. In the oil-bearing sandstone there is much less illite and chlorite, much more kaolinite, and more mixed layers. A detailed analysis of certain formations shows that these modifications did not take place because the oil had migrated in before that could happen. One can consider that the invasion of oil blocked the alteration of the three-layer minerals. Thus we have

190   CLAYS OF MARINE SEDIMENTS

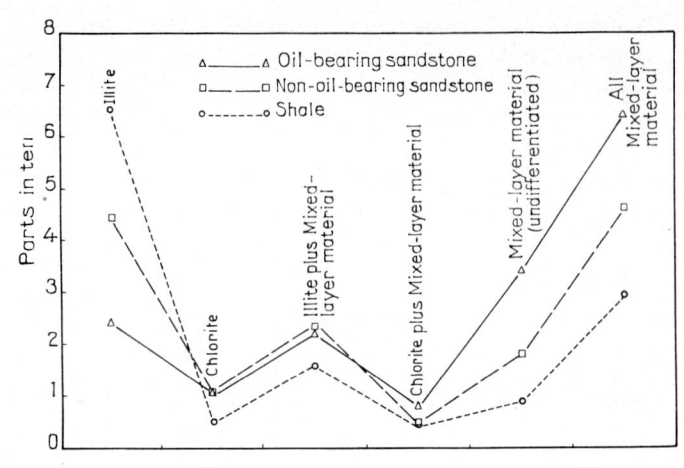

FIG. 30. — *Average clay mineral composition of shales, oil-bearing sandstones, and non-oil-bearing sandstones of the Aux Vases Formation in the Mississippian, Illinois, U.S.A.* (after SMOOT and NARAIN, 1960).

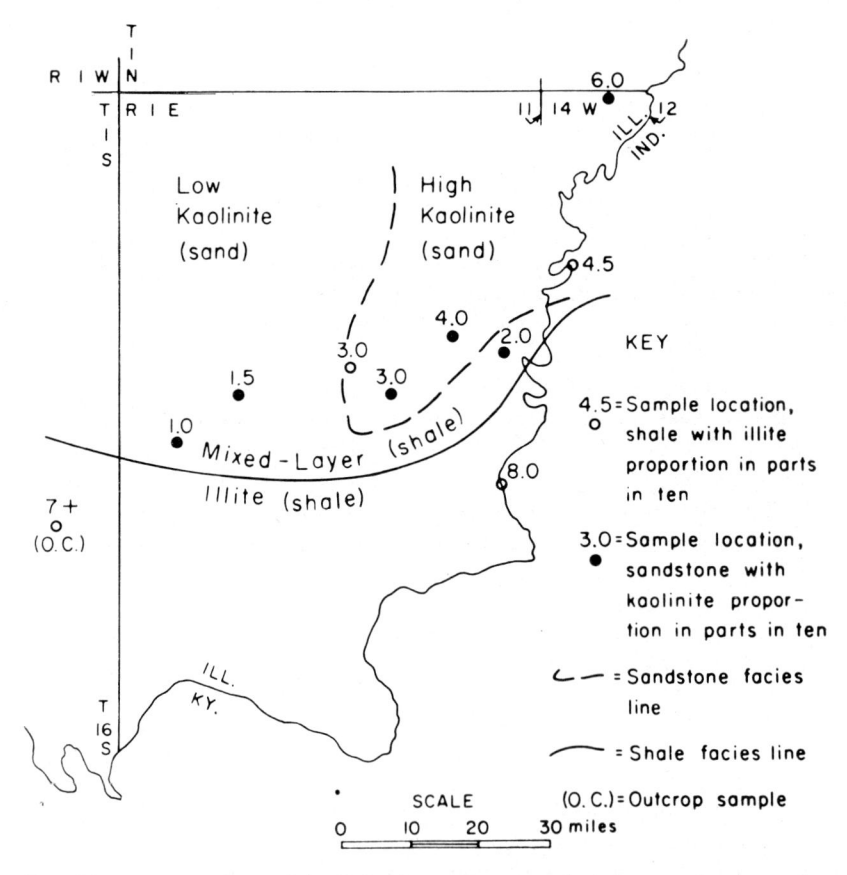

FIG. 31. — *Proportions of kaolinite in sandstone and shale samples from the Degonia, Clore and Palestine Formations of Mississippian age, Illinois, U.S.A.* (after SMOOT, 1960).

at our disposal a valuable but still imperfect tool for the study of the emplacement of the beds. This is the research path that we have followed (KULBICKI and MILLOT, 1960) in the study of the petroliferous sandstones of the central Sahara. This example will be taken up again in the following chapter in the study of the great sedimentary series.

SMOOT (1960) studied the clay minerals of the fossil series not only in vertical section, but also in a horizontal series. He carried out this study for five epochs of the Chester Series in the Mississippian of Illinois. For these five epochs, everything gives the impression that the clay minerals are lined up in a regular suite from the deltaic deposits to the open sea. In the ancient paleogeography the clay minerals spread out in approximatively concentric zones from north to south. Each zone carries the name of its characteristic mineral: *high kaolinite, low kaolinite, mixed layers, illite*. One sees this suite unfold in this order from the shores to the open sea. Figure 31 presents the map of clay zones for the Degonia, Clore and Palestine formations; this can be compared with the idealized scheme of SMOOT (fig. 32). SMOOT shows that the distribution of clay minerals is essentially contemporary with deposition, and he discusses the origin of such a distribution. After having shown that diagenesis had not played a role here, he considers that it is a question of the rate of sedimentation and the environ-

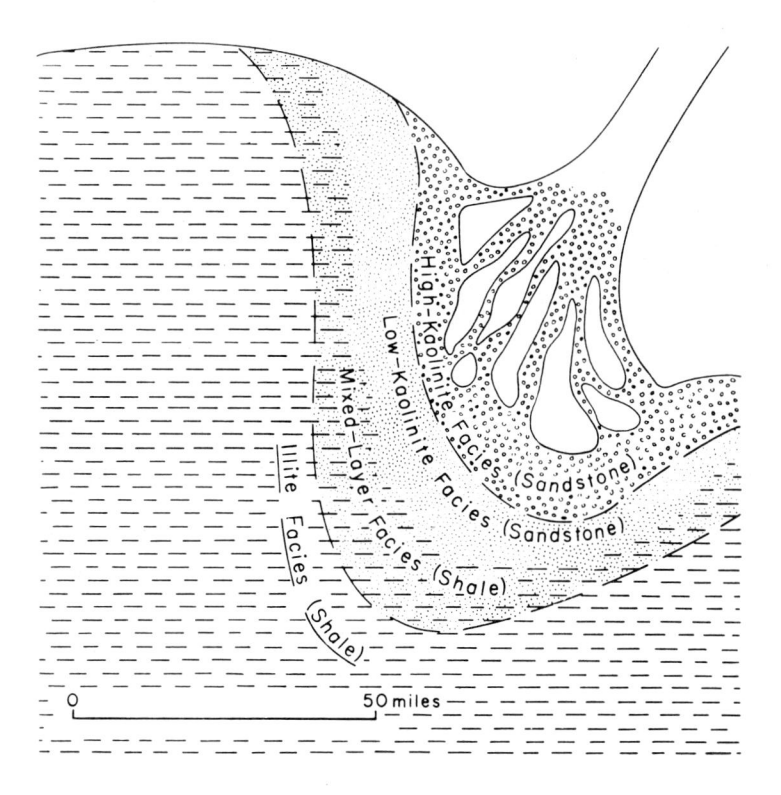

FIG. 32. — *Idealized distribution of clay minerals as a function of facies* (after SMOOT, 1960).

mental conditions. The weathered minerals, which are deposited slowly, migrate far out and are transformed by recrystallization into illite, which is the well crystallized mineral that one finds off shore. On the other hand, kaolinite is deposited as large crystals along with sands close to the coast where, moreover, it will be able to grow only by diagenesis (see also figs. 47 and 48, pages 252 and 253).

We have just encountered several cases in which diagenetic evolution of permeable sandstone horizons increases the amount of kaolinite. This can be observed under the microscope in the appearance of accordion-form or vermiform kaolinite. But this is not always the case. We have already cited the work of MILNE and EARLY (1958) who have found that chlorite is more abundant in sandstones than in the shales that are associated with them. This concerns the clays of the permeable horizons of the Late Miocene of Los Angeles; figure 29 shows the considerable development of chlorite by diagenesis. WEAVER (1959) cites the works of BURST (1958) and QUAIDE (1956), which are unpublished theses. These authors have shown that chlorite is more abundant in the sandstones than in the shales of the Pliocene of California and of the Eocene in Texas. This has also been found in a number of Paleozoic and Mesozoic sandstone horizons in petroleum drill holes in the Sahara. For porous rocks we must foresee various possibilities of evolution by diagenesis either toward kaolinite, as in the series studied by GLASS (1958) or by SMOOT (1960), or toward chlorite, as we have just seen, or toward illite (KULBICKI and MILLOT, 1960), as we shall see. This depends on the nature and conditions of circulation of solutions during diagenesis, and we still know almost nothing about this subject.

Conclusion

The rapid study of detrital sedimentation of sandstone and shale provides us an overall view of the clays that characterize this stage of sedimentation.

The major phenomenon is inheritance. It is certain that during sandy sedimentation, the principal character of which is particulate, the clays as well as the sands are inherited from the continents.

Modifications and nuances can be registered upon this material.

— In the first place, those of the environment. In truth, the example of coal-measure sedimentation is not surprising because lessivage from the coal swamp will certainly have an effect. This is a chemical effect that is peculiar to and originates from the sedimentary environment; it will definitely modify inherited material. In opposition to this are the *transformations shown to take place from the shore line to the open sea* in one of the preceding examples. This is important because, if diagenetic action is really negligible, it confirms that which has been discovered by the study of present-day marine clays. We see the confirmation of a progressive reconstitution, under the influence of marine waters, of well crystallized minerals such as illite and chlorite, in contrast to the degraded, open, and interstratified minerals that are delivered by erosion.

— *Neoformation plays no role* in this detrital sedimentation because of its very character. The velocity of deposition, the avalanche of inherited particles, and the nature of the solutions leave no place for the phenomena of chemical sedimentation.

— *Diagenesis*, on the other hand, comes into play, and everyone agrees in according it an important role. Solutions that are squeezed out of the sediments during lithification and solutions that emigrate after this consolidation model and shape the silicates. The result of these evolutions is in some cases kaolinization, in others chloritization, in still others illitization, or else the degradation of minerals into mixed layers.

— Finally, surficial weathering masks the samples taken from outcrops with a veil of deception.

In the course of these successive experiences the most faithful witnesses of the past are the shales. Their impermeability guarantees to a great extent their conservatism. In order to have a good idea of the original composition of a "clay-sandstone" series, the following can be recommended: "Study the argillites and shales; they are the best witnesses of the original deposits, and, if there has been modification, they will have been the least affected."

III. — THE MARNO-CARBONATE SERIES: FROM ARGILLITES TO LIMESTONES

In the normal sedimentary series of GOLDSCHMIDT and of LOMBARD, as defined in Chapter IV, argillites succeed sands, and limestones succeed argillites. We are in the domain of transition between detrital and chemical sedimentation. This transition occurs in many cases by the intermediary of an alternating marno-carbonate series. The problem is to know whether the transition from the detrital to the chemical environment modifies the allure of sedimentation.

1° *Early works: the clay fraction of limestones*

In the beginning are three works, all oriented in the same direction and tending to show the predominance of illite in limestones.

GRIM, LAMAR and BRADLEY (1937) studied the composition of 35 calcareous rocks of Paleozoic age in Illinois. Illite is characteristic and strongly predominant in most cases. At that time, GRIM and his collaborators had studied the weathering of loess in Illinois and had noted that illite was not stable under the influence of weathering. These observations led the authors to think that the illite in the limestones was authigenic, resulting from transformation of weathered minerals (beidellite and others) from the continent.

MILLOT (1949) studied 25 limestone and marne samples from the Mesozoic series of the Paris Basin. He found there always a predominance of illite, accompanied by chlorite and mixed layers, except in the case of the chalk, in which montmorillonite is pure. In the limestone samples the three-layer clay minerals total 80 to 100 %. Thus kaolinite is quite subordinate. But if this composition is compared to that of marne and especially to the weakly calcareous, very pyritic black argillites, one sees the amount of kaolinite increase. MILLOT thought, at that time, that the calcareous environment could ensure the neoformation of illite.

ROBBINS and KELLER (1952) studied the clay fraction of 27 limestones and dolomites of Paleozoic to Recent age in the stratigraphic series of the U.S.A. Illite is predominant in all samples except two that are dominated by montmorillonite; kaolinite is always subordinate. For the authors two reasons can explain these facts; either there is a transformation into the three-layer minerals within the carbonate-rich environment, or kaolinite flocculates more easily upon reaching alkaline waters.

So, in the first place, the calcareous environment seems to control the nature of clays and favor three-layer minerals.

2° Continuing inventory on limestones and marnes

The inventory of the clay fraction of limestones and marnes has been rapidly accumulated. Results that we could obtain by tens fifteen years ago are obtained today by thousands within the same time. Our information is thus much more abundant, and the sedimentary series analyzed are more varied.

For the limestones, RIVIÈRE (1953) emphasized that the most common mixture in sediments was that of illite and kaolinite, and that this mixture is widespread in limestones as well as in all environments of sedimentation. Marine clays and the clays of marine limestones seem to him to be detrital like the others. Likewise, WEAVER (1958, 1959) noted, on the basis of several thousands of analyses, that illite was the predominant mineral in a great number of limestones and dolomites. He pointed out correctly, however, that other components are commonly present and are even predominant in some cases, whereas in all other sediments illite is the major constituent of the clay fraction. Therefore, it seems that no advantage could be accorded to this mineral in limestones more than in other sediments, and especially in argillites and shales.

In marnes, shales and argillites of marine origin, the same variety appears. Many marnes are known with illite, or illite and kaolinite, but there are also those with montmorillonite, chlorite, or attapulgite. Black clays and marnes rich in organic matter and pyrite do not always contain a great amount of kaolinite. WEAVER (1958, 1959) reports that he knows of many that contain little or no kaolinite, being composed to the contrary mainly of illite and illite-montmorillonite mixed layers. MAUREL (1959) confirmed that the black Jurassic marnes from the Paris Basin are composed of illite and kaolinite, but that those from the Alps, deposited in a geosyncline and submitted to tectonic stress, contain illite and chlorite. BOSAZZA (1948) has even described marine sediments with kaolinite in South Africa. This was also found by RIVIÈRE (1953) in the Gothlandian argillites of the Sahara and since then confirmed by petroleum research.

Thus, the inventory period emphasizes for us the great variety in the composition of the clay fraction of the argillo-calcareous series. The reason for this is simple and important. In this marine domain that has hardly yet taken on a chemical character, products that are inherited from the continent are very abundant. Statistical research on formations of all ages and from all countries emphasizes the diversity of these sedimentary supplies.

There remains the problem of knowing if a marine environment, characterized by calcareous sedimentation, can act upon a given supply either by slight modifications or by more important transformations. Here, it is necessary to change our approach. No longer is the statistical accumulation of results important; precise cases must be followed step by step. And this can be effected in two ways:

— By the study of clay minerals as a function of the paleogeography and facies of a single formation;

— By the study of marno-carbonate alternations.

3° *Variations of clays with paleogeography and facies*

The paleogeographical studies of the argillo-calcareous series are still not very numerous. The reason is easy to understand; it is necessary to have at our disposal many bore holes across a formation, from its edges to its center, and there must be some reason to do this.

MERRELL, JONES and SAND (1957) studied the argillo-calcareous Paradox Formation in the Pennsylvanian of Utah (U.S.A.). They reconstructed the variation in sedimentation as a function of the shape of the basin. They found variations in heavy minerals and in the amount of carbonate. Moreover they noted a progressive change in the clay minerals from the margins toward the center. The suite of the clay minerals is the following: montmorillonite with illite, chlorite with montmorillonite, and illite and chlorite. This suite shows the striking disappearance with increasing distance of the most fine-grained and stable mineral and the increase of illite and chlorite. The only interpretation is that of transformation of the inherited material under the influence of marine waters.

PETERSON (1962) carried out his study as a function of the lithologic facies and microfacies of a carbonate series. He studied the Upper Mississippian of Tennessee (U.S.A.). By means of microscopic studies, he defined a certain number of lithofacies in calcareous and dolomitic rocks of that series. Then he characterized each of these microfacies by their mineralogical assemblage of carbonate and silicate minerals. He found five mineralogical assemblages:

1. Dolomite, corrensite;
2. Calcite, dolomite, corrensite;
3. Calcite, corrensite;
4. Calcite, chlorite-vermiculite;
5. Calcite, chlorite-vermiculite, a mixed layer of montmorillonite and a 10 Å mineral.

The important clay minerals present here are, therefore, a regular montmorillonite-chlorite mixed layer, a vermiculite-chlorite mixed layer and an irregular montmorillonite. PETERSON pointed out a clear connection between these mineral facies and the lithological ones; for this reason he could assert that magnesian solutions had not percolated through the column after sedimentation. This takes a determinant role away from possible later diagenesis. In contrast, the genesis of the mineralogical assemblages apparently took place during or immediately after sedimentation. It is the saline marine water and the interstitial water buried with the sediment that ensure the equilibrium of magnesium with respect to carbonates as well as to silicates. Montmorillonite-chlorite mixed layer (corrensite) and vermiculite-chlorite mixed layer are the result of the fixation of Mg^{2+} ions between the silicate layers. Moreover, this fixation is not indifferent; it depends on the changing conditions of the environment within this calcareous series.

Likewise, TOOKER (1962) studied a series principally of carbonates from the Pennsylvanian of Utah (U.S.A.). It includes alternating calcareous, dolomitic and sandy units. The clay fraction was studied as a function of the different facies; along with illite and chlorite was found a regular 29.4 Å illite-montmorillonite mixed layer, similar to corrensite, but less well defined by the higher order diffraction peaks. But the distribution is not random:

Fossiliferous limestone:	illite predominant, and chlorite
Sandy limestone:	illite, regular mixed layer
Cherty limestone:	illite, regular mixed layer
Dolomitic limestone:	chlorite, regular mixed layer, illite
Calcareous quartzite:	regular mixed layer, illite
Dolomitic quartzite	chlorite predominant, illite
Quartzite:	kaolinite, chlorite, illite

The assemblage of clay minerals is connected to rock type, and this must correspond to equilibria with the major constituents of sea water during sedimentation.

We find here a type of sedimentation very similar to that we shall see develop in the hypersaline facies; it is *the transformation of the inherited clay minerals under the influence of the environment.*

4° *Study of the alternating argillo-calcareous series*

There is one point upon which the measurements show no ambiguity, namely the systematic variation of the clay mixture in alternating series. It is a question of sedimentary series in which marne and limestone alternate tens and hundreds of times.

Such series were studied by MILLOT (1949) in the Muschelkalk, the Lias, the Jurassic and the Cretaceous of the Paris Basin. The clay fraction is composed of illite and kaolinite. The illite is generally accompanied by small amounts of chlorite and mixed layers. In the discussion of the results, it is the total of the micaceous minerals that is contrasted to kaolinite.

In these Mesozoic sediments of the Paris Basin, we were able to show that, in the slightly calcareous, very pyritic black marnes, the percentage of micaceous minerals was 50 to 60 %. In the more calcareous and still pyritic gray marnes, we find 60 to 70 %. In the marny limestones that alternate with the preceding facies, the micaceous minerals make up 80 to 100 % and kaolinite 0 to 20 %.

Thus, in a series of black marnes and marny limestones one observes an oscillation of the clay fraction that is richer in micaceous minerals in the marny limestones than in the black or gray marnes. Analogous results have been obtained by PARJADIS DE LARIVIÈRE (1959) in the cement beds in the neighborhood of Grenoble, and by BERNARD (1959) in the marno-calcareous rhythmic sedimentation in the Triassic of the Cévennes. In the laboratory such oscillations were also found in many alternating series.

An explanation must be given for this phenomenon, but the least that can be said is that it is still very difficult to comprehend today.

— Formerly I had explained (MILLOT, 1949) the greater percentage of kaolinite in the black marnes, relative to more calcareous beds, by the existence of organic matter and sulfate-reducing bacteria in the black mud, the presence of which was revealed by pyrite. An acid milieu developed which could neoform kaolinite. In reality, this has not been demonstrated; to neoform kaolinite, as in a soil, would require lessivage and, if it has indeed been observed that shells of Foraminifera are partly dissolved in such a medium, it is not easy to conceive of a hydrolysing lessivage that would lead to the synthesis of kaolinite.

— Another pathway is that of the differential sedimentation. The initial mixture of clay particles in the calcareous environment was perhaps different from that of the clay particles in the black mud environment. This is not certain because it is still not well known whether the rhythms of these sedimentations are regulated by continental events, such as tectonic or climatic rhythms, or by rhythms of the deposit itself, in the manner of the rhythms of the molasse or of coal measures (BERSIER, 1953, 1958). In the second case, the supplies should not be different in the alternating units of the sequence.

— Another pathway consists of considering not that kaolinite increases in the beds of black marne, but that micaceous minerals develop in the calcareous beds. Transformations of degraded minerals lead to illite and chlorite. But all this is hypothetical and remains to be demonstrated.

— A fourth and final pathway is that of preferential diagenesis in the calcareous horizons by means of the mechanisms pointed out in the clayey sandstone series.

5° *Diagenesis in limestone*

Diagenesis certainly occurs during the evolution of calcareous muds to carbonate rocks. Numerous studies have been concerned with diagenetic evolution during which calcite, dolomite, silica in the form of chalcedony or quartz, and sulfates play a part and successively replace one another.

For the silicates, diagenesis is less well known. However, the development of chlorite in many iron-rich limestones is spectacular. Whether this is true or false chlorite (chamosite), we shall see the modalities of these diagenetic neoformations during the study of iron ores. Many chloritic limestones show under the microscope that the chlorite is the result of a history similar to that of iron ores.

G. Lucas (1955) carefully observed diagenesis within the nodular limestones of *ammonitico-rosso* facies. There is silicification and neoformation of green, pleochroic chlorite. Debrenne (1958) showed the transformation of Archaeocyathus from the Moroccan Cambrian into 14 Å chlorite. It developed from the periphery to the center of these fossils, preserving faithfully all the anatomic details.

Talc has been identified twice in Infracambrian dolomitic limestones in the Congo of equatorial Africa (Bigotte, Bonifas, Millot, 1957) and in the southern Sudan in west Africa (Millot, Palausi, 1959). This talc constitutes either the core of oolites, or concentric envelopes with a foliated and concentric structure, or the framework of stratified dolomites. Formation of the talc preceded the dolomitization and silicification. Since an initial sedimentary origin seemed improbable, we suggested a development by diagenesis.

Thus, here are outstanding cases of evolution of silicates by diagenesis in limestones. Let us note well that these are rocks more porous than clays and, consequently, they are traversed during the transformation of sediments into rocks by solutions squeezed out of the original muds. Here then exists a laboratory whose considerable activity is recognized in the reworking of sulfates, carbonates and silica. It is impossible that silicates and clay minerals would be insensitive to this. Their evolution by late-stage diagenesis must be considered highly probable, in as much as such evolutions have been well demonstrated in sandstones. Let us add that the solutions percolating through limestones are perforce alkaline and we know enough to deduce that transformations and neoformation of three-layer minerals, mica, and chlorite will be most likely under these conditions.

These phenomena of diagenesis in limestones, which are more permeable than the shales that are associated with them, would be by themselves sufficient to explain the differences in clay composition of the alternating series. But, in a more general way, they would explain this tendency of limestones to be commonly richer in three-layer minerals than other sediments.

Conclusion

The overall view of argillo-calcareous sedimentation leads to the comprehension of the modes of genesis of clay minerals:

Inheritance is the fundamental mechanism. It is probable that many marnes and limestones contain simply the detrital clay minerals that they inherited from surrounding continents. Let us add that, if these clay minerals are stable, such as kaolinite, well crystallized illite, and well crystallized chlorite, nothing can happen to them in the normal marine environment.

Transformations are tenuous mechanisms, difficult to grasp, and necessary to be demonstrated in each case. During sedimentation whose chemical character is still modest, they will be weak, whereas, if the waters are charged with magnesium and silica, we move toward clearer transformations of the degraded continental supplies.

Post-sedimentary diagenesis still remains to be explored. The solutions squeezed out of calcareous muds and ulterior circulation have a good chance of developing a new and efficient chemistry.

We are left with the neoformations. In the argillo-calcareous series, which is still very detrital, they are inaccessible or tenuous. But, if the environment becomes rich in silica and magnesia with the possible presence of alumina, in addition to lime, we come to the realm of alkaline chemical sedimentation properly speaking, and we shall study it in the next section.

IV. — ALKALINE CHEMICAL SEDIMENTATION WITH ATTAPULGITE, SEPIOLITE AND MONTMORILLONITE

The attapulgite- and sepiolite-bearing sediments can be lacustrine, marine or hypersaline. It is in the marine series that these sediments are the most widespread; therefore, they will be studied here. We shall see that these deposits are often mixed with montmorillonite.

In 1949 all the known deposits with attapulgite and sepiolite were lacustrine; therefore, I had considered that these minerals were typical of lacustrine alkaline chemical sedimentation. But from 1952 on, attapulgite and, more rarely, sepiolite have been identified in many marine deposits. Therefore, the question deserved to be entirely reexamined, as has been done in later works (MILLOT, RADIER, BONIFAS, 1957; SLANSKY, CAMEZ, MILLOT, 1959; MILLOT, 1961, 1962).

1° *Marine attapulgite and sepiolite*

CAPDECOMME (1952), then CAPDECOMME and KULBICKI (1954) showed that the mineral typical of the base of the phosphatic formations of Senegal was an aluminous and ferriferous palygorskite. The age is Eocene, as attested by marine fossils, and the clay comes from a typical alkaline marine sedimentation with limestones and phosphates.

There are other marine attapulgites (MILLOT, 1953; RADIER, 1953; MILLOT, 1954); first, in a Coniacian argillite of the Mesozoic of Gabon studied at the request of REYRE (DEVIGNE and REYRE, 1957), that had attracted attention because of the trace of an ammonite in a sample of drill core; then, the papyraceous argillites of a great number of collections from the Eocene of the Gao Basin (Sudan, Mali) in which attapulgite is mixed with montmorillonite.

At the same time, studies in the laboratory of RIVIÈRE on samples from several phosphate basins in Africa showed the presence of attapulgite associated with montmorillonite and glauconite (RIVIÈRE and VISSE, 1954; VISSE, 1954, 1954). One can also deduce from the works of BENTOR (1952) and YAALON (1955) the occurrence of attapulgite-bearing argillaceous deposits associated with limestones and chert in the Eocene of Israel.

JEANNETTE, MONITION and SALVAN (1956) studied the clays of the deposits in the phosphate basins of Morocco. They identified clays in which palygorskite is predominant at the top and at the base of the phosphatic series. These facies are connected with calcareous sedimentation, but they are also associated with chert in sheets or nodules and with montmorillonitic clays, abundant in phosphatic series.

Marine attapulgites were also found by ELOUARD (1959) in the Lower Lutetian of southern Mauritania and northern Senegal, then in the Tertiary basin of Dahomey and Togo by SLANSKY (1959) in the Ypresian. Likewise, SLANSKY identified, in the Sangalcam bore hole, Senegal, marine beds bearing sepiolite mixed with attapulgite, 500 km from the border of the basin (fig. 57).

It is important to note that sepiolite has been described in Keuper marls in England (KEELING, 1956). In Morocco, sepiolite has also been found to be very abundant in the Lower Permo-Triassic, whereas it is attapulgite that has developed in the Permo-Triassic above the basalts (LUCAS, 1962). We shall reexamine these sections in connection with the hypersaline facies, but it is interesting to note a kind of indifference of the attapulgites and sepiolites with respect to salinity, since they are found in lakes as well as in normal and hypersaline seas.

Finally, attapulgite has recently been identified in Lutetian and Stampian marine horizons of the small Nummulitic basins of Lower Brittany in the Loire-Atlantique (ESTÉOULE-CHOUX, 1962).

Thus, the attapulgite-bearing marine beds have a great extension that exceeds by far the previously described lacustrine attapulgites. But the conditions of sedimentation are very analogous: association with limestones of chemical origin, with chert that formed contemporaneously in sheets or nodules, and with phosphatic series.

2° *The clay fraction of chalks*

The Turonian chalk of Senonches, Eure-et-Loir (France), had shown a clay fraction composed of montmorillonite (MILLOT, 1949). Since this date, if one makes exception of littoral detrital chalks, which contain sand and littoral clay minerals, all the samples of chalk studied in the laboratory show the presence of montmorillonite. The percentage of this mineral is small, and it is difficult to extract it. But the clay fraction, however small it may be, is composed 100 % of montmorillonite. MILLOT, CAMEZ and BONTE (1957) have given new examples.

Meanwhile, studies on chalks of Germany were carried out by HEIM (1957). The major part of the extracted clay fraction was montmorillonite. HEIM explained this montmorillonite as having been carried from the continents neighboring the chalky seas. The montmorillonite, like the other detrital minerals would have come from south of the Harz.

We have oriented ourselves in an entirely different way (MILLOT, RADIER and BONIFAS, 1957). We considered that the climate of the Late Cretaceous was lateritic on the continents and was not able to furnish detrital montmorillonite. We considered that the chalk was a typical biochemical limestone in which nodules of chert developed and phosphatized beds were common. Finding once more the conditions of sedimentation of attapulgite and montmorillonite of the chemical basins of Africa, we thought that the montmorillonite of the chalk was neoformed, as we shall explain.

3° Origin of attapulgite and montmorillonite in sediments

Attapulgite cannot be reworked from soils. In fact, all soils upon attapulgite-bearing parent rock show rapid weathering of this mineral.

This is the case in particular for the soils in Aquitaine (CAMEZ, 1962; FRANC DE FERRIÈRE, 1961) and for the black soils in Senegal (PAQUET, MAIGNIEN and MILLOT, 1961). The same information can be obtained from the works of MUIR (1951) and YAALON (1955). Attapulgite is thus unstable on the surface and cannot be of pedogenetic origin.

It is excluded, of course, that attapulgite beds be the reworked products of previous beds, in the first place, because attapulgite beds are rare and cannot be in a position to supply the clayey lenses in a chemical series, and secondly, because sedimentation of the attapulgite-bearing argillite and marne is free of detrital particles and occurs in the form of interstratifications in deposits of a very dominantly chemical character, such as carbonates, chert, phosphates, etc.

The conclusion is simple. Since attapulgite cannot be inherited from soils or rocks, it is neoformed in the course of alkaline sedimentation.

The problem of montmorillonite is more difficult. Montmorillonite-bearing soils do exist. Montmorillonite-bearing rocks are common, and volcanic glass is well disposed to be transformed into montmorillonite in an alkaline medium. But in the sediments we are examining here, there is no trace of detrital inheritance from soil nor rock, and no trace of volcanic heritage. Montmorillonite is associated with attapulgite or sepiolite and seems, like them, to be neoformed.

The occurrence of cristobalite in many montmorillonite-bearing sediments (GRÜNER, 1940; FOSTER, 1953; KONTA, 1955; BRINDLEY, 1957) seems to be a proof of this hypothesis, especially, when there is no trace of volcanic heritage, as in chalks (DUPLAIX, DUPUIS, CAMEZ, LUCAS and MILLOT, 1960). Cristobalite and montmorillonite apparently are neoformed in chalks.

4° Origin of the elements necessary for this neoformation

A special kind of alimentation is required for chemical sedimentation that is very rich in silica, lime, magnesia and possibly phosphate but, in contrast, very impoverished in alumina and iron.

In 1952, I sought the origin of these elements in the residue of solutions that had traversed basins characterized by an acid lessivage that generates kaolinite. But this view was too short sighted. According to the conception of J. DE LAPPARENT (1937), kaolinite has been reworked from soils that formed under heavy vegetation, and the products of lessivage, rich in silica, alkalis and alkaline-earth elements, have gone to supply sedimentary basins. AUFRÈRE (1937), likewise, attributed such an origin to the silica of sea water.

This led me in the following direction (MILLOT, 1954): "I explain this peculiar sedimentation more readily by the fact that it was peripheral to the completely leveled African continent. LAFFITTE (1949) has explained that this vast, flat landscape, very different from the landscape that is familiar to us, gave rise to peculiar sediments.

Whenever a continent without relief permits the removal of dissolved material only, sedimentation around it becomes overwhelmingly chemical and biochemical. When the supply is rich in lime, silica and magnesia, the chemical sediments are rich in limestone and chert, and we see magnesian clay minerals appear."

Thus, the features presented in common by certain horizons of basins in Senegal, Togo, Dahomey, Gabon and Sudan appear to be the result of differential lessivage of the African continent with the evacuation of bases and silica. An identical interpretation has been given for the Australian palygorskites by ROGERS, MARTIN and NORRISH (1954). All these facts and fragmentary hypotheses have been incorporated into the vast synthesis of ERHART (1956). But, whereas he interpreted all clay minerals as reworked from soils, one can find here a typical example of clays neoformed in sedimentary basins. However, it certainly is a case of solutions released by continental weathering that prepared these sedimentary neoformations.

5° Geology of landscapes involving attapulgitic and montmorillonitic sedimentation

African basins (fig. 58). — On the flanks of the African continent are littoral basins, such as those of Senegal, Togo-Dahomey, Nigeria, Cameroun, Gabon, etc. There are also interior basins, among which the most complete in the Eocene is the Sudan-Niger basin. Some Eocene horizons, which vary little from basin to basin, include alkaline chemical sedimentary deposits: limestones, chert, argillites or marnes with attapulgite and montmorillonite phosphates. This sedimentation coincides with periods of a great stability of the continent impeding detrital sedimentation, and it coincides also with a great intensity of lateritization. TESSIER (1950), in fact, was able to demonstrate stratigraphically the Lutetian age of the phosphatic lateritoids in Senegal. BORDET (1951, 1955) revealed the occurrence in the Hoggar of lateritic soils fossilized by lava flows that he places at the base of the middle Tertiary. RADIER (1957) reconstructed the climatic and paleogeographic history of Sudan and pointed out that the siderolithic formations post-date the beds with attapulgite: oolitic facies, kaolinitic clays, then sands. Similarly in Dahomey, SLANSKY (1959) showed that the Terminal Continental Series was a siderolithic facies covering the alkaline sedimentation with attapulgite of the Eocene.

French Tertiary basins. — In France, the frequency of attapulgitic horizons during the Tertiary seems to result from the same phenomena. They are due to differential lessivage of the Hercynian massifs, which are covered with a great thickness of kaolinitic weathered products.

But sedimentation remained lacustrine for the simple reason that the base level was too low, and that the country, during this period, was covered only by great expanses of fresh water. This does not prevent the attapulgitic, siliceous and calcareous sedimentation from taking place in the great lakes of Aquitaine, the Berry, and the Paris Basin. However, in southern Brittany, the marine base level is reached and the beds with attapulgite are marine (ESTÉOULE-CHOUX, 1962).

The phenomenon is reproduced during the Oligocene, as shown by the sepiolitic lacustrine beds of the Stampian of Salinelles and Sommières in the Gard (MILLOT, 1949) and by the Stampian marine beds of Saffré in the Loire-Atlantique (ESTÉOULE-CHOUX, 1962). This indicates that lessivage still took place during the Oligocene; thus it must be deduced that soils with a lateritic aspect and the corresponding climates were still able to prevail during this period. Recent information in this direction is provided by the Miocene lake with attapulgite of Le Locle in Switzerland (KUBLER, 1959).

Chalky sedimentation. — One can go further. We have proposed the consideration of chalk deposition in France under the same view point (MILLOT, RADIER, BONIFAS, 1957).

We have been able to point out already that lateritic climates had persisted during all the Cretaceous Period on the emerged uplands of France. Solutions evacuated by intense weathering reached the sedimentary basins. The chalk with its almost pure biochemical sedimentation, with its chert, its phosphatic levels, and its montmorillonite, seems to be the result of this differential lessivage. The percentages are different because the carbonates dominate, but qualitatively the appearance is similar. It is true that the clay that can be extracted from chalk is hardly abundant and is composed only of montmorillonite, but it is also true that petroleum drilling in southeastern France has found attapulgite in the Upper Cretaceous. The sedimentological relations between attapulgite, sepiolite and montmorillonite constitute a very interesting geochemical problem.

Conclusion. Sedimentary sequence and geochemical sequence

The main conclusion from the studies on *attapulgitic* and *sepiolitic* sediments, and on *montmorillonitic* sediments of probably the same nature, is that these *minerals are fully neoformed during sedimentation*. This case is rare enough to be emphasized. There is considerable interest in these silicates that appear to be formed completely within the environment, in the same way as carbonates, phosphates, chert and all other chemical deposits.

SLANSKY (1959) has been able to show the existence of sedimentary sequences in these sedimentary formations of Africa. As one goes toward the open sea, one sees, little by little, the progression of montmorillonite, attapulgite and possibly sepiolite. This sequence is also found, of course, in vertical section; from the bottom to the top of a bore hole it develops in the above-mentioned order, i.e., in positive sequence when there is a transgression and in negative sequence when a regression takes place. This sedimentary sequence is also a geochemical sequence (MILLOT, ELOUARD, LUCAS, SLANSKY, 1960). In fact, montmorillonite is more aluminous than attapulgite, and the latter is more aluminous than sepiolite. Consequently, the more magnesian a mineral is, the farther from coasts it develops, and the more aluminous it is, the closer to the detrital supplies it is. Here all the consequences of these correlations can be anticipated and they will be examined in detail in Chapter VIII concerning the great sedimentary series, and in Chapter XI concerning the mechanisms of genesis.

V. — GLAUCONITIC SEDIMENTS

Glauconie is typical of the marine environment. This fact belongs in the alphabet of every geologist. In addition, glauconite is the ferric homeotype of illite. Thus, it is a question of a clay mineral by definition, and its origin is marine; accordingly, glauconie is of concern to us in this chapter.

The works on glauconie are very numerous, and a complete study of the problems that it poses could fill an entire treatise. That is impossible here, and a very valuable documentation will be found in the main works that I shall cite. A complete historical bibliography is given by VALETON (1958). My aim will be rather to situate the present state of research on glauconie, emphasizing that it would be better to say on "the glauconies". It is important for me to make understood that there are some glauconies from the geological point of view and some glauconies from the mineralogical point of view, and that the genetic mechanisms can be approached today only by taking into account the variety of the deposits and the variety of their nature.

1° *Varieties of occurrence of glauconie in sediments*

It is always necessary to refer to the admirable chapter that CAYEUX, in 1897, devoted to the glauconie of sedimentary formations of northern France. Almost all the modes of deposition of glauconie were enumerated therein and confirmed in the subsequent works of CAYEUX (1916, 1932), as well as in the great number of more recent geological works.

The main kinds of occurrence of glauconie are the following, according to CAYEUX:

1° Fillings of the tests of various organisms, and above all, of Foraminifera (COLLET, 1908);

2° Pseudomorphs of siliceous spicules of the Porifera;

3° Partial epigenesis of calcic tests of mollusks, brachiopods, echinoderms and Bryozoa;

4° Grains of glauconie unrelated to, and not in the form of, organisms (see 9° and 10°);

5° Coatings on detrital minerals, such as quartz, feldspars, etc., and on secondary minerals, such as calcite. Possibility of epigenesis in the cleavages (glauconitization of feldspars);

6° Diffuse pigmentary glauconie in siliceous or calcic cements;

7° Globular glauconie formed by replacement in opal;

8° Coatings on and veinlets through phosphatic nodules;

We can add to this enumeration :

9° Epigenesis of mud balls or fecal pellets (TAKAHASI and YAGI, 1929);

10° Transformation of biotite (GALLIHER, 1935). This occurrence was discovered on the Californian coast. The biotite is altered from its edges and then becomes completely green; it swells and then takes on the aspect of vermicules or accordions analogous to those of leverrierite. Next, the original cleavages disappear, and the vermicule becomes a grain with a cryptocrystalline texture in which no trace of the original biotite is identifiable. GALLIHER has even observed this phenomenon in tests of Foraminifera, which burst under the influence of the swelling biotite that is being transformed. Since that time this mode of formation has been frequently observed, especially by CAROZZI (1951) who showed that this glauconitization of biotite was very common in the Cretaceous sediments of the subalpine chains and of the Jura.

Glauconie is developed, therefore, in numerous ways, either inside the tests of living organism, the state of which at the time when the phenomenon was taking place is not known, or by epigenesis of the most diverse kinds of crystals, or by epitaxic growth on silicates. One will have to keep this variety in mind when we come to the genetic interpretation.

2° *Mineralogical varieties*

Glauconies. — The diversity of glauconitic deposits is doubled, according to modern studies, by a mineralogical diversity. To put it another way, the term "glauconie" includes several things. This was made known by the very important works of BURST (1956, 1958) which were presented in Chapter I. Today a geologist must know, when he recognizes glauconie in a rock, that he has the four following possibilities in front of him.

1° Glauconite, which is a well defined mineral with a regular structure, that of an illite in which ferric iron partially replaces aluminum (1 M glauconite);
2° Disordered glauconite, equivalent to the disordered illites. The structure of the layers is the same, but the stacking of these layers is disordered (1 Md glauconite);
3° A glauconite-montmorillonite mixed layer. This is the equivalent of the illite-montmorillonite mixed layers in which the ferric iron has become important;
4° A variable mixture of minerals in which illite, montmorillonite, chlorite, or various mixed layers can be identified. The total aspect remains that of green clay minerals, but it is not known whether one part or another yet merits the name "glauconite".

Glauconitic illites or aluminous glauconies. — In addition to this variety, there are also minerals intermediate between illite and glauconite. Such minerals have been mentioned three times, but in these cases in lacustrine horizons.

In 1954 JUNG described very pure clay minerals of the Oligocene lacustrine basin of Salins that were ferric illites. The proportion of Al in the tetrahedral sheet was one in ten. The percentage of trivalent ions in the octahedral sheet was 80 % $(Al+Fe^{3+})$ relative to 20 % for Fe^{2+} and Mg. But Al and Fe^{3+} can exchange with one another. The three cases studied by JUNG showed three intermediate stages between aluminous illites and glauconies from the standpoint of quantity of ferric iron. The values are the following:

	Aluminous illites	Ferric illites	Glauconies
$\dfrac{Fe^{3+}}{Al+Fe^{3+}} \times 100$	6 to 16	22, 42, 57	68 to 72

JUNG thought that these illites whose composition was intermediate between illite and glauconite were neoformed; this is plausible since we still do not recognize such minerals among the products of weathering.

In 1958 KELLER described a glauconitic mica in the continental sediments of the Morrison Formation. In this mica

$$\frac{Fe^{3+}}{Al+Fe^{3+}} \times 100 = 43$$

a value that permits us to place this mineral half-way between illite and glauconite. Its origin is attributed to the transformation of volcanic ash in a lacustrine basin under special conditions.

In 1961 NICOLAS reported a small basin in Brittany that must have been lacustrine and where he found a glauconie poor in ferric iron. The use of chemical analysis is difficult because this clay mineral is mixed with kaolinite. However, this may be a third case of continental glauconitic mica.

Celadonite. — It is appropriate to mention also what was known from the work of HENDRICKS and ROSS (1941) about the chemical composition of glauconite and celadonite. The latter is a mineral resulting from the weathering of rocks similar to basalt. Most commonly it fills the vesicular cavities of lava, but it can also replace olivine directly in some cases. One can consider that the weathering solutions of the basalts were solutions containing the magnesium and the iron, along with silica, necessary for the neoformation. But the main point is that celadonites are, in reality, glauconites. The lattice is the same. Celadonites are, on the average, richer in magnesia and poorer in alumina, and thus richer in iron than glauconites, and the Fe^{3+}/Fe^{2+} ratio is smaller. But it is a question of the same mineral with minor variations in composition; the two different names are retained only for the commodity of working in geological situations that are so entirely different.

As a function of this diversity I have proposed a greater precision in the terminology for glauconies in order to avoid confusion. In his vocabulary the geologist will have to distinguish between the three following terms:

1° *Glauconie* is the common name given upon simple microscopic study to green micaceous clay minerals, identified as such, and distinguished as much as possible from chlorite, whatever their arrangement may be in grains, as pigment, in trails, epigenetic, or in accordion form.

2° *Glauconite* is the name of the mineral, correctly defined by modern methods, that corresponds to an ordered or disordered dioctahedral micaceous lattice. The other, interstratified or mixed, cases will be described according to their characters. One term will be provided for the ferriferous illites or aluminous glauconites, and another for *celadonite*, which is more magnesian and more ferrous since it is a product of weathering.

3° *Glauconitite* will be the name reserved for rocks in which glauconie, the predominant mineral, imposes its name upon the rock, e.g., sandy glauconitite, phosphatic glauconitite.

3° *The conditions of genesis*

Since glauconie was first described by BRONGNIART (1823) all observations have confirmed that its origin is marine.

Exceptions are very rare. I have reported two of them (MILLOT, 1949): in the Sannoisian lagoonal beds of Pechelbronn in Alsace and in the Lettenkohle of Damelevières in Lorraine. But both of these formations have mixed characteristics, and it cannot be excluded that they may have been subjected to marine influences. In any case, the lagoonal environment can certainly be characterized, at least seasonally, by concentrations compatible with the genesis of glauconie. The observations of the Russian authors DIADTCHENKO and HATUNTCEVA (1955) are still more interesting; they discovered continental glauconie in residual deposits. The chemical formula differs little from the usual formula for glauconie. One can see progressive weathering of the feldspars and other silicates of the original rocks into this neoformed product, glauconie. We can add here the original phenomenon of the weathering of basic crystalline rocks into celadonite; this phenomenon is continental.

Beside these few exceptions, the environment of genesis of glauconie is marine. Most frequently it has been possible to identify their development with marine facies (CLOUD, 1955). In general, this marine environment is sublittoral, typical of the continental shelf and of epicontinental seas. Commonly this is an agitated environment, as shown by the rock facies, the presence of phosphatic nodules, the granulometry of the sand grains, etc. The glauconitic grains themselves are often calibrated by granulometric sorting. Even though this facies is common, the enumeration above leads to the observation that some glauconies occur in environments more difficult to define. The glauconies of replacement necessitate the study of micro-facies rather than large-scale environmental conditions. In addition, many glauconie-bearing rocks presently include pyrite, the witness of reducing conditions.

The great variety of occurrences of glauconie has brought forward a great variety of propositions concerning its conditions of formation. Some hypotheses take very much into account the agitation of environment, others the occurrence of organic matter, others an oxidizing environment, others a reducing environment, others the necessary presence of biotite from which glauconie would form, etc.

All these directions of research are valuable because, in fact, glauconie occurs in all these situations. What ought to be looked for, however, are the factors common to all these cases.

A question that has preoccupied the specialists very much is that of ferrous iron and ferric iron. In reality, iron occurs in both forms in the glauconite crystal, and the Fe^{3+}/Fe^{+2} ratio falls between 4.1 and 6.2 (HENDRICKS and ROSS, 1941). Thus, one must look for an environment that is oxidizing enough to permit the clear predominance of the ferric ion, and this coincides well enough with geological descriptions of the agitated, sublittoral environment which is conducive to oxidizing conditions. But many glauconitic rocks, including glauconitic sands or sandstones, phosphatic or not, show the frequent occurrence of pyrite, evidence of energetic reduction due to sulfate-reducing bacteria. There are several pathways out of this quandary.

— The first pathway of oxido-reduction has been followed by GALLIHER (1935) and HENDRICKS and ROSS (1941). Genesis of glauconie and genesis of the reducing environment would be concomitant. The reducing environment created by anaerobic bacteria reducing sulfates in the mud is not in equilibrium. Some places, where ferrous sulfide would appear, would function as hydrogen-acceptors, whereas in neighboring areas the trivalence of iron could occur by means of electron transport.

— The second pathway spilts up the sedimentological history into two stages (MILLOT, 1949). In the first stage, the environment is agitated and oxidizing, and glauconie can develop. In the second stage, anaerobic bacteria grow within the deposited sediment, the environment becomes reducing, sulfide is formed from sulfates, but the previously formed glauconite lattice is insensitive. This scheme applies well to certain detrital formations, such as green sandstone, or Gault (pebbles, phosphatic nodules, rolled bones, etc.), which are nevertheless pyritiferous. But it is difficult to comprehend in the case of post-sedimentary diagenesis in numerous muddy sediments described by CAYEUX (1932). It also does not hold much interest for the genesis of celadonite in basalts that are being weathered.

In fact, it seems that this type of reconstruction is premature in view of our knowledge of the muddy environment, on the one hand, and the growth of crystals, on the other. Moreover, we know that the mobility of iron in the hydrosphere is, in most cases, guaranteed by humic or silicic complexes. What happens with our scheme in face of such difficulties?

For these reasons it seems that we must confine ourselves to the fundamental problem as follows:

— Glauconite is a dioctahedral clay mineral, similar to illite. The average formula of 32 glauconites of HENDRICKS and ROSS (1941) is the following:

$$(OH)_2 (K, Ca_{1/2}, Na)_{0.84} (Al_{0.47}, Fe^{3+}_{0.97}, Fe^{2+}_{0.19}, Mg_{0.40}) (Si_{3.65} Al_{0.35}) O_{10}.$$

An average formula was given for illite by GRIM, BRAY and BRADLEY (1937):

$$(OH)_2 K_{0.58} (Al_{1.38}, Fe^{3+}_{0.37}, Fe^{2+}_{0.04}, Mg_{0.34}) (Si_{3.41} Al_{0.59}) O_{10}.$$

Aluminum is thus partly replaced by iron, but in quite diverse proportions because it varies (octahedral aluminum) from 0.12 to 1.27 ions per formula expressed as Si_4O_{10}. Iron is, above all, in a ferric state, with a Fe^{3+}/Fe^{2+} ratio of about 5. Magnesium varies between 0.26 and 0.52 ions per formula in terms of Si_4O_{10}, but 20 out of 32 analyses show it to be limited between 0.35 and 0.45. Thus, we have here a dioctahedral mica that is ferric, aluminous, magnesian and, naturally, potassic, and that must form like all other micas.

— To form like all other micas means to be *formed by the growth of crystals*. Today one can no longer speak of gels and colloids with respect to glauconie because these very widespread notions correspond to a former epoch of our knowledge.

— This can be the growth of entirely new crystals; this will be called neoformation. Or it can be by reorganization of preexisting micaceous lattices, intact or altered; this is called transformation. Or it can occur by means of epitaxic growth between layers of a biotite, or in the cleavages of a silicate. In all cases, the conditions for growth are brought together, and the mineral develops; there is a unity of mechanism.

— This is the way that glauconie forms. In the case of the occurrence described by GALLIHER, it is obvious: between the layers of trioctahedral biotite that is being weathered there is a build-up of layers of the dioctahedral glauconite, which is stable in an aqueous environment. It finds the necessary elements in the biotite itself, as well as in the surrounding marine environment. In other cases, the mechanism will be the same, but the growth will be guaranteed at the expense of or by contact with the numerous clay minerals in the sediments, which we have inventoried: illite, chlorite, vermiculite, various mixed layers. In still other cases, this will be a neoformation, as is evident in the case of celadonite within vesicles of basalt. Finally, this neoformation can utilize other minerals, silicates or not, as guides for the organization of its lattice: feldspar, quartz, peridot, calcite, silica, etc. When the lattice grows externally or in cleavages, one speaks of coatings; when the lattice develops within minerals, one speaks of epigenesis.

— In the most frequent case, which is that of transformation of preexisting clay minerals into glauconie, we find the illustration of these effects in the distinction of BURST (1956, 1958): in its earliest stages, the transformation leads us to the mixture of minerals of group 4; with greater perfection it gives rise to the (10-14 Å) mixed layers; still more advanced, it yields an assembled but disordered glauconite; brought to its final stage, it produces well crystallized glauconite, which can, moreover, also be attained by neoformation. *The categories of glauconites identified by X-ray diffraction seem to us to represent the different stages of glauconitization.*

— The inverse of this mechanism corroborates the hypothesis. Twenty-seven years ago, COLE (1942), studying the weathering of green sands in Australia, described the transition from glauconie to montmorillonite. A careful study made today would show us the progressive interstratifications.

— The elements necessary for this growth and reorganization will be supplied by hosts or by solutions: silicon and aluminum by the hosts, on the whole, and magnesium and potassium rather by the sea water. The problem of iron remains. We do not know, however, in what measure the structures act on elements with several valence states, and the majority of the silicates contain iron in its two states (chamosite, cronstedtite, chlorite). Thus, it is the question of an equilibrium between a structure in the process of growth and the oxidation-reduction potential of the environment. We have no information on this subject.

Conclusion

The glauconie is a mineral that forms either by neoformation, or by transformation, or by diagenesis in marine sediments. In the present state of our knowledge, it is not necessary to look for a precise and narrow mode of formation that calls upon a certain organism (Foraminifera or coprolite), or a certain mineral (biotite or illite), or a certain more or less hypothetical chemical environment.

It is necessary to keep up the discussion, to know the limits of what we know and what we do not know. It must be understood that this "clay" mineral, so similar to others, has all the possibilities of forming like the others, i.e., by the growth of crystals. We all know that montmorillonite can occur in soils by evolution of volcanic glass, by hydrothermal action upon various silicates, by sedimentary neoformation, or by weathering of clay minerals. In all these cases, the conditions necessary for the stability, thus for the growth, of this structure are brought together, and we are beginning to know them and to reproduce them.

Likewise, in the case of glauconite, this crystallochemical structure will form by neoformation or transformation, whenever the siliceous, aluminous, magnesian, ferruginous and potassic solutions have the necessary concentrations. This is realized from the basaltic vesicle to the test of Foraminifera, but sea water seems to be the eminent and favorable environment. It is here that we must direct our research in order to define the conditions of equilibrium between the solution and the growing crystal.

VI. — THE CLAY MINERALS OF IRON ORES

The study of oolitic sedimentary iron ores constitutes a vast problem; it interests us here only because clay minerals named chlorites are found among the main constituents of these ores.

I shall choose as a fundamental example the oolitic iron ores or *minette* of Lorraine in France. However, scientific work has been carried on abreast in all countries and I shall refer to it in the course of our discussion.

1° *Mineralogical identifications*

Before the time of X-rays and of the modern mineralogical identifications, the petrography of the Lorraine iron ore was described mainly by CAYEUX (1922) in a treatise that is fundamental because of the number of precise descriptions given by the author. We can pick up the problem where he left off. Chlorite or iron silicate is identified under the microscope by its colors which range from yellowish green and dark green to greenish black. It occurs in two principal positions; in the oolites of which it forms, in many cases, the concentric aureoles, and in the cement where it has various aspects. from inherited, concretionary aggregates to large, homogeneous, palmate areas.

However, in 1949 the apparent simplicity of the chlorites of iron ores had disappeared owing to the concurrent works of specialists of many countries. In France, ORCEL, CAILLÈRE and HÉNIN (1949) showed that the clay minerals extracted from iron ores were of two types: 14 Å clay minerals (chamosite from Chamoson, bavallite from Bas Vallon, berthiérine from Hayange) and 7 Å minerals (for instance, berthiérine from Uttange). In Great Britain, BRINDLEY (1949) pointed out that the "chlorites" of iron ores, which the Anglo-Saxons were used to calling chamosites, were 7 Å minerals constructed after the model of kaolinite or serpentine. In Germany the same results were obtained by HARDER (1951) from the Liassic iron ores north of Göttingen, where he distinguished chamosites with a lattice of kaolinitic type and chamosites with a lattice of chloritic type. These works resulted in a differentiation within the great group of iron ore chlorites, but also in an irritating question of nomenclature. Certain chamosites from Chamoson are chlorites, and the majority of the chamosites of iron ores are of the kaolinite-serpentine family. Therefore, in France under the influence of CAILLÈRE and HÉNIN, we call these ferromagnesian clay minerals *berthiérine*.

The work was carried on successively by BRINDLEY (1951) and by CAILLÈRE and HÉNIN (1952, 1959). Analogous research, presented in STRAKHOV (1957), was conducted in Russia.

The clay minerals of iron ores are classified in the following way:

1° *7 Å clay minerals.* — They are called *berthiérine* in French and chamosite in English. They belong to the kaolinite-serpentine series; the number of the atoms present in the three octahedral positions is closer to 3 than to 2; thus one can speak of an isotype of serpentine, but the ions in this octahedral layer are Mg, Fe^{3+}, Fe^{2+}, Al. These *berthiérines* occur in the iron ores of Mesozoic age.

2° *14 Å clay minerals, or true chlorites.* True chlorites, characterized by their fixed basal spacing of 14 Å and by the harmonic series of reflections, occur in oolitic ores, but almost only in the Paleozoic ores that have undergone advanced diagenesis.

3° *Swelling 14 Å clay minerals, or swelling chlorites.* — These are swelling chlorites discovered by STEPHEN and MACEWAN (1951) in the Keuper marnes and also found in Mesozoic iron ores. It is well known that these are chlorites whose brucitic layer is very incomplete, filled in enough to ensure the stability of the 14 Å reflection upon heating, but not enough to bind the layers, which remain expandable under the influence of polyalcohols.

Two points concerning this nomenclature need to be clarified. The use of the term "pseudo-chlorite" has been proposed. It is used by some authors to designate sedimentary chlorites that are not true chlorites from a mineralogical point of view; this term, thus, includes the berthiérines and the Anglo-Saxon chamosites. It is used by other authors to designate the swelling chlorites; this gives rise to confusion. In addition, the term "leptochlorite" was proposed in the treatise of STRAKHOV (1957) to designate, without any definite mineralogical meaning, the ferriferous chlorites of sediments. Now, this term "leptochlorite" had been defined by BRINDLEY (1956) to designate ferrous chlorites whose iron was partially oxidized; this is a second source of confusion. Our knowledge of sedimentary chlorites is still too crude to permit us to give specific names to these minerals. This gives the illusion of names of mineral species, whereas the criteria have not been assembled for specific identification. The best practice is to define by their characteristics the chlorites we are speaking of each time that we have the possibility to do it.

2° Conditions of formation

Mineralogy is ahead of geological analysis, and it is not possible at present to know if the different categories of clay minerals identified in iron ores have different origins. However, the fact that true chlorites in France are found only in Paleozoic ores that have undergone a Hercynian history leads one to think that they are products of advanced diagenesis. We are still far from the general metamorphism that gives rise to crystalline rocks, as GUEIRARD (1957) has shown in the case of the collobrierite from Provence in which oolites have been transformed into pale pink garnets.

To show how the conditions of formation of chlorite in oolitic ores are considered, we shall examine the history of the interpretation of the French deposits in Lorraine. But, in the course of development of these hypotheses, there was, of course, interaction with foreign research. Among those that I know, I shall cite the following. GRÜNER (1922) and HALLIMOND (1925) considered "chlorites" to be authigenic. TAYLOR (1949) studied the sequences in English deposits and showed the order of the paragenesis, in which chamosites appear late. In Germany, CORRENS (1947) and HARDER (1951, 1957) showed the diagenetic character of chamosite formed from original ferruginous oolites. The physico-chemical mechanisms were envisaged by KRUMBEIN and GARRELS (1952); they expressed the limits of stability of hematite, siderite and pyrite in terms of pH and Eh. Finally, Russian authors have demonstrated the diagenetic sequence of iron ores (TOTSCHILIN, 1952; THEODOROVITCH, 1954; STRAKHOV, 1957).

Let us examine the suite of hypotheses about the French ore of Lorraine. CAYEUX (1922) considered that "chlorites" were one stage of the successive diageneses to which oolites were subjected. The starting point would have been calcite, then siderite or iron carbonate, then chlorite or iron silicate, then brown hematite or iron peroxide. The oolites were subjected to this evolution, either complete, partial or shortened, within their generating environment. Then they were transported, sorted, and deposited by currents, and there the cement, in its turn, was subjected to similar diagenesis.

BICHELONNE and ANGOT (1938) followed an analogous outline, but they did not think that oolites were originally calcareous. They admitted that they could have formed directly, either in the form of siderite, or in the form of chlorite. Further oxidation would have given rise to hematite.

CAILLÈRE and KRAUT (1954-1957) pushed mineralogical analysis of ores quite far ahead. On the subject of genesis, they concluded that diagenetic evolution occurred *in situ*. The elements became organized and differentiated chemically within the sediment. Here chlorites are once again diagenetic.

BUBENICEK (1960, 1960, 1961) used simultaneously the important observations of CAYEUX and his predecessors, modern mineralogical identifications, and his own petrographic and analytical observations on sedimentary iron ore sequences. The initial phenomenon is littoral sedimentation of well sorted arenites that include fragments of shells, grains of quartz, and *oolites of iron peroxide*. The stratification is cross-bedded, truncated toward the more littoral zones where the arenites become coarse and shelly, and pass seawards to more clayey deposits. In a vertical succession, it is a question of a series of twelve negative sequences beginning with clays and ending with the arenite of the ore. The diagenetic phenomena can be summarized as following: concretionment of the calcite that made up the shell fragments and *reduction of the iron*, which appears first in the form of *chlorite*, then of *siderite*, and more rarely of *pyrite*.

We see that there is unanimity. *The "chlorites" of iron ores are the products of diagenesis*; the diagenesis acts on the original iron oxide of the sediment.

3° Origin of the iron

The problem of origin of the iron has always intrigued geologists because nature at the present time does not provide familiar examples of such accumulations.

In the beginning, one hundred years ago, MEUGY (1869) described the marine basin in which the ore formed as a kind of bay whose shores abounded with ferruginous springs; the deposits of these springs were mixed with those coming from the open sea. However, after 1875, GIESLER was talking of littoral precipitation of iron carbonate, simply dissolved in fluvial waters by the lessivage of surficial formations; oolites would have been formed by wave agitation. The continental origin of the iron of oolitic ores was adopted by many authors, and especially by CAYEUX (1913) at the end of his long study. Taking up the abbreviated expression of Marcel BERTRAND, who said that every mountain range has its red sandstones, its flysch, etc., CAYEUX added, "Every mountain range has its belt of iron ore." Speaking as a geologist, CAYEUX wanted not to evoke high reliefs, but to show that the history of sedimentation during an orogenic cycle includes the formation of iron ores.

It seemed necessary, however, to understand why iron had accumulated in such a considerable quantity, and there were two possibilities for this: accumulation due to marine effects and accumulation due to continental effects. CAYEUX (1913) followed the first direction and envisaged bacterial action. GRÜNER (1922) followed the second, and he suggested that humid, probably tropical and subtropical climates with an abundant vegetation gave rise to rapid weathering of rocks and to dissolution of great quantities of iron and silica.

Many American authors in the last forty years have followed this tradition. HOUGH (1958) interpreted the alternating sedimentation of chert and iron ore in American Precambrian deposits. Seasonal physico-chemical variations in a large lake are held to be responsible for the precipitation of the sediments from continental materials released by a climate of tropical type.

It was by thinking about the geochemical cycle that I arrived at the same opinion on the origin of the iron (MILLOT, 1949). In the geochemical cycle iron ores are situated at the limit between detrital and chemical sedimentation; this must correspond to the epoch of peneplanation of uplands. Now, if the climates are favorable, the peneplaned continents are subjected to lateritization, and iron is stored up. I considered that, as a result of disturbance of equilibrium, these reserves were responsible for the supply of iron to the oolitic ores. This says nothing about the difficult mechanism of utilization of the iron by the sea, but an hypothesis was proposed to explain its abundance.

Considering the role of pedogenesis in geological phenomena, ERHART (1956) arrived at his biorhexistatic scheme and pointed out the alternation of storing up and reworking of continental materials. According to this author, iron ores result from continental accumulations in soils, as was explained in presenting his fresco in Chapter IV. But the problem is not so simple, and a recent work of ERHART (1961) illustrates this. German deposits from the northern Harz present him with two types of ores: oolitic ore and detrital ore. The latter can be coarse, conglomeratic, contain pisolites, concretions, etc. To explain these two types of deposits, ERHART proposes the functioning of biostatic and rhexistatic periods. During the biostatic period, ferruginous solutions are moved seaward through swampy plains: in the sea these solutions nourish oolitic ores and, at the same time, iron is concentrated in emerged areas. During the period of reworking, the continental iron will be reworked into detrital deposits. One can grasp, once again, the bond that exists between continental weathering and the sediments of the surrounding seas.

Conclusion

To conclude, the problem of the chlorites of iron ores, as well as the problem of the origin of the iron in these ores can be considered from a geochemical point of view. In order not to encroach upon a later chapter dedicated to geochemistry, this conclusion will be brief.

— Geochemical calculations by means of the isovolumetric method (MILLOT and BONIFAS, 1955; BONIFAS, 1959) confirm that all the iron and alumina are not stored up on continents during the phenomena of lateritization. In fact, horizons in which the structure is preserved commonly show a volumetric loss of iron and alumina relative to the parent rock. Moreover, soils in equatorial countries are strongly *lessivé* and not encrusted with iron, thus requiring that iron be removed. Thus it is certain that iron can form a part of the solutions of lessivage, just as it can certainly be stored up in ironcrusts (*cuirasses*). Everything depends on the conditions of weathering.

— When iron takes part in continental lessivage, it becomes mixed in peripheral basins with alkaline chemical sedimentation, the functioning of which we have already examined. Thus it becomes the companion of evacuated silica, magnesia and lime and joins together with those compounds to give rise to sedimentary deposits: the American deposits with alternating chert and ore, and the carbonate or chloritic deposits of Lorraine. Let us remember that these "chlorites" are siliceous and magnesian. Moreover, during weathering the solutions of lessivage become charged with elements of GOLDSCHMIDT's third group, phosphorus-sulfur-arsenic-vanadium..., which can easily migrate in the state of large soluble anions. This is the geochemical aspect of marine oolitic iron ore, and this is why it is phosphatic.

— When iron forms a part of the continental reserves, it is carefully separated from all the soluble elements, whether they be the large alkaline-earth ions or the small peroxidized elements. Iron retains as its companion only free alumina and, as the case may be, manganese, nickel,..., as well as alumina silicate in kaolinite. This reworked lateritic iron becomes the siderolithic iron or *minerai de fer fort* whose purity and metallurgical qualities, owing to the lack of "poisons" such as phosphorus, arsenic, selenium, etc., are well known.

— *Therefore, the petrographic as well as the metallurgic properties of our ores depend only on the pathway that the iron followed in the course of lateritic lessivage:*
Indirect pathway, leading to high-grade iron ores in which iron is "free" and "clean";
Direct pathway, where iron rejoins a great number of its companions of silicate rocks. Not only is it "impure", but the immediate effect of diagenesis will be to take away its freedom to reduce it and to engage it in new arrangements: chlorite, siderite, pyrite.

— Here, we return to the "chlorite" of iron ores, formerly called iron silicate. And that is what it is; *chloritization is a diagenetic neoformation* that, upon burial, puts iron and magnesium back into the form of a ferromagnesian silicate.

VII. — THE HYPERSALINE FACIES

1° *General survey of the hypersaline facies*

Vocabulary. — The term most used in French to designate the hypersaline facies is "lagoonal facies" (*faciès lagunaires*). Now, rigorously speaking, lagoons can be either brackish, i.e., hyposaline, or hypersaline. This creates a difficulty of principle because if we want to order the sedimentary environments in a logical way as a function of salinity of the waters, we must consider the following stages: freshwater lakes, brackish lagoons, the sea, and hypersaline lagoons. However, everyone knows that whenever French-speaking geologists say "lagoonal facies" it is traditionally understood that it is a question of hypersaline facies. Everyone also knows that the hypersalinity can occur in lakes as well as in bodies of water dependent on the sea. Although the term "lagoonal facies" is not misleading in common language, it seems that the expression "hypersaline facies" (*faciès sursalés*) is preferable; it does not evoke a special kind of geography.

Paleogeography. — From the geographic point of view, the conditions of hypersalinity can occur in numerous landscapes. TWENHOFEL (1939) has classified them very well. Excluding the deposits of springs and pedogenetic crusts, one finds continental deposits, such as hypersaline lakes, *chotts*, playas, *sebkhas*, etc., and basins communicating with the sea, among which the least open are lagoons (hypersaline). An intermediate stage is realized in interior seas, which are large lakes or small seas, according to individual opinion.

The most famous older work on the paleogeographic conditions of formation of deposits is that of OCHSENIUS (1877). Re-reading it today makes us modest as we see the insight of the older authors. The goal of OCHSENIUS was the study of the potash deposits of the German Permian. In this connection, he outlined the scheme of a lagoon represented by a gulf, a bay, or a part of sea separated from the open sea by a bar. This bar was low or penetrated by a channel. Thus, at the same time the renewal of the water and its concentration by evaporation were guaranteed. The example of Kara Bogaz or Atschi Daria, "as large as the principality of Hesse", was already invoked, as well as the first works of ABICH (1856) on the Caspian Sea and its dependencies. In 1897, ANDRUSSOV showed that a current migrated during all seasons from the Caspian Sea into the lagoon, and he described the variety of salts that precipitated there. A modern study of DZENS-LITOVSKY (1956) has increased our knowledge of this Kara Bogaz, which is at the same time the model salt-bearing basin and the typical hypersaline lagoon.

Across the years opinions oscillated around this initial scheme, but today everyone accepts it with the necessary modifications. Some authors have envisaged a continental environment, and especially WALTHER (1900) who exerted a considerable influence. For him, the term "lagoon" evoked only hyposaline environments, and he proposed both a continental and desertic origin for saline deposits. This opinion is reborn periodically in the literature, but it applies only to special cases.

The excellent discussions by FINATON (1934, 1935, 1937) and by DEICHA (1942, 1942) have shown the variety of the cases possible and the reason for oscillating series in salt-bearing formations. They take up again the study of the famous bar that was so often a source of objections to the theory of OCHSENIUS. They showed that this bar could be replaced by a shallow shelf or even by distance, that is, a long reach of shallow water; the lagoonal environment is not necessarily an isolated environment, but it is differentiated by the equilibrium between evaporation and the introduction of water into the basin (DEICHA, 1942). The paleogeographic reconstructions of SLOSS (1953) demonstrated that the great evaporite-bearing basins were dependencies of the sea and described the lithological sequences that enclose the salt deposit. In the salt-bearing basin of the Upper Silurian of Michigan and Ohio, BRIGGS (1958) gave evidence of the bar and the communications seawards. In addition, the works of Russian authors, presented by STRAKHOV (1957), as well as those of RICOUR (1960), arrive at the same conclusions. The great salt-bearing deposits are dependencies of the marine environment in places where a constraint on circulation permits the evaporation and concentration of salts.

Climate. — Evaporation is necessary. It is sufficient that this evaporation be greater than the incoming water in order for concentration to start. The latter will be hastened if evaporation increases, or if the waters coming into the basin are saline (sea water, continental lessivage of saline deposits) rather than fresh (river water, rain water). Speaking of evaporation, many authors have called upon desert climates. The term "desert" is certainly strained simply because it hardly evokes the immense aquatic expanses, either continental or connected with sea, that are indispensable to ensure hypersaline deposition. It is not necessary that the climate be desertic; it is sufficient for it to include a dry season conducive to evaporation.

Reconstruction of the past climate has to be made for each hypersaline basin. In the case of the Oligocene potash basin in Alsace, QUIÉVREUX (1935) reconstructed the flora and the insect fauna from beautiful deposits that he found in sterile beds of the potash formations. Paleobiology indicates a relatively temperate climate in which the tropical insects of the Eocene leave only a vague souvenir. The mean temperature must have been about 18 to 20 °C, whereas it is 18.1° at the present time at Algiers and 10.5° at Mulhouse (Alsace). The climate of the area around this lagoon that supported almost no aquatic fauna must have been characterized by warm, dry summers and humid, mild winters. We have also examined, in the previous chapter, the features of Permo-Triassic climates; if our reconstructions have some value, these also were not desert climates.

Thus, prudence is always necessary in the case of ancient climates and especially for those that presided over the great hypersaline deposits.

2° Clay minerals in hypersaline facies

Early works. — The inventory of clay minerals deposited in lagoonal sediments is not ancient. I believe that it is correct to situate its beginning in the works of J. DE LAPPARENT (1937, 1937). Unfortunately, there was confusion during that period when our methods of identification of mineral structures were not as perfect as today. J. DE LAPPARENT had studied earth from Attapulgus, Georgia (U.S.A.) and that from Mormoiron (Vaucluse, France). On the basis of X-ray patterns, he had defined an alumino-magnesian mineral whose basal spacing was approximately 10 Å. Since the bravaisite, "sericite-like mineral", and *Glimmerton* of that epoch also had a 10 Å basal spacing, the whole was gathered into one great family, that of attapulgites, in which glauconie also came to be placed. At this point genetic considerations enter in. J. DE LAPPARENT had correctly pointed out that the richer the argillaceous sediments were in what he called "attapulgite", the more accentuated was the saline character of the environment of formation.

If we do not take into account the former confusion between attapulgites and micaceous minerals, a fertile field remains. To be faithful to J. DE LAPPARENT, one can express it in our present language and say: "The more saline the character of the environment of formation of clays, the greater will be the development of micaceous minerals."

The next work is that of MILLOT (1949). In spite of the progress achieved in identifications after ten years, the results are not essentially different. Twenty-five samples of sediments called "lagoonal" were studied; twelve were collected from hypersaline facies in the Triassic of eastern France; three were collected from the hypersaline facies in the Alsatian Sannoisian, and ten from more or less lacustrine hypersaline facies in Tertiary formations, principally from the Paris Basin. In this last group, five samples showed attapulgite, sepiolite or montmorillonite. They form a part of the neoformed sediments of an alkaline chemical character and have already been studied.

For the twenty sediments from the large hypersaline basins, the result was constant: illite, chlorite, mixed layers, and in some cases montmorillonite. This assemblage is united under only one term, "the micaceous minerals", and these constitute 90 to 100 % of lagoonal deposits.

After long comparisons with other environments of sedimentation, and in the state of our knowledge at that time, I was led to the affirmation that lagoonal sedimentation leads to the neoformation of 100 % micaceous minerals. And I concluded as follows (MILLOT, 1949, p. 304):

1° The argillaceous sedimentation differs according to the conditions of sedimentation;

2° There is reworking within the sedimentary basin of material imported or inherited by it;

3° The physico-chemical equilibria of the environment condition the nature of the neoformed argillaceous phase.

Today we see to what extent the differences between neoformation and transformation were obscure in the mind at that time. Let us examine the more recent works.

The Permian and Triassic in Germany. — FÜCHTBAUER and GOLDSCHMIDT (1956, 1959) studied the Zechstein formation in Germany. They analysed in great detail the clay fraction of each facies in the salt-bearing series of the Zechstein, and they have shown that the clay assemblage was different for the principal facies:

— The *argillaceous facies* is characterized by muscovite and chlorite. Here the term "muscovite" designates that which we call illite. In two-thirds of the samples, muscovite is the most abundant, and in the others the two minerals occur in equal proportions;

— The *carbonate facies* shows a much more important development of muscovite, which occurs alone in 17 of 26 samples. In the others it is accompanied by quite subordinate chlorite;

— The *sulfate facies*, represented by anhydrite, contains a variety of minerals: muscovite, chlorite, corrensite, various mixed layers, and talc. The latter must be considered typical of this facies, in which it was the most abundant and the most representative;

— The *salt facies* (sodium and potassium salts) contains, above all, muscovite and chlorite.

The conditions of the genesis of these principal minerals were discussed, as well as those of the accessory minerals that were found occasionally: serpentine, pyrophyllite, montmorillonite. The results were the following:

— Muscovite was considered to be a detrital mineral;

— Chlorite was considered to be a transformed mineral because the variations in the proportion of muscovite to chlorite between the argillaceous facies cannot be considered to be variations in the detrital supply;

— Talc occurred in sedimentary beds interstratified in the series; it was considered to be neoformed in the sulfate sediment at an early stage of diagenesis;

— Serpentine, pyrophyllite and a part of the montmorillonite and of its mixed layers were neoformed, whereas another part of the montmorillonite could be related to volcanic phenomena.

The Germanic Triassic was also minutely studied in Germany. FÜCHTBAUER (1950) showed that the clay fraction of the Muschelkalk was composed principally of illite and that chlorite plays only a minor part. Then LIPPMANN (1954) discovered in the Keuper and then in the Röth (Upper Buntsandstein) a regular mixed-layer mineral to which he gave the name "corrensite" in honor of CORRENS, the master of the German school. A series of different interpretations have been given of the structure of this mixed layer (BRADLEY and WEAVER, 1956; MARTIN-VIVALDI and MACEWAN, 1960). But, in his recent review of this subject, LIPPMANN (1959) defined corrensite as a regular chlorite-montmorillonite mixed layer. This discovery is very interesting because corrensite turns out to be more and more abundant in Triassic hypersaline sediments.

The Keuper marnes in Great Britain. — It is necessary here to recall that it was in Great Britain that HONEYBORNE (1951) discovered the swelling chlorite in the Keuper marnes. The mineralogical analysis of this mineral was advanced by STEPHEN and MACEWAN (1951). Swelling chlorite is an imperfect chlorite, stable on heating, but swelling with glycol. BRINDLEY (in BROWN, 1961) gave an interpretation of it in which he showed that this mineral is a chlorite whose interlayer brucite sheet is not complete, but represented only by pillars set at intervals. Therefore, the layers of this chlorite are not bound strongly enough to oppose the swelling action of polyalcohols, but they are prevented from collapsing below 14 Å by the presence of the pillars. According to all that is known about the variegated marnes of the European Keuper, there is no doubt that this swelling chlorite is but a stage in the transformations between corrensite and chlorite under the influence of magnesian waters.

Let us recall, likewise, that KEELING (1956) discovered a sepiolite in the Keuper marnes from the Midlands. Here we have a magnesian silicate neoformed during sedimentation of hypersaline character.

The hypersaline sediments in the U.S.A. — HARRISON and DROSTE (1960) studied the argillaceous horizons of the gypsum deposits of southwestern Indiana (U.S.A.), which are of Mississippian age. These deposits are part of a calcareous and dolomitic series that includes several masses of anhydrite and gypsum. The clay fraction is composed of illite, chlorite, and illite-montmorillonite mixed layers, but it never contains kaolinite. The authors thought that these minerals came from detrital material carried into the sedimentary basin.

QUAIDE (1958) studied the salt ponds surrounding the south end of San Francisco Bay. The hypersaline muds contain illite, chlorite, montmorillonite and mixed layers and are reworked from marine and non-marine sediments of the surrounding hills, as well as from the soils that cover them. Weak transformations can regenerate mica and chlorite from mixed layers. Essentially, however, the swamp environment, which is very reducing and hypersaline, did not provoke any major changes in the sediments of the bay. The only changes are the opening of chlorite and alternation of micas in the bottom muds of the acid swamps.

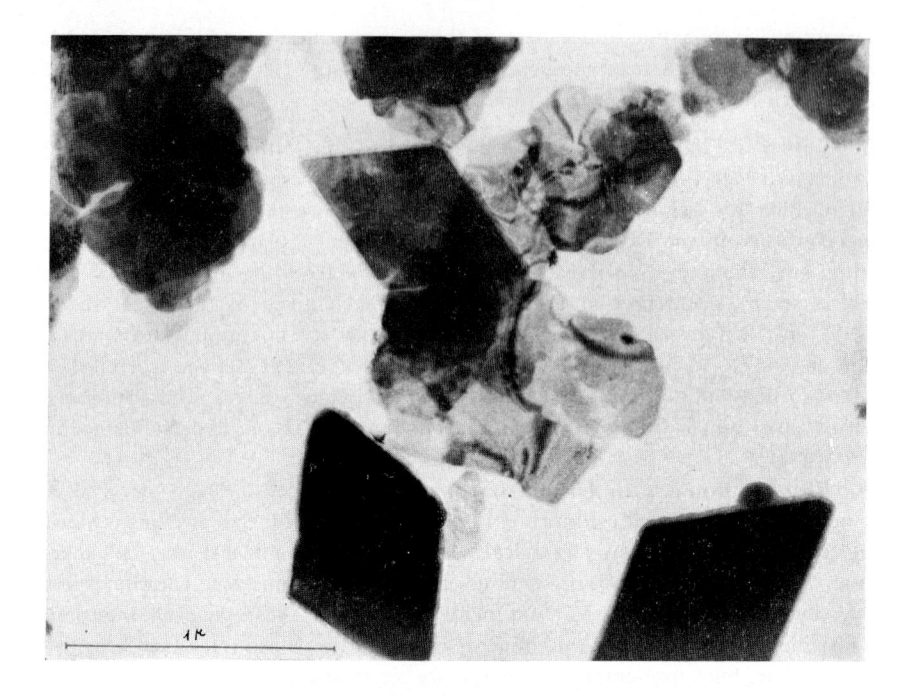

FIG. 33. — *Electron photomicrograph of particles 1 μ and smaller of chlorite extracted from the Triassic shales of Asif n'Aïr Mohamed, Demnate Atlas, Morocco* (photo by MATHIEU-SICAUD).

GRIM, DROSTE and BRADLEY (1960) studied the thin clayey beds in the potash deposits of the Permian of New Mexico (U.S.A.). The evaporite facies contain along with the sodium and potassium salts, magnesite and dolomite. The clay minerals are represented by predominant mixed layers and a small percentage of illite and chlorite. These mixed layers form a natural sequence of montmorillonite-chlorite and vermiculite-chlorite mixed layers, and the regular montmorillonite-chlorite, called corrensite, is very distinct. The authors thought, correctly, that it was highly improbable that chlorite had been introduced into the sedimentary basins to serve as parent material for these specimens; degraded products would be more probable parent materials. Starting with these open minerals, the preferential fixation of the Mg^{2+} ions between the layers leads to an increasing number of "chlorite" layers alternating with the others up to the point of recrystallization of a true chlorite. But, in the hypersaline environment studied, the most stable, thus the most common, structure is corrensite.

This is the same result that PETERSON (1961) found in a comparative study of the mineral assemblages of evaporites of Germany and of New Mexico. We have seen only the abstract of this work, but it can be seen that the clay minerals are represented principally by corrensite, vermiculite-chlorite mixed layers and the mixed layers of montmorillonite.

Thus, not only do we see once again the abundance of these "chloritic minerals" in the hypersaline facies, but also a convergence between formations of the American and European continents.

The Permo-Triassic of Morocco. — The Permo-Triassic clays of Morocco seemed peculiar to the early workers because of the extraordinary development of very beautiful, well crystallized chlorite mixed with illite or pure, in some cases (MILLOT, 1954; JEANNETTE and LUCAS, 1955). A complete study of clay sedimentation in Morocco was made by LUCAS (1962).

In one region, that of the High Atlas, the sediments are very thick, fine grained and composed of red marnes and clays with thin beds of gypsum and salt. This is the Keuper facies that invades all the series. Of 24 identifications described by J. LUCAS for the region of the High Atlas of Demnate, 20 contain more than 50 % chlorite, and 9 of these contain more than 80 %. It is an extremely well crystallized chlorite. The companion of this chlorite is illite that is similarly very well crystallized.

There are some samples in which the chlorite is pure. Figures 33 and 34 show the aspect under the electron microscope of these transformed chlorites that come from the hypersaline variegated marnes of the Permo-Triassic of the Atlas of Demnate.

An entirely different aspect is presented by the Moroccan and coastal Mesetas where the Triassic is much thinner. A typical section is that of the Chabet el Hamra, tributary of the Nefifik wadi in the region of Casablanca. The distribution of clay minerals is given by the Table XIII.

This table shows the very clear difference between the two red argillaceous Triassic formations separated by the basalt flow. The lower series shows only illite in the most sandy horizons. Then rather irregular chlorite appears. Next, there is a good development of sepiolite, which can form 80 % of the mixture. The upper series is dominated by attapulgite, which can, in many cases, constitute more than half of the fine fraction

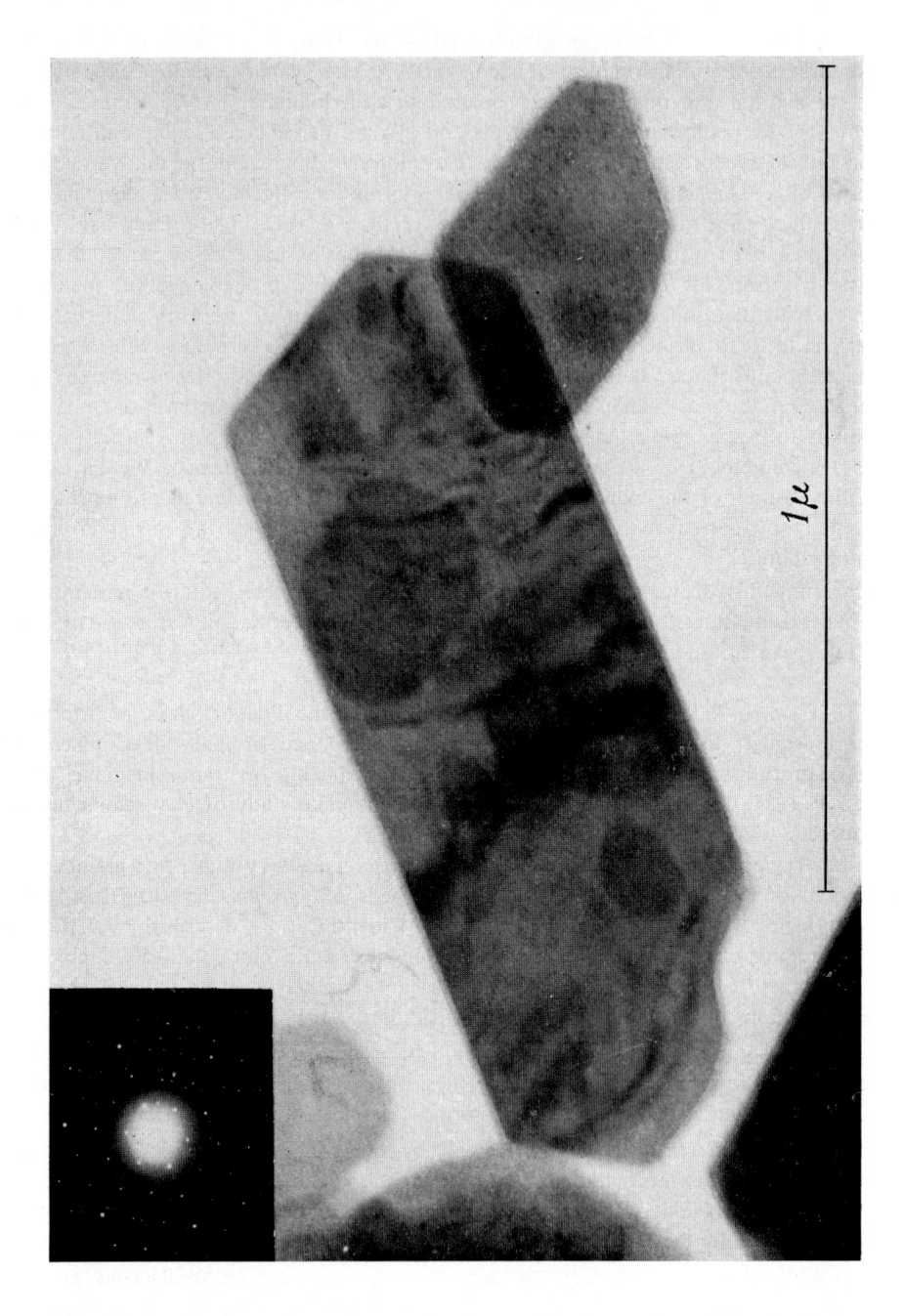

Fig. 34. — *Electron photomicrograph of particles 1 μ and smaller of chlorite extracted from the Triassic shales of Aït Oufad n'Tirhli, Demnate Atlas, Morocco. Corresponding electron microdiffraction pattern* (photo by EBERHART, Laboratoire de Minéralogie, Strasbourg).

TABLE XIII. — TYPICAL SECTION OF THE TRIASSIC OF MOROCCO IN THE EPICONTINENTAL ZONE (CHABET EL HAMRA). VARIATIONS OF CLAY MINERALS (after LUCAS, 1962)

	Sample number	Illite	Sepiolite	Attapulgite	14 Å
Basalt horizon	18	3		7	Trace
	17	3		6	1
	16	5		4	1
	15	5		4	1
	14	4		5	1
	13	1		8	1
	12	5			5 (very beautiful chlorite)
	11	2	8		Trace
	10	5	4		1
	9	7	2		1
	8	7			3
	7	7			3 (C_M)
	6	7			3
	5	10			—
	4	10			Trace
	3	8			2
	2	10			—
	1	10			—

of rock. It must be noted that this mineral appears only above a gypsum horizon formed half by illite and half by a very beautiful chlorite.

In this way, LUCAS (1962) was able to distinguish in Morocco a geosynclinal area and a coastal area. In the latter, the succession of the series with sepiolite and the series with attapulgite is the rule; their companions are 14 Å mixed layers formed of layers of the chlorite, swelling chlorite and montmorillonite. On the other hand, in the geosynclinal series, which is now incorporated in the chain of the High Atlas, the minerals are predominantly chlorite and illite, themselves coming from the mixed layers of the borderlands and characterized by a high degree of crystallization.

Here we see the combination of a mechanism of inheritance of weathered micaceous minerals, neoformation of sepiolite and attapulgite in the epicontinental zone, and transformation into well crystallized mica and chlorite in the geosynclinal zone.

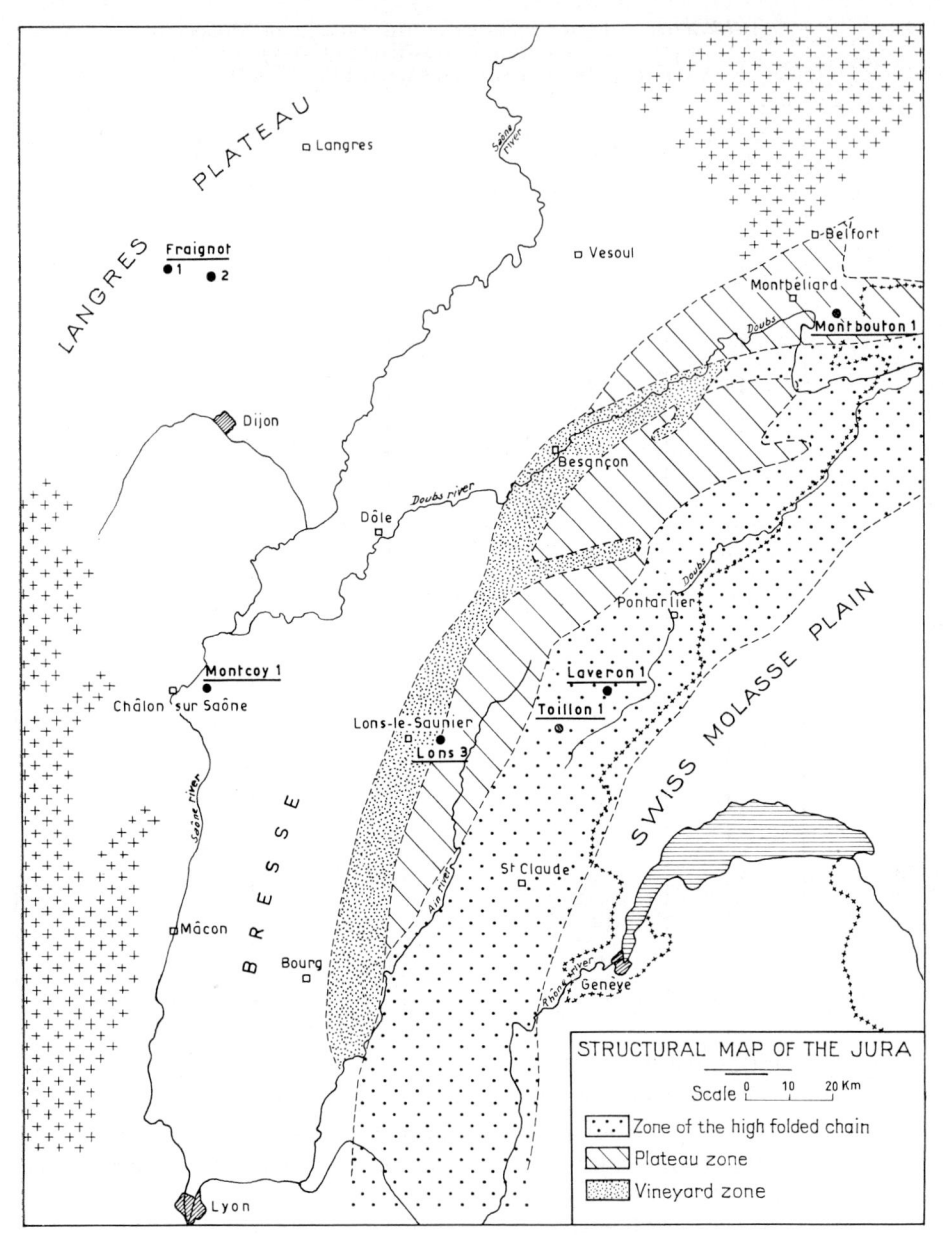

FIG. 35. — *Structural map of Jura. Location of the boreholes relating to the study of the Triassic* (after LUCAS and BRONNER, 1961).

The Triassic in France. — LORRAINE AND THE PARIS BASIN. — Study of the clay fraction of the Triassic in France began with the study of 12 samples collected in Lorraine (MILLOT, 1949); 7 were collected in the variegated or gray marnes of the anhydrite group (Middle Muschelkalk), 1 in the Lettenkohle of the Upper Muschelkalk, and 4 in the iridescent marnes of the Keuper. These rocks are composed of illite, chlorite and mixed layers that the methods of that time allowed me to classify with hydrobiotites-vermiculites. However, the study of Triassic clay sedimentation in Lorraine and in the Paris Basin was taken up again in greater detail by LUCAS (1962). Here we see once again degraded illite and mixed layers on the borders and the development of corrensite and even of chlorite in the zones of maximum subsidence in the center of the basin. This example is described in the following chapter to illustrate the paleogeographic and vertical variations of sedimentation (figs. 53 and 54).

THE JURA. — LUCAS and BRONNER (1961) and LUCAS (1962) have studied the clay sedimentation in the Triassic basin of the Jura by means of the analysis of 600 samples from 7 deep bore holes (fig. 35). The thickness of the formations can exceed 1 000 meters in the thickest part of this sedimentary basin, which had a form elongate southwest-northeast, similar to that of the present range. The Triassic contains its usual three formations, Buntsandstein, Muschelkalk and Keuper, but the Buntsandstein is rather limited. Moreover, the variations of facies are considerable. The borders of the basin tend to be sandy, and the center tends to be invaded by the facies of salt-bearing, iridescent marnes with limited, thin carbonate beds. Thus, it is only between the borders and the center of the basin that the typical trilogy of the Germanic Triassic, i.e., sandstones-limestones-iridescent marnes, is best recognized. Moreover, the calcareous facies and the facies of the iridescent marnes are charged with thin beds of gypsum and masses of halite.

The distribution of the clay minerals is not indifferent; kaolinite is not very important and is limited to the sandy basal and border facies. Everything leads us to think that this kaolinite is a detrital mineral that accompanied the sandy sediments.

Illite occurs at all levels and in all the drill holes studied. However, it is more abundant on the borders than in the center of the basin where its place is taken by the minerals described below. Moreover, it is open or weathered in the basal sandstones, in the peripheral sandstones, and in the Upper Keuper. It is better crystallized in the main body of the Triassic.

The 14 Å mixed layers are very abundant in the Muschelkalk and Keuper. They are mainly mixed layers of montmorillonite, chlorite and swelling chlorite. Their abundance and the regularity of their interstratification increase regularly from bottom to top and from the periphery to the center of the basin. In this central zone, corrensite, the regular chlorite-montmorillonite mixed layer, becomes predominant, especially in the thin salt beds.

Chlorite is never abundant and in many cases is involved in mixed-layer structures. Whenever chlorite is individualized, it occurs either in the peripheral sandstones or in the fine-grained argillaceous facies of the top and of the center of the basin. But in these two cases the chlorites are different; in the detrital facies, the chlorite is characterized by broad, weak peaks in X-ray diffraction, whereas in the fine-grained facies of the Keuper it is very well crystallized. The former is weathered and detrital; the latter is indicative of beautiful recrystallization.

This salt-bearing basin, therefore, shows transformations of its clay minerals during its history, and these transformations are not minor. In the Permian and the sandy

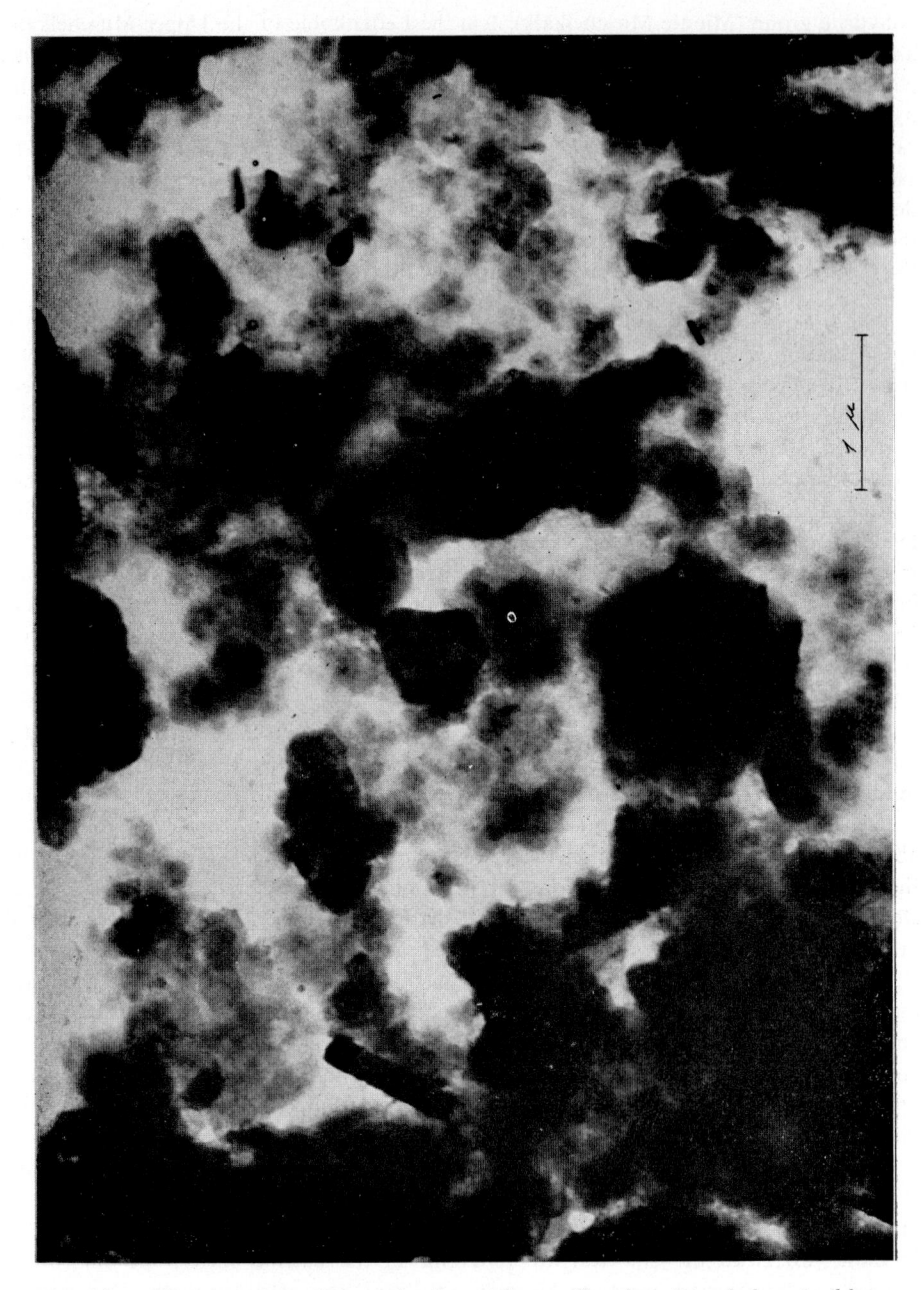

Fig. 36. — *Electron photomicrograph of particles smaller than 1 μ of the mixed-layer clay mineral corrensite extracted from the Triassic shales of Laveron, Doubs, France* (photo by EBERHART, Laboratoire de Minéralogie, Strasbourg).

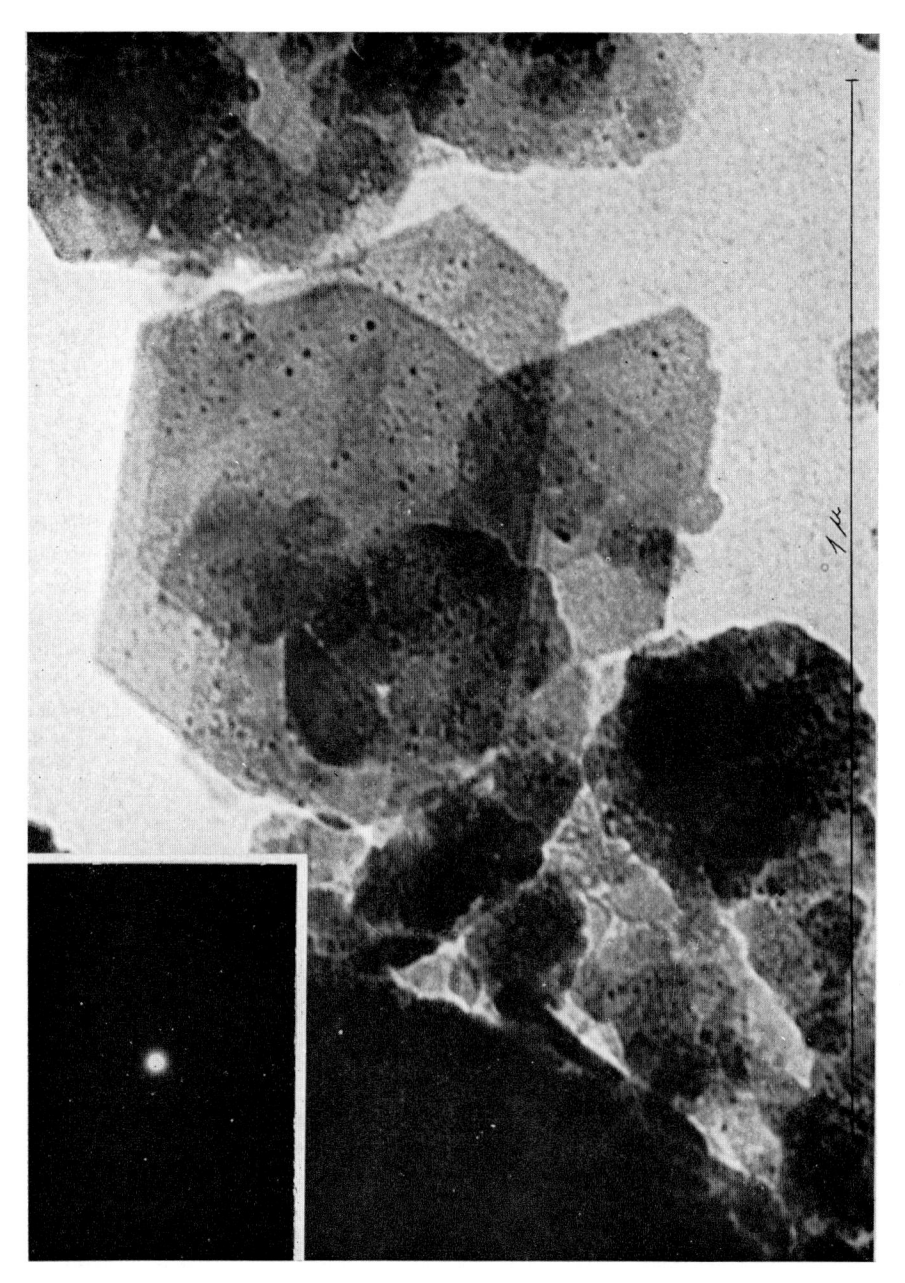

FIG. 37. — *Electron photomicrograph of the fraction smaller than 1 μ of chlorite extracted from the Triassic shales of Laveron, Doubs, France. Corresponding electron microdiffraction pattern* (photo by EBERHART, Laboratoire de Minéralogie, Strasbourg).

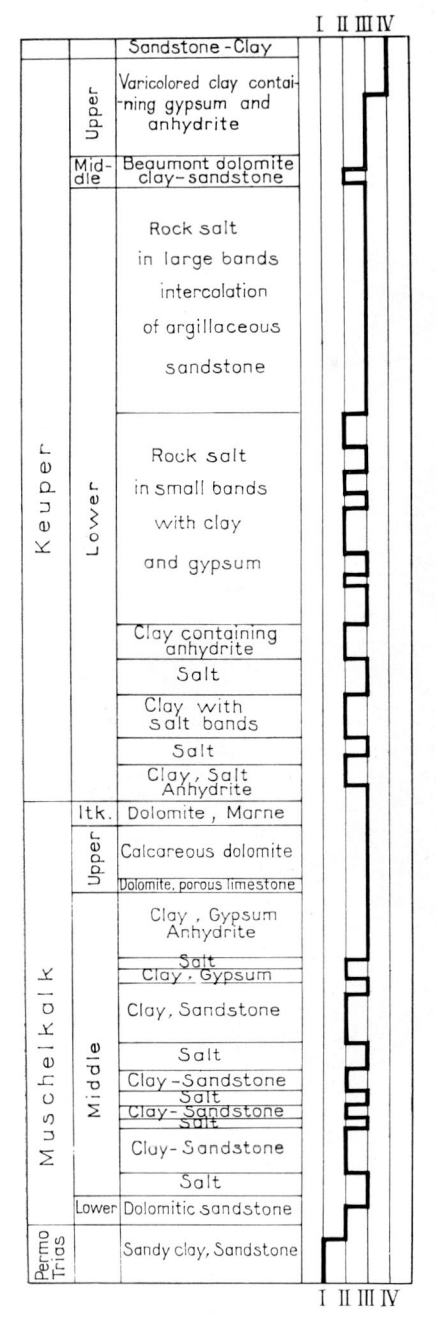

Buntsandstein, one commonly finds 70 % open illite and 30 % very irregular mixed layers, and this mixture is also found throughout the Triassic in the border facies. In contrast, in the Lettenkohle and in the Keuper, the well crystallized illite can be decreased to 40, 20 or even 0 %, while 60, 80 or 100 % of the clay fraction is composed of regular mixed layers and of well crystallized trioctahedral chlorite (figs. 36 and 37).

The most important changes relative to the littoral sediment supply occur, therefore, in the center of the basin in the upper part of the series, there where the chemical sedimentation of the Keuper is most typical. In fact, the Laveron drill hole shows repeated alternations of red clays and salt in the Keuper. These alternations are copied by those of the clay minerals: corrensite in the salt and chlorite in the marne. Lucas (1962) discussed the interpretations. It cannot be a question of variations in the sediment supply because the latter is constant on the shore line. Post-sedimentary diagenesis is likewise not responsible because it would be common to the whole basin and would tend to wipe out the variations. Lucas pointed out that it was a question of transformation.

— The sediment supplied is composed of illite and very irregular mixed layers. They are weathered products from the continent accompanied by a little kaolinite.

— The transformations toward the center and the top of the Triassic show a regularization of the mixed layers that gives rise finally to well crystallized illite, to corrensite and to chlorite.

Thus there appears to be a connection between the clay minerals and the facies.

FIG. 38. — *Borehole of Laveron, Doubs, Jura. Variation of clay minerals as a function of facies:* corrensite in the salted beds, chlorite in the argillaceous beds (after Lucas and Bronner, 1961). I. Illite; II. Chlorite+Illite; II. Corrensite+Chlorite+Illite; IV. Different mixed layer clay minerals+Chlorite+Illite.

— Chemical sediments: well crystallized illite and regular mixed layers;
— Coarse detrital sediments: degraded illite and irregular mixed layers;
— Hypersaline chemical sediments: corrensite in the salt and chlorite in the fine clays of the salt-bearing facies.

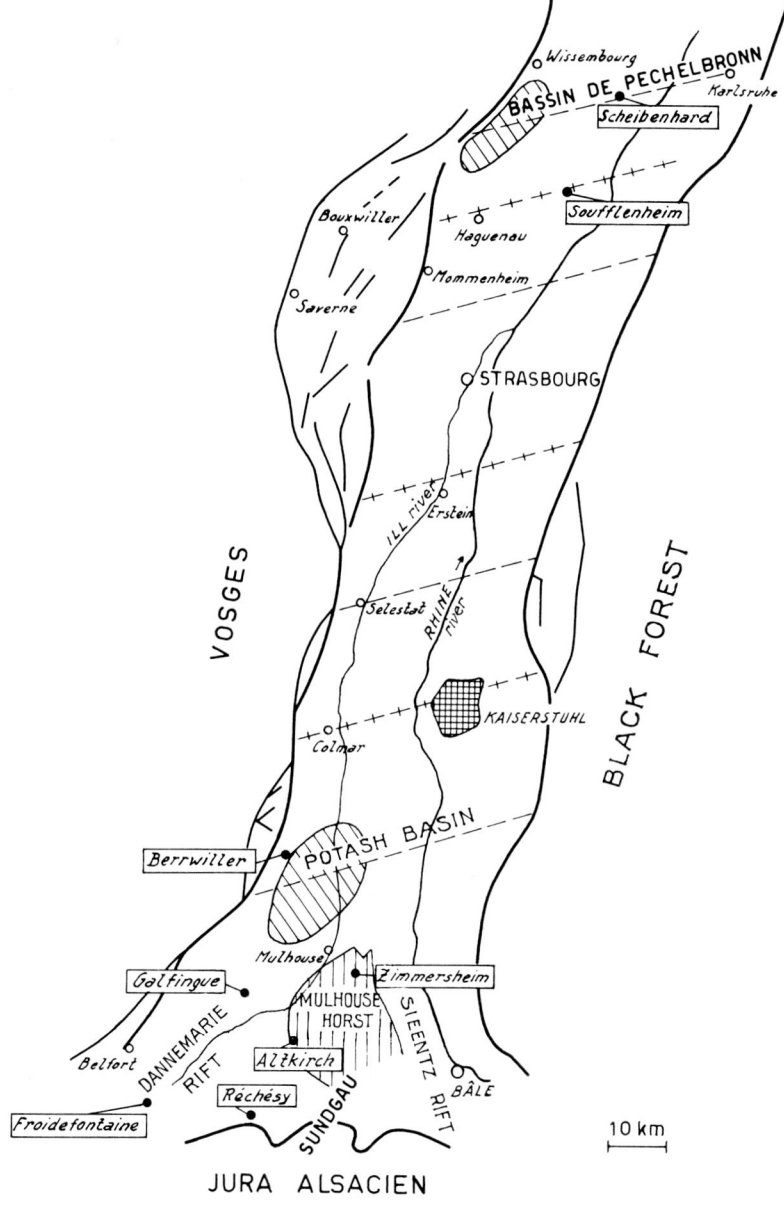

FIG. 39. — *Diagrammatic structural map of the Plain of Alsace, with locations of boreholes the sections of which are shown in figure 40* (after SITTLER, 1962).

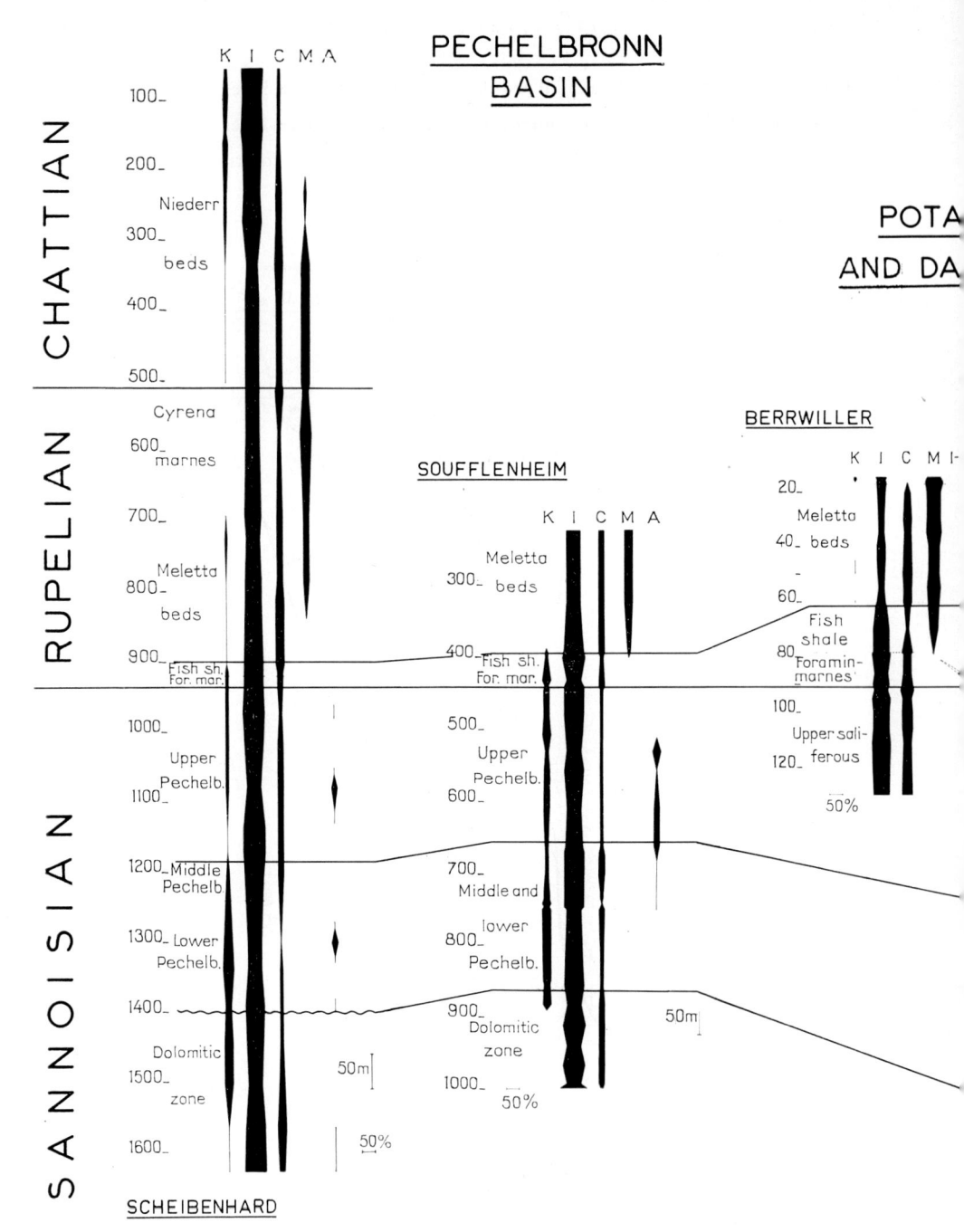

FIG. 40. — *Variations in clay minerals in the Oligocene series of Alsace based on several boreholes* (after SITTLER, 1962).

K   = Kaolinite
I   = Illite
C   = Chlorite
M   = Montmorillonite
A   = Attapulgite
I-M = Illite-Montmorillonite mixed-layer
M-C = Montmorillonite-Chlorite mixed-layer

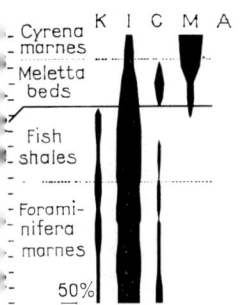

SUNDGAU AND MULHOUSE HORST

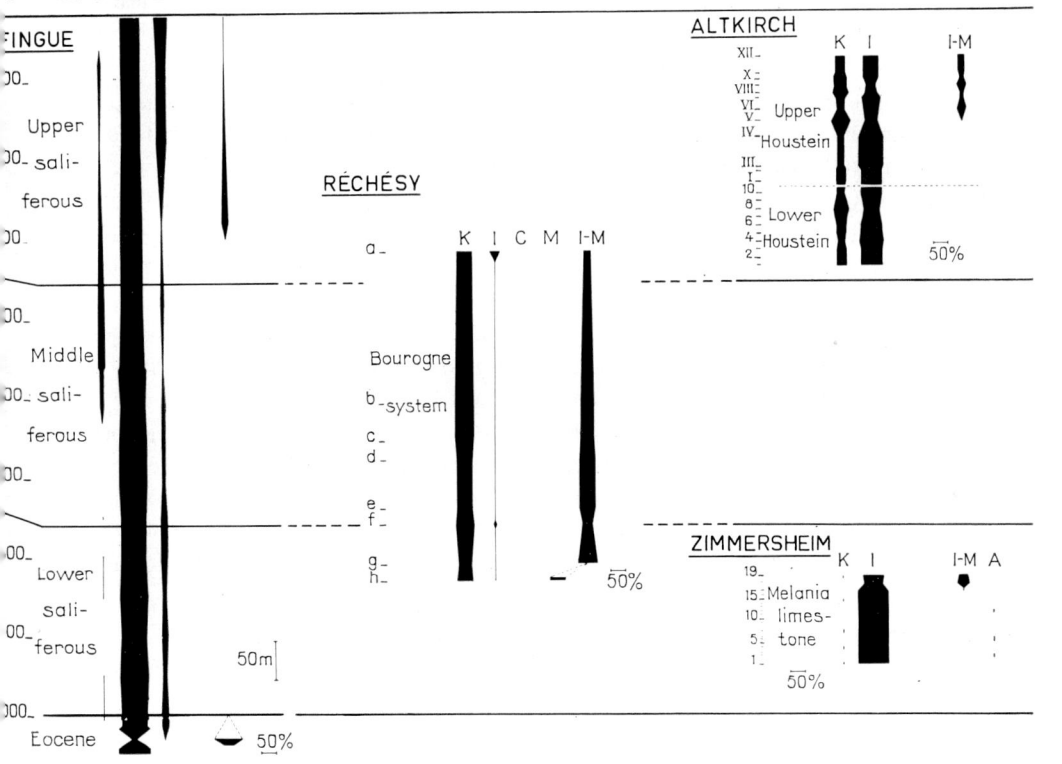

Figure 38 shows a schematic section of the Keuper in the Laveron (Doubs) bore hole and the variations of the clay minerals with the salt-bearing facies.

The study of the Triassic of the Jura is the best presently known example for showing the transformation of sedimentary clays under the influence of the sedimentary environment. Not only is there transformation of the degraded illites of the borders into well crystallized illite and chlorite and into corrensite, but also this transformation is differential, according to the sedimentary facies.

THE SALT-BEARING OLIGOCENE IN ALSACE. — The Oligocene series in Alsace is salt bearing, especially in the Sannoisian formations that include the important potash deposit of Mulhouse.

The nature of the clay minerals in the shales associated with the potash salts has been studied by MILLOT (1949), CAILLÈRE and HÉNIN (1950) and ATAMAN and WEY (1961). It is a question of a pure illite accompanied by a very small amount of mixed layers.

SITTLER (1962) undertook the systematic study of the variations in the clay fraction of the Oligocene series by means of the analysis of several hundred samples collected in the field and from cores of deep bore holes in the different basins. The map in figure 39 gives the location of the sections studied, the principal ones of which are shown in figure 40.

The overall view on clay sedimentation in this basin is the following:

— Kaolinite is certainly detrital. It is abundant in the border facies and practically absent in the potash basin. It is missing altogether in the Middle and Upper Rupelian.

— Illite is also of detrital origin. It is more open on the borders than in the center of the basin and is pure in the potash horizons. One must think of an aggradation based on weathered illites of continental origin.

— Chlorite does not occur in the border facies or on the Mulhouse horst. It is abundant in the potash basin. It is the product of the transformation of degraded minerals of the borders under the influence of solutions of the salt-bearing basin.

— Montmorillonite characterizes the Middle and Upper Rupelian, there where kaolinite is lacking. It is preceded by mixed-layer minerals of the illite-montmorillonite type and seems, therefore, to be the product of degradation of the continental illites, a much more advanced degradation than in the Sannoisian.

— Attapulgite is encountered in the absence of montmorillonite in the lacustrine Eocene and in the Sannoisian, but never later. It reveals the persistence of chemical lessivage, strong enough to supply silica and magnesia, but insufficient to give rise to kaolinization of the weathered products at that time.

These works show once again, in an attenuated but clear way, the essentials of that which has been acquired through the study of hypersaline basins:

— Sedimentary supply of degraded illites to the shores.

— Transformation, by means of aggradation, into well crystallized illite and chlorite under the influence of the K^+ and Mg^{2+} of the salt-bearing basin.

— Neoformation of attapulgite during periods of chemical weathering.

3° Conclusion

The study of the clay fraction of hypersaline sediments is very instructive. Those studies show that the nature of the clays in "lagoonal sediments" or "evaporites" is controlled by at least three mechanisms: inheritance, transformation, and neoformation.

Inheritance is obvious. It is the means by which the original material comes into the basin. Given that the climate favored the strong evaporation indispensable for the production of saline concentrations, arid seasons must have played an important part. This is the principal reason for which illite is the main mineral in the continental supplies. It corresponds to mild weathering that destroys or damages chlorite, that does not yet develop kaolinite, and that delivers great quantities of degraded illites. This is, in fact, what is observed; the subordinate kaolinite remains tied up in the detrital border facies.

Transformations appear here in all their fullness. The rhythmic variations of clay assemblages with the saline and non-saline facies are one indication of them. Another indication is provided by the variations in the composition of the clay fraction from the borders to the center of the basin and, above all, from the basal beds to the facies of chemical sedimentation typical of the Triassic. It is by means of transformation from the degraded and "gaping" micaceous minerals that not only illite, but also mixed layers develop. In them chlorite layers have increasing importance up to the point where corrensite, swelling chlorite, and chlorite itself form. These three minerals can constitute almost all of the clay fraction of the sediment.

Neoformations are evident. They give evidence of the fate within the basin of the sedimentary supplies that are no longer in the solid, but in the dissolved state. These neoformations take place either during sedimentation or in the mud after deposition; they are spectacular, whether it be a case of attapulgite, sepiolite, talc or even, for the sake of curiosity, pyrophyllite and serpentine.

The role of *late-phase diagenesis* is unknown. In any case, it is not possible to attribute to it the transformation of illite into mixed layers and then into chlorite. In fact, late-phase diagenesis applied to the sedimentary column would result in masking or obliterating the variations due to paleogeography and to the alternations of facies. This is not the case.

Much systematic is still necessary, but it appears, in the case of hypersaline sediments, that the marked chemical character of sedimentation is efficacious. It certainly acts on the inherited minerals to transform them and on the inherited solutions to ensure neoformations of new products.

Thus, inheritance is the prime mover of hypersaline clay sedimentation, but it can be slightly or completely modified by the environment itself to such a degree that transformed or neoformed clays appear.

CONCLUSIONS ON CLAYS OF MARINE SEDIMENTS

Marine sedimentation, thanks to its extensiveness and to the numerous studies it has inspired, provides us with a great variety of results and considerable documentation for the studies of genesis. The mechanisms enumerated are the following:

Detrital heritage of clay minerals
 from the continents by direct erosion,
 from continental weathered products,
 from pedogenetic neoformation.

Transformation of the inherited material under the influence of the environment:
 minor transformations, more or less apparent;
 major transformations, resulting in a new mineral association.

Sedimentary neoformation in seas characterized by a strongly marked chemical sedimentation.

Post-sedimentary diagenesis
 modifying the accumulated sediments, only in a subtle way,
 modifying the products deposited by transformation or neoformation.

That which is provided to the geologist in the form of an argillaceous rock is the summation of the effects of these mechanisms; moreover, it is an algebraic sum because some of the mechanisms act in the opposite sense. The problem consists not in discovering what is the principal factor in each case, still less in asserting it, but in demonstrating it.

CHAPTER VIII

EVOLUTION OF THE CLAY FRACTION IN SOME GREAT SEDIMENTARY SERIES

In the three preceding chapters the distribution of clays was studied with respect to each type of weathering and for each sedimentary facies. That documentation was segmented and did not correspond to any historical perspective; it was analytical in style and represented a short or myopic view of the inventory of clay minerals on the surface of the crust.

In the present chapter, in contrast, there will be an attempt to incite discussion. It will no longer be a question of facies, but of sedimentary series, series corresponding to one or several geological stages, or even to a system. Over such a long period sedimentation changes. Sequences pile up and geochemical sedimentary series unfold. Clay minerals are sensitive to this history, and they form a basis for definition of its successive stages.

Among the great sedimentary series of which the evolution of the clay fraction is already well known, four examples have been chosen:

I. The Cambro-Ordovician sandstones of the central Sahara.

II. The Carboniferous of the central U.S.A.

III. The Triassic of Morocco and France.

IV. The Tertiary basins of western Africa.

I. — THE CAMBRO-ORDOVICAN SANDSTONES OF THE CENTRAL SAHARA

1° *Regional situation*

The Cambro-Ordovician sandstones of the central Sahara form the reservoir rock of the French petroliferous deposits of Hassi-Messaoud and Hassi el-Gassi, which are exploited at present. These deposits are located on the western border of the Great Eastern Erg (fig. 41). Numerous studies have been made on these sandstones by geologists and petrographers of oil companies. To present the evolution of clays and silicifications, I shall use the summaries of KULBICKI and MILLOT (1960, 1961), STÉVAUX (1961), VATAN (1962).

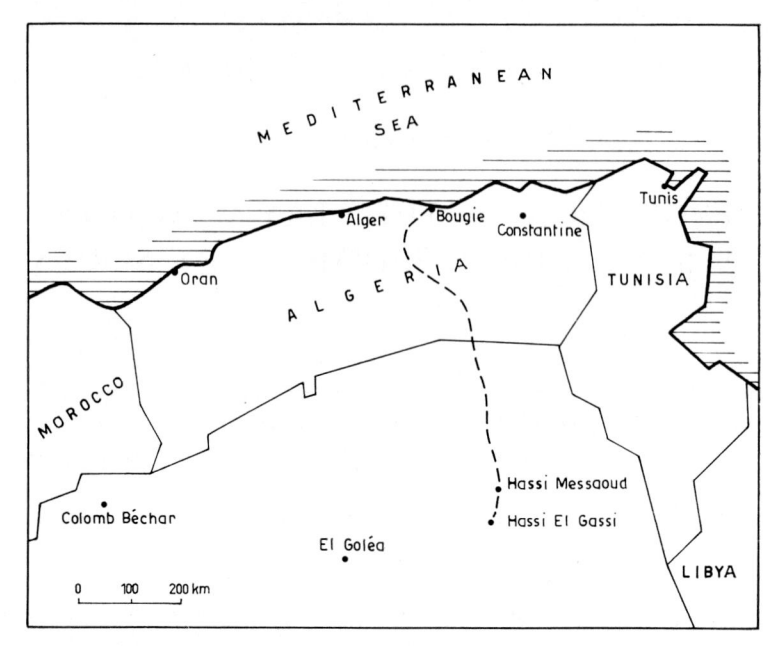

FIG. 41. — Schematic map of Northern Africa. Locations of the oil deposits of Hassi-Messaoud and Hassi el-Gassi in the Sahara.

2° The geological section

The Cambro-Ordovician series of the central Sahara includes the following units:

1. The sandstones from Hassi-Messaoud, about 150 to 600 meters thick in different sections. These sandstones constitute the "Cambrian" of the petroleum industry.
2. The lower transitional zone, intermediate between the "Cambrian" quartzitic sandstones and the succeeding black shales.
3. The black shales commonly containing glauconite and carbonates (dolomite, siderite). They develop to thicknesses of 100 meters and more.
4. The upper transitional zone, intermediate between the black shales and the upper quartzites.
5. The upper quartzites. They also can exceed 100 meters. They become silty near the top and are charged with carbonates, salt and anhydrite.

Formations 2, 3, 4 and 5 form the "Ordovician" of the petroleum industry.

Here we are interested in the sandstones of Hassi-Messaoud - Hassi el-Gassi that constitute formation 1. The irregularity in their thickness is related, in the first place, to the fact that they are commonly eroded at the top; however, this irregularity is related above all to the irregularity of the underlying crystalline basement. A diagrammatic section of these formations is shown in figure 42.

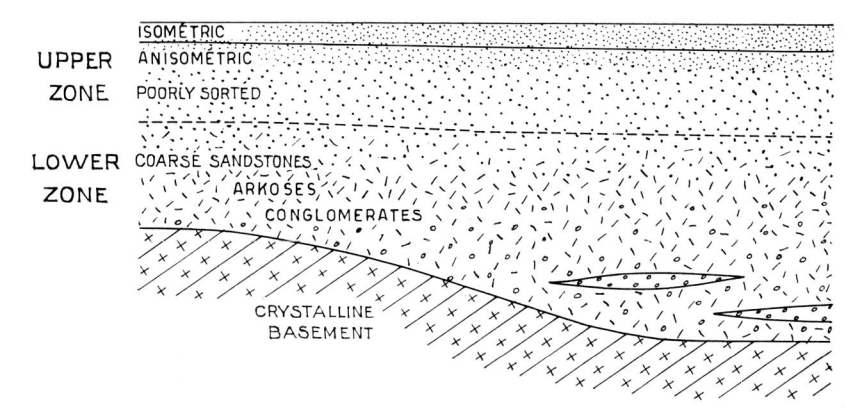

FIG. 42. — *Diagram of the variation in sandstone sedimentation at Hassi-Messaoud and Hassi el-Gassi in the Sahara.*

The series of the sandstones of Hassi-Messaoud includes the following members:

a) The basal or lower series, which is coarse, conglomeratic and begins commonly with *arènes*. It is this lower formation that has a variable thickness (175 to 450 m).

b) The upper series, which is much more regular. Its thickness everywhere is almost constant and equal to 200 m, except where it is eroded at the top. It includes:

— A coarse-grained sandy series, characterized by poor granulometric sorting (100 m);

— An intermediate quartzitic sandy series in which fine to medium grains are associated with coarse elements of quartz; it is called the "anisometric zone" and ends with a coarse horizon that gives rise in all bore holes to a "granulometric peak" (50 to 75 m);

— A very quartzitic sandy series, with well-sorted fine grains, called the "isometric zone" (25 to 55 m); quartzification here attains such a great amplitude that this zone is called in the field the "quartzitic flagstone" (*dalle quartzitique*).

Thus, the unequal thickness of the sandstone series of Hassi-Messaoud - Hassi el-Gassi stems from its coarse lower formation. This is commonly the case in the great detrital piedmont formations. The depressions on the old continental surface are filled up by the initial supply of the coarsest products. When these depressions are filled, sedimentation can occur over great distances. The material is shaped by being moved and reworked across vast pediment slopes that are traversed by seasonal streams.

The same disposition is also found in the Permo-Triassic sandstones of the Vosges, where there was filling up of the depressions of the pre-Permian peneplain by coarse deposits of Permian and Lower Triassic age, and then regularization of the granulometry and of correlations across great distances (PERRIAUX, 1961).

Each bore hole is systematically studied in the oil company laboratories. For each sample collected the granulometry and the amount of quartz, feldspars, siderite, and clay are measured. Each clay fraction is analyzed: mica, illite, chlorite, kaolinite. Moreover, the quantities of trace elements are determined. All these measurements, together with palynologic and paleontologic methods, permit correlation between the bore holes, as well as the reconstruction of the evolution of sedimentation and environment. Sedimentation of the sandstones of Hassi-Messaoud - Hassi el-Gassi has been demonstrated to be of continental origin, the marine transgression coming with the black Ordovician shales.

3° Petrography of the lower coarse series

Since the basal arkosic and coarse sandstones are the least transported and evolved, they are the ones that give the best idea of the original material, while the upper beds, more worked and altered, are less recognizable.

The rock is coarse and principally arkosic, i.e., composed of quartz, feldspars and mica. The fragments have accumulated without granulometric sorting. Fragments of lydian stone or micaschist and micro-pebbles are mixed together. The whole is angular. The cement is composed of sericite-illite and of crystals of siderite, dolomite, iron oxide and pyrite. No alimentation or rearrangement of particles is observed; the deposit is massive, isotropic and has a very poor permeability.

For the bore holes studied by VATAN (1962), the description is as follows: "The lower part of the formation is characterized by the presence of abundant and commonly sericitized feldspars. Siderite occurs in notable quantity. The quartz grains are always angular. The hydromica cement obviously originates in the decomposition of micas and feldspars. In fact, one observes micas becoming frayed and passing into the micaceous clay of the cement, while X-ray diffraction shows the weathered feldspars to be composed of illite and orthoclase. Rapid erosion gave rise to thick arkosic deposits, the lower half of which has preserved its original character."

4° Petrography of the upper series

Since 1959 the petrography of the bore holes of Hassi el-Gassi has been studied with great care by KULBICKI, ESQUEVIN and ELLOY. The main features of these formations are summarized here.

General aspect of the facies. — The upper zone results from much less brutal sedimentation. The alternations of the cross-bedded sandstones and silty beds are explained by fluviatile deposition on a low gradient piedmont. Sediments are reworked several times at the mercy of the regime of the currents and reprisals of erosion. During this winnowing the forms become rounded and the material is sorted by size, leading finally to the isometric zone.

Contrast between the clays of the sandstones and those of the silts. — In a single horizon, the nature of clays differs from the sandstone itself to the intercalated beds of silty clay. One sees a sandstone-shale contrast similar to those of

FÜCHTBAUER (1955), GLASS, POTTER, SIEVER (1956) and SMOOT (1960) mentioned in connection with sandy series. In the shales or thin silty beds, the clays are composed of a mixture of mica, illite, kaolinite and, moreover, chlorite commonly occurs. In contrast, in the sandstones, kaolinite and illite are found, and we shall see that this illite is secondary. Shale and silt being much less permeable than sandstone, they give the least modified image of the original material. When thinking of the nature of the original cement in the sandstones, one must keep in mind the nature of the silts.

Occurrence of the kaolinite in the sandstones. — The kaolinite in the sandstones occurs in the form of accordions or vermicules. This type of crystallization is found in all cases in which micas have been altered to kaolinite, and this has been described by many authors (TERMIER, 1890; ROSS and KERR, 1931; J. DE LAPPARENT, 1934; VATAN, 1939; J. DE LAPPARENT and HOCART, 1939; SCHULLER, 1951; KULBICKI, 1953; FÜCHTBAUER, 1955; KULBICKI and VETTER, 1955). Thus, we can consider that this kaolinite is secondary and comes from the evolution of a preexisting micaceous material because, as noted by many authors, this kaolinite would have been unable to withstand transport in this form. It can even be easily reconstructed that the great fan-shaped masses are the products of the kaolinization of detrital micas. In contrast, the vermicules are probably the product of transformation of small flakes of sericite that correspond to the preserved clay fraction of the silts.

In plate I, the kaolinitic facies are shown.

— Kaolinite, developed on detrital micas, the layers of which it spreads out in the shape of a fan (photos 1 and 2);

— Kaolinite, having completely separated the layers, gives rise to a large accordion (photo 3);

— Kaolinite developed as vermicules agglomerated in the interstices of the quartz grains (photo 4).

The micro-pebbles of quartzite-kaolinite. — On the scale of a thin section, micro-pebbles of quartzite with a kaolinite cement well crystallized in vermicules are often found either in sandstones or in the intermediate micaceous silts. This intraformational reworking of the kaolinite micro-pebbles, in evolved or non-evolved sandstones and in micaceous silts of a very different nature shows that the formation of the kaolinite, as well as the silicification, is very ancient and contemporaneous with deposition. The reworking that occurred was able to include some debris of kaolinized and silicified formations.

Illitization of the kaolinite. — In all cases, and they are numerous, in which illite and kaolinite are found together in the thin sections, kaolinite always forms islets surrounded by illite. Initially the illite develops in thin threads along the contacts of the quartz grains and in cracks in the rock, i.e., in the zones of possible circulation. Then, it begins transforming the kaolinite, which it digests little by little. Never in the course of study of 10,000 thin sections carried out in the S.N.P.A. ([1]) laboratories has

[1] Société Nationale des Pétroles d'Aquitaine.

another reciprocal situation of these two minerals been observed. This results in illite and kaolinite presenting astonishingly similar aspects of facies and size, owing to the fact that illite always develops on kaolinite crystals whose shapes it assumes. Plates I and II show these dispositions. Photos 4, 6, 9, 10 and 11 show the development of illite in cracks and fissures of the rock. Photos 5, 10 and 11 show a partial illitization of kaolinite within large crystals. Photos 6, 7 and 12 show a complete illitization of kaolinite in large vermicules. Photo 8 shows a complete illitization of kaolinite in small vermicules. Some of the photos show the sequence of the events; a detrital micaceous grain is squeezed between two grains of quartz. On the free extremity of the mica, kaolinite develops, separating the layers into fans. Then the neighboring kaolinitic cement is completely illitized, but this illitization reaches only the edge of the fan. It seems that this type of observation demonstrates the three stages in the history of the clay minerals in these rocks: mica, kaolinization of mica, illitization of kaolinite.

Relations with oil and saline waters. Analysis of the bore holes (fig. 46). — The petroleum reservoir of the deposits of El-Gassi is constituted approximately by the upper 200 meters of the sandy formation. Not all the wells are producing, but a correlation, at least provisional, appears to exist between the nature of the clays in the sandstones and the occurrence of oil.

— In the non-producing wells, of which F and partially D are examples, the saline water of the deposit fills up the entire sedimentary column, and there is little kaolinite; the clays that cement the sandstones are composed of illite all the way to the top of the reservoir.

— In the oil wells, kaolinite is distinctly predominant in the productive zones. It yields to illite progressively downwards in the zones where the amount of water in the deposit increases, except in well G, which is sterile and kaolinitic. This bore hole turns out to be situated in a very low structural zone, and the sandy beds are several hundred meters lower in altitude than the neighboring reservoirs of Hassi el-Gassi. They would have never been traversed by mineralizing fluids, and the former kaolinization persists.

PLATE I. — *Diagenetic evolution in Cambro-Ordovician sandstones from the central Sahara.*

FIG. 1. — Detrital mica showing bright colors of birefringence. Between the layers, pale gray kaolinite ($300 \times$).

FIG. 2. — Fanned-out extremity of opened detrital mica. Interlayer kaolinite of gray color ($240 \times$).

FIG. 3. — The detrital mica is almost completely transformed into kaolinite ($240 \times$).

FIG. 4. — Another aspect of kaolinite within a mass of vermicules ($150 \times$).

FIG. 5. — Illite replaces kaolinite between the layers of mica during a second diagenesis ($300 \times$).

FIG. 6. — Complete illitization of opened, previously kaolinized micas ($240 \times$).

1

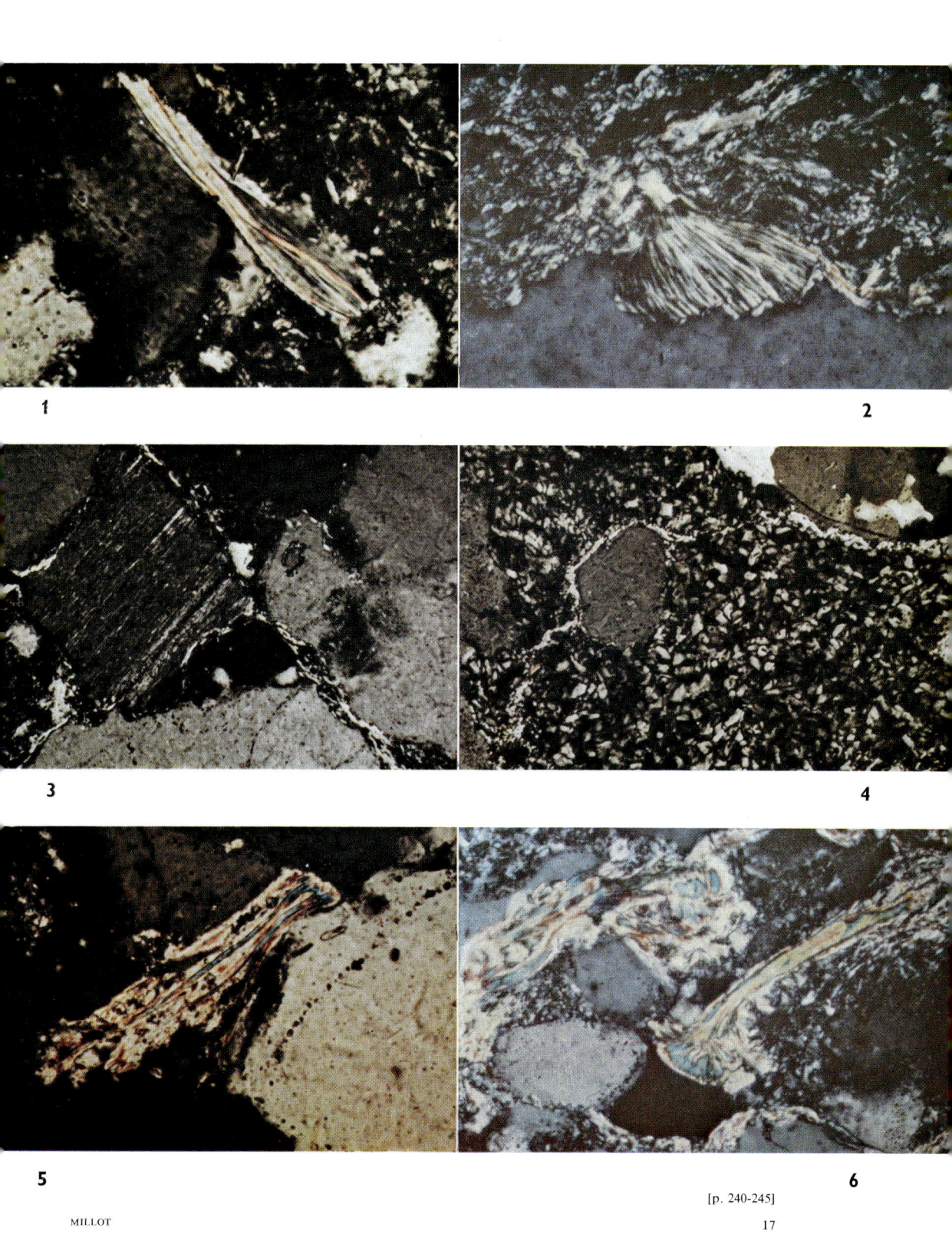

2

3

4

5

6

[p. 240-245]

17

PLATE II. — *Diagenetic evolution in the Cambro-Ordovician sandstones from the central Sahara.*

FIG. 7. — Complete illitization of a previously partly opened and kaolinized mica (240×).

FIG. 8. — Evolution of kaolinitic vermicules into vermicules of illite during a second diagenesis (300×).

FIG. 9. — Illite develops along the cracks in the sandstone and between the quartz and the cement (240×).

FIG. 10. — Illite forms along the edge of a large crystal of kaolinite, the latter being neo-formed on a mica (240×).

FIG. 11. — Illitization progresses in the cracks of the rock, on the edges of the crystal and between its layers (240×.)

FIG. 12. — Complete illitization of large, previously kaolinized crystals (240×).

7

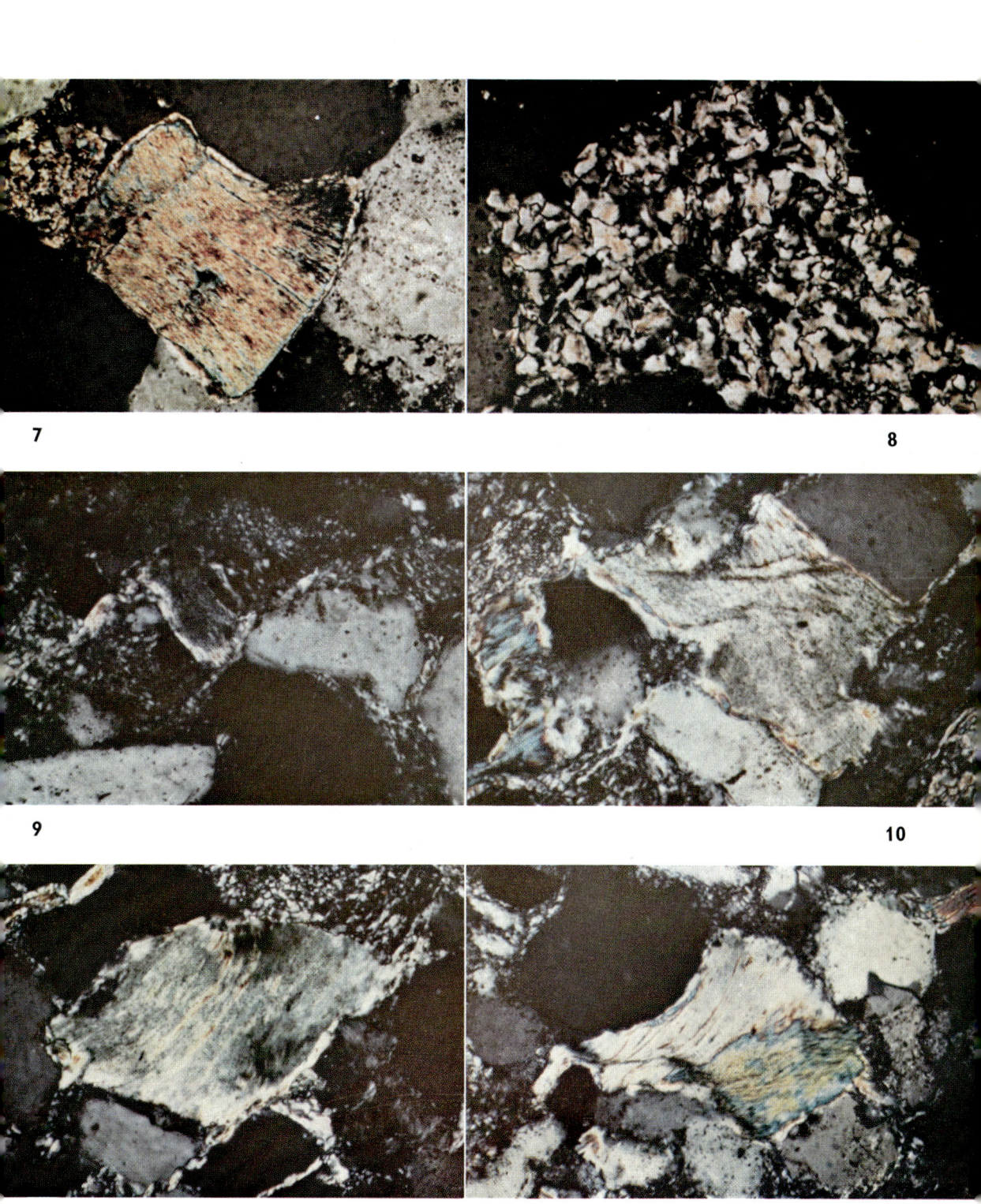

8

9

10

11

12

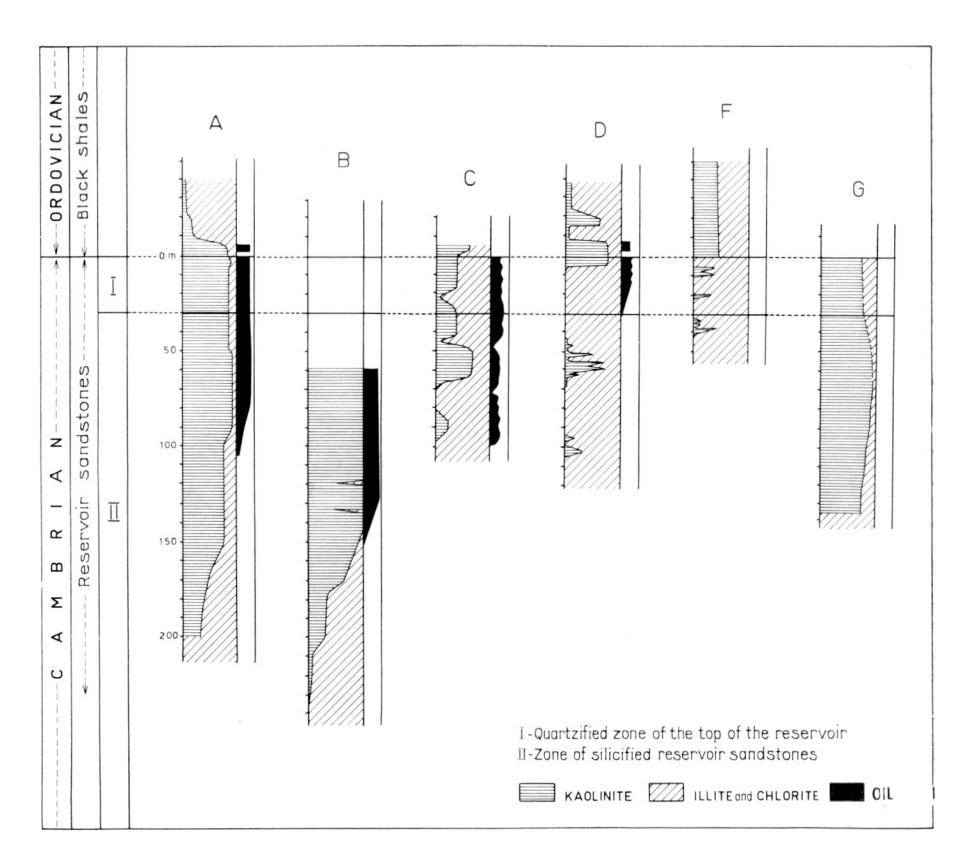

FIG. 43. — *Variations in the argillaceous fraction and in the oil content of some Saharian boreholes* (after KULBICKI and MILLOT, 1960).

— Thus, an indirect connection appears between the occurrence of oil and of kaolinite; the present hypothesis is that the oil protects the kaolinite from the action of saline waters which everywhere have the power of neoforming illite. This protective role of the oil recalls the works of CHEPILOV, ERMOLOVA and ORLOVA (1959) which show the cessation of diagenetic mechanisms upon arrival of oil; this gives information on the age of the migration. Likewise, FÜCHTBAUER (1955) has given evidence of the absence of the neoformation of quartz in reservoir rocks, whereas such neoformation occurs in sterile beds. Lastly, SMOOT (1960) has shown a greater amount of kaolinite in oil-bearing beds than in sterile ones; he attributed this fact to the cessation of diagenesis. Here we come upon a mechanism that will be valuable in the study of petroleum migration and in establishing the chronology of these displacements.

5° *Reconstruction of the history of the sandstones of Hassi-Messaoud - Hassi el-Gassi*

The initial coarse sedimentation. Paleoclimatology. — The material deposited during the first period has been described. Being rich in feldspars, micas and sericite from weathering, it gives evidence of the rapidity of burial of the reworked arkose. This burial created a reducing and alkaline environment that gave rise to the neoformation of siderite and pyrite and to the continuation of the sericitization.

These coarse sediments from the "hills" on the peneplain filled up the basins. The material, therefore, was not transported far and rapidly accumulated in these lowlands.

An attempt can be made to reconstruct the climate presiding over original weathering. It is a question of limited arenization permitting hydrolysis of ferromagnesian minerals, releasing iron, and hydrolysis of plagioclase, feeding the sericitization, but it hardly goes beyond that. This can be attributed to a climate that was not aggressive, or to rapid erosion, or to both. In any case, it is certain that, with such low relief, it was not a question of a tropical humid climate because these climates never provide preserved feldspars nor abundant mica. It seems that warm and alternating climates should be called upon, like those we reconstructed in the case of the great red-bed series. The humid period sets off the hydrolysis and sweeps away the *arènes*. The dry period inhibits vegetation and fixes the iron.

It seems difficult for the moment to use the information that can be drawn from present-day weathering of the basement rocks beneath the Cambrian sandstones. The latter are accessible in outcrops, where they show intense kaolinitic weathering. By drilling, likewise, progressive kaolinization of the basement granites has been reported. This is a contradiction of the preceding facts and can be interpreted in two ways. Either this weathering was terminal and coincided with the time when the sandy series buried the topographic relief, no longer permitting rapid erosion; kaolinization could have occurred before burial. Or this alteration took place long afterwards under the influence of circulating ground water; this would have been "crypto-weathering". Such is the opinion expressed by BIROT, CAPOT-REY and DRESCH (1955) and BIROT and PINTA (1961). The reason, without doubt, is that one does not see why so loose a material would not have taken part in the detrital sedimentation of the epoch.

Be that as it may, we encounter here moderate weathering giving rise to arkose that came from the Precambrian topography of the area.

Evolution of the sedimentation. Kaolinization. Silicification. — After the depressions of the peneplain had been filled up during the Saharan Cambrian, the sediment came from a much greater distance and underwent much more intense modifications. It was reworked across wide piedmont surfaces of low gradient. Feldspars disappeared, grains became rounded, and calm zones permitted the deposition of clayey silts. Zones of more rapid flow produced sands mixed with large flakes of mica and cemented by various clay minerals, the approximate composition of which is given, with a minimum of deformation, by those very silts.

During that time the transgression was beginning, and well before its waters invaded the area and deposited the black shales, it raised the base level insensibly, decreased the flow velocity, made the courses of the streams even or meandering, allowed only limited transportation of sediment, which was reworked an indefinite number of times until the isometric zone was formed, marking the end of the exclusively continental regime. It was during this second period of sandy sedimentation that kaolinization and silicification took place. We are assured of this fact by the witness of the micropebbles; they correspond to the continual reworking by erosion of the sandy areas that were being transformed. The waters, ceaselessly renewed by the flow from upstream, were more and more pure and more and more acid since they were leaching a material from which cations were quickly removed by hydrolysis of the feldspars. They flowed through the permeable formations either as ground water during the dry season or as fluvial water during the humid season. The large micas were kaolinized into fans; the sericite flakes were kaolinized into vermicules; some silica was removed and ensured quartzification by the mechanism that we shall study in the next chapter.

VATAN (1962) reconstructed this evolution in the same way. Observing that the upper sandy series had a kaolinitic cement, he emphasized: "The transition from the illitic clays, rich in silica, to kaolinite, poorer in silica, must be accompanied by release of silica. This, therefore, explains the fact that there where is no hydromica cement there is no secondary silica, and that, in contrast, secondary silica is accompanied by kaolinite." VATAN quantified these transformations. Calculated with alumina constant, the epigenesis of hydromica to kaolinite releases 20 % silica and 15 % alkali. The silica gives rise to quartzification, whereas the alkalis are removed; this loss is apparently responsible for the porosity of the reservoir.

Emplacement of hydrocarbons. Persistence or illitization of kaolinite. — The sedimentary series has been buried since the Ordovician transgression and during a long Paleozoic and Mesozoic history. At still poorly defined epoch, posterior, of course, to the emplacement of the overlying impermeable Triassic beds of all the reservoirs, the oil-bearing deposit was formed.

The silicified and kaolinized sandstones serve as reservoirs. There where the oil collects, kaolinite is protected and persists. In contrast, there where marine water enters in, illite develops on a large scale from the kaolinite, as the observations suggest. This transformation affects the large kaolinites formed from detrital micas, as well as the small kaolinites derived from small sericite flakes.

This is diagenesis under the influence of migrating saline water. The ease of the transformation can be understood if one thinks not only of the saline concentration of the solutions, but also of the temperature.

The photographs of plates I and II show different stages of this illitization.

Evolution of boron. — Study of the trace elements has been made and that of boron is especially instructive. Boron occurs in concentrations close to 60 to 80 ppm in the lower series, which is rich in mica. This amount drops to 20 ppm after kaolinization, while remaining at 60 or 80 ppm in the silts that correspond to the slightly deformed original material. Let us note that the secondary illitization of kaolinite does not change the concentration of 20 ppm. In the black "Ordovician" shales the amount rises and oscillates between 85 and 120 ppm. This is an indication of the marine transgression.

6° Conclusion

This is an attempt to reconstruct the history of the sandy series of Hassi-Messaoud. It is based on innumerable measurements of the workers at the research center of the Société nationale des Pétroles d'Aquitaine in Pau (Basses-Pyrénées, France). Because it is only an attempt, it is fragile and asks only to be modified by new observations and new arguments.

As now developed, based on our most recent knowledge of detrital series, of silicification, and of the evolution of clay minerals, it teaches us several things:

1. The clay fraction of the sandstones of Hassi-Messaoud has undergone a three-fold history:

    — It has a *detrital origin*, like that of the formation including it.

    — It has undergone *early diagenesis*, penecontemporaneous with sedimentation, by kaolinization of micaceous debris and by silicification of detrital quartz under influences of a climatic character.

    — It has undergone *late-phase diagenesis* by illitization of the preceding kaolinite under the influence of underground migrations of saline water.

2. The *kaolinization* is an evolution due to the *circulation of acid water* in a piedmont area. It is tied to *silicifications*; these are two parallel diagenetic mechanisms.

3. *The entry of oil into a reservoir rock blocks diagenesis therein.* In contrast, its absence or its departure *permits* the high temperature *saline waters to illitize kaolinite.*

4. *The argillaceous cement of the sandstones is the living part of these rocks.* Its study provides information about their history, which can involve several episodes.

II. — THE CARBONIFEROUS OF THE CENTRAL U.S.A.

1° *The American Carboniferous*

Three great regions must be distinguished within the American Carboniferous.

— The Carboniferous of the Atlantic border (Newfoundland, Nova Scotia, Appalachian Mountains) is analogous to the Carboniferous of the European Hercynian and Caledonian mountains, very thick and commonly folded, with a thick, detrital Mississippian (Dinantian) and a productive Pennsylvanian (Westphalian and Stephanian).

— The Carboniferous of the central states is analogous to that of the Russian platform. It includes a Mississippian composed of an alternation of limestone and shale corresponding to the Tournaisian, Visean and Namurian, and a Pennsylvanian preserved in four great coal-bearing basins (Michigan, eastern interior basin, western interior basin and Texas).

— The Carboniferous of the American West, i.e., of the Rocky Mountains and Pacific coastal ranges, is much folded, distinctly marine and devoid of coal; it would be comparable to that of the Urals.

It is in the Carboniferous sediments of the center that much work has been done on clays because of coal and, above all, petroleum research.

2° *Study of argillaceous sediments of the Carboniferous of Oklahoma*

(WEAVER, 1958, 1960)

Regional situation and general section. — Figure 44 shows the regions studied. Principally these are the Anadarko basin and the Ardmore basin in western Oklahoma. Farther to the east is the Ouachita geosyncline and farther west, the basement rocks of the Wichita Mountains and their prolongation in the Criner Hills.

Figure 45 shows diagrammatically the distribution of clay minerals throughout the Upper Mississippian and Lower Pennsylvanian series of southern Oklahoma. The section trends NW-SE and shows the distribution in four major facies the history of which WEAVER (1958, 1960) was able to reconstruct as a function of the orogenic history of these regions. In the shales, the clay minerals are the reflection of tectonic activity and of the zones of alimentation.

History of the argillaceous sedimentation. — During the Meramecian, the Ouachita geosyncline was receiving illite and chlorite from crystalline basement rocks and the foreland of the basins received, above all, illite which probably came from the reworking of older deposits (Lower Mississippian and Devonian) in which this mineral is predominant.

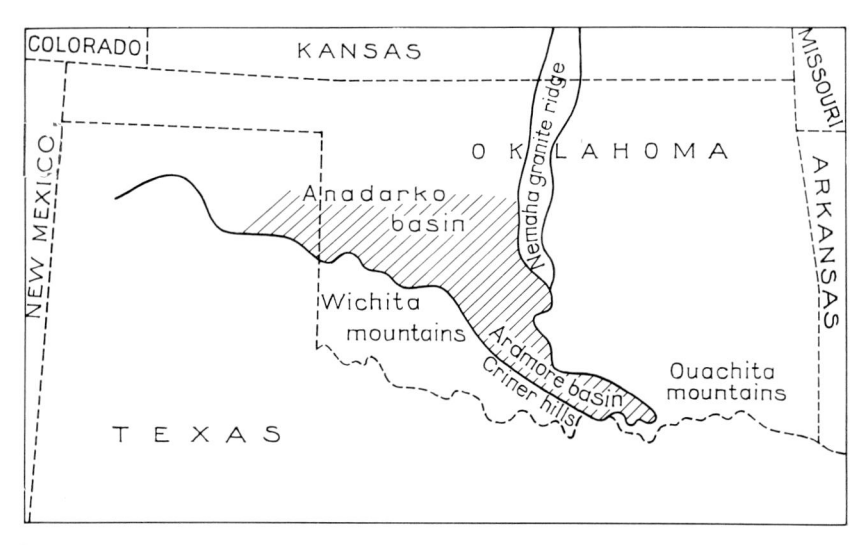

FIG. 44. — *Diagrammatic structural map of the sedimentary basins and of their source area during the Mississippian of Oklahoma, U.S.A.* (after WEAVER, 1958).

At the end of the Meramecian, considerable volcanism developed in the southern part of the Ouachita geosyncline. The volcanic ash is preserved in the southern Ouachitas, but in the north it has been transformed into montmorillonite. The Caney beds in this region are very rich in montmorillonite in their terminal part.

The Upper Mississippian corresponds to the Chester and Springer formations grouped

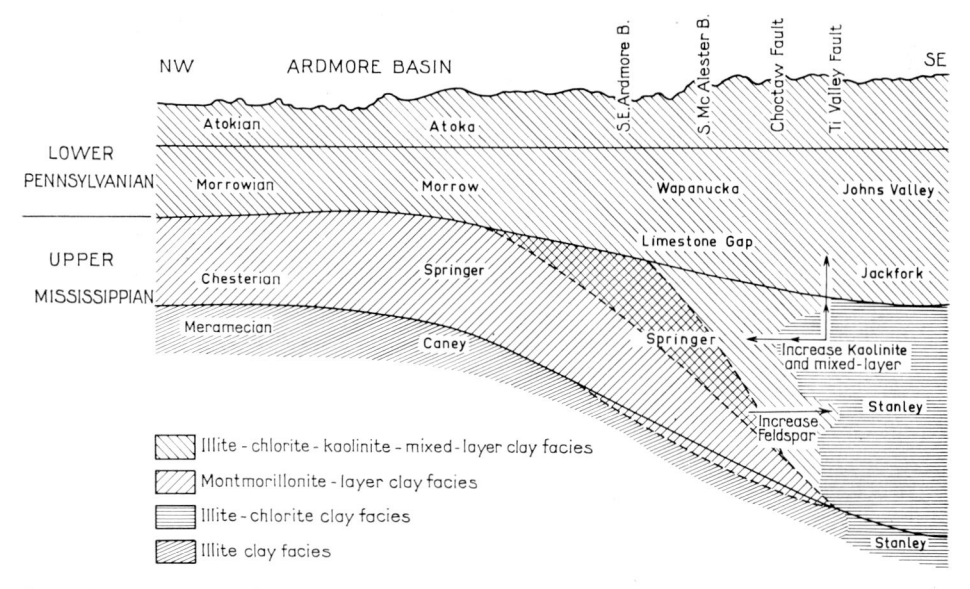

FIG. 45. — *Generalized cross section through southern Oklahoma, showing distribution of major clay mineral facies* (after WEAVER, 1960).

together in the Chesterian Series. During this epoch, a southern source of supply provided the Anadarko basin and the western Ardmore basin with clays of montmorillonitic type and with sandstones of orthoquartzitic type. During that time, a metamorphic zone of supply, situated to the SE, sent a series of illitic and chloritic clays as well as sandstones of graywacke type into the Ouachita geosyncline. A tongue of this illitic-chloritic facies spreads all the way into the eastern Ardmore basin. At the contact between these two facies a mixture is found with less montmorillonite than in the east and more montmorillonite than in the west.

Before the end of the Mississippian Period the basin was little by little filled up, and thus a glauconitic sandstone was able to cover over the greater part of its surface. The southern zone of supply bearing montmorillonite was hardly active any longer. But the southeastern zone was rejuvenated. Illite and chlorite coming from this zone were poured into the Ardmore and Anadarko basins all the way to the west and gave rise to a mixed zone about 200 feet thick everywhere in the basins; this period is not drawn on the diagrammatic section.

Then, the old Wichita Mountains and the Criner Hills began to rise, while being covered periodically with ash and tuff. They provided the second supply of altered montmorillonitic clays from the Upper Mississippian beds. Along the southern faulted flank of the Ardmore basin are found zones of shales with rather pure montmorillonite contained within a conglomeratic series which is the reflection of reactivation of the border faults with periodic discharge of detritus northward into the basin.

In the Pennsylvanian System the first formation is the Morrow Formation, which covers the Chester and Springer Formations. They present a mechanical mixture of the underlying and overlying beds. This produces a mixed facies in which (2 M) illite is abundant and is accompanied by mixed layers, kaolinite and chlorite. This permits the correct placement, during drilling, of the limit between this formation and the underlying formations which are richer in montmorillonite.

The Atoka Formation is rather similar but contains more mixed layers. This is due to an orogenic phase that separates the Morrow from the Atoka and that gave rise to a new supply of degraded minerals from the south. Farther northwest in the Anadarko basin, a distinct chlorite-vermiculite mixed layer appears in the Atoka Formation and is very characteristic of the Des Moines Formation which follows. It seems to come from a zone of supply farther to the west.

In the Ouachita geosyncline, the chloritic and illitic sedimentation continued throughout all the Mississippian (Stanley Formation). In the Pennsylvanian, the Jackford shales contain a (2 M) illite and mixed layers along with small proportions of chlorite and kaolinite. The Johns Valley shales contain a similar mixture, but it is much richer in chlorite and kaolinite. The Atoka facies becomes similar to those of the western basins.

On the whole, during all the history of these epicontinental basins and of the nearby geosyncline, there was a conflict of influence between two categories of sediment supply: the sediments of metamorphic origin which accumulated more readily in the Ouachita geosyncline, and the sediments of montmorillonitic character which came from the south and were dispersed in the Ardmore and Anadarko basins. But each domain tended to encroach upon its neighbor. WEAVER compared this situation to that which we know at present in the southeastern U.S.A. The great plain of the Gulf of Mexico evacuates montmorillonite which is, in part, of volcanic origin, while the region of the Appalachian Mountains with its crystalline basement sends forth illite, chlorite and also kaolinite. These materials characterize sedimentation in the Gulf of Mexico and on the Atlantic coast, but interact in the northeastern part of the Gulf of Mexico.

Stratigraphic and petroleum interest. — These studies on the clay fraction of the Carboniferous shales of the Oklahoma basin facilitated correlations during oil-well drilling. In particular, equivalence in age of the Chester and Springer Formations could be established by the same clay mineral suites. One could also distinguish the upper, Morrow and Atoka, formations from the underlying ones, and that had not been possible by microscopic methods. Within the Springer Formation, one was able to make subdivisions that are very useful for petroleum prospecting. The productive shales are, in fact, in the upper part of the Springer Formation. When the latter is reached by drilling after crossing a fault, it is very important to know if the drill is above or below the productive zone. This avoids regrettable confusion in the course of drilling; for example, drilling was stopped in error when Des Moines beds had been taken for those of the Morrow Formation, and those of the Morrow for those of the Caney Formation. Clay minerals permit one to locate himself and to avoid such errors.

FIG. 46. — *Schematic cross section in southern Oklahoma showing electric log correlations* (upper diagram) *and clay mineral correlations* (lower diagram) *between two sections approximately 15 miles apart* (after WEAVER, 1960).

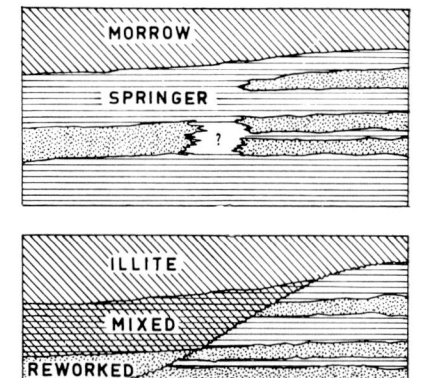

An example of correlation with geophysical methods (electric resistivity) has been given by WEAVER (1960). Figure 46 shows the results obtained by the electrical method and by the clay mineral method. This gives evidence of the unconformity of the Morrow beds on the Springer beds. And although the reworked layer at the base of the Morrow Formation is probably reworked from the Springer beds, the two are not equivalent.

Lastly, there is a correlation between the amount of oil and of clay minerals of montmorillonitic type or similar, swelling mixed layers in the sedimentary rocks. This relation is noted in Carboniferous sediments in Illinois as well as in Oklahoma, where it is the Chester beds that are richest in oil and in minerals with variable basal spacing. This correlation was presented by WEAVER (1960) for numerous formations, and it is also noted by Russian workers. This is explained by the quantity of water of adsorption and also by the high pressures that are necessary to expel this water of adsorption from the swelling minerals. If, in addition, clay minerals play a role by means of surface effects in the evolution of hydrocarbons, one can conceive that it would be those with the most developed internal surface that would facilitate the genesis of oil. These questions are being studied now and can be of great importance in petroleum prospecting.

3° *Study of argillaceous sediments of Illinois*

Very exhaustive studies have been made on Carboniferous sediments of the eastern interior basin within the state of Illinois. These works have already been summarized in Chapter VII in connection with the study of the clay fraction in detrital series.

The Mississippian System is represented by the Chester Formation, and the study of the clays was carried out by SMOOT (1960). Several effects were pointed out:

— The evolution of the clay fraction of sandstones in comparison to that of shales. The former are richer in kaolinite and mixed layers, the latter are richer in well crystallized illite and chlorite.

— The statistical abundance of kaolinite, which is greater in the oil-bearing sandstones than in the barren sandstones, the arrival of the oil being able to prevent some diagenesis (fig. 43).

— The horizontal distribution of clay minerals in the ancient paleogeography. Figures 47 and 48 show an example of this distribution. Figure 47 shows the distribution for the Cypress Formation, and figure 48 for the ensemble of the Paint Creek, Bethel and Renault Formations. One sees the nature of clay minerals vary with distance from the source; toward the south mixed layers evolve into illite in an environment that is clearly marine (see also figs. 31 and 32, p. 190 and 191).

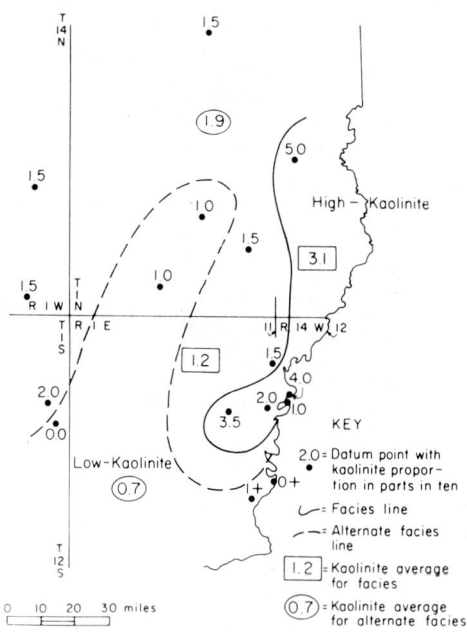

FIG. 47. — *Proportions of kaolinite in samples from the Cypress Formation* (Mississippian age) *in Illinois, U.S.A.* (after SMOOT, 1960).

The Pennsylvanian System was studied by GRIM, BRADLEY and WHITE (1957),

GLASS, POTTER and SIEVER (1956) and SIEVER (1958). Its clay fraction is constituted by a rather complex mixture of illite, kaolinite, chlorite and mixed layers. The main results have already been given in Chapter VII; they show the following principal features:

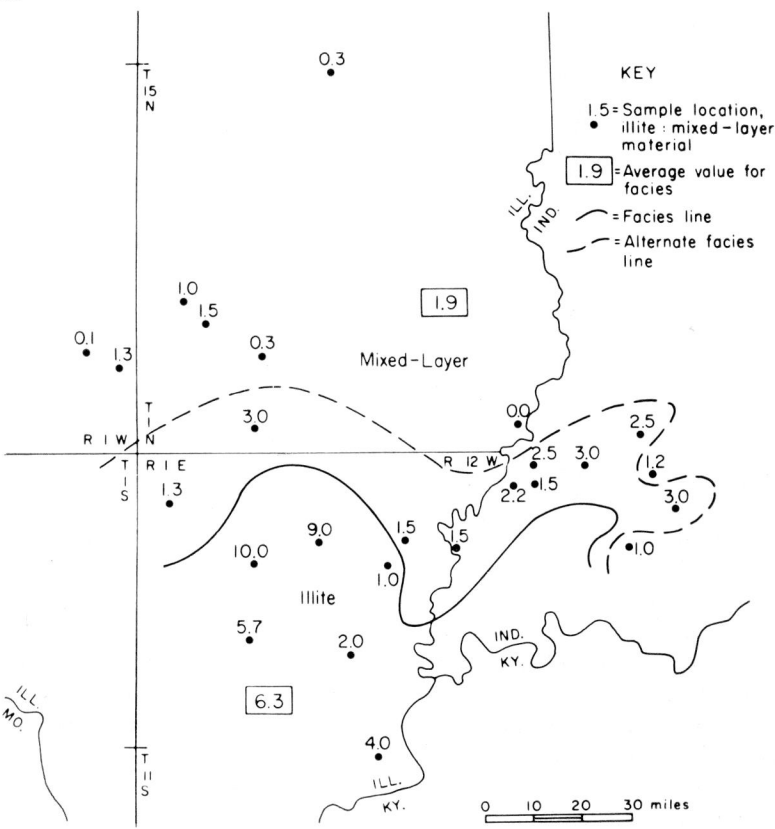

FIG. 48. — *Variations of the ratio of illite to mixed-layer clay minerals in the Paint Creek, Bethel and Renault Formations, Mississippian of Illinois, U.S.A.* (after SMOOT, 1960).

— The samples collected from outcrops are richer in kaolinite and mixed layers than core samples. This results from weathering.
— The amount of kaolinite in sandstones is higher than that in shales. This kaolinite is commonly in the state of a fragile aggregate, in the form of "books" or vermicules. This results from diagenesis. The quantitative comparison based on 60 samples is given in figure 49.
— The strata underlying coal beds (underclays) are richer in kaolinite and mixed layers than the overlying strata; this results from weathering contemporaneous with the coal-producing swamp.
— The clay fractions of shales depend on differences in the supply of detrital minerals, the shales alternating with quartzitic sandstones and graywackes. This is a result of heritage.

These examples studied in detail show the respective roles of heritage, of modifications occurring during sedimentation, of weathering immediately subsequent to deposition, of late-phase diagenesis and of recent weathering.

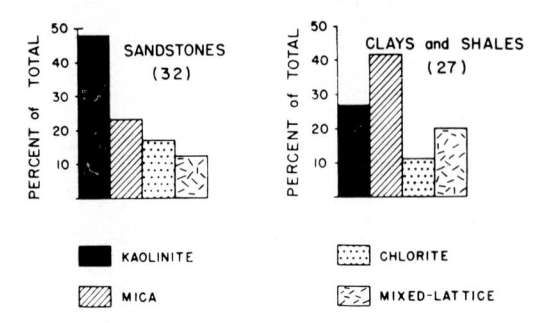

FIG. 49. — *Histograms showing contrasting average clay mineral composition of basal Pennsylvanian outcrop sandstones and of clays and shales in Illinois, U.S.A.* (after GLASS, 1958).

4° Study of sediments associated with fireclays in Missouri

KELLER, WESCOTT and BLEDSOE (1954) described the deposits of the Cheltenham Formation in the Pennsylvanian of Missouri. These deposits include beds with diaspore and fireclays and present important variations of facies.

The Cheltenham Formation of Missouri covers an erosion surface on Paleozoic (including Mississippian) dolomitic and calcareous formations. This surface has been deeply worked by dissolution and transformed into karst. The Pennsylvanian deposits are in the form of a wedge thinning southwards against the Ozark dome and thickening northwestwards. Erosion has reworked a considerable part of the Pennsylvanian deposits, which still persist only here and there as remnants on the flanks of the dome, as well as in karstic depressions. The cover beds are constituted by more recent Pennsylvanian formations, thus dating the deposits to an emergence between Mississippian and Pennsylvanian time.

Going from southeast to northwest on the Ozark dome, one sees in succession a region characterized by diaspore and flint clays (hyperaluminous fireclays) with deep karstic pits, then four successive, concentric zones of flint clay, semi-plastic and semi-flint clays, refractory earths, and finally shales. Figure 50 shows this distribution. The conditions of formation of each zone have been deduced from observations of structural geology and paleogeomorphology, and from characteristics of the sediments formed in a series of marshes and swamps that supplied sediment to the solution basins and the karstic depressions. The original material of the earlier sediments was principally illitic.

The diaspore- and boehmite-bearing deposits had formed in a non-marine environment. Intense lessivage by acid waters charged with organic matter converted the parent rocks and clay minerals into aluminous minerals and well crystallized kaolinite (fig. 51).

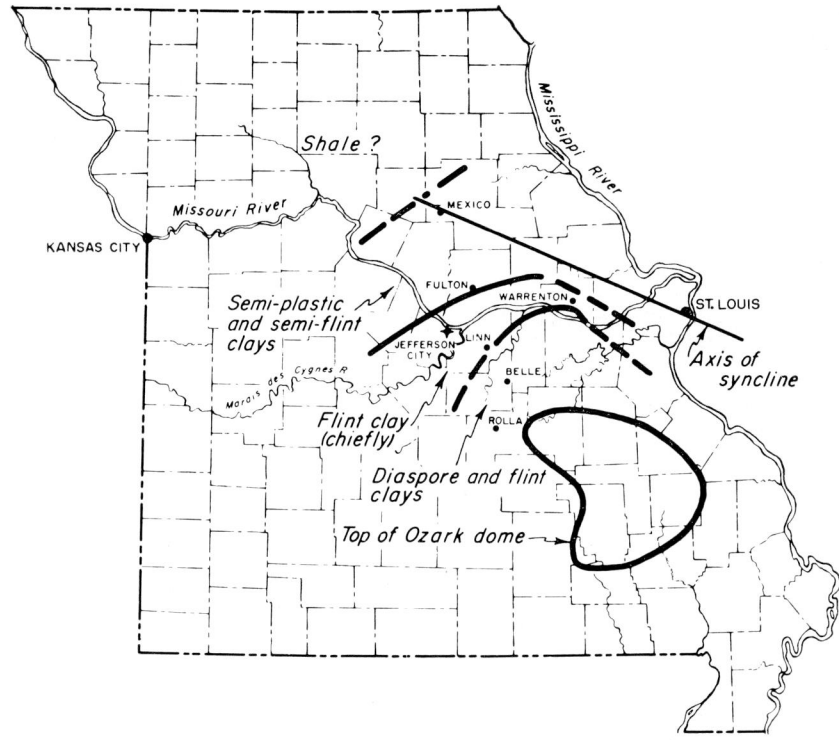

FIG. 50. — *Map of Missouri, U.S.A., showing the distribution of the different facies of the Cheltenham Formation* (after KELLER, WESCOTT and BLEDSOE, 1954).

Toward the Cheltenham sea, on the flanks of the uplands, more moderate lessivage led to flint clays.

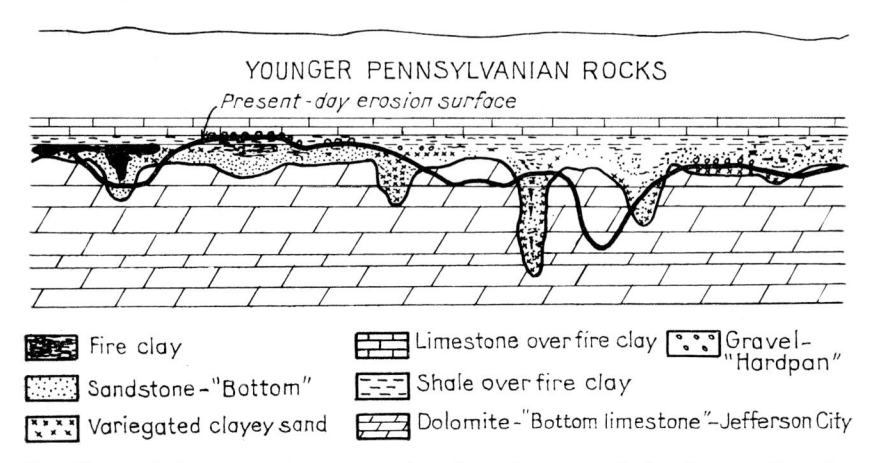

FIG. 51. — *A diagrammatic cross section through a part of the diaspore-flint clay district of Missouri* (after KELLER et al., 1954).

In the zone of semi-flint and semi-plastic clays, illite becomes abundant. It is thought that the environment was less acid, or alternately acid and alkaline. It is possible that the swamps were occasionally brackish, inhibiting thus the transformation of illite into fireclay mineral. It is also likely that the more rapid deposition decreased the time for lessivage prior to weathering of the parent illite into clay of the kaolin type (fig. 52).

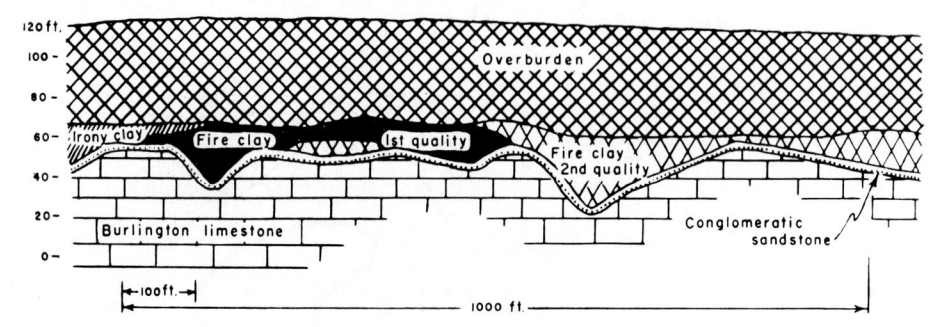

FIG. 52. — *A diagrammatic cross through a semi-plastic and semi-flint fire clay pit, Missouri, U.S.A.* (after KELLER et al., 1954).

The shales situated to the northwest of the region are illitic and are interpreted as a typical marine deposit.

This interpretation of progressive transition of the environments of formation from the continental to the marine implies that the clay minerals were in equilibrium with their environment and that they formed, *grosso modo*, where they occur at present.

Here we find once more the conditions of formation of a siderolithic facies with variations in intensity of lessivage and transport. There is direct action of the environment, but one will note that this environment is an environment of lessivage with a continental character.

5° Conclusions

Studies on American Carboniferous sediments are quite numerous. Here I have presented only a few examples. The principal lesson that can be deduced from them is the following:

1. Mississippian and Pennsylvanian sedimentation had a *predominantly detrital character*. In an orogenic period, each depressed zone received from the emergent massifs its lot of detrital products. The detrital supplies changed as the surrounding landscapes were being deformed.

2. Within the whole mass of detrital material, nuances can be observed. In particular, if a suite of weathered minerals is delivered to the sea, the kaolinite is easily deposited along the coasts with the detrital products, whereas the weathered minerals reach to open sea. There *transformations can take place* under the influence of the marine environment leading to the reconstruction of well crystallized illite from mixed layers.

3. Lastly, if a zone emerges from the sea and is subjected to humid tropical conditions within a continental regime, *strong weathering* begins, leading to aluminous products rich in aluminum oxides and kaolinite, with a gradation of facies as a function of the intensity of lessivage.

III. — TRIASSIC ARGILLACEOUS SEDIMENTATION IN MOROCCO AND IN FRANCE

(LUCAS, 1962)

J. LUCAS has presented a synthesis of his studies on Triassic argillaceous sedimentation in Eurafrica (JEANNETTE and LUCAS, 1955; LUCAS and BRONNER, 1961; MILLOT, LUCAS, WEY, 1961; LUCAS, 1962). This synthesis concerns Triassic basins in Morocco, the Jura, Lorraine and the Paris Basin, complemented by data on the Triassic of the Sahara, Spain and the Pyrenees.

1° *The Triassic of Morocco*

The results of the works on Morocco have been presented in Chapter VII concerning hypersaline sedimentation. The broad outline of the results obtained is given here. LUCAS distinguishes two regions: an epicontinental region, that of the Moroccan Meseta and of the Casablanca basin, and the geosynclinal region of the High Atlas. In the epicontinental and salt-bearing basin, an attapulgite series succeeds a sepiolite series. These two minerals are not alone; they are accompanied not only by illite but also by 14 Å minerals in which layers of chlorite, swelling chlorite and montmorillonite are interstratified. Since the detrital supplies are composed of degraded micaceous minerals, one must think of the addition of two mechanisms: a neoformation of sepiolite, then of attapulgite, from solutions and, during the same time, a transformation of the micaceous debris into 14 Å mixed layers.

In the geosyncline, the evolution involves only transformation and leaves no place for neoformation. But the transformations are spectacular. Starting with the micaceous debris found on the borders of the basin, one attains in its center true trioctahedral magnesian chlorites with perfect crystalline shapes, as shown on figures 33 and 34. This chlorite is accompanied by similarly well crystallized illite. The transformation leads to both of these end members in parallel fashion.

2° *The Triassic of the Jura*

Here, too, a detailed exposé was given in Chapter VII. The Triassic argillaceous sedimentation in the Jura corresponds, once again, to that of a basin of strongly marked subsidence of geosynclinal type. The series has a thickness of 70 m on the margins (bore hole of Fraignot, Côte d'Or) and of about 1 300 m in the center (bore hole of Laveron, Doubs). The sediments supplied to the basin are illitic, and accessorily chloritic, degraded minerals. It is possible to see two evolutions, one from the borders to the center of the basin, the other from the bottom to the top of the series. In the both cases, we see a continuous progression from the degraded illite, by way of illite-montmorillonite mixed layers, to the organization of a regular mixed layer, which is the corrensite of LIPPMANN (1955), and then to well crystallized chlorite (fig. 36 and 37). The evolutions are less strong than in Morocco, but one sees corrensite develop preferentially in the saline facies and chlorite in the clays interstratified between the salt beds. The regular alternation of clay minerals with the alternating Keuper facies emphasizes the role of the environment in these transformations (fig. 38).

3° The Triassic of Lorraine and the Paris Basin

Twenty-two deep bore holes have been studied in Lorraine and in the Paris Basin. The clay minerals identified are essentially illite and 14Å chlorite-montmorillonite mixed layers. The mixed layers are most commonly irregular, but they tend in places toward corrensite. True chlorite is rather rare and little abundant. Lastly, neoformed attapulgite was encountered. Figure 53 shows the situation of the bore holes studied across the Triassic of Lorraine and of the Paris Basin, such as reconstructed by

1 Zone of Triassic outcrops
2 Undivided Trias
3 Upper Keuper
4 Schilfsandstein (Middle Keuper)
5 Clastic dolomite (Middle Keuper)
6 Rock salt of Lower Keuper
7 Upper Muschelkalk, calcareous-dolomitic
8 Middle Muschelkalk with sandstone faci
9 Middle Muschelkalk with anhydrite faci
10 Lower-Middle Muschelkalk
11 Vosgian sandstone

FIG. 53. — *Western boundaries of the principal Triassic formations in Lorraine and in the Paris basin* (after RICOUR, 1959).

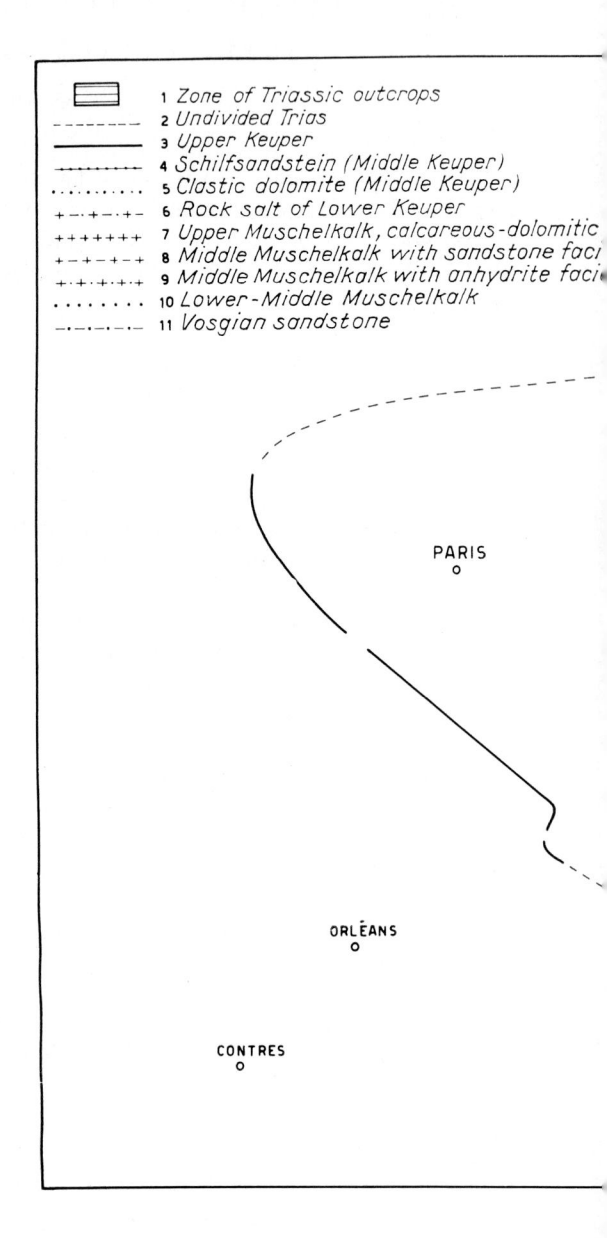

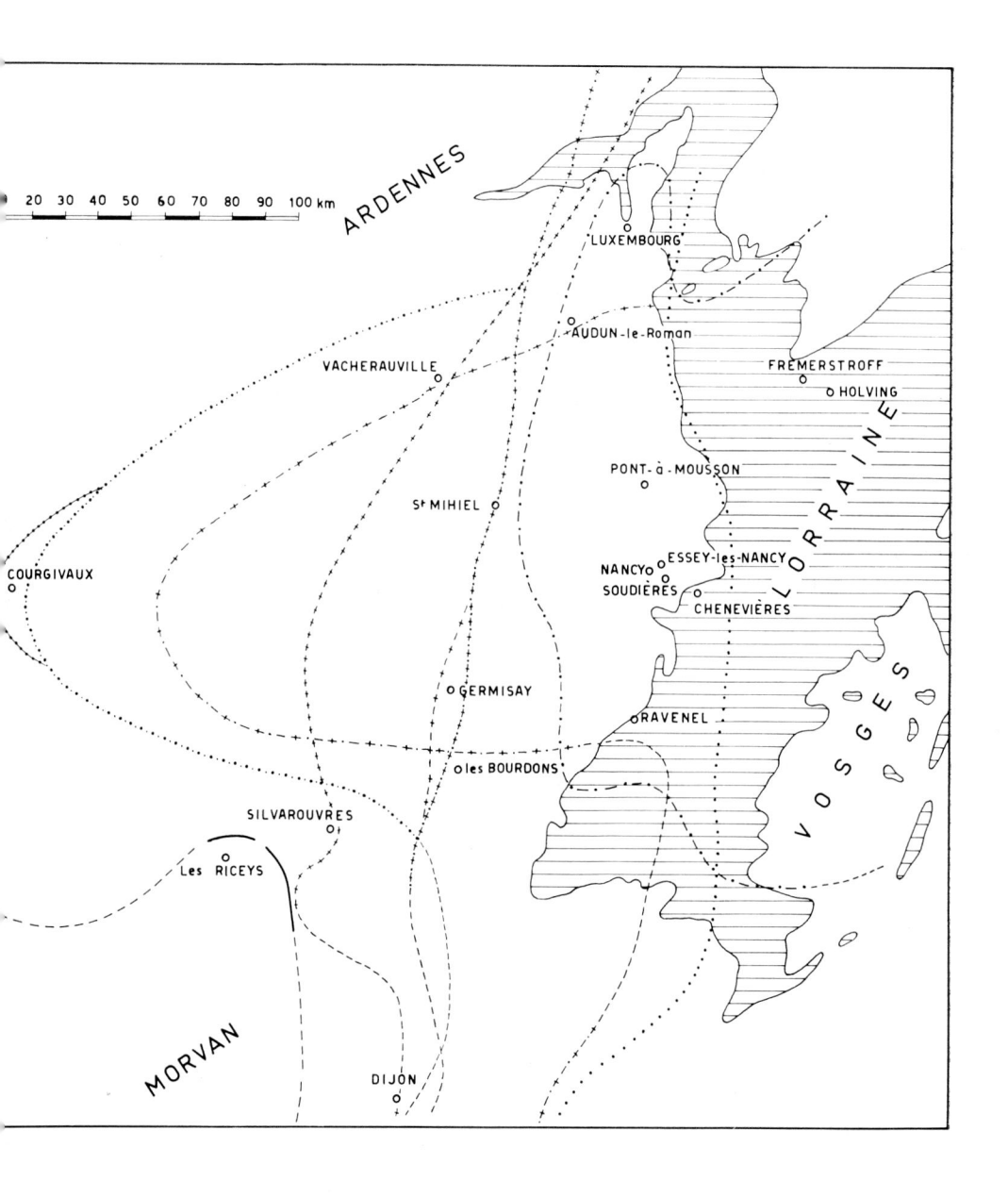

RICOUR (1959). In all the bore holes one sees a slow variation in the overall composition of clays. In the lower sandy levels, open illite is dominant by far, accompanied by a small percentage of degraded chlorite and by little kaolinite. As one rises through the Triassic series, one sees the illite decrease at the expense of 14 Å minerals. The bore hole of Ravenel (Vosges), shown in figure 54 emphasizes this slow variation. Other bore holes show the same drift, but jerkily. At the top of the Lower Keuper and at the bottom of the Upper Keuper, the quantity of minerals in the 14 Å complex present in the mixture can reach 80 % of the clay fraction in contrast to 5 to 10 % at the bottom. In the thickest parts of the Triassic of the Paris Basin 70 % of the mixture is composed of an almost regular chlorite-montmorillonite mixed layer, close to corrensite, and even a well crystallized chlorite (bore hole of Courgivaux, Seine-et-Marne). There are horizontal variations of same amplitude that correspond to these vertical variations. At Riceys (Aube) and Silvarouvres (Haute-Marne), which were on the borders of the Triassic Paris Basin, the different horizons of the Triassic are scarcely recognizable because they are very detrital. Open illite characterizes the entire section. Then, going toward the north or the center of the basin by way of Les Bourdons (Haute-Marne), Germisay (Haute-Marne), Saint-Mihiel (Meuse), one sees the 14 Å mixed layers become more frequent and more abundant. There is an increase in regularity that parallels the increase in abundance. At Saint-Mihiel, one finds up to 70 % mixed layer very close to corrensite. At Vacherauville (Meuse) and Audun-le-Roman (Moselle), the situation is similar. Lastly, at Courgivaux (Seine-et-Marne) mentioned above, which is the westernmost, being situated in the center of the basin, corrensite and a well crystallized chlorite are present.

Attapulgite has been found locally in the bore holes of Fremerstroff (Moselle) and Holving (Moselle). This point is very important; not only because it shows the similarity of this Triassic sedimentation with that in England and Morocco, but also because this mineral is neoformed and is evidence of the active presence of products in solution, especially silica and magnesia.

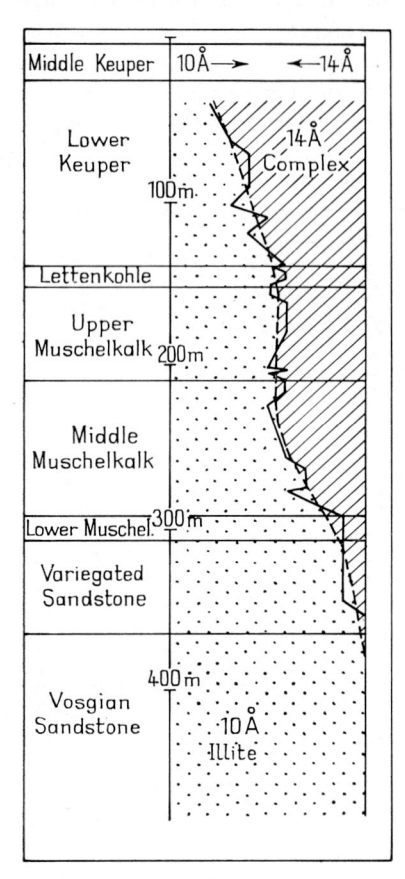

FIG. 54. — Section of the borehole of Ravenel (Vosges). Variations in clay minerals (after LUCAS, 1962).

4° Comparison with neighboring basins

In this itinerary of research, going from Morocco to Lorraine, comparisons are possible with neighboring basins.

In the Sahara, the Triassic is very detrital at its base and is characterized by illite. It passes rapidly to the salt-bearing facies and the illite is replaced by mixed layers of corrensite type.

In Spain, in the coastal Catalan basin, one observes the same phenomenon as in the Paris Basin: illite characterizes the base of the Triassic, and chlorite-montmorillonite mixed layers occur in the upper levels. Going toward the Pyrenees, the Triassic series thickens and well crystallized chlorite appears and can become dominant. Once again, the zones of great subsidence, as in Morocco and in the Jura, are conducive to growth by transformation of a magnesian chlorite, accompanied by well crystallized illite.

Conclusions

The work of J. LUCAS (1962) is very important *because it demonstrated with certainty the existence of transformations of silicates during sedimentation.* It is not a question of true neoformation, except in the case of palygorskite, because no mineral is formed entirely from solutions. It is also not a question of heritage because the particles inherited by the basin are illitic and degraded. It is a question of an intermediate mechanism; the basin inherits minerals that are open, leached of their cations, and in the process of expansion. With the help of the environment, well defined minerals such as illite or chlorite, or minerals in the process of being organized, are shaped.

In many cases, the result is hardly recognizable relative to the detrital material received at the shores of the sedimentary basins.

Furthermore, these constructive or positive transformations are observed in *space* as well as in *time*; it is a question of the environment of deposition. In fact, it suffices if sedimentation diverges from a detrital character and orients itself toward a clearly chemical character. This is what the Triassic basins achieved in two ways.

— During Triassic time, the chemical character of environments was accentuated from the basal sandstones to the salt-bearing formations at the top. Transformations of clays were able to succeed each other in time.

— But at a given epoch, the inherited particles made their way slowly from the sandy, hyposaline coasts to the middle of these shallow Triassic seas where ions accumulated. Along the way, the particles are completed, organized, and take on a new aspect. Transformations were able to succeed one another in space.

It will be said that this is described for Triassic basins and that this phenomenon must be characteristic of the Triassic. It is certain that Triassic sedimentation oriented clay minerals toward magnesian forms because we see the occurrence by transformation of irregular chlorite-montmorillonite mixed layers, corrensite and magnesian chlorites themselves and by neoformation attapulgites and sepiolites. But transformed and well crystallized illite is also organized. Moreover, each epoch has its own chemical character. The Cenomanian transgression favored glauconite, and continental lessivage of the Eocene favored palygorskite. Other environments will orient transformations after their own fashion and, particularly, toward well crystallized illite. Here we find the key to the differences between sediments of *different environments.* As was said only a short time ago, the environment can shape the inherited material in its own likeness.

IV. — CLAY MINERALS IN THE TERTIARY BASINS OF WEST AFRICA

During Upper Cretaceous and Eocene time the African continent was subjected to transgressions that invaded its borders and penetrated sometimes even deeply into its most depressed zones. The study of the evolution of the clay fraction during the Tertiary will be presented here according to recent works on the basin of eastern Sudan (Mali), on the basin of Senegal-Mauritania and on the basins of Dahomey-Togo and the Ivory Coast. Figure 55 shows these various geological basins.

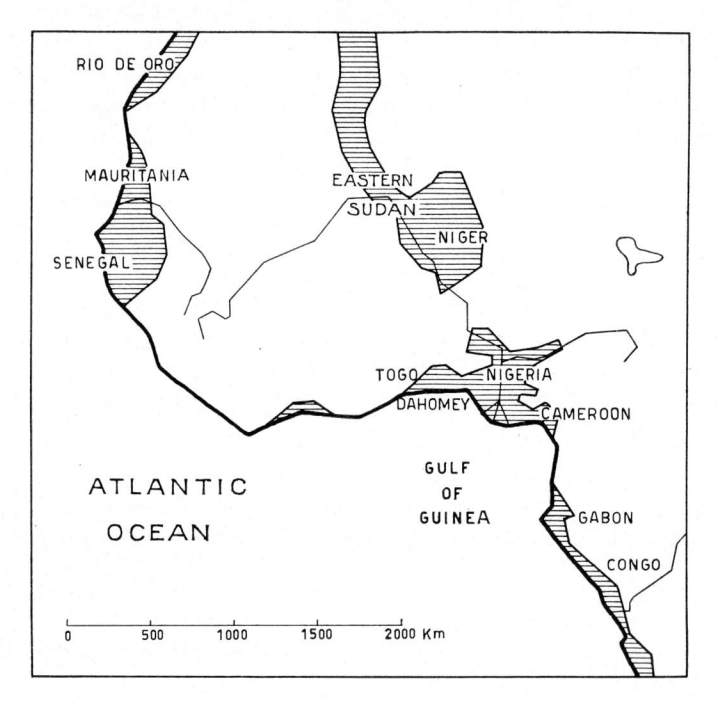

FIG. 55. — *Cretaceous and Tertiary basins of Western Africa.*

1° *The eastern Sudan basin*
(RADIER, 1953, 1957)

The history of the Gao basin in eastern Sudan can be summarized in five stages. A geological section is given in figure 56.

UPPER CRETACEOUS. — Marine filling of the zone of subsidence.
During the Upper Cretaceous the straits of Gao functioned as a subsiding rift valley, as shown by geophysical evidence. Only the uppermost 250 m of the filling is known.

Sedimentation was rhythmic and the following deposits alternate: micaceous and glauconitic sandstones, dark green or chocolate-colored micaceous clays, molasse and breccias resulting from the reworking of earlier formations. Near the top, deposits are finer grained: black pyritic, lignitic and bituminous argillites. These are facies typical of the filling of a deep rift marginal to a shield. The clay fraction of the samples is composed of 80 % montmorillonite, 10 % chlorite, and 10 % mica.

DANIAN, LOWER EOCENE. — Reef limestone.

The sea rose out of the rift and spread on to the crystalline platform. In the axis of the rift, the sedimentation remained molassic, micaceous, glauconitic and montmorillonitic. On the platform it was chemical, calcareous, characterized by reefs, and always montmorillonitic.

The interpretation of these facts can be made if one postulates the existence on the neighboring continents of semi-arid soils favorable to montmorillonite, which was reworked with the constituents of the molasse.

MIDDLE EOCENE. — Interior sea in the process of disappearing.

As the limestones become more and more clayey, they form chalky marnes with montmorillonite (50 to 60 %) and attapulgite (20 to 50 %). The neoformed character is accentuated and attapulgite increases. One arrives at papyraceous marnes in which attapulgite (40 to 100 %) and montmorillonite (0 to 40 %) constitute the clay fraction. These attapulgite-bearing strata contain beds, banks and lenses of phosphatic gravels and an extreme abundance of coprolites and bones of vertebrates. Chemical sedimentation of alkaline character sets in.

UPPER EOCENE, OLIGOCENE. — Continental lacustrine regime.

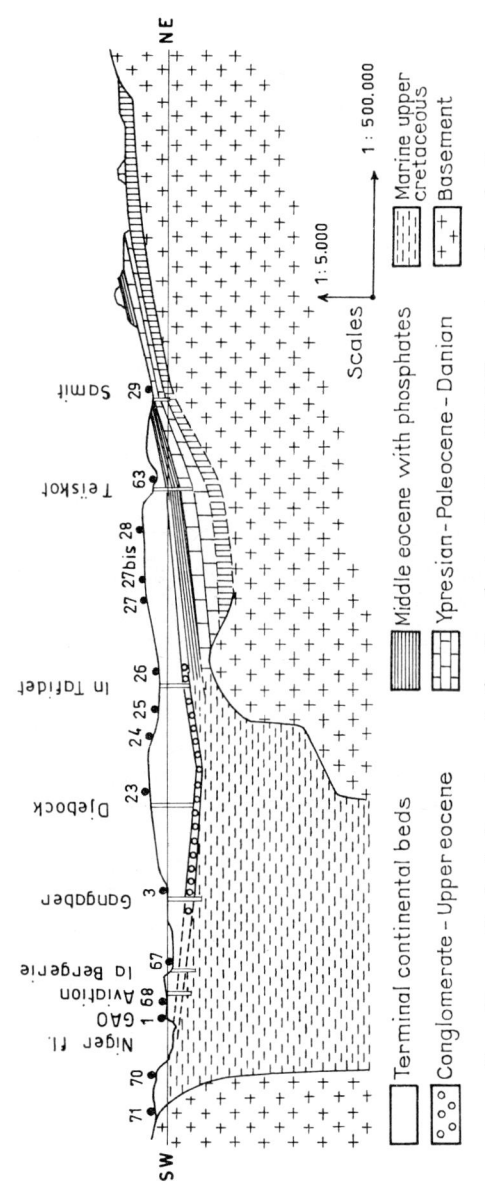

FIG. 56. — *Geological section of the Gao basin. Section SW-NE passing through Gao, Eastern Sudan, Mali, Western Africa* (after RADIER, 1957).

We pass from the marine to the lacustrine regime; the clay fraction is split between attapulgite, montmorillonite, and kaolinite, which was present previously only in traces. Argillites lose their fissility, become variegated, and include thin beds of gypsum and a horizon of cherty gaize about 0.5 to 3 m thick. Then kaolinitic clays come in, along with horizons of argillites, thick in some cases, with ferruginous oolites. This is the transition from chemical sedimentation to the renewal of erosion.

POST-OLIGOCENE. — Siderolithic regime, Terminal Continental Series.

The regime of demolition, begun previously, is accentuated and is produced at the expense of lateritic material that had accumulated on the continents during the preceding periods. With the occurrence of kaolinitic clays, sands, sandstones, ferruginous horizons we are in an obviously siderolithic regime; this is the Terminal Continental Series of African geologists.

RADIER (1957) interpreted this series as the product of the succession of alkaline chemical sedimentation during a period of large-scale lateritic weathering on the continents and of siderolithic sedimentation by reworking. This is the biorhexistatic series of ERHART (1956).

2° The basin of Senegal-Mauritania
(TESSIER, 1952; ELOUARD, 1959)

The central part of the basin of Senegal was the object of the fundamental study of TESSIER (1952). Hydrogeological and petroleum works furnished complementary information through drilling (SLANSKY, 1959). Briefly, this is the series:

— The Maestrichtian is, above all, composed of sand and clayey sand.

— The zone of transition from the Cretaceous to the Tertiary is clayey marne and is capped by an important horizon of zoogenic limestone with a Paleocene fauna.

— The zoogenic limestone is overlain by a phosphatic, glauconitic level, then by papyraceous clays and marnes from the final Paleocene.

— The Ypresian is still of clayey marnes and papyraceous, with calcareous and phosphatic intercalations.

— This papyraceous series is capped by the marno-calcareous horizon of N'Gazobil, which marks the limit between the Ypresian and the Lutetian.

— The Lutetian becomes calcareous, but it is invaded by phosphatic sedimentation in which papyraceous marnes still are found. This phosphatic sedimentation was interrupted, in the course of the sedimentation, by emergences (TESSIER, 1950) that has the ability to transform calcium phosphates into aluminum phosphates or "phosphatic lateritoids", which are the result of a kind of autochthonous bauxitization.

Thus, the attapulgitic sedimentation is encountered first during the Paleocene-Ypresian, and then during the Lutetian. Let us note that in the bore hole of Sangalcam (fig. 57), situated to the west of Thiès in a zone of intense subsidence, the results of analysis of the clay fraction were the following (SLANSKY, 1959):

— In the upper part of the Cretaceous, kaolinite dominates and is associated with illite and montmorillonite.

— At the extreme limit of the Cretaceous, it is montmorillonite that dominates, associated with kaolinite.

— In the Lower Paleocene, montmorillonite occurs alone with traces of illite.

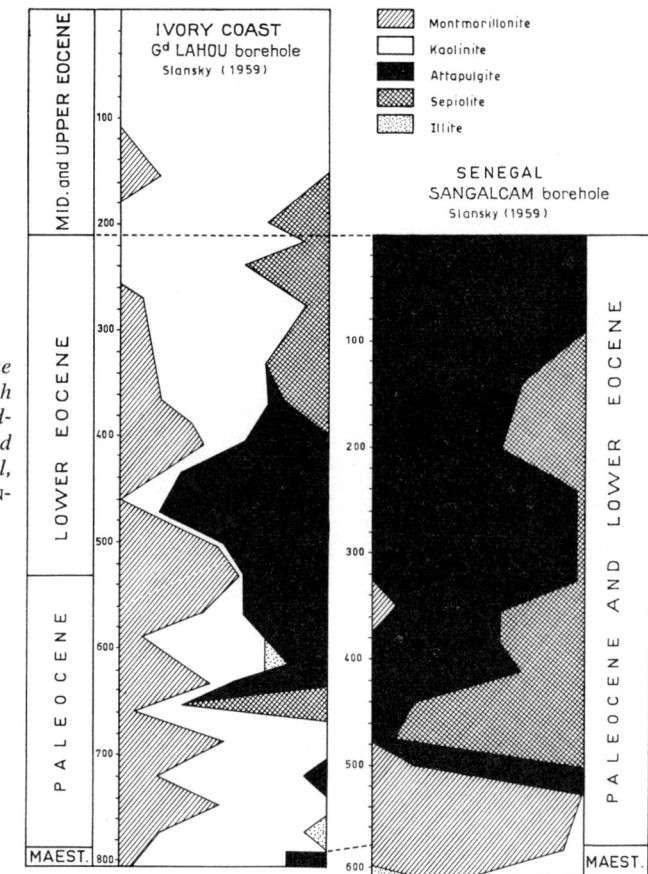

FIG. 57. — *Evolution of the argillaceous fraction through the boreholes of Grand-Lahou in Ivory Coast and of Sangalcam in Senegal, Western Africa* (after SLANSKY, 1959).

— In the Upper Paleocene, attapulgite appears. At first associated with montmorillonite, it immediately joins with sepiolite to constitute the main part of the clay fraction up to the top of the Ypresian. The thickness of the attapulgite-sepiolite bed attains 500 m here, and for 475 m attapulgite and sepiolite together represent 100 % of the clay fraction. Within this long series, sepiolite has two maxima, corresponding to the Upper Paleocene and to the Ypresian. This double develop-

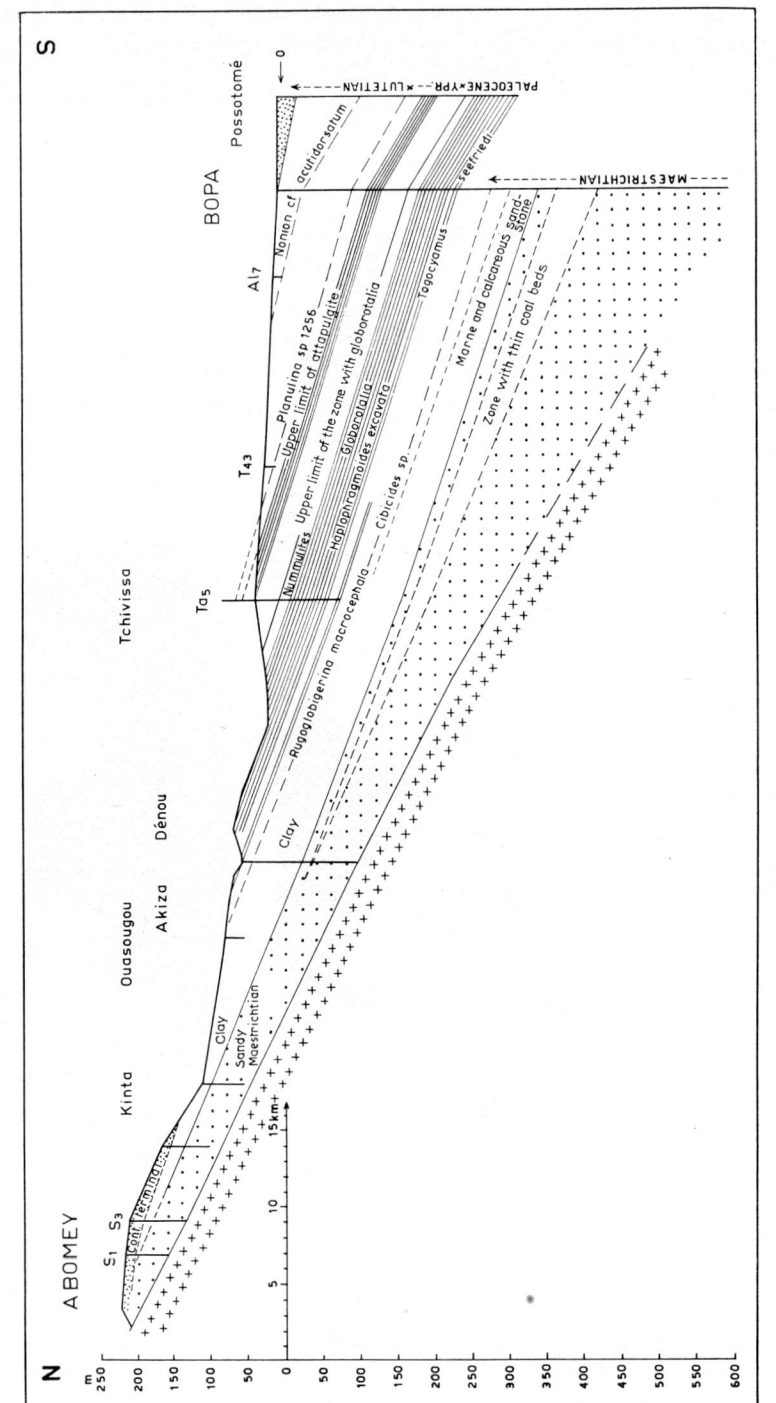

Fig. 58. — North-south geological section of the basin of Dahomey, Western Africa (after SLANSKY, 1959).

ment of the sepiolite is to be compared with the rhythms of attapulgite and sepiolite in the Ivory Coast. This comparison is presented in figure 60.

ELOUARD (1959) has studied Eocene sedimentation in the northern part of the Senegal basin at the present-day mouth of the Senegal River and in the territory of the Islamic Republic of Mauritania. In these regions, only the Lutetian transgression covered the Precambrian basement. Two different kinds of sedimentation followed one another.

— Chemical sedimentation in an alkaline environment, resulting in limestones, dolomites, cherts, and attapulgitic and montmorillonitic argillites. These are the epicontinental deposits of Lutetian age.
— Sedimentation of siderolithic character, with sands, sandstones, iron oxides and kaolinitic argillites; these are the detrital deposits of the Terminal Continental Series.

In addition to these vertical evolutions of the clay fractions, ELOUARD made important observations on the horizontal distribution. At 65 km from the borders of the basin, the clay fraction contains only attapulgite. As one approaches the border, attapulgite decreases and montmorillonite increases. Very close to the coast, its place is taken by kaolinite, a detrital clay mineral introduced with the sands.

3° *The basins of Dahomey-Togo and of the Ivory Coast*
(SLANSKY, 1958, 1959)

SLANSKY studied the Upper Cretaceous and Tertiary basin of Togo and Dahomey, which is but the western extremity of the great sedimentary basin of Nigeria. A geological section of the basin of Dahomey is presented in figure 58.

— During the Maestrichtian, the sea progressively invaded the basin, which passes from a detrital, sandy facies to a clayey facies. The base is continental, then becomes brackish, and then really marine. The clay fraction is kaolinitic.
— During the Eocene, the sedimentation took on a chemical character; there was a transition from sandstones to calcareous sandstones, then to zoogenic limestones, which are in some cases phosphatic and glauconitic. Towards the middle of the Upper Paleocene, papyraceous argillites came in, then disappeared, and came in again during the Ypresian. This is expressed in figure 59, which represents the evolution of the clay fraction in the bore holes of Bopa and Hetin Sotta in Dahomey. One sees that the Eocene presents a clay fraction with attapulgite and montmorillonite, including two major peaks of attapulgite during the Paleocene and Ypresian. During the Lutetian, the sedimentation evolved considerably; attapulgite disappeared, and montmorillonite dominated by far at first, then it was relayed by kaolinite, that which announced the arrival of sands. At the base of the stage, the agitated phosphatic sedimentation still prevailed, then it disappeared.
— The Terminal Continental Series is unconformable on the Lutetian. It is uniquely detrital: sands, sandstones, kaolinite, and ferruginous horizons. Here we are again in the presence of the siderolithic facies, which became coarser and coarser as the regression was accentuated.

SLANSKY confirms, in addition, the horizontal evolutions noted by ELOUARD. The farther from the coasts one goes, the more the detrital kaolinite decreases in favor of the clay minerals of chemical origin, montmorillonite and attapulgite. But, moreover, attapulgite increases toward the open sea at the expense of montmorillonite. This permitted SLANSKY (1959) to define the clayey sedimentary sequences that we shall come back to later on.

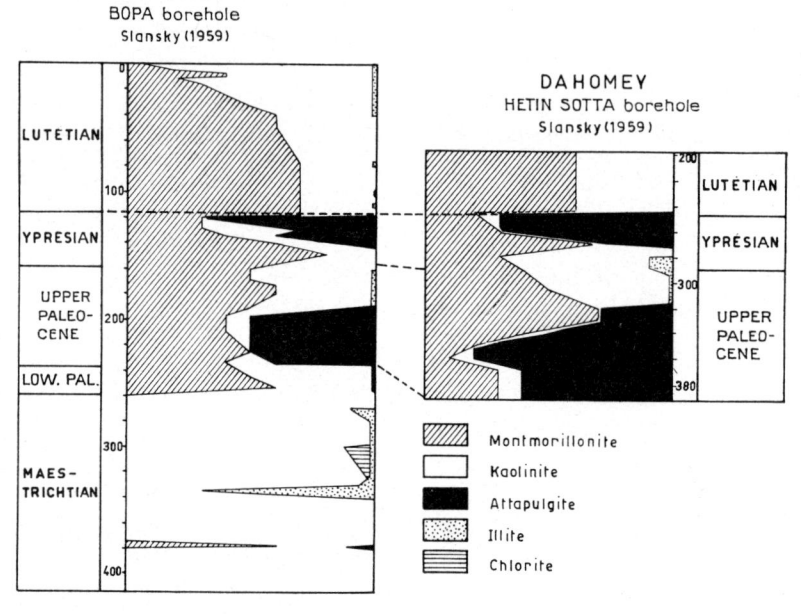

FIG. 59. — *Evolution of the argillaceous fraction through the boreholes of Bopa and Hetin Sotta in Dahomey, Western Africa* (after SLANSKY, 1959).

In the Ivory Coast, a small, littoral Tertiary basin is seen in outcrop, and the sedimentation there is comparable. A deep hole was drilled at Grand Lahou, and the study of the clay fraction is shown in figure 57. In this case, one visualizes an invasion by attapulgitic chemical sedimentation in the following manner:

— Chemical sedimentation began earlier because attapulgite already made its appearance during the Maestrichtian.
— Chemical sedimentation was prolonged later.
— Attapulgite is accompanied by sepiolite, which is more magnesian.
— The sum of attapulgite+sepiolite passes through three maxima, in the Paleocene, in the Ypresian and towards the Lutetian. The first and the third of these maxima depend heavily on sepiolite.
— Chemical sedimentation disappeared in the Upper Eocene and gave way to the formation of kaolinite and to the siderolithic facies of the Terminal Continental Series.

Comparisons with sedimentation in Senegal, 2 000 km distant, and with that in Dahomey, 750 km away, show the amplitude of these variations in sedimentation peripheral to a vast continent.

4° Age of the attapulgite- and sepiolite-bearing formations in West Africa

Stratigraphic correlations are not always easy at a distance of several thousand kilometers because of the frequent poverty of and the variations in faunas. But identification of Foraminifera and revision of macrofaunas have made some progress, and the synchroneity will be examined here to the extent that the documentation has been published.

It is a question of comparing the epochs in which the development of attapulgite has been observed at present. For purposes of comparison the data have been added for Morocco from JEANNETTE, MONITION, ORTELLI and SALVAN (1959), for Niger from GREIGERT (1961) and FAURE (1961, 1962) and for Gabon from REYRE (in DEVIGNE and REYRE, 1957). All the identifications, except those concerning Morocco, were made in the Institut de Géologie at Strasbourg, France.

These comparisons are presented schematically in figure 60.

In Morocco, the attapulgite-bearing beds are reported from the base of the Maestrichtian and the Ypresian.

In Senegal, if we take into account only field observations, we see three developments of the attapulgite-bearing beds in the Paleocene, Ypresian and Lutetian. If the results are added from the bore hole of Sangalcam, which unfortunately does not cross the Lutetian, one sees palygorskites invading all the clay fraction, with two sepiolite maxima in the Paleocene and the Ypresian.

In the Ivory Coast, after one or two appearances in the Lower Paleocene, attapulgite and sepiolite develop continuously up to the Upper Eocene, with three maxima in the Upper Paleocene, in the Ypresian and Lutetian.

In Dahomey-Togo there are two attapulgite-bearing horizons, the Upper Paleocene and the Ypresian. It is known that analogous horizons occur in Nigeria.

In Sudan, attapulgite begins in the Ypresian, then expands in the Lutetian, although SLANSKY reports revisions of a fauna that would permit making the series older, but we must wait for complementary information.

In Niger, attapulgite-bearing levels have been identified in the Turonian and at the limit between the Turonian and the Senonian. In western Niger, the Eocene beds of Sudan continue with the same probable age.

Lastly, in Gabon, attapulgite has been identified in the Coniacian. It is very likely that it will be found also in the chemical Eocene series.

On the whole, one sees the attapulgite- and sepiolite-bearing formations develop in the basins peripheral to the African continent in accordance to rhythms that commonly are contemporaneous and, in all cases, similar. The three privileged periods are the Paleocene, Ypresian and Lutetian. Progress in the analysis of geological sections will be instructive for the confirmation and the tightening of these correlations. Already we must look for a large-scale explanation for such a phenomenon.

270 EVOLUTION OF THE CLAY FRACTION

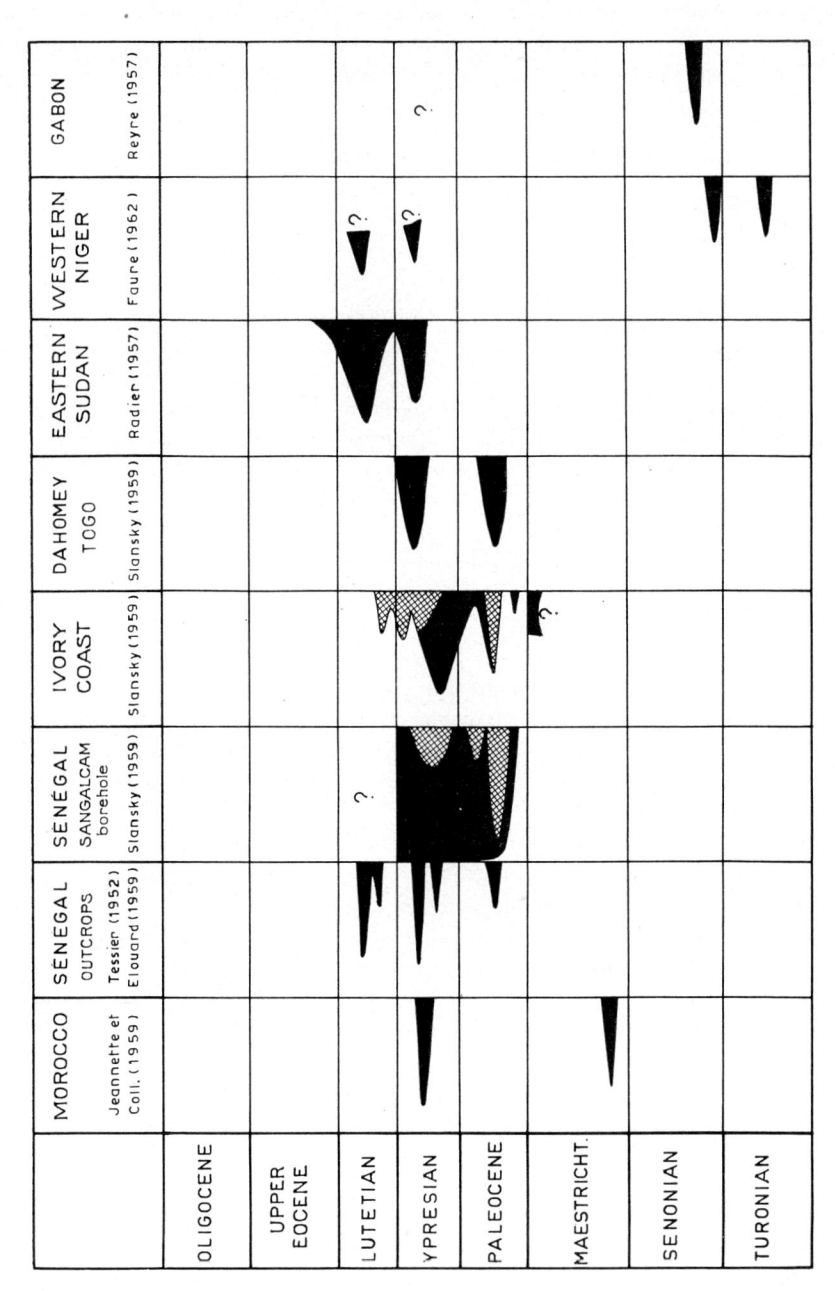

FIG. 60. — *Comparative evolution of the argillaceous fraction in the Tertiary basins of Western Africa* (Attapulgite black, sepiolite cross-hatched).

5° Attapulgite-bearing facies and associated facies: antagonism and paragenesis

Attapulgite and, in some cases, sepiolite occur in series with an alkaline chemical character: carbonate rocks, chert, tabular flint, calcium phosphate, glauconie, montmorillonite. These associations were very frequently evoked by authors who have studied these formations. First, CAPDECOMME (1952), then CAPDECOMME and KULBICKI (1954) in Senegal, then RADIER (1953, 1957) in Sudan, MILLOT (1953, 1954) in Sudan and Gabon, VISSE (1953, 1954) in several African basins, MILLOT, RADIER and BONIFAS (1958) for the ensemble of this sedimentation.

But all these facies of chemical origin within the African phosphatic and attapulgitic series are not randomly mixed. JEANNETTE et al. (1959) and SLANSKY, CAMEZ, MILLOT (1959) studied the paragenesis. Phosphates and glauconie are intimately mixed; they testify to conditions of intense agitation and form together. In contrast, attapulgite and phosphates are separated in alternating beds; the attapulgitic marnes are poor in phosphates, except when it is a question of gravels or reworked oolites; the phosphatic beds contain no attapulgite. This results in the fact that attapulgite and glauconie are separated, as has been verified; even if the series contain both minerals, it is always in separate levels. Montmorillonite is much more tolerant; it has been seen that it is the common companion of attapulgite; it is also the principal mineral of the cement of phosphates. Lastly, flint, by definition, alternates with the other facies.

Here, we can grasp the hesitations of the sedimentation between several facies that are associated in series, but are not mixed together in the same sedimentary level. On the scale of the sedimentary environment, we find ourselves facing not only mineral paragenesis, but also antagonism.

6° Sedimentary sequence of the alumino-magnesian clay minerals

Regular ordering of the clay minerals in the Eocene basin of Mauritania was noted by ELOUARD (1959). As one moves away from the former shore lines, the detrital kaolinite is replaced by the montmorillonite-attapulgite couple of chemical origin. But the proportion of attapulgite increases with distance from the coast. SLANSKY (1953) corroborated this fact by his observations in Dahomey, Togo, and the Ivory Coast and studied this clayey sedimentary sequence. Recent analyses of the fissile clays in the Ypresian of Senegal showed that they are composed uniquely of attapulgite, associated or not with montmorillonite. In contrast, in the bore hole of Sangalcam, which is farther from the coast and corresponds, above all, to a more continuous, thicker, thus more remote sedimentation, sepiolite occurs and can even form 90 % of the clay fraction. Moreover, it shows two maxima at the times of the two transgressions.

The complete sedimentary sequence is now known (SLANSKY, 1959; MILLOT, ELOUARD, LUCAS and SLANSKY, 1960) and presents itself as follows: kaolinite-montmorillonite-attapulgite-sepiolite.

Its first member is detrital and littoral, the following three are chemical and farther and farther removed from the coasts.

This sedimentary sequence, in the sense of LOMBARD (1953), which has been described in the horizontal plan, is found again in bore holes in vertical section. Whenever we see in a bore hole, for instance in the Paleocene, kaolinite, kaolinite-montmorillonite mixtures, montmorillonite, montmorillonite-attapulgite mixtures, attapulgite, and attapulgite-sepiolite mixtures following one another, we have traversed a positive sedimentary sequence, equivalent to a transgression. The opposite vertical succession characterizes regressions. The examples studied above even show, in the course of the great Eocene transgression, two withdrawals: between the Paleocene and the Ypresian and between the Ypresian and the Lutetian. The rhythms of the alumino-magnesian clay minerals register these comings and goings very well. The chemical sedimentation obeys the displacements of the shore lines.

7° *Geochemical sequence*

If this sequence is examined from the point of view of the minerals, one sees a succession of three different mineralogical types: dioctahedral montmorillonite, attapulgite, and sepiolite. In each of these types, aluminum and magnesium can be exchanged, but the tolerance of this exchange is limited and different for each type. All chemical analyses show that montmorillonites are characterized by a high Al/Mg ratio, attapulgites by a Al/Mg ratio hardly greater than 1, and sepiolites by a low Al/Mg ratio. These differences in composition correspond to differences in crystalline structure.

If this sequence is examined from the point of view of its total chemical composition, one observes the continuous decrease in the amount of alumina, as well as a parallel increase in the amount of magnesia.

Thus it can be said that the sedimentary environment varies regularly from the coasts toward the open sea, but that neoformation gives a discontinuous mineralogical response to a continuous chemical variation. So long as aluminum is important, it orients the neoformation towards montmorillonite. When aluminum decreases, montmorillonite is relayed by attapulgite, and when magnesium occurs alone, sepiolite forms. The response to the continuous chemical variations is a continuous variation in the mixtures of minerals of fixed composition. Exactly the same rules apply as in crystalline rocks.

Lastly, it can be noted that the most aluminous minerals are the most littoral and the most magnesian are the most remote. Leaving aside the case of kaolinite, which is littoral, it must be noted concerning the montmorillonite-attapulgite-sepiolite sequence that in a sedimentary basin it is alumina, the least soluble element, that becomes exhausted in the direction of the open sea. Here we found what the syntheses will show us: the privileged role attributed to alumina in neoformations because of its exceptional affinity with respect to silica.

8° The geochemical cycle in the African Tertiary basins

This montmorillonitic, attapulgitic, sepiolitic sedimentation within facies of alkaline chemical character is too general on the periphery of the African continent to have local causes. The cause must be sought on a continental scale. Since this sedimentation of carbonates, phosphates, flint and magnesian clays is chemical, it has its source in lessivage of the African continent itself (MILLOT, 1954; RADIER, 1957; MILLOT, RADIER, BONIFAS, 1957).

Large-scale lessivage requires humid tropical climates, and it can be proved that the African continent was, in fact, subjected to lateritizing climates at that time.

TESSIER (1950) has demonstrated that phosphatic lateritoides are interstratified in the Lutetian series in Senegal. Since they are the result of intense lateritization of emerged calcium phosphates, they testify to the reality of these climates. BORDET (1951, 1955) has revealed the existence in the central massif of Ahaggar of lateritic soils fossilized by volcanic flows, which he dates as earliest middle Tertiary. RADIER (1957) has carefully analyzed information from paleobotanists, and in particular that of AUBRÉVILLE (1949, 1949), on the Tertiary forest floras; in the Eocene the African flora was equatorial and capable of protecting the thick lateritic weathering products.

GREIGERT (1961) and FAURE (1961, 1962) have brought us direct proofs of intense lateritic weathering of the Eocene. In fact, in Niger where widespread and very deep weathering is preserved, it can be shown that such weathering corresponds to areas that were not covered by the Eocene sea. Weathering into kaolinite, alunite, bauxite, etc., attacked all formations of Senonian age and older in these regions that remained emerged. They are never visible either in outcrops or by drilling in the regions that were subjected to the transgression. Moreover, neither Eocene nor post-Eocene deposits have been affected by this weathering. Thus, we have at our disposal the proof that the large-scale weathering was really Eocene and was the source of material for the lateral chemical sedimentation, as expressed in the opinions of MILLOT (1954), RADIER (1957), ELOUARD (1959) and SLANSKY (1959).

Thus, it is possible to describe the geochemical cycle that took place in the peripheral Tertiary basins of Africa.

1° In the Upper Cretaceous, the sediments are sandy clay, molasse, and characterized by montmorillonite in Sudan and Niger. The majority of the authors have attributed the latter to the reworking of weathered products and soils in regions subjected to tropical climates of the Eocene. This is one way of looking at it which would correspond to a great development of montmorillonite-bearing soils at that epoch. The studies on the clay fraction of soils permit us to enliven the discussion a little. Montmorillonite-bearing soils in a tropical climate with a dry character are the result of the recombination, in the poorly drained depressions, of silica, alumina, and magnesia released by hydrolysis during the humid, warm seasons. This recombination can be produced in depressions where molasse accumulates as well as in continental depressions. There is no difference in nature between them.

Thus, in the absence of determinative arguments, we shall say that during the Cretaceous erosion supplied the sedimentary basins with a clayey sand material of molassic character. This occurred under a dry tropical climate, confirmed by the paleobotanists. This climate was conducive to hydrolysis, lessivage of silica and bases, and, in the absence of strong lessivage, neoformation of montmorillonite, either in continental depressions, or in marine depressions where all the material finally was collected.

This line of reasoning is all the more defensible in as much as the Upper Cretaceous in Dahomey, the Ivory Coast and Senegal is kaolinitic. This shows that the intense weathering must already have begun, and the sporadic appearances of attapulgite in the Cretaceous corroborate this interpretation.

2° During the Lower and Middle Eocene, a great humid period with tropical forests prevailed. Weathering intensified and provoked considerable accumulations of lateritic products in emerged areas: kaolinite, ferruginous crusts, bauxite, lateritoids, alunite, etc. This gigantic weathering evacuated dissolved products in great quantities to the sedimentary basins either lacustrine or marine according to the base level. These products were immediately engaged in chemical sedimentation, of which the principal facies are carbonates, flint and tabular silica, phosphates and the alumino-magnesian clays, montmorillonite, attapulgite and sepiolite.

The history of sedimentation, the transgressions and the regressions can be reconstructed by means of the variations in these deposits, observed by examination of the littoral facies where the detrital supply persisted (kaolinite), examination of the epicontinental facies where dolomitization, silicification and phosphatization were common, and examination of quiet-water facies farther from the coasts (attapulgite- and sepiolite-bearing argillites). This was the epoch of biostasy.

3° After Middle Eocene time, at a date that varies with the province, regression and erosion started up again on a continent with a radius of several thousands of kilometers. All the material stored up on the continent during the preceding period was delivered to transportation and collected in basins. This was the rhexistasy of ERHART (1956), and RADIER (1957) showed very well that, *grosso modo*, the continental cover deposits accumulated upside down in the basin of Gao. In fact, the succession shows the following significant dominants: beds with ferruginous oolites-kaolinitic clays-sands and sandstones. This represents the statistical reworking of crusts (*cuirasses*), lithomarges, and initial zones of arenization.

This Terminal Continental Series was announced in the sedimentary basin by a very sensitive precursory sign, the invasion of detrital kaolinite. This phenomenon was produced at different epochs, according to the tectonic evolution of the different regions. Thus the Terminal Continental Series began in the Upper Eocene, in the Oligocene, or in the Miocene, in different cases.

V. — OVERALL VIEW OF THE EVOLUTION OF CLAYS IN THE GREAT SEDIMENTARY SERIES

Evolution of the clay fraction has been followed in four great sedimentary series. All of them represent several tens of millions of years, and these four series were chosen expressly to show the reader four principal styles in the interpretation of the history of silicates during a geological period. When arranged in stratigraphic sequence, these four examples illustrate the phenomena in an indifferent order, which by chance is not so illogical.

1° *Diagenesis in Cambrian sandstones of the Sahara*

The Cambrian sandstones of the Sahara illustrate, naturally, the *mechanical inheritance* of minerals coming from Precambrian uplands. But they illustrate, above all, the phenomenon of *diagenesis* or, better, of *diageneses* because two diageneses can be reconstructed. The mechanical heritage is completely disguised by these deceptive diageneses, and all the more so because detrital micas and illites have been kaolinized, then again illitized. But the convergence of the terms hides some differences. The initial micas and the terminal micas have neither the same aspect, nor the same composition, nor, if the isotopic methods are to be trusted, the same age. These evolutions are connected to migrations of the water supply, thus to the entire dynamics of a region. In the beginning, continental weathering spread over the piedmont of the Precambrian uplands. Lessivage of this piedmont kaolinized the silicates. Then much later, the underground circulation of oil and saline water led to the reappearance of the micaceous lattice. The chronology of the deformation of the region is that of hydrodynamic evolution and mineral evolution. It is seen that the latter can serve, for their part, to date the former.

2° *Heritage in the Carboniferous series of the U.S.A.*

The American Carboniferous series illustrate chiefly *mechanical heritage*. Without doubt, transformation during deposition, weathering in the coal measures and diagenesis can produce shades of difference in the result, but the bulk of the material seems to be really detrital, having come from the surrounding emerged massifs. Here there is direct intervention of the *tectonic history* of the American platform. The rising massifs provided the subsiding basins with their own substance, and the trajectories followed by these detrital minerals permit us to reconstruct the ancient paleogeography.

3° *Transformation in Triassic clays of Eurafrica*

The Triassic series of Europe and North Africa illustrate transformation. Without doubt, heritage is responsible for the initial supply, and neoformations complete it. But, by means of a peculiar chemical character, which is chiefly potassic and, above all, magnesian, *this inherited material is transformed and reconstructed into new and very well crystallized minerals, especially illite and chlorite*. Reconstruction of the climates and the geographies that would explain such large and widespread basins has not yet been achieved. Terms for comparison with this great post-Hercynian epoch will have to be sought in the variegated series that post-date the Alpine chains of the

Orient or Asia, or in older formations. This was an epoch of tectonic calm, of which our still agitated continents give no image at present.

4° *Neoformation in the Eocene of West Africa*

The Tertiary series in Africa show, with a considerable amplitude, phenomena of *sedimentary neoformation.* We cannot consider them as curiosities of puny character. The cumulative thickness of tabular silica can reach some tens of meters, phosphatic beds constitute mineral deposits in which one can move about in galeries, beds with 100 % palygorskite have a thickness of 500 meters in Senegal. It is a question of another style of genesis for silica and silicates, *synthesis in environments of a marked alkaline chemical character.* This epoch favorable for chemical sedimentation was again an extraordinarily calm epoch during which the immobile continents were dissolved of the major part of their substance. But towards the close of the Eocene, agitation set in again, climates changed, there was erosion, and the siderolithic facies invaded everywhere. At the same epoch, Europe was going through an analogous history. We can grasp only the appearance of phenomena, and once again the causes reside in the dynamics of the Earth's crust. It is understandable that Alpine orogenesis provoked the reworking of the weathering mantles in Europe, but we must agree that its effect is just as clear and occurred at the same time in Africa. The rhythms of sedimentation can define the age of these oscillations.

To study a sedimentary series requires the concurrence of all methods, since each of them throws light on one part of the past. Paleontology and field studies are complemented by sedimentology. Today, the study of silicates on the Earth's crust adds to that which can be obtained from the study of carbonates and detrital sediments. *Silicates are sensitive to continental climates, to the "climates" of seas, and to underground environments. They are degraded and disappear, but they can also be transformed by means of aggradation. They can also be reborn from the natural solutions.* But it will be noted that these evolutions require the slow action of the water and that the active orogenic periods do not allow it. Thus, *periods in which tectonism is active will furnish preferentially detrital and inert argillaceous materials,* whereas *periods of stability will ensure evolution of chemical character on the continents and in sedimentation.*

CHAPTER IX

SILICIFICATIONS, FLINT AND GROWTH OF CRYSTALS

Introduction

Clays and silicifications. — It may seem surprising to study siliceous rocks and silicifications in a treatise on clays. In fact, I hope that the reading of this chapter will show how necessary it is. Here are the principal reasons. Since clay minerals are silicates, the architecture and development of their lattice are chiefly controlled by the assemblage of oxygen tetrahedra with a central silicon ion. Now this fundamental elemental structure is that of crystallized silica in its three forms: quartz, cristobalite and tridymite. Many rocks of pedogenetic or sedimentary origin are siliceous, that is, formed chiefly of silica tetrahedra. It is obvious that these siliceous formations formed at low temperature, under conditions of the hydrosphere which are the conditions of the formation of clays; moreover, they are frequently closely associated with the latter. Thus, looking at clays and siliceous formations on a structural scale, one sees the similarity in their natures. In clays, the tetrahedra are arranged in a two-dimensional planar layer; in siliceous formations, they are arranged spatially in a three-dimensional lattice. This is the fundamental difference for the specialist in crystalline structures, but for the common people, is it not only a difference between two types of occurrence of silica? This chapter is designed to show what this difference represents for the geologist.

Silicifications in the hydrosphere. — Silicifications occur in nature anywhere from great depths to the surface. Metamorphic, hydrothermal and volcanic silicifications will not be considered here, but only silicifications in the hydrosphere.

These silicifications, in the surficial domain, present, moreover, several aspects. The first aspect is that of climatic silicifications related to pedological phenomena and prevailing especially under warm climates. Another aspect is provided by silicifications affecting the sediments themselves, either very early in their history, as in the case of flint, or during their transformation into rocks.

In a general way, we in geology have been *wrapped up* in traditional teachings on silica and silicifications that are difficult to overcome. One of the tenacious chapters in this affair is the question of colloidal silica in natural waters and its possible flocculation. This is a result of the considerable repercussions of works carried out by chemists on colloidal silica around the beginning of this century. Geologists adapted these important discoveries to their problems, but this merits revision today.

There is much reason to believe that in the domain of silicifications the chemistry of colloidal silica does not apply or applies only rarely. It is necessary to call upon the new knowledge which chemists and mineralogists have at their disposal today. One must consider how this new knowledge can take the problem of surficial silicifications into account. This work is attempted here; it cannot avoid having a "bookish" aspect. But the naturalist cannot be master in the physico-chemical disciplines; he tries only to be open to their teaching and to consider how the approach to natural phenomena can be made otherwise than in a manner exactly contradictory to the present state of knowledge of chemists and mineralogists.

Limits of this study. — First of all two remarks must be made ; they limit the scope of the arguments that will be presented here. It is the question of ageing and of the role of biological chemistry.

Numerous observations show that siliceous neoformations age. A single example shows the strictness of this rule: all Paleozoic flint is now quartz, and we have no reason to suppose that they were not previously similar to their more recent successors. Therefore, all logic on silicifications is suspect because the petrographer when observing a stone does not know the evolution undergone since its formation. There is a distressing but inevitable degree of uncertainty here. We shall see, however, that this difficulty does not obscure the entire problem because silicifications that are definitely contemporary differ according to the environment of formation of the material replaced. The attempt must, therefore, be made.

On the other hand, the presentation given here takes into account only the mineral chemistry. Now, works that are already numerous have given us information on the role of organic chemistry in the mobility of natural silica. Silico-humic complexes are the best example. It is very likely that biochemical factors play a role in the transport of silica, and probably in its mineral organization. One could express the same prudence concerning the ferri- and ferro-silicic complexes. The attempt being made here does not take this into account, and that increases its provisional character.

These two restrictions being made, one can present an interpretation of silicifications in the hydrosphere in terms of mineral chemistry. I shall take up again my previous publications (MILLOT, 1960, 1961), and add to them, in the course of the four following sections:

I. Chemical data: silica in nature.

II. Mineralogical data: chalcedony and opal.

III. Geological and petrographical data: climatic silicifications and sedimentary silicifications.

IV. Silicification and growth of crystals.

I. — CHEMICAL DATA.
SILICA IN NATURAL WATERS

1° *Silica is a mineral macromolecule*

Silica is a mineral macromolecule whose behavior in water is controlled by the laws of the polymerization and depolymerization. The solubility of silica in water has been studied in Chapter III. The lessons that we have been able to learn from the study of natural silica are the following:

1. Silica occurs in some cases in the *crystalline state*; it has then a periodicity in the three spatial directions that corresponds to the linkage of SiO_4 tetrahedra. It is a question of a three-dimensional distribution of these tetrahedra such that each oxygen ion is shared by two tetrahedra, corresponding to an overall formula of SiO_2. The three possible crystalline forms are *quartz, cristobalite* and *tridymite*.

2. In other cases, there is *no periodicity* in a high polymer composed of disordered tetrahedra; these are the *amorphous silicas*. Siliceous gels are an example of the latter; on the other hand, opal gives evidence of a more or less well-marked degree of organization.

3. In water, a depolymerization equilibrium is established, which can be represented by the following equilibrium:

$$(SiO_2)\,n + 2\,n\,H_2O \rightleftharpoons n\,Si(OH)_4.$$

In equilibrium with silica, water contains *the $Si(OH)_4$ monomer*, which is the monomolecule of silicic acid. *These monomers are free and in true solution.*

4. In presence of amorphous silica, the depolymerization equilibrium is obtained in two ways, as shown in Chapter III (fig. 8): either from amorphous silica put in the presence of water, or from solutions supersaturated with silica. In the first case, monomers leave the amorphous silica to go into solution. In the second case, there

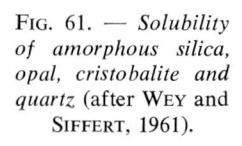

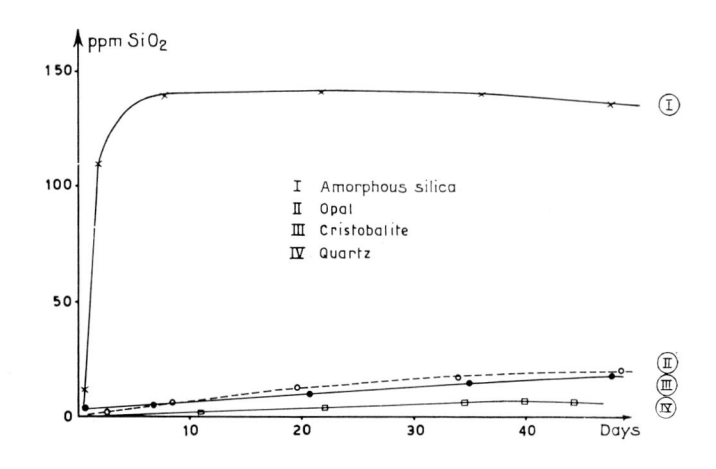

Fig. 61. — *Solubility of amorphous silica, opal, cristobalite and quartz* (after Wey and Siffert, 1961).

are polymerization and deposition of amorphous silica. *Equilibrium is attained at 20 to 25 °C for a concentration of 120 to 140 ppm, or mg/l, of silica; this limiting value is called the solubility of the amorphous silica.*

5. *The depolymerization equilibrium in the presence of crystalline forms of silica is much lower.* Measurements are very difficult to make because of the slow release of molecules and because of the sources of error that can appear over a period of months. SIEVER (1957) and KRAUSKOPF (1959) gave some values. The solubility of quartz appears to be 7 ppm; that of chalcedony is hardly higher. Diatomite powder (opal) gave 22 ppm after two years. WEY and SIFFERT (1961) present solubility curves obtained during two months (fig. 61). The solubility of quartz was 7 ppm, that of cristobalite was less than 20, and that of opal about 20. WEY and SIFFERT showed that it was very probably a question of true solubility and not of incomplete dissolution because of the slowness of the release of the monomers. The crushing of quartz makes the surface of the grains amorphous, and the solubility increases. But once the residue is washed and put into equilibrium with water, it gives again the initial solubility of quartz; thus, this solubility would be really characteristic of the mineral. It appears, therefore, that the solubility of the crystalline forms of silica is six or twelve times lower than that of amorphous silica. One can well understand that irregularities of the lattice or of the surface of the small particles have an influence on this solubility.

6. It will be noted that the *solubility curve* for opal in figure 61, given by WEY and SIFFERT (1961), is *very close* to those of the *crystalline forms of silica* and very different from that of amorphous silica. In my opinion opal is cryptocrystalline, as we shall see later on. Anyhow, it is reasonable to suppose that the solubilities vary according to the degree of crystallization of the natural forms of silica, but the main point is that they are lower, and greatly lower, than the solubility of amorphous silica.

7. *The solubility of silica is the same in sea water as in fresh water* (OKAMOTO et al., 1957; KRAUSKOPF, 1956, 1959). It was even experimentally shown that sea water is a good depolymerizing agent for colloidal silica; this is due to its low concentration of dissolved silica (OKAMOTO et al., 1957). Metallic cations have no effect on the solubility of silica except Al^{3+}, and, under particular conditions, Mg^{2+}, as has been observed in Chapter III. Let us note that Al^{3+} has very low solubility for ordinary values of pH and very low abundance in sea water. As for Mg^{2+}, it can act only at very high pH values or in high concentrations, conditions that are not common in natural environments.

8. From this it can be deduced, and this is fundamental for our subject, that *a solution saturated with $Si(OH)_4$ silica (120 to 140 ppm at 20 to 25 °C) is supersaturated with respect to crystalline forms of silica and is capable of ensuring their growth.* Moreover, the difference between the solubilities of amorphous silica and of the crystalline forms of silica is such that every natural solution containing 40, 60 or 80 ppm of silica in true solution is unsaturated relative to amorphous silica, but it is supersaturated relative to quartz and silicates. These natural solutions are thus capable of ensuring the growth of such structures in a direct way. This corresponds to the

laws of thermodynamics. It is certain that the experimental crystallization of quartz from solutions under normal conditions of temperature and pressure could not be obtained (SIEVER, 1957), but this may result from the slowness of reaction, as well as from the presence of foreign ions, as we shall see.

It will also be noted, parenthetically, that natural solutions containing silica in true solution are likewise supersaturated with respect to silicates, clay minerals or not, and are capable of nourishing them; we shall come back to this question with regard to synthesis and neoformations in Chapter XI.

9. Likewise, from the thermodynamic point of view, when it is question of true solution, *all slow precipitation of silica must occur in the form of crystallized silica* and not in the form of amorphous or colloidal silica, or of silica gel because the crystalline forms of silica are the least soluble. The colloidal forms of silica can be anticipated only in cases in which solutions supersaturated with silica are suddenly induced to flocculate.

10. Lastly, amorphous forms of silica or silica gels, characterized by their disorder, can be transformed into crystalline silica only if energy is provided to them. There is, therefore, no possibility of direct transition from an amorphous silica to a crystalline silica. This would be true in the case of a completely amorphous gel maintained in *statu quo*. But nature is never homogeneous nor permanent. If one considers a gel of amorphous silica containing the germs of crystals, there can be under favorable conditions growth of the crystalline nucleus at the expense of the gel by the intermediary of interstitial solutions. The highly soluble amorphous silica supplies free monomers of $Si(OH)_4$ tetrahedra of the interstitial solution, and these are capable of feeding the growth of the crystalline forms.

2° *Geochemical consequences*

These chemical data must be constantly present in the mind of the geochemist in order to interpret the facts of nature. Traditional teaching that we received is entirely modified by these data, and we must reexamine completely our geological reasoning.

The state of silica in natural waters. — All recent measurements on the concentration of silica of natural waters, i.e., those taking advantage of modern colorimetric methods, give values lower than 120 to 140 ppm at 25 °C, thus lower than the solubility of amorphous silica. This is true of soil water, springs, rivers, lakes and seas, as well as in connate water, i.e., the interstitial water of muddy sediments. Information on these measurements is given in the works of SIEVER (1957, 1962), GOLDBERG and ARRHENIUS (1958), KRAUSKOPF (1959), BIEN, CONTOIS and THOMAS (1959), DAPPLES (1959), and ROUGERIE (1961). The concentration of silica in sea water is very low: 0.1 to 5 ppm for surficial waters, and 5 to 10 ppm for deep waters. In the interstitial water of marine muds the concentration can reach 20, 40, 60, 80 or even 100 ppm. In ground water the concentration is quite variable, depending on climate and weathering; it is about 10 to 30 ppm and can reach 60 ppm.

Traditional teaching tells us of colloidal silica in natural waters. It can be deduced from the preceding facts that the *normal state of silica in natural waters is in true solution.*

The influence of pH variations. — *The silica of natural waters in true solution has a constant solubility* for all the pH values that do not exceed 9, that is, for the acid and alkaline conditions encountered in soils and sediments.

It is currently admitted in our traditional teaching that silica is insoluble or only slightly soluble in an acid medium (pH 4 or 5) and more and more soluble between pH 4 and natural alkaline pH values. *This is an error.* It is no longer possible to call upon pH variations to account for the solution or precipitation of silica in natural environments.

Influence of the metallic ions. — The metallic cations dissolved in water have no influence on the solubility of silica, except aluminum, which is, moreover, insoluble or only slightly soluble under natural pH conditions. *This implies that salts in sea water have no effect on the solubility of silica.* Our traditional information indicates that the cations of sea water have a flocculating effect on silica. This is incorrect. The silica of natural waters cannot flocculate upon arrival in sea water because it is not in a colloidal state, but in solution, and because the cations of sea water have no effect on this solubility.

We shall mention only the effect studied by BIEN *et al.* (1958-1959). Freshwater charged with material in suspension experiences a decrease in its low concentration of silica in solution as a function of chlorinity. Here we enter into the domain of exchange and of reactions between the inherited mineral particles and the sea water itself.

Necessity of a new hypothesis to explain silicifications. — The main problem of geochemists is to explain the silicifications of all origins, pedological or sedimentological. Let us say right away that this problem cannot be resolved as it has been approached up to now:

— By the flocculation of colloidal silica that does not occur in natural solutions;

— By the action of pH or metallic cations because these factors have no influence on the solubility of monomeric silica in natural solutions.

Thus, one is led to the search for a new hypothesis to explain silicifications. I shall present here the hypothesis of crystal growth (MILLOT, 1960; MILLOT, LUCAS and WEY, 1961).

This new hypothesis is the following:

— Since we know that natural waters contain tetrahedra of $Si(OH)_4$ silica in true solution;

— Since we know that these solutions, which are unsaturated with respect to amorphous silica, are easily supersaturated with respect to the crystalline forms of silica in concentrations of 20, 40, 60 or 80 ppm;

— We shall state that the crystalline forms of silica are nourished by the growth of crystals from the free tetrahedra in the water.

There is nothing in this hypothesis that is not commonplace for the most diverse kinds of neoformed minerals: carbonates, phosphates, silicates.

Proposition: *Silicifications, flint, cherts, etc., are the products of the growth of silica crystals from the natural solutions containing monomeric silica in solution.*

II. — MINERALOGICAL DATA. CHALCEDONY AND OPAL

If our knowledge has been much improved in the domain of chemistry, the same is true in that of mineralogy with respect to what is usually called chalcedony and opal.

1° *Chalcedony*

The essential information about chalcedony has been obtained by petrographers working with the polarizing microscope. In this way chalcedonite, quartzine and lutecite have been distinguished. LACROIX (1893-1913) maintained the use of these terms, "on condition, of course, that they are not considered to be distinct mineral species, but to be various types of occurrence of the elemental lattice of rhombohedral quartz." MICHEL LÉVY had explained the variations in the birefringence along the fibers by a twisting of the (n_g) negative acute bissectrix around the axis of the fiber, to which it remains perpendicular. LACROIX had given evidence of the presence of opal in the fibrous varieties of quartz.

X-ray diffraction has confirmed that chalcedony is composed essentially of quartz. The first opinions seem to have come from WASHBURN and NAVIAS (1922), RINNE (1924) and SOSMAN (1927). But the essential study is that of CORRENS and NAGELSCHMIDT (1933).

According to CORRENS and NAGELSCHMIDT, chalcedony can be represented by an association of quartz fibers alternating with thin layers of opal. This fibrous construction can be observed under the polarizing microscope, but the optical examination does not allow an appreciation of the thickness of the fibers because it is not possible to distinguish fibers from packets of fibers. X-ray analysis examines this texture in detail.

Each quartz fiber is composed of a piling up of quartz crystallites. The C axis of each crystallite is perpendicular to the long direction of the fiber around which the crystallite can rotate. As a visual image one can consider a high pile of match boxes with every box slightly offset relative to the preceding one by a rotation around the axis of the pile. This is the explanation of the aspect of chalcedony fibers under the microscope; the birefringence varies along the fiber which takes on a moiré luster. Every fiber is constructed so as to be very slightly offset relative to its neighbor since the migration of the moiré luster observed between crossed nicols is very similar from one fiber to the next. A more serious displacement can occur from one packet of fibers to the next.

The physical properties of chalcedony are rather different from those of quartz: lower density, appreciable water content, lower index of refraction. To account for these properties, Correns and Nagelschmidt called upon opal that would form thin layers between the fibers. Calculations specify the probable amount of opal and account suitably for the density and the water content. Application of Wiener's formulae relative to the form birefringence of an alternating quartz-opal ensemble permits us to account for the value of the indexes.

In a very detailed analysis of this problem, Donnay (1936) approved of the appeal to the phenomenon of form birefringence to solve the problem of chalcedony. But he noted that, even though the hypothesis of the opal as second constituent of chalcedony after quartz is plausible, any one other product with indexes lower than those of quartz would have the same result.

Since the time of these works one has been able to show, as will be seen later, that the majority of opals take advantage of some crystalline arrangement similar to that of cristobalite. Many authors have not been able to give evidence of the blurred peaks of cristobalite on X-ray diagrams of chalcedony (Midgley, 1951; Folk and Weaver, 1952; Tovborg, Jensen et al., 1957; Drenck, 1959). These various authors were able to calculate the size of the quartz crystallites that constitute chalcedony; they evaluated their size to be about 400 to 1 000 Å, but they did not find any trace of opal.

The study was undertaken by electron microscope by Folk and Weaver (1952). They obtained beautiful pictures by the replica method. Cherts of the "microcrystalline quartz facies", according to the distinction of Keller (1941), turn out to be composed of polyhedral fragments about 1 to 2 μ in diameter that are mutually interlocked. In contrast, cherts of the "chalcedony facies" did not show their fibrous disposition under the electron microscope, but only a profusion of bubbles disposed in rows. These bubbles correspond to the brownish zones visible in chalcedony under the microscope. Their weakly negative relief indicates that they are not filled with air. Folk and Weaver decided in favor of water that would keep in solution the Al, Fe, Na, Mg, Ca ions that could be measured. These bubbles would explain the density, the water content, and the indexes of chalcedony all at the same time. In these works it is no longer possible to study the fibrous structure of chalcedony.

Pelto (1956) reconsidered the problem of chalcedony in another way, by applying to it the arguments used to explain the structures of the boundary between grains in metals. He reexamined the fibrous structure studied under the microscope and by X-ray diffraction, and he proposed a description of the boundaries from fiber to fiber in chalcedony. The degree of disorder near the boundary between two crystalline structures increases with the difference in orientation of these crystals. When the difference in orientation between two grains is small the atoms on the boundary can be brought into the lattice by an elastic distortion requiring a small amount of energy. This distortion can extend to a considerable distance from the boundary. In contrast, if the difference in orientation is great, the lattice can scarcely be improved, and the boundary is sharp.

The case of the boundary between the fibers of chalcedony apparently is of the first type. The crystallites situated between the adjacent fibers are united by a relatively thick transitional structure composed of a set of dislocations and islands of poor adjustment. On the boundary of the bundles of fibers the dislocations can be so close to each other that an intermediate lamellar zone acquires properties similar to those of a liquid. Where the join is poor, the fit is made by the impurities, foreign ions in the water, a part of which can be related to silica. This mineralized water could also fill in the pores observed under the electron microscope by FOLK and WEAVER. This interpretation by PELTO permitted him to account for not only the diverse physical properties of chalcedony, but also its chemical reactivity and the suppression of the calorific phenomena close to the inversion point of quartz.

BRAITSCH (1957) reconsidered this question of the presence of opal in chalcedony. After study by X-ray diffraction and the optical methods, which confirmed the works of CORRENS and NAGELSCHMIDT, he showed that it was a question of definition of language. As long as X-ray diffraction gives only two blurred bands, and no recognizable lines of low-temperature cristobalite, it is better to abide by the denomination of opal.

Hildegard HOSS (1959) studied the chalcedonies of culm. She obtained by electron microscopy very beautiful replicas of well oriented and crystallized chalcedonies in which the parallel structure is shown. To characterize opal in these chalcedonies, she used the test of ENDELL (1948), who obtained the stages of the crystallization of cristobalite by heating of diatoms. In the heated chalcedonies, HOSS saw, in addition to the quartz peaks, the 4.05 Å reflection typical of cristobalite, whereas the original material presented only the very beginnings of crystallization.

All these works concur, if one is willing to conceive that there exist here stages between great disorder and the organization into cristobalite. What we must remember is that chalcedony is composed chiefly of quartz, but that there are places where the silica is disordered and associated with water and cations. The degree of disorder is variable. Now we are going to see that it is this disorder in the organization of silica that is typical of opal. Thus, the most modern techniques permit the interpretation and the verification of the works of petrographers such as LACROIX, CORRENS and NAGELSCHMIDT.

2° Opal

Before the time of investigations by means of X-rays, opal was considered amorphous. However, MALLARD (1890) had already reported the existence of a fibrous-lamellar variety of silica: lussatite.

LEVIN and OTT (1932) were the first to prove the existence of real crystallinity in some fifteen samples of opal. The patterns presented diffraction bands rather than clear peaks, thus making interpretation difficult. They identified the reflections of high-temperature cristobalite. In the following year LEVIN and OTT (1933) discovered opals the patterns of which included also the reflection of low-temperature cristobalite. Likewise, DWYER and MELLOR (1934) identified high- and low-temperature cristo-

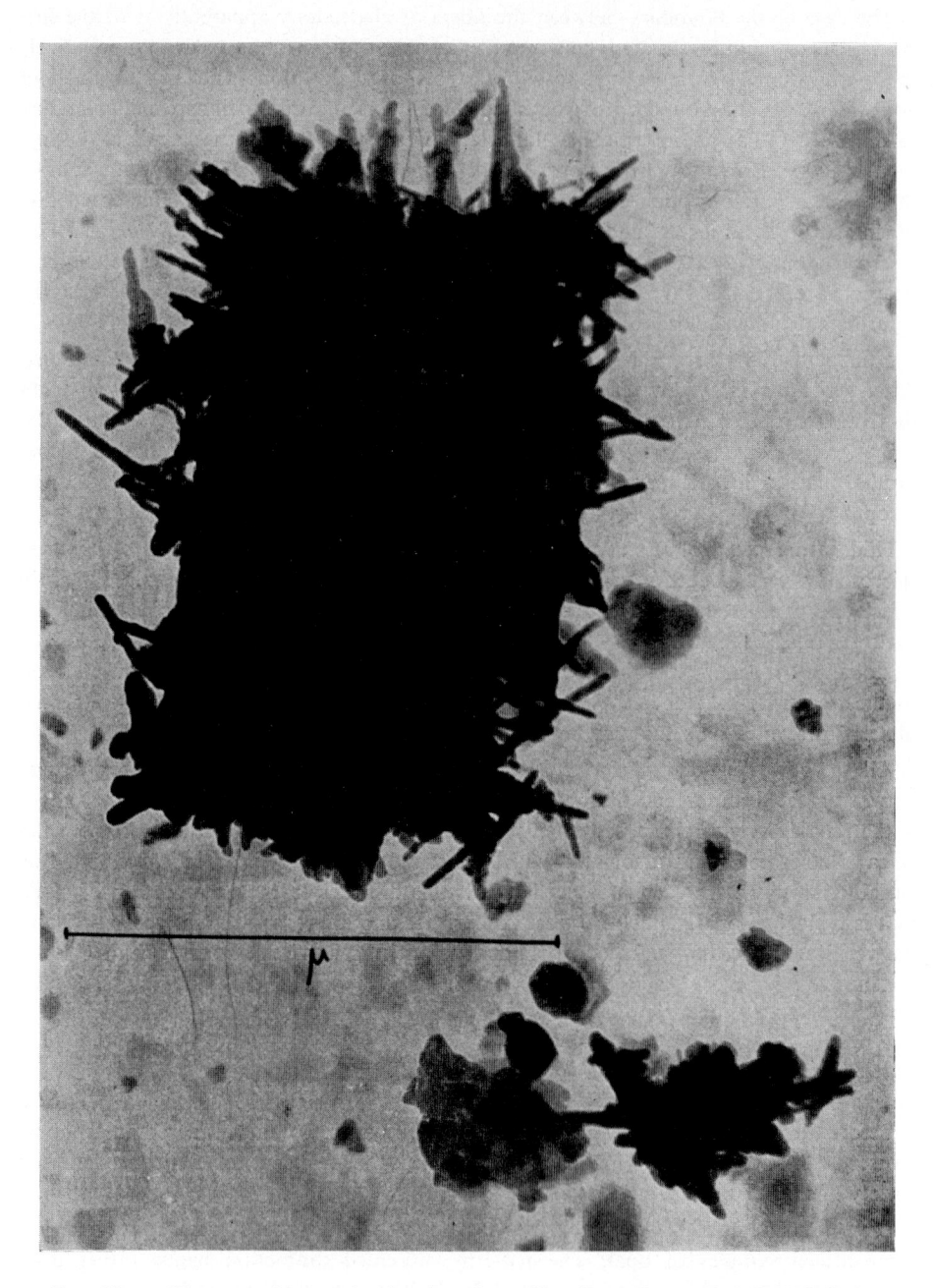

286 SILICIFICATIONS, FLINT AND GROWTH OF CRYSTALS

Fig. 62. — *Electron micrograph of pulverulent silica from Berry, France. Needles or tubes of opal* (after Pézerat, 1961).

balites, as did RAMAN and JAYARMAN (1953) who described alternating layers of these two crystalline forms.

FLÖRKE (1955) explained that the patterns attributed to high-temperature cristobalite must correspond to broadened peaks of low-temperature cristobalite and to the absence of certain peaks. This broadening of peaks and the absence of some of them apparently is explained by disorder in the lattice and by the small size of the crystallites. FLÖRKE described the disordered lattice of opal-cristobalite as a unidimensional disorder, and he explained this as follows. The layers of SiO_4 tetrahedra would be normally ordered within their own plane, but the stacking of these layers would be disordered, either with a periodicity of two (hexagonal arrangement corresponding to tridymite) or with a periodicity of three (cubic arrangement corresponding to cristobalite). In fact, FLÖRKE showed that in some "opals" the (200) reflection of tridymite practically merges with the principal peak of cristobalite (111). Thus, there are no opals in which the low-temperature cristobalite is associated with high-temperature cristobalite, but there are some opals in which cristobalite and low-temperature tridymite are more or less disordered.

BRAITSCH (1957) studied the textures of opals in great detail. He showed that in the case of lussatite the direction of the (111) disorder was perpendicular to the direction of the ($1\bar{1}0$-$10\bar{1}$-011) fiber; the elongation is positive. In contrast, in lussatine the elongation of the fiber and the direction of the disorder are confused. Pseudolussatite adds textural disorder to the preceding structural disorder by means of aggregates of crystallization domains that are oriented differently.

FLÖRKE continued his work: his exposé, during the *Journées internationales d'études* in 1960 in Brussels, threw a brilliant light on these problems. In the surficial zone of a crystal, coordination between atoms is disturbed. In the case of silica lattices this disturbed zone is estimated to be 10 Å in thickness. The volume of the disturbed zone is equal to 0.6 % for a 1 μ crystal and to 48.8 % for a 100 Å crystal. Very small particles would thus behave quite differently from the large ones; the disordered zone in very small particles represents an appreciable part of their volume. All kinds of bonds can occur between the water and the impurities. One arrives at the state that I shall call here "cryptocrystalline", which is neither the crystalline state nor the amorphous state, between which there is, moreover, no limit. Thus one must conceive of opals with various degrees of organization.

It is indispensable, likewise, to note the role of cations in the structures of the various crystalline forms of silica, such as BUERGER (1954) has described to us. WEIL (1926) had already shown the favorable influence of impurities on the crystallization of cristobalite. But BUERGER noted that the analyses of natural cristobalite and tridymite always gives a proportion of elements other than silica and oxygen; these elements are chiefly alkalis and alkaline earths. Cristobalite and tridymite represent structures more open than that of quartz and these structures allow the accommodation of large cations, while electrostatic equilibrium is ensured by Al ions in tetrahedral position.

Reciprocally, a crystal that includes these elements as it grows cannot have the structure of quartz, but only that of cristobalite or tridymite. These structures are stable below their temperature of formation because they are filled up by foreign ions.

At low temperature the lattices are victims of distortions, but they cannot attain the lattice of quartz, which is the stable form of crystalline silica at ordinary temperatures. This is very important in the interpretation of silicifications by crystal growth (MILLOT, 1960).

Other observations confirm this accumulation of facts; SWINEFORD and FRANCKS (1959) studied the opals in Neogen formations of Kansas (U.S.A.). They distinguished two varieties of opal. The first named the "low cristobalite tridymite type" corresponds to the descriptions of FLÖRKE, with bands in the proximity of the reflections of cristobalite and low-temperature tridymite, and containing various cations. The second variety was called "diatom type"; it presents only wide and diffuse bands around 9.8 Å and between 4.9 and 3.4 Å, corresponding to a very low degree of crystallinity.

PÉZERAT (1961) studied samples from the Berry of pulverulent opal associated with illites and siliceous nodules. Upon analysis this silica, considered amorphous up to that time, turned out to be an irregular cristobalite-tridymite structure according to the scheme of FLÖRKE. Moreover, one observes under the electron microscope that this opal powder is composed of spherulites in the form of needles or tubes (fig. 62). These needles or tubes are 200 Å thick and 1 000 to 3 000 Å long and show, therefore, that a very elongate organized facies corresponds to a disordered organization. The interest of this example is that it shows an aspect of opal in the form of free needles instead of a massive and non-dispersable aspect as in the majority of opals.

Thus, natural opals present a range of organizational possibilities. The best organized variety appears to be composed of very small particles in which layers similar to low-temperature cristobalite or low-temperature tridymite alternate in unidimensional disorder. The other varieties are still more disordered, so much so that they present only rings by X-ray diffraction. One will note that there is a causal relation between this smallness of the particles and their disorder.

3° *Geological consequences*

The petrographic and geological consequences of these new mineralogical data are important.

— Flint of the "microcrystalline quartz" facies is simple; it is a question of assemblage of many small isodiametric quartz grains into polyhedra, the size of which is commonly 1 μ.

— Flint of the "chalcedony" facies is different; we must remember that chalcedony is a fibrous assemblage of quartz crystallites with serious irregularities in the lattices at the boundaries between fibers. It is in the boundary area between the fibers that silica is apparently disorganized in the company of water and mineral cations; this leads to the notion of "opal" in the broadest sense of the term.

— As for natural opals, or more exactly the majority of them, they are not amorphous; they are partially organized. The arrangement of tetrahedra is similar to that of cristobalite or possibly of low-temperature tridymite with partial disorder. This disorder favors the presence of water and foreign ions and limits the size of particles. These particles are grouped into a texture that is fibro-lamellar in some cases, or in aggregates (BRAITSCH, 1957). Under the microscope, this aggregated texture produces a statistical isotropy that has misled generations of observers concerning the real nature of opal.

III. — GEOLOGICAL AND PETROGRAPHIC DATA

1° *Climatic silicification*

The climatic silicification connected to weathering will be distinguished from sedimentary and post-sedimentary silicification. Numerous workers in all countries have been concerned with this question. It is the problem of mechanisms that interests us, and for this reason only the fundamental or recent works will be cited.

Silicification in the present Kalahari Desert has been studied for a long time. KALKOWSKY (1901) has already given detailed descriptions of the silicification, allied in some cases with dolomitization, of the various rocks that crop out in this great desert. STORZ (1928), in a fundamental and very well illustrated treatise, showed the whole variety of siliceous neoformations and presented an overall view of the problem of climatic silicification in the memoir of his teacher KAISER (1926) on the diamond-bearing deserts of South-West Africa. He showed relationships with weathering phenomena developed under more humid climates of the African Tertiary and made comparisons with the *meulières* (millstone grits) of the Paris Basin. Helpful information is found also in Germany in the book of WALTHER (1900). From other countries, one can mention the works of HUME (1925) on Egypt, of FERSMANN (1926) on the desert of Karakum, of KAISIN (1956) on the weathering of limestone, of SWINEFORD and FRANKS (1959) on Neogene silicification in Kansas.

In France, we find that GOSSELET had already reported in 1888 the Tertiary transformation of sandstones into quartzites in the Paris Basin by means of what he called *métamorphisme atmosphérique*. CAYEUX (1929) described the microscopic aspects and all the characteristics of this phenomenon. Since that time, the following works have been added.

ALIMEN (1936, 1944) studied the silicification of sands into the Fontainebleau sandstones, then the dissolution of the quartz of the Stampian calcareous sandstones. CHOLLEY (1943) studied the meulierized surface of the Miocene in the Paris Basin. According to him, the *meulière* is a surficial silicification corresponding to a humid tropical climate with a well marked dry season that gives rise to sheet floods and to phenomena of slow percolation. VATAN (1947) interpreted the silicifications in the form of flagstones and the formation of *meulière* in the Paris Basin as surficial phenomena. CAILLEUX (1947) described quartzose concretions of pedogenetic origin. AUZEL and CAILLEUX (1949) examined silicifications in the northern Sahara and called upon sheet flow and climatic and subaerial phenomena. ALIMEN and DEICHA (1958) studied the meulierization of the Pliocene hammadas of the northwestern Sahara and the phenomena of solution and silicification to which they attributed a climatic origin.

I shall reexamine here a summary of the facts observed in the last fifteen years by the geologists in Saharan Africa. These facts have already been brought together recently (MILLOT, RADIER, MULLER-FEUGA, DÉFOSSEZ and WEY, 1959; MILLOT, 1960). They will serve as examples and as arguments.

Silicification in the Gao basin (Sudan) RADIER (1957). — The observations were presented in 1949 and 1950 by KIKOINE and RADIER, and then reexamined in detail by RADIER in 1957.

a) QUARTZITIC FLAGSTONES. — Quartzites form rather regular flagstone pavements over great areas in the Sahara. In some cases, the assemblage is more or less chaotic because of subsidence of an underlying unconsolidated bed. Under the microscope, one can observe all the intermediate steps from quartzitic sandstones to quartzites; peripheral growth of the quartz grains is seen and, in different cases, a cryptocrystalline cement of chalcedony or of iron oxide. These quartzitic flagstones do not belong to a specific stratigraphic horizon. They overlie formations of various ages: whitish sands of the Maestrichtian-Senonian, pale or yellowish sands of the lower Paleocene, whitish sands of the Terminal Continental Series, recent alluvial sands and chiefly those of the floodplain of the Niger River. These facies have never been found in the numerous well and drill-hole sections that have been studied. The only probable interpretation is to consider these quartzites as a surficial modification of sandy sediments.

b) SILICIFIED LIMESTONES OR "TERRECHT" FACIES. — The calcareous deposits that crop out in circular pattern around the Adrar des Iforas present a cuesta landscape; the structural limestone surfaces of the Danian and Ypresian beds crop out as wide plateaus bordered by abrupt cliffs. The surface of these plateaus is sculptured by erosion into a kind of lapies, which was called *Terrecht* (stone sponge) in the Tamashek language. The Terrecht I corresponding to the spongy Danian surface with cutting *arêtes* and bristling with sharp spines has been distinguished from the Terrecht II which gives rise to rounded forms, pierced by channels, having the aspect of skulls, and belonging to the Paleocene-Ypresian. Under the microscope, the limestone appears silicified; the matrix of the rock is composed of islands of residual calcite and of areas of quartzite in a fine-grained mosaic. These silicified limestones have never been found in bore holes or wells where stratigraphically-equivalent formations are represented by limestones or marny limestones. There is a close dependence between outcropping, silicification, and the Terrecht facies. The proposed interpretation is based on the hypotheses of CHOLLEY (1943) for the origin of the *meulière* in the Paris Basin. We arrive at the reconstruction of a climatic phenomenon of pedological character.

Silicifications in the Fezzan (eastern Sahara) MULLER-FEUGA (1952). — The desert outcrops in the Sahara of the Fezzan provided excellent examples for petrographic analysis; 110 thin sections were studied. An evaluation of these analyses leads to the following distinctions.

a) QUARTZIFIED SANDSTONES. — Quartzified sandstones are common on the tabular surface of the summits, plateaus, and hammadas. They are surficial and modify the underlying rock through a thickness that increases with the permeability of the rock.

The outlines of the quartz grains are distinct. The main characteristic of these sandstones is growth of the grains, with a tendency towards polygonal forms.

Opal, with one exception, was not found in these samples. Such facies occur also in bore holes penetrating the thick "Intercalated Continental" Series. After some discussion, it turns out that the beds that were reached are fossil surfaces.

b) OPALIZED OR FERRUGINOUS SANDSTONES. — The opalized or ferruginous sandstones have been sampled at the foot of slopes. They have never been found in drilling, and only once on a plateau. Quartz grains show general traces of corrosion; they have not grown, and they are edged with ferruginous or opalized cement. In the latter case the opal has the tendency to be organized into chalcedony.

c) ARENACEOUS LIMESTONES WITH HALOED QUARTZ GRAINS. — In the lowlands dominated by the plateaus there are arenaceous limestones on the surface that have undergone a curious evolution. The quartz grains are strongly corroded. Their outline is surrounded by a thin envelope of clay minerals, then by an envelope of opal whose variable thickness is in many cases equal to the diameter of the grain. This opal tends to crystallize into chalcedony and even into quartz. In addition, the opal invades the cryptocrystalline calcite cement in some of the samples. In other cases, the corroded quartz grains are surrounded by halos of calcite. These microcrystalline halos are multiple in some cases and stand out clearly against the cryptocrystalline cement of the rock.

d) DISTRIBUTION IN THE LANDSCAPE (fig. 63). — An example will illustrate the distribution of these weathered facies. In the Cherguya, near el-Bder (NE of Fort Leclerc), we find ourselves at the edge of a plateau dominating a foreland that is slightly rolling: the lowland. The sandstones forming the cliff have not been transformed.

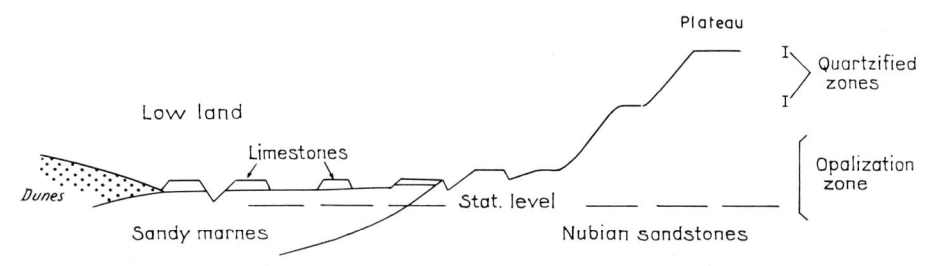

FIG. 63. — *Distribution of silicifications in the Cherguya, northeast of Fort Leclerc in the Fezzan.* Quartzification on plateaus. Opalization in the lowlands (after MULLER-FEUGA, 1952).

On the plateau there are quartzified sandstones. At the foot of the cliff, at altitudes relatively close to the water table, opalized sandstones are found. In these opalized rocks the silica of the quartz grains is dissolved, but it is found in the form of opal in the cement. Calcite gives way, but it reappears in some cases in halos around the quartz grains. The latter, which are corroded, are surrounded either by these halos of calcite or by a thin border of micaceous clay minerals or by a halo of more or less chalcedonic opal. Here we can comprehend the variations in solutions during neoformation the growth of which they ensure. It is possible to affirm that the solutions are alkaline and that they oscillate around delicate equilibrium values for silica and calcite, as well as for silicates, but we cannot be more precise. We are in the domain of epigenesis. MULLER-FEUGA, therefore, deserves great merit for having shown us the relation between the various forms of silicifications in the landscape. The high plateaus are subject to quartzification, whereas depressions and ground water give rise to the formation of chalcedony and opal.

Silicification in the Great Bend of the Niger (Sudan) DÉFOSSEZ (1958). —
Analogous phenomena as well as subterranean silicification of great amplitude have been studied in the Saharan regions of the Great Bend of the Niger.

a) SURFICIAL SILICIFICATIONS. — Whether it be a question of the sandstones of Ydouban or of Hombori Douentza or more recent primary sandstones, numerous specimens collected on the surface show traces of the growth of quartz grains and transition to quartzite (DÉFOSSEZ, 1958). The same surficial quartzification has been observed in the southern extensions of these formations, in the southern part of Sudan and in the northern part of Upper Volta, by PALAUSI (1958).

Dolomitic limestones are also affected by this surficial silicification. In the region of Tin Simaman, Tin Mazurel, In Kafen Amman, and Tin Akoff, the dolomite changes on the surface to a black and red, scoriaceous, vesicular chert, very much marked by eolian erosion. Under the microscope, the carbonates remain as traces, especially in ghosts of oolites; fibrous or spherulitic chalcedony is prevalent with secondary quartz in some cases.

On the other hand, silicification in the depressions takes on another aspect. Thus, in the bottom of the pond of Guéderon, northeast of Gondo, a distinct section is observed:

— 1 meter of stratified sandstone composed of ferruginous sandstone alternating with shale that has been silicified into opal;
— 1 meter of breccia silicified either into opal or into fibrous chalcedony being transformed into quartz;
— 3 meters of sandstones with chalcedonic cement, accompanied by a few thin beds of opal.

Here we see once again the principal facies that we observed previously.

b) SUBTERRANEAN SILICIFICATIONS BY GROUND WATER. — Ground water of the Gondo basin, which is utilized by the people, has been carefully studied. Here it is found that the most diverse stratigraphic horizons which have been encountered in wells or drilling are marked by strong silicification. This is particularly noticeable in the zone of fluctuation of the water table; it can also occur at depth, but never in dry horizons.

In the Irma bore hole, jointed carbonate rock abounds with fissures filled with opal. The silicification penetrates the carbonate mass and isolates the dolomite crystals.

In wells at the Gondo, dolomites collected at the level of the water table are completely silicified and colored blue, red, white or brown. The trace of former oolites is found in the form of silicified ghosts. Fibrous chalcedony is the principal component and the secondary quartz appears.

It happens in some cases that the water table washes the basal detrital facies of the Terminal Continental Series which covers the dolomitic series of the Gondo. In this case the clays and sands of this series are abundantly silicified. The clays are transformed into brightly colored, porcelainized rocks. In thin section, the isotropic matrix turns out to be composed of opal in which some detrital and micaceous particles float. The sands are consolidated into silicified sandstones with an opal cement.

Here we have the spectacle of strong silicifications at the level of a water table under a permeable covering, fed by the Sudanian rains.

Silicification in the Zemnour Noir (Northern Mauritania) SOUGY (1959). — SOUGY studied the residual buttes of the Upper Cretaceous which are remnants of the border facies of the transgression of that period. He studied both their stratigraphy and their petrography. "Posterior to its deposition, this Aidate formation underwent the following transformations: emergence at a time that is difficult to define at present, and then silicification, all the stages of which can be observed from simple opal cement to a quartzite mosaic, with an intermediate stage with chert nodules (replacement of the sandstone by chert). In a general way, silicification, zero at the base, intensifies toward the top of the formation. It is maximum in the terminal quartzitic bed, which may be a quartzite mosaic."

These observations on climatic silicification are very briefly summarized here. However, they lead to an appreciation of the influence of two principal factors that orient this phenomenon.

First is the question of the morphology of the landscape. Quartzification occurs especially easily on the surface, whereas in basins and within the ground water mass more disordered forms of silica, i.e., chalcedony and opal, develop preferentially.

Secondly there is the question of parent rocks. Even under ordinary conditions one sees that carbonates prefer silicification into chalcedony and clays silicification into opal. The nature of the bedrock submitted to the action of solutions has an influence on silicification; this point will be verified in the following paragraph.

2° *Flint, cherts and silicification of sedimentary rocks*

Here we are concerned with silicification affecting sediments after their deposition, whether it be immediately after deposition, or during a more or less belated history after burial. We find ourselves confronted here by a gigantic bibliography of several thousand works, which it is absolutely impossible to think of summarizing. However, in France, it is necessary to mention the incomparable work of CAYEUX (1929). It seems that the essentials of his observations and of the coincidences that he gave evidence for still persist.

One can also refer back to examples of the silicification of oolitic carbonate rocks given by G. LUCAS (1942) and KIESLINGER (1954), as well as to the studies of silicification given by SIEVER (1959) and other American authors in a treatise on silica in sediments.

Every geologist knows a great number of examples of silicifications affecting siliceous carbonate, argillaceous, or organic sediments. I shall take the liberty to mention here an example observed by ELOUARD (1959); it is an interesting case.

Silicification in the Lutetian of southern Mauritania. — Many phenomena of silicification have been observed by ELOUARD (1959) in the Lutetian formations in the northern part of the sedimentary basin of the Senegal River in southern Mauritania. These silifications are penecontemporaneous with sedimentation, since they are also found as pebbles in the immediately overlying formations.

The limestones and dolomites completely silicified or transformed into flint are numerous. The silicification occurred by means of epigenesis of the carbonates into fibrous chalcedony. Quartz is only accessory.

Arenaceous limestones can be transformed into cherts. The latter show detrital quartz grains and, in some cases, phosphatic nodules, both being coated with a cement of secondary chalcedony. The presence of many silicified fossils testifies to the calcareous origin of the rock. Here we encounter arenaceous limestones or calcareous sandstones, silicified into chalcedony or cryptocrystalline quartz.

The ferruginous sandstones of the loop of the Gorgol River are brick or salmon red; they are associated with quite quartzitic sandstones. These facies are mutually transitional. Silicification leading to the quartzitic sandstone took place by removal of the ferruginous halos followed by growth of the quartz grains that had been thus purified. Neoformed quartz developed upon the nucleus and assumed the same orientation. Then the grains came into contact with each other, and thus a mosaic quartzite formed.

The sandstones of the Badalli wadi are quartzitic sandstones with haloed quartz grains. But these sandstones contain small nodules of about 2 by 1 cm, black, hard, and of heterogeneous aspect. Under the microscope these nodules show the debris of silicified organisms (lamellibranchs, echinoderms) immersed in a chalcedonic cement. The contact between the quartzitic sandstone and the cherty nodules is an insensible transition in which one can observe quartz grains without overgrowths and surrounded by chalcedonic cement. Everything indicates the past existence of a sand with calcareous nodules in which the silicification followed two different paths according to the material replaced.

The quartzitic sandstones of Kaedi are composed of detrital quartz grains, of grains with overgrowths, and of a chalcedonic cement. Locally the chalcedony forms masses of the size of a quartz grain. One can even find overgrown quartz grains coated with chalcedony.

These different cases show how silicification evolves differently according to the material attacked (ELOUARD and MILLOT, 1959). If the material is a carbonate, calcareous or dolomitic, it is chalcedony that develops in abundance. Quartz, if it occurs, is only cryptocrystalline and subordinate. If the material is a sand or a sandstone, it is quartzification that develops in most cases, except if cement is calcareous giving rise again to chalcedony.

Lastly, the sandstone of the Badalli wadi evolves in two different ways; the calcareous nodules are transformed into chalcedony and the sandstone into quartzite. This evolution, which differs according to the facies in the same rock, is one of the most valuable points made by ELOUARD.

Once again we see that the material offered for silicification seems to have a role in the result obtained.

Overall view on silicification in sediments. — The aim pursued here is to search for an explanation of the phenomena of silicification. It is thus necessary to understand precisely the essentials evoked above in connection with the works of CAYEUX.

a) It seems that the first essential point is replacement. When a sandstone is transformed into quartzitic sandstone, into quartzite, into chalcedonitic or opaline sandstone, silica replaces a "primordial cement", in the words of CAYEUX. Whenever the original material is a tree trunk, a fossil, a Foraminifera, an oolite, or simply a mineral, it is again a question of replacement since the microscope reveals delicate structures that have been silicified. Similarly, one could mention valuable observations that give evidence of this replacement in gaizes and clays. It is a general phenomenon.

Concerning flint, it will be noticed that it is more abundant in calcareous than in dolomitic rocks. Moreover, if Ca and Mg decrease from the periphery to the core of the flint, the Ca/Mg ratio also decreases (CHILLINGAR, 1956). Replacement seems to prefer the calcite, but it is not known whether Mg is expelled or if it finds a niche in the siliceous framework.

Let us state clearly that these phenomena of replacement or epigenesis prohibit an appeal to a possible phenomenon of flocculation of colloidal silica. There is no hope of obtaining such detailed silicification by coating the mineral or organic structures with a silica gel, even if one could succeed in explaining its presence. In contrast, it is necessary to envisage a mechanism of permutation, which will be called epigenesis or pseudomorphism, as the case may be.

These silifications are, of course, the effects of silica in solution in interstitial waters. It is well known that oceanic waters are very poor in silica because measurements indicate only a few ppm or mg/l. In contrast, interstitial waters give quite variable values, which can rise as high as 100 ppm while still remaining below the solubility threshold of amorphous silica. The reason resides in the fact that the silica of sea water, like that of lakes, is used in great quantities by siliceous organisms: diatoms, Radiolaria, Silicoflagellata, siliceous sponges. But in the mud, the silica of the tests of these dead organisms can become available again and contribute silica to the connate waters. The silica in organisms is thus only a detour; coming from the continents, it is stored up in the siliceous tests, but it can be released again after their death.

b) A second point that seems essential concerns *the nature of the material offered for replacement.* CAYEUX (1929) approached this question at the end of his treatise where he permitted himself to rise above the patient inventory of facts. An important although not completely determinant reason for the variety of silicifications must be looked for in the host material itself.

— *Siliceous features in a calcareous environment develop chalcedony in a typical way.* Opal and quartz are subordinate.

— In contrast, *opal is characteristic of the siliceous impregnations that have an appreciable occurrence in clays.*

— *Quartzine finds a favorite domain in the epigenesis of gypsum.* But attention must be called to the development of bipyramidal quartz grains in marnes rich in sulfate accumulations: anhydrite or gypsum.

— Let us add that *quartzification develops in sands and sandstones* whenever the cement does not form an obstacle.

Thus, we shall note that in the sedimentary silicifications, early or not, *the composition of the material silicified exercises some influence on the genesis of the different modes of occurrence of silica.*

IV. — SILICIFICATION AND GROWTH OF CRYSTALS

It is now possible to attempt an explanation of all accumulated facts. I shall do this by rejecting the traditional flocculation of silica and by presenting the hypothesis of genesis of natural silica by the growth of crystals.

In reviewing our position, we see that this is where we are.

Natural waters are true solutions of monomolecular $Si(OH)_4$ silica. They are insensitive to pH variations. They are unsaturated with respect to amorphous silica, but easily supersaturated with respect to the more or less regularly crystallized forms, such as quartz, chalcedony, and opal. They can, therefore, ensure the growth of the latter.

An interpretation of silicification in the hydrosphere is thus possible by the application of knowledge acquired of the growth of crystals. This line of reasoning has already been used by DEICHA (1945) in the case of the crystallization of gypsum. Concerning silica and neoformed quartz in Tertiary sediments of the Paris Basin, he wrote: "the perfection of the crystal form, the development of free bipyramidal crystals, and the limpidity suggest the idea of slow crystallization in a medium close to the physico-chemical equilibrium. By this character, this medium seems to be the opposite of that which provided deposits of silica in the form of less stable varieties... The deposition of unstable or metastable substances from supersaturated solutions is common; the lesser stability of such substances is, in fact, an indication of higher solubility... In the case of silica, the infinitesimal solubility of quartz permits us to think that supersaturation prevails when silica is supplied to water by the decomposition of silicates." In brief, we find here the complete orientation that must be followed in the crystal-growth hypothesis.

Experimental works on the growth of crystals, such as presented for instance by DEICHA (1946), KERN (1953, 1961) and DEKEYSER and AMELINCKX (1955), provide guide lines for geologists without underestimating the complexity of the phenomena. Slightly supersaturated solutions, poor in impurities, can nourish monocrystals starting with a nucleus, adding step by step the ions necessary for regular stacking. In contrast, more concentrated solutions induce a proliferation of nuclei and give rise to twinned and polycrystalline structures. Sudden changes of temperature and concentration accentuate this disorder. Likewise, "impurities" or forcign ions can also disturb the regularity of the growing structures.

Let us transpose this information to silicification, while noting at the same time that this transposition is somewhat risky because the bonds in silica are partially covalent. But, in a naturalistic approach to such a problem, it is valuable to point out the coincidences.

1° *Quartzification*

1. Climatic quartzification prevails on Saharan surfaces. They are indicated by the growth of grains. A silica supply is necessary. This silica cannot come from afar because the phenomenon takes place on upland surfaces. Rain water runs across the surface, percolates into the soil or a surficial loose material, and traverses, at most, the first few meters of the outcrop. In any case, its history is short. The waters are "fresh", charged with very few ions, but they can acquire rapidly, by dissolution of silica or hydrolysis of silicates, a concentration of silica higher than 10 ppm. With the help of the temperature difference between the surface and the interior of the rocks, supersaturation with respect to quartz is reached. Grains leached by these acid and aggressive solutions act as nuclei. On their surfaces regular growth starts, and the former lattice increases and is continued. The halo is orientated on the grain itself.

2. In the transformation of the sedimentary sands and sandstones into quartzites the same mechanism can be invoked. Everybody knows that water circulating in pure sandstones is very soft, the intensity of percolation being sufficient to remove the cations. Within these sands and sandstones, this soft water ensures growth of the quartz grains, which develop halos.

3. *What is essential here is to have available "clean" solutions, poor in cations, but somewhat charged with dissolved silica so that the growth of quartz is maintained.*

2° *Silicification into chalcedony*

1. Climatic silicification has shown us the evolution to chalcedony in the low areas, and especially in the carbonate facies. In this case, the waters have a more prolonged history. Either by percolation through the rocks, or by accumulation in lowlands, or by residence in the ground water reservoir, the concentration of various elements in solution increases. The water can attain higher concentrations of dissolved silica; such solutions are common in nature. They are capable of nourishing siliceous neoformations if nuclei exist. Moreover, these nuclei are not lacking, even if they cannot occur spontaneously. Numerous particles and surfaces can play this role: thin flakes of minerals, clay particles, and without doubt a great number of crystalline surfaces. The stacking of silica tetrahedra begins, but numerous causes of disorder come into play: multiplication of the nuclei, inadaptation of their lattice to the regular development of the quartz lattice, and the presence of foreign particles and ions. Incapable of a regular development, quartz is organized into long spiral fibers separated by disordered boundaries where water and cations are trapped. One obtains thus fibrous chalcedony with all the variety of cases found in nature.

2. Within sediments the phenomenon presents itself similarly. The water enclosed in the sediments is likewise charged with diverse ions because its migration is controlled by its imprisonment or because it circulates with ground water. Thus, it is alkaline and likely rich enough in silica to induce silicification such as is observed. Let us note here that carbonates are particularly predisposed to be transformed into chalcedony; one day

it will be necessary to explain this coincidence and to see whether the presence of calcium ions or the calcite lattice plays a role.

An intermediate case occurs between the easy development of quartz and the development of fibrous chalcedony; this is the case of microcrystalline quartz. When the regular growth of the macrocrystals is hindered, quartz can still develop into a micropuzzle of microcrystalline quartz. This is a common stage in flints.

3. *If microcrystalline quartz flint and chalcedonitic flint are different facies, they do present a common character,* that of the organization of quartz in small sizes varying between about one micron and 500 Å. *This will be the microcrystalline domain.*

3° Silicification into opal

1. Inventory of the facts has shown us how silicifications of climatic origin in the Fezzan and Sudan readily take on the form of opalization, and this occurs chiefly in clays. In this case, the neoformations are even more disturbed than in the preceding cases. We know today that the opals of petrographers are irregular organizations in which a lattice similar to that of cristobalite or tridymite is discernible. Moreover, the irregularities are added together. The linkage of the SiO_4 tetrahedra is disturbed by the intervention of large alkaline or alkaline earth cations; this leads to structures similar to cristobalite, but at the same time this structure is filled in and prevented from evolving spontaneously into quartz. The siliceous sheets pile up with that unidimensional disorder described by FLÖRKE. Lastly, the very limited development of crystallites gives rise to surficial disorders and to that misleading, statistical isotropy. We are on the boundary of the crystalline state.

2. In sedimentary rocks where opalization is frequent, a similar mechanism can be invoked. Construction of the lattices is seriously restricted, and the results show products intermediate between chalcedony, chalcedonic opal, and cristobalite opal such as we have studied.

In both cases, the waters governing these neoformations have a long history because of their circulation as ground water or through sedimentary formations. These are waters rich in cations and perhaps richer in silica that give rise to the very imperfect organization on numerous nuclei. Particularly curious is the preferential transformation of clays into opal. The role of clays in the orientation of siliceous neoformations will have to be explained by the influence of the ions present, of the siliceous framework of the replaced clays, of the size of clay minerals, or by any other factor that would play a pilot role in this domain.

3. *Such will be our third case,* which could be called "penecrystalline", if the term existed; *we shall be content with the word "cryptocrystalline"* that testifies very well to the character of this crystallinity that is so discreet that it remained hidden from petrographers for a century.

4° Formation of silica gels in nature

There remains the direct formation of silica gels in nature. The question is poorly known and sure observations are rare. It is necessary to reexamine the observations in the light of our new knowledge about silica.

The first research pathway is the examination of weathering horizons and soil profiles in which the "amorphous material" or "aluminosilicic gel" mentioned by many authors occurs. Solutions in soils having never revealed a concentration of silica higher than the limit of solubility, it is highly unlikely that flocculation of siliceous or aluminosiliceous gels can be invoked. It is a question of residues of crystalline structures with their cations removed by hydrolysis, or residues of leached glassy structures if the parent rock is volcanic. Thus, these are but stages in the degradation of crystalline and glassy silicates; they are not neoformed gels. Two pathways are possible for their evolution: either weathering increases and these amorphous products themselves are dissolved, and this is probably their usual destiny; or conditions are united that are conducive to neoformation within the mass of these silicate residues, and they can nourish the solutions that ensure the neoformation.

The second research direction is that of direct precipitation of silica from solutions that become supersaturated. Places in which the dissolved products can accumulate are necessary, and evaporation can be one of the principal factors. In the great lacustrine depressions under warm climates we know that this silica feeds extensive diatom blooms. But the surplus silica seems to be able to flocculate, as described by ERHART (1943) in the central delta of the Niger River in the Sudanian basin and by ERHART (1953) and SIEFFERMANN (1959) in the Chad basin. One can also report the observation of DENISOFF (1957): the sand of the bank of the Congo River apparently is cemented at the level of emergence of the water table by a silica gel whose physico-chemical character is not defined. For the submarine environment, H. and G. TERMIER (1952) cited several examples of beds of silica gels observed on the bottom, outside the bar of the Mississippi River and south of Queensland, Australia. Such contemporary cases must be studied to find out the composition of these gels, their charge, and their possible evolution.

On this subject, it is of interest to mention the experiment of BISQUE (1962). He reacted a solution of less than 50 ppm SiO_2 with a porous, clayey dolomite. He obtained silicification by the creation of silicified bonds between the clay particles. This is not a replacement but a kind of polymerization in which the clay particles are incorporated. Here is an interesting itinerary to follow up.

But for the moment, flocculation of gels in nature is poorly known, and we know nothing of their fate. Their ageing could orient them toward poorly organized opals.

5° Importance of the factors of silicification

We see, therefore, that silicification in the hydrosphere presents a whole gamut of possibilities with increasing disorder. Before summarizing what has been understood from a genetic point of view, it is necessary to see clearly the factors determining silicifications. Two have been evoked:

— The nature of the circulating solutions;
— The nature of the material replaced.

These two factors are not completely independent variables because the nature of the solutions depends partly on the material through which they percolate. A siliceous sandstone maintains the poverty in various cations of the waters that traverse it; a limestone charges these waters with calcium ions.

However, it is also seen that the effects of the same water body can be different on limestones and on clays. One sees that sulfate rocks, although very rich in calcium ions, can give rise not only to microcrystalline silica, but also to beautiful crystals of quartz. The latter are, for instance, the famous bipyramidal quartz grains of the Triassic resulting from epigenesis of sulfates. They are twinned, filled with inclusions, but, however, "perfectly quartz". If, under the influence of similar solutions, the effects are different, this is not due to the solutions but to the material itself. It can only be a question of the lattice or of the structure of the mineral or organic material replaced. In other words, the periodicities, or one or the other of the periodicities, or such and such an element of the structure guides the neoformation that occurs by epigenesis.

We are ignorant in this domain. The only glimmer concerns the clays, which are readily silicified into opal. The tetrahedra are organized in ring-shaped tetrahedral sheets, which are not very different from the ring structures occurring in cristobalites. Therefore, one can conceive that the clays, with their cations largely removed, except for silicon, offer the beginnings of rings capable of forming the layers of cristobalite and tridymite that FLÖRKE described to us as stacked up but disordered in their stacking. Moreover, the aluminum ions in tetrahedral position would persist in part, giving the framework a charge capable of retaining the foreign ions that are always present. Therefore, the small size of clay particles, the ions they contain, and the organization of their tetrahedral sheets appear to me as appropriate guides for the elaboration of opals, such as they are understood today.

Replacement carried out step-by-step in minerals of larger size can permit us to understand the edification of particles of greater diameter. In the case of the epigenesis of calcite, the absence of Al^{3+} ions would allow the edification of the quartz lattice, but, being disturbed by the Ca^{2+} cations, it would give rise only to the microcrystalline facies. Lastly, leached and clean quartz grains would offer to solutions ideal nuclei for the regular development of quartz itself.

V. — CONCLUSIONS

The variety of silicification phenomena in the hydrosphere seems to be interpretable by the phenomenon of the growth of crystals.

Chemical and mineralogical knowledge accumulated in recent years cannot allow an appeal to the flocculation of silica in any case as an explanation of the phenomena of silicification in the hydrosphere. This is impossible from the three-fold chemical, mineralogical and petrographic point of view.

On the other hand, the phenomenon of the growth of crystals, which ensures the neoformation of so many minerals of weathering and sedimentation, can account for the variety in the observed phenomena. The ensemble will be examined in the following series of propositions:

1. The growth of quartz grains is the obvious witness of the orderly growth of a crystal from dilute, clean solutions. It leads to quartzification.

   This is the domain of the regular growth of macrocrystals.

2. The restricted development of quartz will lead either to the microcrystalline quartz facies, composed of small quartz grains interlocked in a mosaic, or to the chalcedonic facies. Here the parent solutions are richer in cations or impurities than the preceding case. The crystalline structures begin to grow, but remain of small size, or else they develop with this peculiar fibrous facies that is typical of a first mode of disorder in crystals.

   This is the domain of restricted microcrystalline growth.

3. A still more disordered stage is next. The parent solutions and the minerals subjected to the silicification phenomenon cannot ensure the regular growth of crystals. Several causes of disorder converge and lead to that which petrographers call opal. These opals contain trapped water and cations and can develop only crystalline outlines, similar to cristobalite, but of very small size, of the order of 100 Å. Many intermediates occur between chalcedonies and chalcedonic opals and opals.

   This is the domain of the very restricted and imperfect cryptocrystalline growth.

4. It is not excluded that, at the limit, solutions very much enriched in silica could polymerize this silica and give rise to true gels. It seems that this phenomenon is rare in the hydrosphere, but study of present-day silicifications is necessary. In addition, it occupies a choice position in the interpretation of silicification of deep seated origin: volcanic emanations, geysers, hydrothermal veins, etc.

   Here one appears to enter into the domain of the flocculation of silica.

5. If suitable cations are present, under conditions that we shall examine later, and in sufficient proportions, they are no longer tolerated, but associated into new structures, those of the clay minerals. This is the reason for such a frequent association of montmorillonite with cristobalite, and of attapulgites and sepiolites with flint, because silica divides itself between liberty and the variety of clay combinations, which develop by themselves in many cases.

   Here we penetrate into the domain of argillaceous neoformation.

In short, the mode of occurrence of silica in growing siliceous neoformations depends on the order that can be guaranteed or on the disorder that must be allowed.

The role of metallic impurities is to increase the disorder of silica, as long as it is being organized upon itself. But when these cations become more abundant, they constrain the silica into another type of order; this is the genesis of clay minerals.

CHAPTER X

GENESIS OF CLAY MINERALS: INHERITANCE AND TRANSFORMATION

Throughout the exposé of facts three principal processes have been invoked to account for the genesis of clay minerals:

— Detrital inheritance;

— Transformations;

— Neoformations.

It is time to examine them *per se* and to study their mechanism.

DETRITAL INHERITANCE OR TOTAL INHERITANCE. DEFINITIONS OF LANGUAGE. — There are two ways of using the term inheritance, a way restricted to the inheritance of detrital particles and a much more general way that can be summarized as follows: "Every natural environment inherits from the past its solid, colloidal, and dissolved constituents." It is the limited definition that will be adopted here, that of the mechanical or detrital inheritance.

As a matter of fact, a weathering horizon works on all the constituents of the parent rock that it inherits. It can fractionate, modify, or dissolve these constituents and ensure neoformations; it only works on its inheritance in order to transmit it to the soil.

A soil works on the material it inherits from the underlying weathering horizon, or directly from parent rocks if the soil is shallow. It accelerates fractionation, transformation, dissolution and undertakes its own neoformations.

A lacustrine, marine, or hypersaline sedimentary environment inherits from the continent particles and solutions upon which it works in its turn. It can leave them inert, transform them, or ensure new syntheses.

The progressively buried sedimentary column inherits from the sedimentation process material that is fresh sediment composed of minerals and solutions. The whole is going to be confined under more recent, overlying sediments, and, little by little, diagenesis will transform it into sedimentary rock through the squeezing out of water, the circulation of solutions, and the increase of temperature and pressure.

At each stage of these transitions an inheritance is transmitted. But here, by convention, the term inheritance concerns only the detrital, particulate supply that remains inert.

We arrive at the following distinction:

1° *Clay minerals will be maintained intact because of their stability:* this is mechanical or detrital inheritance.

$2°$ *Clay minerals will be modified by the environment because of their instability:* this is transformation.

I. — MECHANICAL INHERITANCE

Several synonymous expressions designate this phenomenon: mechanical inheritance, detrital supply, detrital origin, allochthonous origin.

Thus, the clays inherited are called by different names: inherited clays, allochthonous clays, detrital clays, clays of mechanical origin.

$1°$ *Clays inherited by weathering horizons*

Weathering horizons receive clay particles by inheritance from the parent rock:

— When the parent rock is sedimentary and in particular argillaceous, all argillites, shales and marnes furnish the weathering process with the clay particles they contain, consequently, with the most varied lots in which all the clay minerals may be present;

— When the parent rocks are crystalline or metamorphic rocks, the biggest part of the supply comes especially from schists, micaschists, gneiss, and accessorily from granites, in the form of micas and chlorites.

$2°$ *Clays inherited by soils*

If the soils are formed directly from the parent rocks because of their shallow development, we return to the preceding problem. If soils characterized by biological activity are developed from a thick weathering horizon, detrital inheritance takes on multiple aspects. In fact, soils inherit:

— Clay minerals inherited by the weathering horizon and neither transformed nor destroyed during that passage;

— Minerals transformed by weathering: biotites transformed into chlorite, chlorite into vermiculite or montmorillonite, muscovites into illite, vermiculite, montmorillonite;

— Minerals neoformed during weathering: weathering of feldspars and aluminous silicates into sericite and, in some cases, montmorillonite; weathering of ferromagnesians into chlorite, perhaps into vermiculite, celadonite, or montmorillonite; weathering of volcanic glass into montmorillonite, or possibly into illite. And in an entirely different context, weathering of silicates into disordered kaolinites, into ordered kaolinites or into gibbsite.

$3°$ *Clays inherited by sediments*

The bulk of sedimentary clays is inherited from the continents. Again the origins multiply:

— By direct supply from the continental bedrock undergoing early and active erosion whenever tectonic or climatic conditions are favorable.

— By supply from weathering mantles, which can accumulate several tens of meters of unconsolidated material (*arènes*, lateritic clays) under favorable climates.

— By supply of soil clays from the drainage basins of rivers and streams to the basin of sedimentation. These clays include not only all the products that the soils inherited and that survived that transition, but also:

— The clays transformed within the soil: transformation of sericite, illite and chlorite into vermiculite, then into montmorillonite, along with the whole suite of mixed layers;

— The clays neoformed within the soil: kaolinite in the strongly *lessivé* soils and gibbsite at the extreme; montmorillonite, on the other hand, in hydromorphic or calcimorphic soils.

4° *Inherited clays subjected to diagenesis*

The deposited sediments are buried and undergo diagenesis. The clay minerals of the sedimentary mass that is going to be transformed into rock contains the totality of the former heritage plus the results of the work of the sedimentation process itself, that is:

— Clays transformed by the sedimentary environment, transformations that we have seen to be either inexistent, discreet or massive as a function of the increasing chemical character of the sedimentation;

— Clays neoformed during chemical sedimentation, which produced sediments with attapulgite, sepiolite, glauconite and, in certain cases, montmorillonite.

And, if a second diagenesis occurs in the sedimentary series in response to a tectonic change, it is going to attack the inheritance mentioned above, augmented by the results of the preceding diagenesis that had already modified, by transformation or neoformation, the products received.

5° *The conditions of inheritance: stability and instability*

If clay minerals inherited by one environment from another are inherited without modification, it is because they are maintained in conditions of stability in their new environment. If these conditions are not guaranteed, transformations take place.

These conditions of stability are different for each structure; their detailed analysis can be made only after the study of transformation, neoformation and synthesis. But indications from nature show very well the limit of such stability.

— Kaolinite is a mineral that forms chiefly in weathering and in soils. It is described as being very stable in the hydrosphere; this is exact, but not without qualification. Within a lateritic profile in which drainage increases and where the water level falls because the discharge is greater than the supply, solutions are going to be strongly desaturated in silica. Kaolinite is then going to be desilicified and transformed into gibbsite. Inversely, during diagenesis, if alkaline salt brines at high temperature come in, kaolinite will be illitized as we have seen in the Saharan sandstones.

Except for these extreme cases, kaolinite is stable in the hydrosphere, in soils and in sediments, and it is the most typical and the most resistant inherited clay mineral.

— Illite and chlorite are common detrital minerals inherited by very numerous soils and sediments. They will be stable during weathering of an overridingly physical character, in poorly drained soils of modest chemical activity, in sediments deposited in alkaline water, and in alkaline diagenesis.

They will be vulnerable in acid environments of lessivage by freshwater, free of cations. This is the case of weathering of a prevailingly chemical character (podzolization, lateritization) and of diagenesis of acid character. They will evolve by transformations of degrading or negative type, chlorite being the most vulnerable.

— The degraded, interstratified, three-layer minerals and vermiculite are minerals that appear in the cycle during weathering. These are stages of the degradation that will be pursued if these minerals are inherited by a degrading, agressive environment of lessivage. In contrast, such unfavorable evolution will be interrupted if these fragile minerals are brought again into an alkaline environment, favorable to their organization; they will have the tendency to reconstitute the original micas and chlorites.

— Attapulgites and sepiolites are minerals neoformed in alkaline sediments rich in silica and magnesia. They will be stable in this environment but very vulnerable in environments of lessivage, and especially in soils, where they disappear rapidly. It will be noted that in the black tropical clays (*argiles noires tropicales*), which are favorable to the neoformation of montmorillonite, attapulgites are degraded and transformed into montmorillonite; this shows their extreme instability outside their genetic environment.

— Montmorillonite is a mineral of many possible origins. It has a double instability. In weathering, montmorillonite is rapidly destroyed as soon as lessivage produces sufficient desaturation of cations and silica. In transformation and diagenesis, montmorillonite clearly has a tendency to fix potassium or magnesium ions and to give rise to micaceous or chloritic lattices.

— Glauconite is also doubly unstable. Formed in marine environments, it does not resist pedological activity of mild weathering and produces iron oxides and various mixed layers. Glauconite does not resist diagenesis any better; it can evolve into sericite, siderite or chlorite.

6° *Conclusion*

Mechanical inheritance is one of the major phenomena of the clay cycle in the hydrosphere. The part played by detrital inheritance in the genesis of soils and sedi-

ments is considerable. It decreases only when the environments are capable of degrading the inherited clay particles or using them in constructions; this is the case of evolved soils and sedimentary environments with a clearly marked chemical character.

This means that many pedological and many sedimentary formations offer conditions of stability and, therefore, of survival to the clay particles. This inoffensive pedogenesis and this detrital sedimentation form the framework of large-scale inheritance. As soon as the environments act on the inherited material, it is modified; this is transformation.

II. — TRANSFORMATION

Transformations of clay minerals can take place during weathering (*Verwitterung, altération*), during sedimentation itself, or during diagenesis and metamorphism. The term *diagenesis* applies to modifications of sediments that occur between sedimentation (syngenetic) and metamorphism (epigenetic). This is the usual meaning in France and Germany (Correns, 1950). In America, authors have taken the habit of calling diagenesis the phenomena contemporary with sedimentation as well as those posterior to sedimentation; this leads to confusion.

On the other hand, in August 1962, the Eleventh National Conference on Clays and Clay Minerals was held at Ottawa (Canada) and the main subject on the agenda was a "Symposium on Clay Mineral Transformations." Here a common language was used.

1° Definitions

Transformations of clay minerals. — Here the expression transformation of clay minerals applies to those changes that modify a clay mineral without altering its two- or three-layered structural type. The most frequent and the best known cases are the transformations within the three-layer clay minerals. The principal examples are the following:

Biotite ⇌ chlorite
Biotite ⇌ sericite
Biotite ⇌ (Tri)vermiculite
Chlorite ⇌ (Tri)vermiculite
Vermiculite ⇌ (Tri)montmorillonite
Chlorite ⇌ (Di)vermiculite
Chlorite ⇌ C-V mixed layer

Illite ⇌ I-V mixed layer
Illite ⇌ (Di)vermiculite
(Di)Vermiculite ⇌ V-M mixed layer
Vermiculite ⇌ montmorillonite
Illite ⇌ montmorillonite
Illite ⇌ chlorite

Transformations of the two-layered kaolinitic type are much less well known.

Kaolinite ⇌ disordered kaolinites (fireclays)
Kaolinite ⇌ halloysite
Halloysite ⇌ disordered kaolinites (?)

A limiting problem can be discussed, that of the transition from a two-layer mineral to a three-layer mineral, for example, illite ⇌ kaolinite, and chamosite ⇌ chlorite. If it could be demonstrated that during the process certain tetrahedral or octahedral layers were used for the genesis of the new structure, it would be possible to speak of a transformation; this is certainly the case in the transformation of chamosite into chlorite by diagenesis. If, on the other hand, it is supposed that the process passes through solutions and that the new minerals are built from the ashes of the former ones, one must speak of neoformation. This is perhaps the case in the kaolinization of illites. At any rate, there must be intermediate stages because, in the process of degradation of lattices on their way towards dissolution, there is no formal limit between transformation and neoformation. However that may be, it is not the problem of limiting cases that interests us at present, but a clear view of the ensembles.

Degradation. Aggradation. — The two terms degradation and aggradation are not part of the French geological vocabulary, although they are standard in English. The second does not even exist in French. But the symmetry of these terms, which cannot be replaced by any others, is so eloquent that it was used by GRIFFITHS (1952). We will adopt them for the clearness of our exposé. Degradation will be the negative transformation in the sense of weathering and of loss of substance. Aggradation will be the positive transformation in the sense of construction and of gain of substance (LUCAS, 1962).

In natural transformations one passes from stable minerals in sedimentary, metamorphic and crystalline rocks, i.e., from micas and chlorites to open minerals with variable basal spacings that are more frequent in weathered materials or in environments rich in water. This will be degradation.

One can pass, in the opposite sense, from gaping, open and unstable minerals with variable basal spacings to clay minerals more and more similar to micas and chlorites. There is no term for it in French; we shall use the transposition of the Anglo-Saxon word "aggradation"; this will be agradation in French.

The following are quite good examples; reading left to right we follow a degradation, from right to left an aggradation.

Degradation ⟶

Illite ⇌ I-Vmixed layer ⇌ vermiculite ⇌ V-M mixed layer ⇌ montmorillonite
Biotite ⇌ hydrobiotite ⇌ (Tri)vermiculite ⇌ (Tri)montmorillonite
Biotite ⇌ chlorite ⇌ C-V mixed layer ⇌ (Di)vermiculite ⇌ (Di)montmorillonite

⟵ Aggradation

Danger of the abuse of language. — It is well known that if the interlayer ions of a Ca-montmorillonite are exchanged under the influence of a KOH solution, the periodicity of the lattice passes from 14-15 Å to 10 Å. In contrast, a K-vermiculite treated with Mg salts passes from 10 to 14 Å. These are not transformations; they are base exchanges in minerals with variable basal spacing.

FOSTER (1954) correctly insisted on the "hypersimplifications" to which one could be led in explaining the transformations of montmorillonite into illite by replacement of the interlayer ions by potassium and magnesium and reciprocally. In the illitic layer the deficit of charge due to replacements of trivalent ions by divalent ions in octahedral position is about 0.22 to 0.32 (number of ions per Si_4O_{10} structure). The deficit of charge due to the replacement of silicon by aluminum in tetrahedral position is about 0.46 to 0.62. The total varies between 0.66 and 0.94; it is this total that is responsible for the charge of the layer being saturated by interlayer cations.

In montmorillonite *sensu stricto*, in which tetrahedral replacements are weak or nil, the total is between 0.30 and 0.50. The highest charge of the montmorillonitic layer is lower than the lowest charge of the illitic layer.

In beidellite, which is a montmorillonitic layer that admits tetrahedral replacements, the total charge of sheet is between 0.50 and 0.67. We barely reach the lowest known charge of illite layers.

Such are the reasons for FOSTER's recommendations. One cannot simply pass from montmorillonite or beidellite to illite by replacement of interlayer ions; the minerals have different lattices.

WEAVER (1958) likewise showed how nomenclature could "magnify" natural phenomena. He takes the example of the degradation of illite into illite-montmorillonite mixed layer, which may by aggradation give again the lattice of illite. When we use the term montmorillonite to designate swelling layers alternating with illite in a degraded stage, one could suppose that "true montmorillonite" is really present. But that is neither proved nor probable. It is a question of layers of mica preserving their high charge but leached of their cations so that the reconstruction is easy. This is not transformation but adsorption on an uncharged layer. The layers of the clay minerals must be considered as detrital; only their interlayer ions are exchanged.

Therefore, our arguments run the risk of being false if we do not know well enough the exact meanings underlying the words we use to designate minerals or mechanisms. Now, what we call "transformation" does not apply to such restricted exchanges but to true evolution of the minerals themselves.

Conventions of language and reality of the transformations. — The language we use here is very precise. It is prescribed by the tools we have at our disposal, which are the only means of access to the realities offered to us.

That which we call layers of montmorillonitic type are layers with a periodicity of 14-15 Å that increases up to 17.2 Å by the action of glycol and decreases to 10 Å upon heating. This information is provided by X-ray diffraction, and we cannot easily obtain information about the chemical composition of these layers. This is regrettable but for the time being inevitable because the natural products are complex mixtures of minerals that in many cases are themselves interstratified and subject to various isomorphic replacements; no chemical analysis can be interpreted without simplifications that take all significance away from these calculations.

That which we call layers of vermiculitic type are layers with a periodicity of 14-15 Å that does not increase with glycol treatment and decreases to 10 Å upon heating. We do not have any more chemical information and we know, still by means of X-ray diffraction, that the majority of these vermiculites are dioctahedral and that they come from tri- or dioctahedral micas or chlorites. In 1948 already, MACEWAN, who was the first to describe such structures in soils, placed vermiculites between quotation marks and added: "*I do not want to give the impression that these structures are exactly the same substances (vermiculites-hydrobiotites) as those so designated by* GRÜNER (1934)."

We could paraphrase it by saying "a structure similar to montmorillonite in its behavior in X-ray diffraction". However, we say "layers of montmorillonitic type" and most often simply "montmorillonite"; this is a convention, it is in general use and understood by everyone. But we do not lose sight of the fact that, as is often the case in physical measurements, we receive information only from the effects. Of course, it is not at all excluded that the so-called montmorillonite is really montmorillonite. It is the same thing in the case of vermiculite, as well as for chlorite whose 14 Å reflection remains unchanged both in presence of glycol and upon heating, although we cannot always make out in the mixture whether this clay mineral is dioctahedral or trioctahedral.

These conventions of language having been established, it remains to be seen whether we are describing only base exchange or true transformations. If we are looking only at exchange of cations between immutable, resistant, detrital layers, our transformations are illusory. This phenomenon certainly occurs in nature, but it is not the only one. As a matter of fact, the examples given in the descriptive chapters as well as in the following paragraphs show that true transformations exist. Let us only mention the most outstanding phenomena: weathering of trioctahedral minerals into dioctahedral minerals, occurrence of colloidal products in soils and in the fine-grained material carried by rivers, destruction by weathering of the brucitic layer of chlorites, increase of heat-stability of chlorite during sedimentation and diagenesis, and formation of trioctahedral chlorites from dioctahedral illites. These are the outstanding facts that indicate that the layers have been affected.

From the thermodynamic point of view, the structure of a mineral cannot change at constant temperature and pressure without variation in its chemical composition. Now, the preceding examples show that the chemical compositions of the clay minerals can be mobilized; first, of course, the interlayer ions, but also the octahedral and tetrahedral layers because observations in nature show fragments of minerals such as allophanes and dissolution of octahedral cations and silica. Emptying or filling in the interlayer space changes the electrostatic equilibrium of a mineral and has repercussions on its deeper layers; the mineral tends constantly to reorganize itself in order to ensure the neutrality of its equilibrium.

This will be shown by the studies on transformations.

2° *Degradation*

The facts. — Examples of clay mineral degradations are innumerable; they have been set forth principally in Chapter V on weathering and soils, and are especially frequent in an environment of lessivage. The most spectacular ones have been pointed out by the Scottish school of MACEWAN (1948, 1948), WALKER (1949, 1950), BROWN (1953), MITCHELL (1955), and by the American School of JACKSON (1948, 1952), GIESEKING (1949), WHITTIG and JACKSON (1955, 1956), and in the numerous works summarized by JACKSON (1959). Similar degradations have been found in France by CAMEZ,

Franc de Ferrière, Lucas and Millot (1960), Camez (1962), as well as in other countries. The ensemble of facts has been described in Chapter V.

The principal transformations are:

— Transformation of biotite into hydrobiotite (biotite-vermiculite mixed layer), into trioctahedral vermiculite and into trioctahedral montmorillonite, with all the intermediate mixed layers.

— Transformation of biotite into chlorite, into dioctahedral vermiculite and into dioctahedral montmorillonite, with all the intermediate mixed layers.

— Transformation of muscovite-sericite into illite, into dioctahedral vermiculite and into dioctahedral montmorillonite, with all the intermediate mixed layers.

— In numerous soil profiles these transformations have advanced far enough for the breakdown of the mineral lattice and the formation of amorphous products called allophanes.

Degradations are characteristic of continental weathering, but they take place also during sedimentation and diagenesis. In sedimentation they are rare because sedimentary environments are generally too saturated to undertake such negative transformations. They would take place in the ensemble of sub-aquatic alterations called "halmyrolyses" by Hummel (1922), and especially during the alteration of volcanic products placed in the unstable conditions of a marine environment. The transformation of biotite into glauconite is a good example of this.

During diagenesis it can happen that percolating solutions are conducive to lessivage and able to degrade clay minerals. Diverse cases have been observed; here are a few examples:

Smoot (1960) showed that mixed layers become more abundant during the diagenesis of American Pennsylvanian sandstones. This is an indication of the degrading effect of solutions upon clay minerals in the sediments.

Transformations of biotite into chlorite or into chlorite and sericite, by diagenesis as well as by weathering, are extremely well known. Biotite becomes whitened, releases its iron and titanium, and yields flakes of alumino-magnesian chlorite and possibly sericite. Beautiful examples are shown by Carozzi (1960). Dunoyer de Segonzac and Millot (1962) show such evolution by diagenesis, parallel to the neoformation of pyrophyllite. The sericitization of glauconite noted by Arbey and Le Fournier (1962) in Saharan diagenetic sandstones can be added to the previous examples.

The mechanisms. — Walker (1949, 1950) described the mechanism of transformation of biotite into vermiculite. This was the application of the experimental works of Barshad (1948, 1949, 1950) and of Walker (1949) on ion exchange in vermiculite, but with modification of the octahedral layers themselves.

The 10 Å reflection of biotite becomes broader. Then diffuse intermediate bands appear between 10 and 14 Å, corresponding to a variable number of 14 Å layers interstratified with 10 Å layers. Then one comes to either a diffuse distribution between the two reflections at 10 and 14 Å or a single reflection at 12 Å that corresponds to hydrobiotite. The latter structure contains nearly the same proportion of biotite layers as vermiculite layers. Finally, the 10 Å periodicity and the mixed layers grow indistinct, and the result is an intensive 14 Å reflection not affected by glycerol treatment.

The following interpretation can be made: there is replacement of the most accessible K ions by water molecules. If this replacement continues toward interior of the mica, the bonds between the layers are loosened and the spacing between them increases. Within the layers themselves modifications are produced: (1) oxidation of ferrous iron to ferric iron, which disturbs or ruptures the lattice, (2) a replacement of some oxygens by hydroxyls and (3) migration of magnesium between the layers to balance the vacant charges. The lessivage continues, and the process is carried on from layer to layer so that a mixed layer evolves as a result of the statistical increase of the number of layers similar to vermiculite.

JACKSON et al. (1948, 1952, 1954, 1959) have multiplied the studies on weathering of clay minerals in soils and have shown evidence of weathering sequences, i.e., the succession of the clay mineral structures throughout pedogenesis. The common series is the following:

$$\text{Mica} \longrightarrow \text{intermediate stages} \longrightarrow \text{vermiculite} \longrightarrow \text{montmorillonite}$$

One sees a decrease in the intensity of the basal reflection of the mica, as well as its broadening and the appearance of intermediate bands between 10 and 14 Å for the untreated sample, and between 10 and 18 Å for the glycerol-saturated sample. The potassium content decreases, the water content and the internal surface area accessible to polyalcohols increase. To reach the beidellite type of montmorillonite one passes through the stage of dioctahedral vermiculite. This evolution can occur only by partial dissolution of the layers, as is shown by the presence of silico-aluminous colloidal residues.

MURRAY and LEININGER (1956) have described similar evolutions in soils developed on tills of Wisconsin and Illinoian age in Indiana (U.S.A.). They emphasized very much the oxidation of iron that was certainly present in the layers of illite and chlorite. This oxidation leads to disorder in the structure, to a rupturing of the crystal, and to progress in weathering.

CAMEZ, FRANC DE FERRIÈRE, LUCAS and MILLOT (1960) and CAMEZ (1962) have studied numerous podzols formed under Atlantic climate. Some soils initially contained only illite, others much chlorite, and others a mixture of the two. This has permitted the description of the two parallel pathways leading to vermiculite through the intermediate stages of illite-vermiculite mixed layer on one hand, and of chlorite-vermiculite mixed layer on the other. It is reasonable to assume that after lessivage of the interlayer ions magnesium will be the first to leave the octahedral sheets in order to balance the charges between layers. Thus, in chlorite, magnesium disappears from the octahedral sheets which become principally aluminous and therefore dioctahedral.

DROSTE et al. (1962) have shown likewise that the transformation of chlorite into vermiculite and montmorillonite was more rapid than that of illite into vermiculite and montmorillonite. The fragility of chlorite depends very much on the nature and rate of the dioctahedral and octahedral substitutions. In fact, this controls in the layer with three sheets the charge deficits that must be balanced by the admission of Al into the intermediate brucite layer. The less abundant aluminum is in the brucite layer, the easier weathering is. The liberation of ions from these interlayer sheets would be in the order of their solubility, i.e., Fe^{2+}, Fe^{3+}, Al^{3+}. Illite begins the evolution described above, but with delay. According to GARRELS and HOWARD (1959), not only a removal of K ions takes place, but also a loss of silicon and aluminum. It is interesting to note that the early weathering of chlorite protects illite against hydrolysis by the increase in saturation of the solutions. This is the line of reasoning that corresponds to the perspicaceous observations of J. DE LAPPARENT (1909). During the weathering of a granite with two feldspars and two micas, the transformation of biotite is delayed; in fact, potassium is abundant in the intergranular solutions; it nourishes the neoformation of sericite but does not bring about degradation of the biotite. On the contrary, in a granite without potassium feldspar, the environment is unsaturated in potassium, biotites are chloritized and the leached K ensures sericitization of the plagioclases.

Thus, by successive approaches, the mechanisms of transformation by degradation can be understood. Obviously they depend on the composition of the solutions and the composition of the minerals.

The more unsaturated the solutions are with respect to a given element, the more quickly this element will be leached from the crystals. In addition, the composition of minerals corresponds to diverse structures and stabilities. GRIM and BRADLEY (1955) have studied the stability of clay minerals from the crystallochemical point of view. They showed that the bond in kaolinite between the tetrahedral sheet and the exclusively aluminous octahedral sheet results in a size and a stability greater than in the three-layer minerals, which, moreover, tolerate quite large substitutions. Among the micas, BASSETT (1960) showed a difference in the location of the K^+ ions in the hexagonal cavities of the mica. In the trioctahedral micas, such as phlogopite, the hydroxyl ions of the layers are perpendicular to the cleavage such that the H^+ protons of the hydroxyl ions are the closest to the K^+ ions, thus having a repulsing action upon them. In contrast, in muscovite the hydroxyl ions of the layers are inclined to the cleavage, and the H^+ protons are farther from the K^+ ions. The result is a better fixation of potassium in its negative environment and an increased resistance of muscovite to weathering.

These studies show the reasons for the decreasing stability of the principal minerals with which we are concerned, kaolinite, muscovite, chlorite, and biotite, and they also show the facility with which degradations are going to take place under identical conditions and in the same order.

Experimental studies. — **Degradation of kaolinite.** — GASTUCHE, DELVIGNE and FRIPIAT (1954) experimentally reconstructed the weathering of kaolinite such as it can be observed by the electron microscope. Polar substances that accept protons, such as nitrobenzene or nitromethane, produce parting and slippage of the layers in large crystals of kaolinite with dilatation of the interlayer space. If the crystals are small, there can be a rolling-up, similar to that of halloysite. Oxalic and phosphotungstic acids attack the internal parts of the lattice and not the edges, giving rise to hexagonal holes.

OBERLIN (1957) and OBERLIN, TCHOUBAR et al. (1957, 1961) have weathered kaolinite by the action of sulfuric acid at pH 2. Electron microdiffraction showed disorders leading to disordered kaolinites (fireclays). The superposition of layers one upon the other gives rise to the *moiré* effect in electron microscopy.

Degradation of three-layer minerals. — CAILLÈRE, HÉNIN and GUENNELON (1949) treated samples of phlogopite with boiling 40 % $MgCl_2$ during 135 hours. Opening of the layers was observed with 10, 12 and 14 Å diffraction peaks. One can deduce from this the formation of structures of the vermiculitic type with mixed layers.

JONAS et al. (1960, 1960, 1961) carried out very interesting studies on the swelling, the size and the degradation of three-layer minerals.

a) MECHANISM OF THE OPENING OF LAYERS. — The opening of the layers of clay minerals is the result of the antagonism of two forces. On the one hand, the crystals are subjected to the adsorptive forces of the molecules of liquid; these forces act as a wedge trying to disjoint the layers. The strength of these adsorptive forces depends on the liquids; it is, for instance, greater for glycol than for water. Let us add that the sum of the adsorptive forces exerted upon the edge of a crystal in order to disjoint the layers is proportional to the perimeter of the particle.

On the other hand, there is the action of the bonding forces between the interlayer ions and the layers themselves. The surface of the layers has a negative charge, and all the more negative as isomorphous replacements of Si by Al are more numerous. The charge of the interlayer ions is positive. Statistically, each negative charge of the layer is neutralized by a positive charge of the interlayer ions. The sum of the bonding forces between the layers and the interlayer cations is proportional to the number of interlayer cations, or to the product of the charge density times the surface area of layer (charge density = charge per unit area of surface). Therefore, this bonding force depends at the same time on the structure and on the size of the mineral.

Expansion occurs when the adsorptive forces of the liquid exceed the bonding forces between layers and interlayer cations.

FIG. 64. — *Relationship between resistance to expansion* (F) *in three-layer minerals and charge density* (d) *for varying particle surfaces* (s). (LUCAS, 1962, after JONAS and THOMAS, 1960).

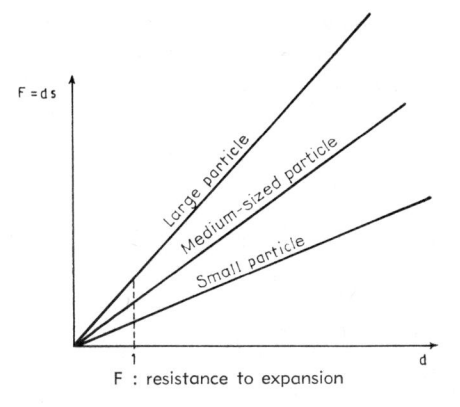

b) RESISTANCE TO EXPANSION. — It is easy to appreciate the influence of structure when it is known that montmorillonite, which contains little tetrahedral aluminum, is easily expanded by glycerol or glycol, whereas mica does not tolerate such expansion.

Figure 64 shows the influence of particle size: on the ordinate the resistance to expansion is represented by the sum of the interlayer K^+ cations, which is equal to the charge of the layer; on the abscissa the charge density of layer is represented.

The wedging forces exerted by molecules of the liquid are proportional to the perimeter of the particles, and the resistance to expansion is proportional to the surface area of the particles; particles of small diameter are going to be much more easily expanded than larger ones. Moreover, the diameter will have little influence on particles with weakly charged layers, but its influence will be great on highly charged particles.

Experimentation confirms all these facts.

JONAS and THOMAS (1960) experimented on particles of the same size but with different layer charges. They showed that expansion increases when the number of interlayer cations, representing the layer charge, decreases. JONAS and ROBERTSON (1960) studied particles of the same composition but of different sizes. They showed that the expansion increases as the size decreases.

c) SIZE AND COMPOSITION IN THE DEGRADATION PROCESS. — JONAS and BROWN (1961) studied the chemical composition of particles as a function of their size. It is well known that on the surface of crystals there is a zone the composition of which is disturbed because of disequilibrium. In micaceous minerals the interlayer zone is particularly disturbed because of the solubility of K^+ and its fragile bonding. The disturbed zone has a constant thickness whatever the size of the crystal may be. When the crystal is large, this zone represents a small percentage of its volume; when the crystal is small, about $1\ \mu$, this zone represents an important fraction of the crystal volume. The smaller a micaceous crystal is, the closer its chemical formula is to montmorillonite because of the modifications of the composition of its edges.

d) CONCLUSIONS. — These studies are very instructive for us because they show that several properties of the three-layer minerals evolve in the same direction: capacity of expansion, size, and chemical composition. Small particles, which are the most fitted for expansion, also are the most subject to lessivage by natural solutions, a process that removes the interlayer ions and leads to a chemical composition tending toward that of montmorillonite.

General view on degradation. — With the aid of observations, interpretative reasoning, and the experiments of JONAS, we arrive at a comprehension of the phenomenon of degradation. The most important point is not to consider clay minerals as immutable when the interlayer ions are removed. Everything shows us the contrary.

When solutions are sufficiently unsaturated in cations to dissolve the soluble elements of clay minerals, they are attacked by three different ways: by the defects, the inclusions and the holes in the crystals, by the oxidation of ferrous iron if present, by the interlayer joints. This brings about a division, then a micro-division of the minerals.

As the size decreases, action on the joints increases, and the release of interlayer cations is accelerated. In the case of biotite, iron, titanium and magnesium reach the interlayer zone; this is "chloritization" of biotite with "exudation" of iron oxides and rutile or anatase into the joints. If the process is maintained, the deserted octahedral sheet becomes organized into a dioctahedral sheet that is dominantly aluminous because of the double intervention of the original octahedral aluminum and the tetrahedral aluminum that left its position. At the same time, the charge decreases and the possibility of expansion appears; this is the weathering of chlorite into dioctahedral vermiculite. If the phenomenon is accentuated, size decreases even more, and the charge as well, and the capacity for expansion increases; montmorillonite is obtained.

In the case of weathering of muscovite or sericite, the mechanism is similar but delayed because of the greater stability of the K^+ ion in its position and because of the low solubility of alumina. Illite already shows an impoverishment in potassium and an opening of its layers; dioctahedral vermiculite represents the stage with the possibility of moderate expansion; montmorillonite, the final stage, reunites the characteristics of small size, high capacity for expansion, and a very "impoverished" chemical composition.

Let us note here that this montmorillonite that occurs at the end of degradation is probably not very different from true montmorillonites; both show the same behavior in X-ray diffraction. Now we understand the reason. It may also be noted that all these degradations occur progressively from layer to layer. That allows us to comprehend the different stages of the process by means of the mixed layers which show a gradual transition from one type to another.

Finally, when the term "end of degradations" is used, it evokes the termination of the crystalline state because these very small structures remain vulnerable if lessivage is maintained, dissolving silica. The crystalline structure then collapses, and allophanes appear, which are simply silico-aluminous gels. Allophanes, moreover, are commonly ephemeral because either they finally dissolve or they nourish kaolinitic neoformations in acid environments.

Degradations of clay minerals are but a moderate form of hydrolysis, and this progressive character is not only the result of the slowness of the attack but also the fruit of the many possibilities of rearrangement that typify the three-layer clay minerals. In this way, clay minerals are quite different from feldspars, which can choose only between stability and dissolution. On the contrary, clay minerals, impoverished first in monovalent, then in divalent cations, maintain their existence in a progressively changing form up to the point where these minute clay minerals are of the order of $0.10\ \mu$ and similar to montmorillonite. Let us note, along with GRIM and BRADLEY (1955), that these clay minerals cannot be other than minuscule because the only minerals that grow in nature are mica, talc and pyrophyllite, but not the intermediate stages, which are fine grained and are therefore clay minerals. Weakly charged montmorillonite is "the most clayey of all clays."

3° Aggradation

Aggradations are transformations opposite to degradations. They develop after weathering and before metamorphism. Their amplitude can be determined from comparison of their starting point with the final result of the process.

At the starting point, we find all the material resulting from degradation and carried by rivers and soils to the sedimentary basin. The example of the Rappahanock River in Virginia shows the nature of the material brought into the sea (NELSON, 1960). The clay minerals identified were kaolinite, illite, disordered illite, dioctahedral vermiculite, 12 Å montmorillonite, 14.2 Å montmorillonite, chlorite, and amorphous, organic and detrital materials.

At the arrival point, we are in the zones of advanced diagenesis, in sedimentary rocks that have completed their lithification. Practically the only clay minerals that can be extracted are sericite and chlorite. All the intermediate modifications are transformations by aggradation.

The facts. — The known facts that demonstrate the existence of positive transformation, or aggradation, are numerous already and are accumulating every day.

Aggradation is very rare in weathering horizons and soil formation. This is understandable because weathering is characterized by lessivage and by environments of degradation. There are, however, horizons of accumulation and hydromorphic soils in which all kinds of aggradations are possible. These aggradations are superimposed on neoformations, and their analysis is not simple.

The first domain in which aggradations are amplified is that of sedimentation. It was in present-day sediments that the phenomenon was defined by DIETZ (1941) and GRIM, DIETZ and BRADLEY (1949). Since that time many studies on the present-day argillaceous sediments in estuaries and on coasts have multiplied the information available on aggradation. These examples were described in Chapter VI on sedimentation in estuaries and in Chapter VII on present-day marine sedimentation. These were the studies of GRIM and JOHNS (1954), GRIFFIN and INGRAM (1955), POWERS (1957, 1959), MILNE and EARLY (1958), JOHNS and GRIM (1958), GRIFFIN and JOHNS (1958), NELSON (1960), PINSAK and MURRAY (1960).

In studying ancient sediments, I was able to show, in 1949, the occurrence of vermiculite and chlorite in sedimentary clays. I came to the notion of natural transitions between micas - hydrobiotites - vermiculites - montmorillonites and chlorites in sediments, and I emphasized that "there is no reason why such transitions should not be possible in nature or that one should not find here and there species of ambiguous character. These species must be considered as products evolving from one type to another... It is possible that the names of species are too precise to encompass a reality that, without a doubt, is in full evolution; these rigid terms, however, are the only ones at our disposal and I insist that a broad meaning be given to them." (MILLOT, 1949). Calcareous and hypersaline sediments show a widespread development of illite and chlorite that I attributed to the influence of the environment. "In summary, the clay fraction in sediments is not only detrital, clastic, mechanical or allogenic because it varies with the environment in which it formed. It is at least partly authigenic. Today we are led to make the following sort of affirmation; in general, argillaceous sedimentation is authigenic by means of reworking of a material that was in large part allogenic." (MILLOT, 1949). It is certain that, when I discovered these variations, I gave them an exaggerated importance *vis-à-vis* detrital inheritance. On the other hand, the notion of transformation was still obscure at that time, so that we had only the alternative of detrital origin and authigenic origin. Transformations represent that intermediate stage that modifies the inherited material according to the environment.

This was the constant opinion of GRIM, supported by his works on present-day sediments and on ancient sediments: GRIM (1953, 1958), GRIM and JOHNS (1954, 1958), GRIM, BRADLEY and WHITE (1957), GRIM, DROSTE and BRADLEY (1960). This was also the result of considerations of the structure and stability of clay minerals (GRIM and

BRADLEY, 1955). Works that give clear evidence of transformations in ancient sedimentary basins require many bore holes in order to compare the paleogeography and the sedimentologic data for a given epoch. Detailed studies are quite recent: MERRELL, JONES and SAND (1957), PETERSON (1961, 1962), LUCAS and BRONNER (1961), LUCAS (1962), TOCKER (1962). All these studies have been presented in Chapter VII on carbonate or hypersaline sedimentation. One sees not only the aggradation of illite, but also that of corrensite and chlorite in sedimentary environments of chemical character. I shall recall only the example described by LUCAS (1962) in the Triassic basins; here the borders of the basin are supplied with clay in which degraded illite dominates; in the center of the basin corrensite or trioctahedral chlorites are found, which are minerals transformed by aggradation (figs. 33, 34, 36 and 37).

Transformation by aggradation is also very important in diagenesis. Moreover, it is often very difficult to distinguish from that which takes place in sedimentation because the two are superimposed. Only paleogeographic studies allow such distinction. The principal phenomenon is the transformation of all the clay minerals into sericite and chlorite: illite, montmorillonite, mixed layers, glauconite, chamosite. The examples have been given in the successive descriptions in Chapters VI and VII.

The mechanisms. — The first interpretation of the mecanism of aggradation is restrictive. It has been clearly defined by WEAVER (1959) and must be taken into account. It certainly acts, in part, on the least degraded minerals. The layers of the three-layer minerals, opened by weathering, are considered as intact. When they reach the sea they admit cations into the interlayer position; this contraction is not a transformation, but a "rejuvenation". This mechanism probably plays a big role on slightly weathered minerals with a high layer charge.

But transformations of the layers themselves also occur to the degree that we have

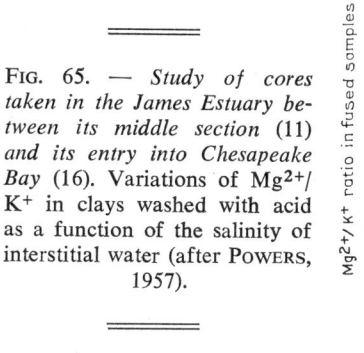

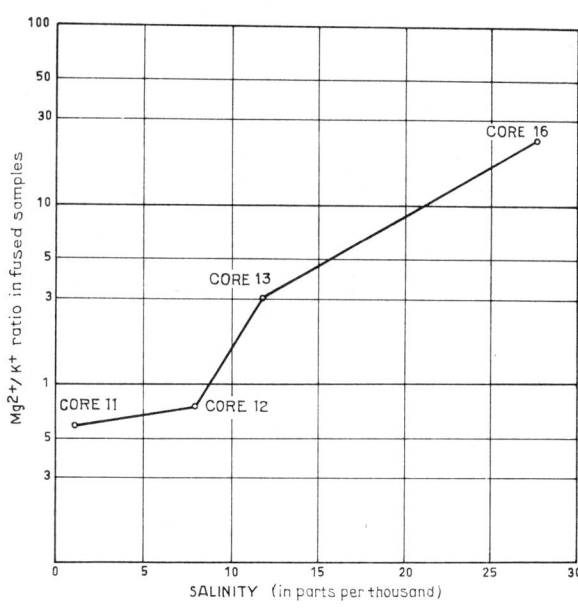

FIG. 65. — *Study of cores taken in the James Estuary between its middle section (11) and its entry into Chesapeake Bay (16). Variations of Mg^{2+}/K^+ in clays washed with acid as a function of the salinity of interstitial water (after POWERS, 1957).*

seen them degraded during weathering. In order to reconstruct the normal minerals of micaceous or chloritic type, it is indispensable that the trajectory opposite to that of degradation be followed. In their discussion of the crystallochemical possibilities of such reverse reactions, GRIM and BRADLEY (1955) invoke the image of a "memory" of the inner structure of the layers, which seeks to reconstruct the past. The image is very suggestive, even if we are obliged to note that this memory fails at times and that the transformed minerals can be of a type quite different from that of their ancestors.

Thus we are able to reconstruct progressive aggradations from the fixation of ions to profound modifications of the lattices.

GRIM and JOHNS (1954) reported an increase of illite and chlorite in the Gulf of Mexico from montmorillonite, which is the principal mineral inherited from the volcanic formations and soils of Texas. Figure 28 shows the evolution of the argillaceous fraction from the river to the sea.

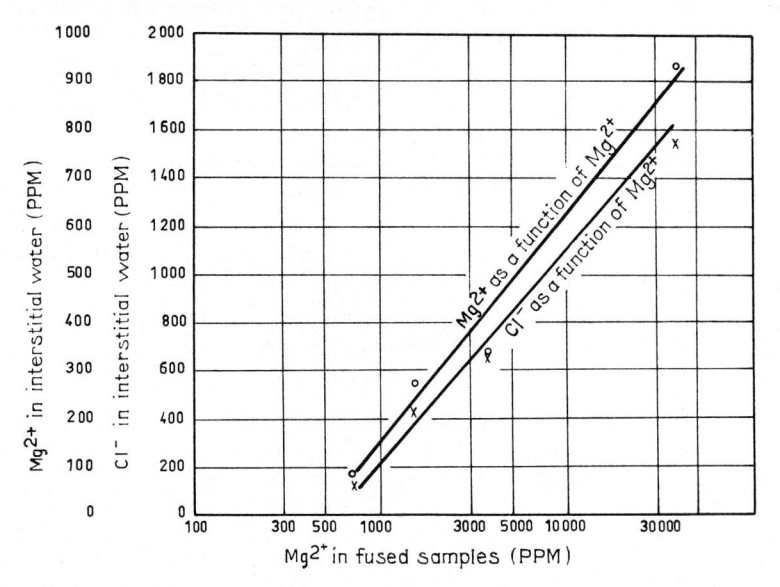

FIG. 66. — *Relation of Mg^{+2} concentration in clays washed with acid to the concentrations of Mg^{2+} and Cl^- in interstitial water.* James Estuary, Maryland, U.S.A. (after POWERS, 1957).

POWERS (1957) made measurements on samples collected between the James River and the sea in Maryland (U.S.A.). He studied at the same time the interstitial waters of muds, the exchangeable ions of the clay minerals, and the composition of the latter after acid washing. The curves given by POWERS (1957, 1959) show the fixation of Mg^{2+} and K^+ in the interlayer sheets and in the lattice itself. In fact, the samples washed with acid in order to extract the exchangeable ions show an exponential increase of the amount of Mg^{2+} in the lattice relative to the amount of Mg^{2+} of the interstitial waters of muds. Figures 65 and 66 show these results.

GRIM, DROSTE and BRADLEY (1960) looked for an interpretation of the considerable development of corrensite (montmorillonite-chlorite mixed layer) in the hypersaline sediments of the Permian of New Mexico. Once the concentration of Mg^{2+} ions in the water and in the interlayer spacings has reached an equilibrium, it is quite probable that an economy of space results from the regular and alternate distribution of these ions. This corresponds to Pauling's rule, which requires that crystalline structures reach neutrality in the smallest possible space. Here we find a reason for the frequency of this regular mixed layer in sedimentary deposits.

LUCAS (1962) went beyond the stage of corrensite and reached the stage of chlorite in the central part of the Triassic sedimentary basins of Morocco, the Jura and the Paris Basin. He frequently found clay fractions that were composed only or almost only of chlorite, as isolated tests had shown previously (MILLOT, 1954). On this basis he could advance the study of this chlorite, which certainly is transformed because it results from degraded illite, and all the intermediate stages were detected. This is a trioctahedral chlorite of the pennine clinochlore type, very rich in magnesium and containing alumina but no iron. Electron photomicrographs (figs. 33, 34 and 36) show the very well crystallized, lozenge-shape clay minerals. All the paleogeographical, paleosedimentological, mineralogical and geochemical arguments converge to show that these clay minerals are trioctahedral chlorites transformed from degraded illites from the borders of the basin. It has already been emphasized that here transformations approach the frontiers of neoformation.

Finally, reorganization of lattices during aggradation leads to an explanation of the behavior of trace elements in clay minerals, and especially in illite. If reworking of the tetrahedral and octahedral layers takes place within the sedimentary environment, certain ions can easily take part in the game. The question has been examined in Chapter VIII. The most significant example is that of boron, which increases with the salinity of the genetic environment and which very probably lodges in the tetrahedra of illite. Trace elements found in greater abundance in marine clays are nickel and lithium, which can enter the octahedral layers, and fluorine and chlorine, which are external.

Next, evolution takes place during diagenesis, after deposition of the sediments. BURST (1959) believed these transformations to be obvious: disappearance of montmorillonite, increase of illite and chlorite, decrease of the capacity for expansion of mixed layers, increase of the heat stability of chlorite. He showed these evolutions in samples collected in the Eocene series of the Gulf Coast. He added, furthermore, that these diagenetic effects do not prevent us from understanding the original variations, which, acting upon a different material, had a different effect.

POWERS (1959, 1959) studied the relative behavior of K^+ ions and of Mg^{2+} ions on the surface and at depth. On the surface the adsorption of Mg^{2+} ions is much greater than that of K^+ ions, although the latter have a lower ionic potential, a smaller ionic radius in solution, etc. But the Mg/K concentration ratio is about 5 in terms of the number of atoms. Fixation of magnesium is preferential. Once buried, the sediment is filled with water that is no longer renewed, and the probability of fixation of potassium increases as the probability of fixation of magnesium decreases. Thus, an "equivalent level" will necessarily be found, below which potassium will, in its turn, be preferentially adsorbed and the quantity of illite will increase noticeably in order to make up the usual composition of fossil sediments. Beyond that we would enter the domain of "dry" diagenesis. No longer is the diameter of the hydrated ions of importance, but it is the crystalline ionic diameter. The Mg^{2+} ions will tend to enter the octahedral layers, and the K^+ ions will tend to replace them in the interlayer positions; we are at the threshold of metamorphism, and we shall take it up again in Chapter XII.

Experimentation. — The first experimental tests on aggradational transformation of three-layer minerals were carried out by CAILLÈRE and HÉNIN (1947, 1948). Treating montmorillonite with 20 % $MgCl_2$ during 40 days at room temperature, they obtained a mineral similar to chlorite and stable upon heating and in the presence of glycerol. When they treated montmorillonite by boiling in a normal solution of KOH during 8 hours they obtained, after three or four successive operations, a diagram showing peaks of illite that remains stable in presence of glycerol.

Agronomic tests on K^+ fixation in soils should be considered as experiments in aggradation. It seems that it was VOLK (1933) who first applied the techniques of X-ray diffraction to the study of this problem. He noted a much greater intensity of the muscovite peaks in the clay fraction extracted from soils that had been treated over a period of fifty years than in reference soils. He attributed this fact to a neoformation of muscovite; today we would say that it was fixation of potassium in the open clay minerals. Since then many authors have studied these facts (GIESEKING and MORTLAND, 1951); however, the most elaborate and demonstrative tests are those of VAN DER MAREL (1954, 1959). Under the effects of potassium salts the layers of open illite, vermiculite and montmorillonite close again by fixation of cations in the interlayer position. The effect is less marked in Dutch marine sediments than in fluviatile sediments; this is a result of natural fixation by marine water. These properties are of great interest for agriculture; systematic studies have been undertaken in order to examine the variations in yield of cultures as a function of the clay minerals in the soils treated with potassic fertilizers (FRANC DE FERRIÈRE, 1956; FRANC DE FERRIÈRE, CAMEZ, MILLOT, 1957, 1958; FRANC DE FERRIÈRE, BLANCHET, MILLOT, CAMEZ, 1960). Soils with kaolinite and chlorite have a low capacity for fixation and for release of K^+ ions relative to soil solutions; montmorillonites are very favorable; vermiculites in podzols and in podzolic soils act as traps for potash, which preferably should be provided in the form of carbonate. These phenomena are interesting for agriculture, but for the subject at hand they have the value of a model because that which occurs in soils charged with potash must also occur in sediments in saline or hypersaline environments. Let us recall that, for many reasons, DEBYSER (1959) was able to call the muddy environment that still has its organic matter a "submarine soil".

In addition to these indirect tests, direct experiments on the action of sea water on clay minerals have been undertaken. WHITEHOUSE and MACCARTER (1958) treated a series of minerals with artificial sea water. Illite and chlorite presented no modifications, but montmorillonite showed diffraction reflections of illite, chlorite and various mixed layers. The tests lasted six months to five years. Through chemical methods structural changes were demonstrated involving modification of the SiO_2/Al_2O_3 and SiO_2/R_2O_3 ratios, which implies rearrangements in the octahedral and tetrahedral layers. Moreover, by X-ray diffraction one could observe the appearance of a peak close to 1.5 Å, which signifies the replacement of the dioctahedral by the trioctahedral configuration of chlorite. By electron microscopy considerable modifications were noted in the original form of montmorillonite, which was fibrous. The authors spoke of reaggregation or of recrystallization or of both together. Organic matter inhibits or retards these transformations. Re-suspension of the material slows down the development of chlorite and favors that of illite. In this study one has at his disposal a great number of quantitative data on experimental transformations.

CARROL and STARKEY (1960) studied the action of sea water on clay minerals and measured the preferential fixation of Mg^{2+} over Ca^{2+}; this is attributed to the greater abundance of Mg^{2+} in sea water. They also noted a release of silica, alumina and iron, but, of course, these experiments that were performed with small quantities of sea water and during only a few months did not lead to notable mineralogical transformations.

SLAUGHTER and MILNE (1960) continued the experimentation of CAILLÈRE and HÉNIN. Precipitates of magnesium and aluminum hydroxides were immediately dispersed and intimately mixed with a montmorillonitic suspension. A structure similar to that of chlorite developed. Likewise, by means of an electrolytic method, YOUELL (1960) fixed brucite sheets between the layers of a montmorillonite and obtained a swelling chlorite.

General view on aggradation. — Aggradations are transformations that reconstruct normal clay minerals from the degraded minerals provided by weathering. They occur under the influence of solutions that are rich in mineral cations and silica in sedimentary environments and under the influence of the successive early and late stages of diageneses.

There is a whole gamut of aggradations that corresponds to the gamut of degraded minerals, from the simple fixation of cations to the complete reorganization of crystals. It proceeds layer-by-layer in a progressive way and results in the variety of mixed layers we observe.

Initially magnesium is commonly fixed preferentially. Later on, it can be removed into the octahedral layer when potassium comes into dominance in its turn. This brings about a new tetracoordination of aluminum, and we approach the micaceous lattice, first in illites, then in sericites.

If, on the contrary, the influence of magnesium is maintained, it enters the octahedral layers while aluminum takes again its tetrahedral position. Having become magnesian again, the clay mineral is trioctahedral, and the increased charge cements the interlayer brucite layer.

The study of these crystallochemical evolutions under the constructive influence of the environment has hardly begun; it holds much instructive information in store for us.

III. — CONCLUSIONS:
INHERITANCE AND TRANSFORMATION BY ADDITION AND SUBTRACTION

The mechanisms of inheritance and transformation account for the origin of clay minerals in weathering, soils, sedimentary environments and zones of diagenesis.

Inheritance. — Every environment inherits solutions and particles from the past.

— *The solutions* may be oriented or not toward neoformation.
— *The mineral particles* are either stable or unstable.
  – *When they are stable* they provide the detrital heritage, properly speaking.
  – *When they are unstable* they are subject to transformation, either through degradation or through aggradation, according to local conditions.

Degradation. — Degradation takes place in the open environment of lessivage, characterized by removal of constituents of the clay minerals.
— *This is, therefore, transformation by subtraction.*
— The choice domain of degradation is found *in the zone of weathering and pedogenesis*. Sedimentary degradations are curiosities (Halmyrolysis). Degradations find a new domain in *certain diageneses*.

Aggradation. — Aggradation takes place in confined environments, or in environments rich in cations, characterized by the entrance of supplementary elements into the clay mineral lattices.
— *This is transformation by addition.*
— The choice domains of aggradation are those of *sedimentation* and *diagenesis*. It is rare and poorly known in weathering horizons and can occur only in the horizons of accumulation in soils. On the other hand, sedimentation and diagenesis reconstruct in part the original material, which will be perfected by metamorphism.

Inheritance and transformation account for the fate of clay particles in the geochemical cycle. *The question has been raised as to whether clays are fine-grained detrital products.* Today it is possible to answer in a more discriminative way. The answer can be presented as follows:

Yes, because the mechanical inheritance, to a great extent, provides material for weathering, sedimentation, and diagenesis.

No, because transformations modify a part of the inherited clays in weathering horizons, in soils and in sediments.

No, because neoformations are added to the former processes, in weathering and in soils, as well as in sediments.

CHAPTER XI

GENESIS OF CLAYS: NEOFORMATIONS AND SYNTHESES

This chapter will bring together and compare our knowledge on:
— The neoformations in nature,
— The syntheses in the laboratory.

I. — NEOFORMATION IN NATURE

Several synonymous terms designate the neoformed clays: neoformed clays, clays of neoformation, clays of neosynthesis, authigenic clays.

1° *Neoformation in weathered products and soils*

Sericitization of feldspars. — In the initiation of weathering we had the opportunity to see the phenomenon of sericitization of feldspars. The conditions of sericitization have been defined (J. DE LAPPARENT, 1909; MILLOT, 1949). It occurs in the lower part of the weathering zone where water circulates slowly through the microfissures of crystals. The release of alkaline and alkaline-earth elements from feldspars maintains an alkaline environment. It is important to note that sericite develops initially only in plagioclases, and, if they are zoned, in the zones that are richest in anorthite. This detail and the presence of small calcite rhombohedra, which can be seen under the microscope, reveal that the environment is calcic. Calcium seems to favor the neoformation of a micaceous lattice into which it does not enter. We know that a number of aluminum silicates are altered into damourite, sericite, pinite and various micaceous products; such is the case of cordierite, andalusite, zoisite, etc. Let us mention that volcanic glass can weather to illite under special conditions of lessivage (SCHLOCKER and VAN HORN, 1958).

Chloritization of ferromagnesian minerals. — At the same time, the ferromagnesian minerals become loaded with flakes of neoformed chlorite. These are not chlorites of retromorphosis, but chlorites of weathering. In the alkaline environment of hydrolysis of these silicates the elements from the ferromagnesian minerals organize into chlorite rich in iron and magnesia. Biotite is the most common example. J. DE LAPPARENT (1909) pointed out that chloritization of biotite is faster in granular rocks without orthoclase; here the K^+ ions are released into solutions poor in potash, and the potash nourishes sericitization. In the opposite case, in rocks rich in orthoclase, potassium is very abundant in solution and biotites subsist longer.

Kaolinization. — We emphasized kaolinization very much in Chapter V. It is not the normal weathering process of feldspars in our climates where the weakly drained environment of hydrolysis remains alkaline. On the other hand, kaolinization develops when lessivage accelerates and the medium becomes acid. This is the case of kaolinitic weathering in the various lateritic profiles. Tropical and equatorial rainfall ensures very vigorous lessivage during the humid seasons; the ions are released by hydrolysis and the electrolytes removed. When a permanent water supply provides the required silica, alumina organizes with it to form kaolinite. The pH is acid, cations are absent, and the kaolinitic horizons can develop a thickness of several meters. In ferruginous, warm-climate soils one finds series of thinner weathering profiles in which kaolinization subsists at first, then decreases gradually toward regions of temperate climate. Kaolinization also develops in those places where there is percolation of humic matter. Thus, bleached horizons are found under peat bogs and similarly are found in beds underlying some coal beds. Not all bleached horizons are kaolinitic; if a horizon is bleached, it is by lessivage of iron. Many of these bleached horizons are kaolinitic, however.

Aluminum sesquioxides. — The crystallization of aluminum sesquioxides is also a neoformation, if we are willing to admit this term to designate the appearance of gibbsite or hydrargillite at the expense of silicates. This phenomenon also occurs in intertropical countries, but only when drainage is excellent so that, in spite of the abundance of tropical rainfall, the water table falls and disappears into the fissures of the parent rock. In such a zone, which is abundantly leached but not immersed, the alumina that is released organizes alone and crystallizes into gibbsite. In the case of this neoformation, lessivage is not necessarily more intense in terms of discharge than in the case of kaolinization, but it does not maintain a high water table in the zone of weathering; this prevents dissolved silica from being delivered for the synthesis of kaolinite.

Here we are, of course, in an acid environment totally desaturated in cations, including the silicon cation.

Montmorillonite in weathered products and in soils. — *Montmorillonite commonly results from weathering of volcanic glass either in situ or pulverized in the form of ash and accumulated in sedimentary beds. This occurs only if the drainage is moderate enough to maintain an alkaline environment. Thus, deposits of primary or secondary bentonites are produced. The generating environment is alkaline, whether it be the effect of plutonic liquids or vapors, or the effect of waters of sedimentary basins.*

Montmorillonite has also been identified as a fugitive phase at the base of lateritic profiles in the zone of initial weathering. This is a case in which the profiles are bathed in water, where the direct evolution of silicates into gibbsite does not occur. We are in the zone of hydrolysis; the component ions of the silicates are released into an environment that is still alkaline. These are the conditions of the genesis of montmorillonite, and it is found there transitorily. Towards the top of the profile, the environment becomes acid and kaolinite prevails. This phenomenon is not very important in terms of the quantities produced since it is discreet and ephemeral, but it indicates how very susceptible clay neoformations are to the conditions of the environment. Even there where kaolinite is formed through several meters or several tens of meters of thickness, there can exist a very limited environment at depth, at the contact with the parent rock, where montmorillonite forms.

Montmorillonite also appears systematically and abundantly *in soils* in which hydrolysis occurs but *in which drainage is weak enough to permit the solutions to remain there while they are concentrated by evaporation.* This is typical of hydromorphic, calcimorphic soils in warm regions. In calcimorphic soils poor drainage and evaporation accumulate silica, alumina, magnesia, and commonly iron, in an alkaline environment rich in Mg^{2+} and Ca^{2+} ions. Minerals of the montmorillonite group develop; this is the case in chernozems, in tropical black earths, and in tirs. Once more we see in the development of montmorillonite the intervention of the alkaline environment and the tendency for accumulation of silica and alkaline-earth cations.

2° *Neoformation in sediments*

Kaolinite. — Cases of certain neoformation of kaolinite in sediments are not numerous, and today we are sure that kaolinite is a mineral of continental neoformation characteristic of weathered products and soils. Moreover, the examples described are much more comparable to phenomena of weathering or diagenesis than to phenomena of sedimentation. Thus, the development of vermicular kaolinites in sands and sandstones is a result of the percolation of acid water through porous sediments. The formation of halloysite in siderolithic pockets is also a recrystallization under the influence of lessivage due to drainage by underlying karst, or the result of the action of acid solutions coming from the oxidation of pyrite. The kaolinitic evolution of *Tonstein* results from the intervention of either biochemical and pedological mechanisms under the influence of humic matter, or the rapid weathering of volcanic silicates in an acid environment. Finally, the majority of kaolinite crystallizations are posterior to sedimentation and belong to the domain of diagenesis.

In summary, *kaolinite is a mineral typical of continental weathering* in which solutions percolate and are purified; *kaolinite cannot develop in sedimentary basins* where solutions accumulate and are enriched.

Illites and chlorites. — Today it seems most likely that *the majority of sedimentary illites and chlorites are inherited*, either in a well crystallized form or in a degraded state that permits further positive transformations. I know of no certain description of neoformation of illite or chlorite during sedimentation. In fact, all the recorded descriptions have characters that show their relationship either to weathering (halmyrolyses), or to transformations by aggradation during sedimentation, or to later diagenesis.

A problem of limits arises here. When much degraded illites or chlorites are found on the border of a sedimentary basin, chemical analyses reveal deficits of several cations. The patterns obtained by X-ray diffraction show very weak and broad peaks. On the other hand, after transformation by aggradation, the peaks are well developed. We have often observed peaks the height of which increased ten-fold, indicating a regularization of the lattice and of the stacking of the layers. In addition, the size of crystals increases considerably, and electron microscopy reveals beautiful crystal forms, as exemplified by a "transformed" sedimentary chlorite from the Moroccan Triassic shown on figure 33. Under these conditions, are we dealing with a transformation or a neoformation?

The problem of limits, and above all of limits of definition, can always give rise to long discussions. However, this much can be said: if, as seen for the Triassic, beautiful crystals of chlorite on the order of $1\ \mu$ are constructed on particles of degraded illite on the order of $0.1\ \mu$, we are dealing with the growth of crystals; this is as close to neoformation as it is to transformation. Here, it is preferable to class this phenomenon with transformations because the beginning of crystalline organization took place on a preexisting crystal lattice, but the idea of categories, which is necessary in the scientific domain, cannot hide the continuity of the phenomena.

In such limited light, one could say that sedimentary environments give evidence of gigantic phenomena of neoformation of illite, chlorite, corrensite, etc., in conditions of alkaline chemical sedimentation, whether lacustrine, marine or hypersaline. This neoformed character would be emphasized by the fact that the free ions of solutions take part in the process, i.e., K^+ and Mg^{2+}, of course, but also iron and silicon.

Nevertheless, since these evolutions with growth of crystals operate on materials that are already clay minerals, although disordered, we shall maintain them in the category of transformations, although we realize how arbitrary are our classifications.

Glauconites. — Glauconites are either entirely neoformed, developed by epitaxic growth on preexisting clay minerals, or the result of the transformation of micas. Here again our categories cut across a continuous and probably mixed reality. Let us say that glauconite commonly is neoformed, that the almost exclusive environment for this neoformation is the marine environment with special supply of iron. It seems that the iron can be provided by ferromagnesian minerals such as biotite or ferri- and ferro-organic complexes, but its dynamics are still poorly known.

Sedimentary attapulgites, sepiolites and montmorillonites. — These minerals are typically neoformed and typical of sedimentary basins of alkaline chemical character in which carbonates, tabular silica, phosphates, etc., develop. It will be noted that dissolved silica is abundant in these environments where, according to different cases, it organizes upon itself (sedimentary flint, cristobalite of montmorillonites) or with aluminum and magnesium. The African basins show us successively, proceeding in an offshore direction, montmorillonite, which is more aluminous, then the alumino-magnesian montmorillonite, and then sepiolite, which is more magnesian. It is in this manner that montmorillonites are situated among neoformed aluminomagnesian clays. In some cases they are mixed with attapulgites and sepiolites, but in other cases they can occur alone in siliceous and calcareous basins; the Cretaceous basins that produced chalk seem to be a good example.

Let us note in addition that these neoformed attapulgites, sepiolites and montmorillonites seem to be indifferent to the salinity of their genetic environment. They occur in calcareous or hypersaline lakes as well as in calcareous or hypersaline seas. The abundance of the sodium and potassium salts has no influence on the mechanism of their growth, but an alkaline environment is necessary.

3° *Neoformation in the domain of diagenesis*

There is a final domain in which neoformations occur, after the domain of continental weathering and that of sedimentary accumulations; this is the domain of diagenesis. This is understood here as the ensemble of evolutions that occur after sedimentation and before metamorphism; it is agreed that these limits are vague.

Silicates that result from diagenesis are numerous. TOPKAYA (1950) has presented an important study on authigenic silicates. It concerns postdepositional authigenesis, i.e., neoformation of diagenesis. Tourmaline, quartz, orthoclase, microcline, albite, oligoclase and muscovite are the principal authigenic silicates described. Let us examine the neoformation of clay minerals during diagenesis.

Kaolinite. — *Kaolinite is a diagenetic product common in the course of evolution of permeable sediments* subject to the circulation of acid waters which ensure underground evolutions similar to those occurring during weathering: hydrolysis of silicates, release of ions, combination of silica and alumina into minerals of the kaolinite family. In the course of study of a detrital series it was noted that, in an argillaceous-arenaceous series provided with a common supply of nutrients, the arenaceous levels are more kaolinitic than the shales (GLASS, POTTER and SIEVER, 1956; SMOOT, 1960).

Microscopic examination confirms these neogeneses by the presence of kaolinites in the form of fans, vermicules, or accordions. The example described in Chapter VII of the series of Paleozoic sandstones of the Sahara illustrates this case; it is figured on plate I (KULBICKI and MILLOT, 1961). These diagenetic neoformations of kaolinite have been described by many authors, and especially by VATAN (1939), KULBICKI (1953, 1954) and KELLER (1958).

Here we find ourselves confronted by the action of underground acid solutions with their cations removed and capable of starting the kaolinization of silicates and, in particular, of feldspars, micas, and clay minerals.

Illites and chlorites. — The development of white mica and chlorite in the course of diagenesis is universal. I say white mica because one observes all the intermediates between a more or less open illite and the muscovite that appears during metamorphism.

Illites and sericites appear under a great number of conditions to which the observer who does not study these questions does not pay attention because we are used to considering micaceous flakes as common minerals. The above-mentioned case of diagenesis in Cambro-Ordovician sandstones showed us a beautiful example of illitization of kaolinite (KULBICKI and MILLOT, 1960). We attributed it to a second stage of diagenesis, at high temperature under the action of saline waters that succeeded oil in its migrations. Moreover, sericitization of kaolinite is the rule in advanced diagenesis because this mineral is no longer found at the level of schists. Sericite can also be seen to develop in arkoses by diagenesis from feldspars. This is a case of silicates that resisted weathering but that gave rise to micaceous clay minerals in the zone of diagenesis. In another way, one commonly sees in sandstones or calcareous sandstones micaceous films that surround the quartz grains with a fine concentric envelope. We have mentioned, concerning silicifications in the Fezzan, MULLER-FEUGA's observation (1962) of successive aureoles around quartz grains in calcareous sandstones; some of these aureoles are micaceous. These micas were certainly neoformed by diagenesis on their siliceous support, and their occurrence together with aureoles of calcite or opal shows that the environment was alkaline.

Diagenetic illitizations, which can be either transformations or neoformations, certainly are much more frequent than we realize today. Waters recovered through oil drilling from the zone of diagenesis have a high temperature and high salinity; these are the conditions of hydrothermal development of the micaceous lattice. The role of these mechanisms has been discussed in connection with the marno-calcareous series; the frequency of illite in marine or lacustrine calcareous deposits may find its origin here, but much work is yet to be done.

Diagenetic chlorites are very common and have even been said to be universal by PUSTOVALOV (1955). He mentioned examples of regional character in diagenetic series observed in Russia. Already J. DE LAPPARENT (1924) had shown the development of chlorite in indurated Devonian rocks of the northern Vosges. He showed the chloritization of the walls of Radiolarian tests, a phenomenon that was effected at the same time as the development of vermiform chlorites within the chalcedonic matrix of these organisms. The objection can be raised that these formations were caught up in the Hercynian orogenic cycle and show evidence of slight epizonal metamorphism. But this is not the case for the chloritizations mentioned in Chapter VII concerning diagenesis of sandstone or carbonate series. In particular, MILNE and EARLY (1958) and WEAVER (1959) showed preferential chloritization of sandstone horizons in Tertiary series that are absolutely not metamorphosed.

If we add to the true chlorites the problem of the false chlorites or chamosites of iron ores, we note in this case a beautiful development of neoformation by diagenesis. These clay minerals are the result of silication of the iron sesquioxides coming from sedimentation. They are not only ferruginous but also aluminous and magnesian. The proximity of calcite and the presence of phosphates show that this neoformation takes place in an alkaline environment. Moreover, in advanced diagenesis these chamosites are transformed into true chlorites, as in the ores of Normandy.

The geochemical cycle of iron will be reexamined in Chapter XII, but it is necessary to emphasize that the natural evolution of argillaceous series by advanced diagenesis consists of incorporating iron into chlorites, which will be the starting point for beginning metamorphism.

KOSSOVSKAIA and SHUTOV (1958) studied the zonation of sedimentary sandstone formations of series of the Russian or Siberian platform and of the geosyncline of the Verkhoyansk arc in eastern Siberia. They described four zones, two of which interest us.

— The first zone is characterized by the lack of alteration of the argillaceous cements. It concerns deposits of the platform and the upper parts of the geosyncline.

— The second shows an evolution of the argillaceous cements. It concerns only the basal horizons of the platform and the periphery of the folded structures of the geosyncline beyond a depth of 6 000 m. This zone is characterized by the complete disappearance of montmorillonite, by the hydromicatization (illitization, sericitization) of kaolinite and by the development of hydromicas and chlorites. It must be added, of course, that if plagioclases are present, they are sericitized, and ferromagnesian minerals are chloritized.

— Zones 3 and 4 belong to metamorphism. In zone 3 chlorite and sericite are included in the zones of growth of the quartz grains engaged in the quartzitic texture. In zone 4 chlorite and muscovite (coming from sericite) show a brush-like habit on the surface of crystals: metamorphic recrystallization has definitely begun.

Sericite and chlorite are the two clay minerals characteristic of diagenesis and all the more so when diagenesis is advanced and is tending toward metamorphism.

Pyrophyllite and talc. — Pyrophyllite of diagenetic origin has been described, as far as I know, only in the lower Devonian sandstone formations of the middle syncline of the Armorican Massif (DUNOYER DE SEGONZAC and MILLOT, 1962). This pyrophyllite represents the deposit that was the origin of detrital pyrophyllite found by LAFOND (1961) in the alluvium of the Vilaine River. It is crystallized in nests within the sandstones where it occupies the emplacement of grains that have disappeared. It is believed that these grains were former feldspars that evolved first into kaolinite and then into pyrophyllite; the intermediate stage seems to be necessary in order to evacuate potassium, which would have led automatically to sericite as is usually the case.

Talc is frequent in hydrothermal deposits, but its presence has been observed in non-metamorphic carbonate rocks of the African Infracambrian (BIGOTTE, BONIFAS and MILLOT, 1957; MILLOT and PALAUSI, 1959). It formed the concentric envelopes of oolites or the clayey matrix of the rock. It had seemed to us to be more likely the result of late-phase diagenesis, but the discovery of talc as an early mineral in the Zechstein (FÜCHTBAUER and GOLDSCHMIDT, 1959) raises the problem again. Whether it be contemporaneous with sedimentation or with diagenesis, this talc in the Zechstein is certainly neoformed; and its association with the sulfate facies does not leave any doubt about the alkaline conditions of its formation.

Attapulgites and sepiolites. — Attapulgites and sepiolites are minerals that are neoformed in sedimentary basins, but they can also result from later diagenesis. DEMANGEON and SALVAYRE (1961) have described a fibrous palygorskite, certainly diagenetic, that had replaced crystals of dolomite by epigenesis. PUSTOVALOV (1955) described diagenetic palygorskites as being very frequent in Russia. BIGOTTE (1956) has found a very magnesian palygorskite in geodes and joints of limestones with talcose oolites in Equatorial Africa. But we are at the limit of the hydrothermal domain where these minerals are rather frequent. It is in this situation that they received the names "asbestos board, mountain wood or mountain cork" (*carton, bois,* and *liège de montagne*) and that they show an asbestos-like aspect.

4° Summary on the neoformation of clay minerals in nature

Neoformation of clay minerals in nature has just been presented in the three domains of surficial weathering, sedimentation and diagenesis. Indications of the conditions of neoformation are convergent and the same terms are always used to describe the natural environments of neoformation.

— *The well-leached (lessivé) environment, which is acid and free of cations, determines the formation of kaolinite.*

— *The insufficiently renewed environment, which is alkaline and rich in cations, determines the formation of the three-layer minerals:* illites, montmorillonites, chlorites, attapulgites. It is the proportions of the elements present that lead to one or the other of these neoformations.

Let us try to determine more closely the reasons for this double orientation by the examination of the coincidences observed in nature.

Lessivage. Environment of lessivage and confined environment. — First the effect of lessivage will be examined. This term is rather vague, but for the geologist and the pedologist it corresponds to the dynamics of waters in the natural environment, and this is essential. We shall distinguish the leached (lessivé) or open environment and the confined environment.

— *The environment is leached when water can pass through it easily, thanks to a natural open exit:* joints in the underlying rock, drainage of ground water, etc. The loss of water is greater than the supply. One speaks of drained, open or leached (lessivé) environment. The dissolved products are evacuated.

— *The confined environment is one in which means of exit are deficient; the water level is maintained constant through evaporation.* But the dissolved products are concentrated instead of being evacuated. One speaks of a closed, confined, poorly drained environment, or an environment of accumulation.

— *The environment of lessivage maintains dissolution and hydrolysis.* As the ions are released, they are evacuated, but the question of discharge is important. If discharge is high relative to the speed of hydrolysis, all the ions are evacuated, and the environment becomes acid. If the discharge is weaker, the ions released are only partially evacuated, and the environment is alkaline. Thus the environment of lessivage is not inevitably acid, but it is an environment that loses a part of its substance. *Neoformations encountered in this domain will be neoformations by subtraction.*

— *The confined environment blocks dissolution and hydrolysis.* As solutions arrive, ions are accumulated; the water level is regulated by lateral drainage that is weaker than the loss by evaporation. Thus, there is no loss of substance, but accumulation, and *neoformations found in this domain will be neoformations by addition.*

EXAMPLES:

— *Environment with strong, acid lessivage.* — Hydrolysis as well as removal of all soluble products are at a maximum. In a crystalline rock everything is evacuated except alumina, which organizes into gibbsite; this is bauxitization. If the lessivage is a little less strong, it allows a permanently high water table, the alumina remains in place and traps dissolved silica; this is kaolinization. The same thing occurs during the kaolinizing percolations of diagenesis. *Gibbsite and kaolinite are neoformations by subtraction within an acid environment.*

— *Environment with restricted alkaline lessivage.* — Hydrolysis is operative, a part of the ions in solution subsists, pH is high. Alumina, which is the least soluble element, organizes with silica and with the cations present in the environment to give rise to three-layer minerals. *Sericitization of feldspars, chloritization of ferromagnesian minerals, evolution of volcanic glass into montmorillonite; these are neoformations by subtraction within an alkaline environment.*

— *Alkaline confined environment.* — Hydrolysis is blocked, but the environment is supplied with ions from outside. These ions are chiefly the alkaline and alkaline-earth ions, which are the most soluble and mobile, but there is also silica in the monomolecular state and perhaps iron and a little alumina. In an alkaline environment these ions combine into *montmorillonites of all compositions, attapulgites, sepiolites, chamosites,* and possibly *siliceous neoformations. These are neoformations by addition within an alkaline environment.*

— *Alkaline lessivage.* — During diagenesis, highly alkaline solutions can migrate as deep ground water, or connate water removed by squeezing. This is alkaline lessivage at high temperature; it is the domain where diagenetic sericites and chlorites develop; these are *neoformations by addition within an alkaline environment.* In spite of their circulation, these solutions are concentrated through the thickness of the sedimentary column; this is another way for them to be "confined"; the result is identical.

— *One sees that it is too simple to relate lessivage to acidity and confinement to alkalinity;* this is an error that makes natural neoformations seem mysterious. That which must be related is *lessivage to subtraction* and *confinement to addition.* In the case of lessivage, the environment will be acid, neutral or alkaline according to its intensity and quality. In the case of confinement, the environment will be always alkaline because the salts of the hydrosphere are salts of weak acids and strong bases.

pH. — The best measurement of these characteristics in a more or less leached or confined environment is the pH. It is direct proof of the proportions of anions of weak acids and of cations of strong bases. The measurement of the pH of solutions within a weathering horizon, within a soil horizon, or within a sedimentary mud specifies the total intensity of the renewal of waters. An acid environment is an environment in which lessivage is sufficient to ensure the removal of soluble basic cations. An alkaline environment is an environment that is poorly drained or that is confined, where basic cations linger or accumulate.

The pH of fossil sediments. — An echo of the original pH may be recorded in sedimentary rocks, which are fossil sediments. To measure it, it is sufficient to pulverize a sedimentary rock and to put it in equilibrium for 24 hours with distilled water. The measurement of pH yielded results that I have presented in statistical form (MILLOT, 1949). The pH of suspensions obtained from argillaceous rocks characterizes the type of clay mineral they contain. Kaolinitic rocks gave an acid pH, whereas rocks containing three-layer minerals of micaceous type gave an alkaline pH. These results are plotted in figure 67. It is not quite the same thing as the abrasion pH of minerals that has been defined by STEVENS and CARRON (1948). What these authors measured was the pH of hydrolysis of pure, ground-up minerals. Here, it is a question of sedimentary rocks; all the minerals and elements they contain are involved in the hydrolysis, and in spite of the antiquity of their history, the echo of former conditions is recorded. I say "echo" because the disappearance of organic matter and many successive evolutions could have modified the initial value. The fossil pH is to the initial pH as a fossil is to an animal. Let us say that it is somewhat different, but it is

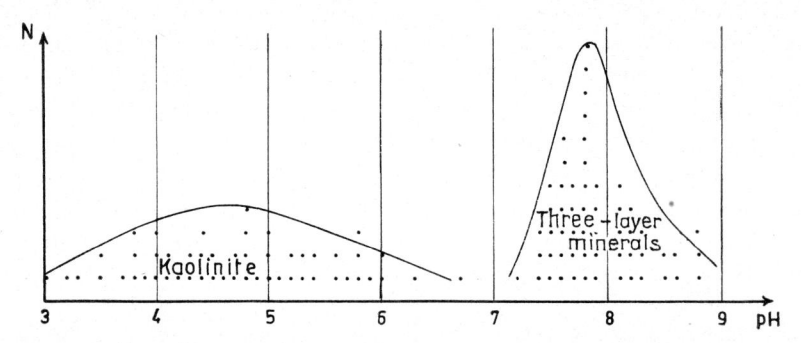

FIG. 67. — *Variation of the pH values of suspensions of argillaceous sediments.* Acid pH in the case of sediments rich in kaolinite; alkaline pH in the case of sediments rich in micaceous minerals (after MILLOT, 1949).

recognizable. Without guaranteeing the initial state and in the absence of totally deforming diagenesis, that which must be rather frequent, one can say: "The internal environment of a rock possesses some of the characters of the original sediment, characters that are fixed, preserved, fossilized, approximately intact, or at least of the same order of magnitude as in the original." (MILLOT, 1949).

Role of alkaline and alkaline-earth cations. — These are the cations of the first group of GOLDSCHMIDT, qualified as "soluble cations". These are especially Na^+, K^+, Ca^{2+}, and Mg^{2+}. These cations *vis-à-vis* weak acids determine the reaction of the environment. Their absence leads to acid reaction. Their presence imposes an alkaline reaction.

— This connection of the functioning of cations and pH, and thus the intensity of lessivage, accounts for the natural coincidences. The absence of cations coincides with the formation of kaolinite and their presence with the neoformation of three-layer minerals of micaceous type.

— In addition, cations enter into the neoformations of three-layer minerals of micaceous type, but this does not occur for kaolinite, which is uniquely aluminous.

Potassium is necessary for the neoformation of illite, and magnesium for the neoformation of montmorillonite and, above all, of attapulgite and sepiolite. Iron and magnesium are necessary for the elaboration of chamosite, chlorites and glauconites.

— It would seem that calcium serves no function, although it is familiar in carbonate neoformations. In fact, many indications allow us to think that it does play a role. In a feldspathic rock undergoing weathering, sericite develops within the plagioclases that contain a certain amount of anorthite, i.e., of calcium silicate. Deposits of kaolin which are hydrothermal develop readily in massifs of alkaline granite, granulite in the French meaning of the term, greisen, etc. But they do not exist in calco-alkaline granites; it is as if calcium had the right of veto. In lateritic weathering, the development of kaolinite is often delayed in the zone of initial weathering as long as plagioclases are not sericitized and hydrolyzed. Sericite is potassic, whereas plagioclases are calcic. As soon as sericitization is complete, kaolinization develops its full amplitude at the expense of alkaline feldspars (sodic and potassic) and of sericite (potassic). One must, therefore, believe that the release of alkaline ions is more tolerant *vis-à-vis* kaolinization than calcium, which seems to poison it.

In the neoformation of attapulgites, sepiolites and montmorillonites, which are neoformations typical of the hydrosphere, the concentrations of sodium and potassium seem to be indifferent since these authigenic minerals form in freshwater as well as in marine or hypersaline water. It is, thus, the concentrations of magnesium and calcium that are required. This is well understood in the case of magnesium, which participates in the structures. But calcium does not seem to be simply a companion of magnesium because it always occurs in the form of calcite or phosphates in formations bearing the aluminomagnesian clays.

The same line of reasoning can be applied to calcimorphic soils, which are calcic as their name indicates and characterized by the occurrence of montmorillonite (chernozems, black tropical soils, tirs). Moreover, in soils in which kaolinization is initiated, such as ferruginous soils or the upper horizons of podzols, kaolinite forms only after desaturation of calcium. It does not seem that alkaline ions are so constraining.

Let us finally say that the very spectacle of marine sedimentation, in its ensemble, is cause for reflection. As long as sea waters are only saline, transformations remain discrete. *But if the chemical character is accentuated, and if carbonate sedimentation begins*, transformations accelerate, growth of crystals takes on an original character leading to minerals such as corrensite or chlorite, or neoformations can take place (attapulgite, montmorillonite, sericite).

— All this is not easy to interpret, but it seems that alkaline-earth ions play a special role in the determinisms we are studying. The ionic potential of alkaline-earth ions, which have a double charge for rather small diameters, is much higher than that of alkaline ions. We shall see that experimenters have studied the effect of electrolytes on syntheses and on the tetracoordination of aluminum, but it is the most frequently a question of sodium; here there is a subject for investigation. For the time being, by means only of the correlation of observations in nature, I conclude as I did in 1949 (MILLOT, 1949, p. 288):

"The role of the nature of the cations present ... is obvious for those which enter into the chemical composition and into the crystalline structure of the minerals themselves ... On the other hand, much less obvious and very important, it seems, is *the role of certain cations in the neoformation of minerals whose structure they do not enter*. This is, in the first place, the case of calcium, followed frequently by magnesium, both being bivalent cations. *A whole group of facts shows that the presence of bivalent cations in an environment favors the formation of micaceous minerals and blocks the genesis of kaolinite.*"

Mineral and organic anions. — Our knowledge of the role of anions in these phenomena is rather limited. Naturally their role is obvious in the regulation of pH. Likewise, very soluble anions accelerate the removal of cations during lessivage and, consequently, the speed of hydrolysis. But the specific role of certain anions on neoformations has also to be studied. Specialists in hydrothermal deposits emphasize the effect of the sulfuric anion in the synthesis of kaolinite and especially of halloysite. Does it act as a strong acid or does it intervene in the coordination of aluminum? Likewise, the role of organic anions must be considered. The favorable effect of humic acids on kaolinization has always been emphasized; research on syntheses by WEY and SIFFERT (1961) will advance this question, as we shall see in the following pages.

Oxidation-reduction potential. — The role of the oxidation-reduction potential is still poorly known. It must play a role, sometimes direct, sometimes indirect, in the neoformations of clays.

— The Eh of solutions regulates the solubility of iron which, when it is free, is mobile only in the ferrous state. It also regulates the stability of ferro- and ferri-humic complexes; in an oxidizing environment, the latter are destroyed, and iron sesquioxide appears.

— Eh intervenes directly in the neoformation of ferriferous clay minerals such as chlorites, chamosites, and glauconite, that contain iron in both oxidation states. CAILLÈRE and HÉNIN (1960) have good reasons to think that iron might enter a clay mineral in a reduced form and then be oxidized, which would not be without consequences for the structure.

— Eh must intervene indirectly because it presides over the conditions of life. The presence of oxygen permits the development of flora and fauna whereas anaerobic conditions develop a special bacterial population. Anaerobic bacteria reduce organic matter and sulfates and give rise in sediments to a new sort of biochemistry, as revealed by the odor of muds, the appearance of colloidal iron sulfides, etc. This cannot help but intervene in neoformations occurring in the muds and, in particular, in the geochemistry of iron ore clay minerals.

Silica and alumina. — Clay minerals are always siliceous and in very many cases aluminous. For this reason, the role of the concentrations of silica and alumina of the genetic environment has been examined. All experience acquired in the last several decades shows that *the SiO_2/Al_2O_3 ratio of the original silicates that supply* *alumina and silica does not intervene in the neoformations of clays.* This follows from the fundamental works of CORRENS and VON ENGELHARDT (1938). The silicate structures are destroyed and silica and alumina are delivered in conditions that depend on the environment and not on the original material. The best confirmation of this is that different neoformations are obtained from one and the same silicate under different conditions.

CORRENS (1939), HÉNIN (1947), EDELMAN (1947) and MILLOT (1949) have studied the role in neoformations of the SiO_2/Al_2O_3 ratio of solutions. This took place at the time when it was established that kaolinite forms in an acid environment and that three-layer minerals form in an alkaline environment. In addition, at the same epoch it was thought that silica was less soluble in an acid environment than in an alkaline one, being well understood that these terms designate the range of natural pH values.

Therefore, in combining the knowledge acquired in 1949, I presented the following line of reasoning (MILLOT, 1949):

a) In an environment from which cations are evacuated, either by weak lessivage in an acid environment or by vigorous lessivage in a neutral environment, the solubility and stability of SiO_2 and Al_2O_3 in solution are such that the neoformation of clays is of kaolinitic type in which the SiO_2/Al_2O_3 ratio is equal to 2.

b) In an environment in which solutions are not renewed, cations are abundant, producing a basic pH; the solubility and stability of SiO_2 and Al_2O_3 are such that the neoformation of clays is of micaceous type in which the SiO_2/Al_2O_3 ratio is greater than 3.

Although the premises and conclusions are exact as they are observed in nature, the reasoning is false. It is based on two errors. The first is the question of the solubility of SiO_2, which was considered at that time to be higher in the natural alkaline environment than in the acid environment; this is erroneous. The second error is the appeal to the stability of colloidal alumina flocculated by cations; it is not a question of flocculation of colloidal alumina; it is a question of the growth of crystals based on Al^{3+} ions.

Today one can present quite a different outline. It is not a question of solubility of silica, which is constant in the range of natural pH values; it is a question of the quantity of silica present in solution. In fact, silica is soluble in the molecular state in natural environments and it can occur in greater or lesser quantities. The outline is the following:

— In an environment of lessivage, the concentration of silica in solution depends on the intensity of lessivage. In an environment of *vigorous lessivage* silica is exported, alumina remains alone and crystallizes as gibbsite ($SiO_2 = 0$). In *more moderately leached* environments silica is present in solutions and alumina is associated with it there to give rise to *kaolinite* ($SiO_2/Al_2O_3 = 2$). In an environment *still less leached* cations persist and silica is more abundant; one comes to *sericite* and possibly to *montmorillonite* (SiO_2/Al_2O_3 greater than 2).

— In a confined environment, the products of hydrolysis accumulate; cations and silica are abundant, alumina is rare because it is weakly soluble and therefore travels poorly. Neoformations are always oriented toward minerals in which alumina occurs in small quantity.

— Thus, the abundance of silica and alumina has a role in the elaboration of clay minerals, but their respective concentrations depend more on the intensity of lessivage or confinement than on the pH of the environment or the amount of cations. *The dynamics of these solutions is the common driving force determining the pH, the concentrations of cations, silica, alumina, and, finally, the neoformations themselves.*

Conclusion. — The dynamics of natural solutions regulates the nature of the neoformations of clay minerals in weathered products, sediments, and rocks in evolution. *This dynamics acts at the same time on the pH, on the amount of alkaline and alkaline-earth ions and on the amount of silica and alumina in the environment of neoformation.*

Naturalistic approaches outline the rules, but the experimental method permits us to give evidence of mechanisms and to know their limits. Let us study the syntheses of clay minerals in laboratory.

II. — LABORATORY SYNTHESES

Experimental syntheses of clay minerals at low temperature are the results of work of only a few investigators. Each school took up the problem in its own manner; therefore, it is convenient to present successively the results and the teaching of each scientific team.

1° Syntheses of clay minerals by Caillère, Hénin, Robichet and Esquevin

Methods. — 1° METHOD OF ELECTROLYTIC ATTACK. — The first encouraging trial was that of electrolytic attack of a magnesium anode by a potassium silicate (HÉNIN and CAILLÈRE, 1947). A 7.2 Å-mineral related to antigorite appears, although deficient in silica. Other trials were attempted with Al, Fe, Mg, Ni, or Zn electrodes and solutions of silicates, aluminates and chromates. A whole series of products have been obtained; they all are composed of hydroxide layers separated by sheets of water and maintained in this position by tetrahedral anions. The interpretation that was given already showed that it was not a question of reaction of the silicate on the already formed brucite, but an association of the silicic anion on brucite that was forming (CAILLÈRE, HÉNIN, ESQUEVIN, 1953). The products obtained were similar to antigorite and characterized by a large deficit of silica; since the method could not be improved, it was abandoned in favor of the following ones.

2° METHOD OF DILUTE SOLUTIONS. — The concentration of the solutions employed is chosen to be of the same order of magnitude as that of solutions of natural drainage

(20 mg/l). Thus, one hopes to approximate the mechanisms of nature. The realizations are the following (HÉNIN and ROBICHET, 1953; HÉNIN and ROBICHET, 1954):

Elements stable in alkaline and acid environments are put separately in a very dilute solution. Then they are combined in a large 5 litre container already containing 2 litre of distilled water maintained at 100 °C.

a) Sodium silicate and aluminum chloride give rise to boehmite, whatever the pH value may be.

b) Sodium silicate and sodium aluminate give rise to boehmite at pH 8 and to a small quantity of a product difficult to characterize, perhaps kaolinitic, at pH 6.

c) Sodium silicate and magnesium chloride give rise to an antigorite at pH 7 and to a montmorillonite at pH 8.

3° METHODS OF NATURAL SOLUTIONS. — In order to approach natural conditions still more closely, the water of drainage of experimental weathering compartments was used. By concentration and evaporation at 90 °C, a montmorillonite was obtained (HÉNIN and ROBICHET, 1954). Already boiling is no longer necessary here. The works of PÉDRO which will be examined later are the confirmation of this trial.

4° METHOD BY ATTACK OF SOLID SILICATES. — Synthesis of magnesian antigorite has been achieved by the method of dilute solutions in a glass sphere. But chemical analysis revealed that the resulting product contained 2 g of silica whereas only 2 mg had been supplied. The excess silica came from the dissolution of the inner surface of the sphere (HÉNIN, 1954). In a whole series of tests performed since that time, the role of the inner surface of the glass sphere has been verified and has been used, even when the sphere was made of fused silica (CAILLÈRE, ESTÉOULE, HÉNIN, 1962). This observation has also opened a pathway to the studies of rock weathering realized by PÉDRO and which will be examined later.

5° SYSTEMATIC STUDIES OF THE DIFFERENT FACTORS AT WORK. — Studies have been undertaken to examine the role of various factors on the synthesis of montmorillonitic products (HÉNIN and ROBICHET, 1955).

a) Influence of the Si/Mg ratio at a given pH value: magnesian montmorillonite forms at pH 8 for SiO_2/MgO ratios between 0.15 and 1.40.

b) Role of electrolytes: $CaCl_2$ prevents the formation of magnesian silicate. Na_2CO_3 contributes to an increase of pH up to 11 and gives rise to an almost amorphous product. NaCl favors the formation of montmorillonite and apparently permits obtaining this structure at a ratio lower than the limiting SiO_2/MgO ratio (0.15).

c) Role of pH: the lowest pH limit below which montmorillonite cannot be obtained is situated near 7.5.

The synthetic products obtained. — The techniques of preparation of clay minerals at low temperature being perfected, syntheses have been carried out using all the cations susceptible of entering the octahedral sheet: Mg, Ni, Fe^{2+}, Fe^{3+}, Zn, Co and Mn^{2+}.

1° MAGNESIAN CLAY MINERALS. — The first clay minerals obtained by synthesis, as has just been seen, were antigorites and montmorillonites. This is important, in consideration of the beautiful development of these clay minerals in nature. This also reveals the ease with which brucite forms from solutions. With high concentrations of magnesia and at pH 7, antigorite is obtained. With increasing concentrations of silica and at pH 8 or 9, stevensite is obtained.

2° NICKELIFEROUS CLAY MINERALS. — By the same procedure CAILLÈRE, HÉNIN and ESQUEVIN (1954, 1956) achieved the synthesis of nickeliferous clay minerals. At pH close to 6, nickeliferous antigorite is easily obtained. At high pH values, nickeliferous montmorillonite is obtained. pH is not the only factor acting; the relative quantities of silica and nickel hydroxide also intervene.

3° FERRIFEROUS CLAY MINERALS. — The synthesis of ferriferous clay minerals has been attempted. However, since no exclusively ferriferous clay minerals occur in nature, a small quantity of magnesium was introduced. Sodium silicate pure or mixed with sodium aluminate, respecting the proportions of Si and Al in the tetrahedral sheet, have been added to the acid solutions of iron and magnesia. Ferriferous saponites, nontronites, ferriferous beidellites, pseudo-chlorites, and chamosites have been obtained (CAILLÈRE, HÉNIN, ESQUEVIN, 1953, 1955; CAILLÈRE, HÉNIN, 1960). The role of the proportion of Si to the sum of the other cations is very clear. Abundant silica gives rise to ferriferous montmorillonites, whereas deficiency of silica yields chamosites, and a strong deficiency of silica gives rise to oxides. pH is also of importance; at pH values above 8, the 14 Å-minerals appear; at lower pH values, the 7 Å-minerals occur. Usually one obtains dioctahedral 14 Å products, whereas chamosite is trioctahedral. The experiments lead to the conclusion that iron enters into the product in its reduced form, but can be oxidized afterwards.

4° ZINCIFEROUS CLAY MINERALS: ESQUEVIN (1956, 1958). — By the same methods, zinciferous clay minerals have been prepared. Solutions of sodium silicate, on the one hand, and of zinc acetate and magnesium chloride, on the other, permitted ESQUEVIN to obtain, at pH higher than 7, a zinciferous montmorillonite of the stevensite type. But a whole series of experiments were run by varying pH, Zn/Si ratio, electrolytes and amount of alumina. It is indispensable that dilution be great, otherwise unlayered silicates, such as calamine (hemimorphite) and willemite are obtained. The presence of electrolytes plays a role of catalyst or inhibitor of neoformation, according to the nature of their anions or cations. Sodium chloride, in particular, is depolymerizing for the silica of the walls of the sphere and has a determining influence on the coordination number of zinc at the moment of precipitation of the hydroxide, which must be layered in order to start the synthesis.

The role of alumina is very informative; its behavior is very different from that of the other cations. A whole series of syntheses has been made at pH 7, 8 and 9 and with increasing concentrations of alumina. Low pH values favor the two-layer minerals of the chamosite type, but so do high concentrations of alumina. At pH 9, aluminum goes preferentially into tetrahedral position, and one passes from the preceding stevensites to zinciferous saponites or sauconites. At pH 8, aluminum goes into octahedral position and at the same time increases in tetrahedral position; zinciferous and

TABLE XIV. — CONDITIONS OF FORMATION OF ZINCIFEROUS CLAY MINERALS

	0	2	4	6	8	10	12
$\frac{Al}{Zn}$	0		$\frac{1}{6}$		$\frac{1}{3}$		$\frac{1}{2}$

[Diagram showing fields of Stevensite, Saponite, Chamosite, Calamine, and Boehmite across pH 6 to pH 9]

aluminous saponites are obtained. At pH 7, the fixation of aluminum is very strong, and chamosites are obtained. The conditions of these various syntheses are shown on table XIV.

5° COBALTIFEROUS CLAY MINERALS: CAILLÈRE, HÉNIN and ESQUEVIN (1958). — Sodium silicate, pure or mixed with sodium aluminate, is put in the presence of cobalt acetate, mixed or not with magnesium chloride. Minerals of the antigorite, chamosite, stevensite and saponite types can be prepared. The behavior of cobalt is similar to that of nickel.

6° MANGANIFEROUS CLAY MINERALS: CAILLÈRE, HÉNIN and ESQUEVIN (1959). — Manganese is supplied in the form of acetate, but one must work in a medium containing neither oxygen nor carbonic gas to avoid the change of valency of the Mn^{2+} ion or the formation of carbonate. The same types of products were obtained as in the preceding case, but they do not have a degree of crystallinity comparable to that of the other synthetic minerals, and the amorphous products are easily obtained. This is attributed to the difficulty in obtaining the hexacoordinate Mn^{2+} ion.

Generalities. — This group of syntheses, carried out successfully over fifteen years of study, allows one to draw some conclusions of general scope (CAILLÈRE, HÉNIN, ESQUEVIN, 1957; CAILLÈRE and HÉNIN, 1961).

1° Clay minerals form by *fixation of silica on a hydroxide sheet* by means of precipitation from *very dilute solutions.*

2° This requires *the hydroxide sheet to be layered itself,* preparing in some way the formation of layers. Now, a layered structure necessitates *hexacoordination of the cation.*

3° If the cations used are classed in order of increasing size,

$$Al^{3+} Fe^{3+} Mg^{2+} Ni^{2+} Co^{2+} Fe^{2+} Zn^{2+} Mn^{2+},$$

one sees that hexacoordination is common to all these cations, but also tetracoordination is frequent for the first two and eight-fold coordination for the last one. It is exactly in the cases of these cations of extreme size that synthesis shows itself to be the most difficult.

4° Thus, the problem comes back to that of how *to induce the appearance of the layered hydroxide in which the cation is hexacoordinate in the presence of silica tetrahedra* that will fix on it. Experimentation shows that, in the case of zinc, whenever the tetracoordinate hydroxide forms, one ends up with three-dimensional and not layered silicates.

5° The proportions of the elements in solution and the pH of the solution control the nature of the neoformed products. A relatively high concentration of cations other than silicon fixes the pH of precipitation at a rather determined and low value (between 7 and 8); two-layer minerals of the antigorite-chamosite type are obtained. In contrast, a low concentration of cations relative to a higher concentration of silica permits precipitation, at pH 8 or 9, of three-layer silicates of the montmorillonite type.

6° *It is very easy to introduce aluminum into the forming minerals.* Aluminum tends to go first into tetrahedral position, then, and only later, into an octahedral position. Reduction of pH promotes the development of this octahedral position corresponding to the hexacoordination of aluminum.

2° *Syntheses by Pédro based on solutions from weathering*

The starting point of these experiments is found in the experimentation of HÉNIN and ROBICHET (1954) on the concentration of solutions of natural lessivage leading to montmorillonite and on the treatment of a silica gel by solutions of magnesium acetate, leading also to montmorillonite (HÉNIN, 1957).

Action on artificial allophanes. — Here PÉDRO used the residue from his experimental weatherings. We know that he obtained, from granite, basalt and andesite, alkaline suspensions containing amorphous material among which the electron microscope shows spherical silico-aluminous globules (PÉDRO, 1958). These products are comparable to allophanes. But even after three years these suspensions had not yet begun any evolution toward the crystallization of a mineral from these amorphous globules. By treating these products with magnesium acetate one obtains, as the case may be, products of the montmorillonite type or of the antigorite type (PÉDRO, 1960, 1961).

Carbonic lessivage of silicate rocks. — If experimental weathering of rocks is carried out in the presence of CO_2, one obtains direct neoformations in the products of lessivage. In fact, in those products CO_2 increases as well as the proportion of the leached cations, and *a montmorillonite appears* mixed with crystallized calcium carbonates. The conditions are close to those in nature since the pH of the suspension is about 8 to 9 (PÉDRO, 1960, 1961).

Experimental lateritization. — The experimental, intense lessivage of crystalline silicate rocks performed in the laboratory by PÉDRO (1958, 1961) has already been described. The solid residue shows that the weathering results in neoformation of iron and aluminum oxides and hydroxides. This corresponds to the complete lessivage of silica and of basic cations and to the complete release of alumina, which readily crystallizes into gibbsite.

The principal lesson of these experiments is to show not only that experimental neoformations are possible, but also to show how the dynamics of lessivage controls the neoformations that can be expected. According to the relative abundance of the elements present in the environment of weathering, neoformations proceed differently. At the extremes one obtains:

— *In an environment rich in cations and silica, a three-layer neoformed mineral of the montmorillonite type;*

— *In an environment of complete removal of silica and cations, the crystallization of gibbsite.*

This comes right back to the natural mechanisms.

3° *Studies of gels and syntheses by Fripiat, Gastuche and de Kimpe*

Belgian specialists have pursued studies on gels and on weathered products from the Congo for several years. The lesson that can be drawn from these studies will be exposed in successive stages, starting with the two summaries that provide the earlier bibliography (GASTUCHE, FRIPIAT and DE KIMPE, 1961; GASTUCHE and DE KIMPE, 1961).

The coordination of aluminum in alumino-silicic gels. — An aluminosilicic gel acts as ion exchanger. The negative electric charge of these gels depends on the concentration of alumina. In fact, alumina is capable of taking on tetracoordination in the tetrahedra, and this is the origin of the negative charge. But this replacement of Al for Si in the tetrahedra is limited because each Al tetrahedron must be connected with four Si tetrahedra. Therefore, the negative charge and the base exchange capacity of a gel reach a maximum for $\frac{Al_2O_3}{Al_2O_3 + SiO_2} = 0.3$. When this limit is exceeded, all the tetrahedral positions available for aluminum in the gel are occupied and sixfold coordination becomes possible. In fact, gibbsite, in which aluminum is hexacoordinate, appears in gels of suitable composition obtained from silicate and aluminate solutions whose pH is lowered to 8. It can be deduced that in a gel the increase of the concentration of alumina favors the hexacoordinate form of aluminum.

Influence of pH and electrolytes on the coordination of aluminum. — The negative charge of an alumino-silicic gel of constant composition increases with the pH of its environment. This amounts to saying that the number of tetracoordinate Al ions increases with pH, or that high pH values favor the tetracoordination of aluminum, low pH values its hexacoordination. These results have been obtained:

— By the study of variations of the base exchange capacity of gels in terms of their pH of formation (fig. 68).

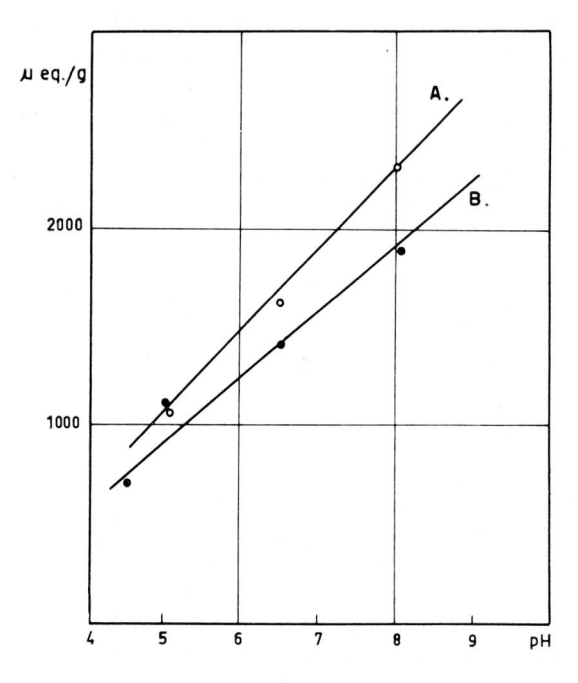

FIG. 68. — *Variation of the base exchange capacity (μeq/g) of silico-aluminous gels as a function of their pH of formation.*
$Al_2O_3/Al_2O_3+SiO_2 \# 0.29$
A: Products formed in an environment saturated with NaCl. B: Products formed in the absence of salts (after FRIPIAT, GASTUCHE and DE KIMPE, 1961).

— By the study, as a function of pH, of the proportion of hexacoordinate aluminum relative to the total amount of alumina in gels; this proportion varies with the variations in the Kα fluorescence angle of aluminum as well as with the displacement of the infrared absorption bands corresponding to the Si-O bond (FRIPIAT, GASTUCHE, DE KIMPE, 1961).

The presence of an electrolyte foreign to the reaction emphasizes the tetracoordination of aluminum, as is verified by the upper curve (A) of figure 68. In contrast, the complete removal of ions by dialysis permits the complete development of hexacoordination and the synthesis of gibbsite (HERBILLON and GASTUCHE, 1962).

Study of the ageing of gels. — The study of the evolution of gels in the course of time shows that ageing induces a decrease of their internal surface area and of their electric charge. This can be followed:

— by measurements of the base exchange capacity;

— by measurement of surface area.

GASTUCHE, FRIPIAT and DE KIMPE (1961) interpreted these results as the effects of the organization of gels. The electron microscope allows these changes to be detected; whereas fresh gels showed agglomerations of fine particles, more evolved gels organize into flakes. On the other hand, infrared spectrography (FRIPIAT, 1960) in the range between 8 and 11 μ shows, by the increase of the frequency of the Si-O vibration, that the rigidity of the structure increases. Upon ageing, gels are organized into two-dimensional structures with an increase in structural rigidity.

Appearance of clay minerals. — Within these ageing gels, the appearance of clay minerals has been detected. These recrystallizations are too scarce to be revealed by X-ray diffraction; therefore, the identification of crystals has been achieved by measurement of the b parameter on patterns of electron microdiffraction.

GASTUCHE and DE KIMPE (1961) noted the appearance of kaolinite when they studied gels prepared at pH 4.5 to 5 and then aged. This crystallization corresponds to the stabilization of the hexacoordinate structure of aluminum, whereas at higher pH values a mica appears. These authors also succeeded in stabilizing a hexacoordinate aluminous gel at an acid pH; and by adding silica very slowly they obtained at 43 °C tubular crystals of kaolinite, which were identified by electron microdiffraction.

All the observations were made under conditions in which the lattice of gibbsite can form (acid pH, desaturation in cations). GASTUCHE, FRIPIAT and DE KIMPE (1961) joined together with CAILLÈRE and HÉNIN in considering that lattices are built from a gibbsite or a "pregibbsite" in which aluminum is hexacoordinate. "In the absence of conditions favorable to the appearance of these nuclei, the three-dimensional polymerization of silica will take place and give rise to mixed alumino-silicic gels."

This appearance of clay minerals within gels can be compared with experiments of other Belgian investigators, GILLIS and DEKEYSER (1961). Gels of silica and alumina are enclosed in small bags made of collodion or cellophane and devoid of inorganic matter. The technique is that of the growth of whiskers. Silica and alumina pass through the wall of the bag. Amorphous products are obtained, but also crystalline products. In an experiment in which the Al/Si ratio in the gel is 4, the initial pH is 5.1, hexagonal crystals are obtained after a duration of 3 months in a humid, saturated atmosphere; these crystals can be observed under electron microscope and give X-ray patterns with peaks at 24-26 Å and 7.2 Å that disappear on heating to 650 °C. They must not be very far from kaolinite, which apparently formed from the solutions from the gel that sweated through the wall of the bag.

General results. — Although the studies of the Belgian authors do not lead, for the time being, to abundant neoformation of clay minerals at ordinary temperatures, they are very interesting because they provide us with information on the internal mechanisms of the evolution of silico-aluminous structures.

— *The increase in the relative concentration of alumina in a gel favors the hexacoordinate form of aluminum.*
— *Acid pH favors the hexacoordinate form of aluminum.*
— *Electrolytes partially inhibit this hexacoordinate form.*

— *The ageing of gels increases the rigidity of their structure and their two-dimensional organization.*

— *High concentrations of silica, high pH values, and abundance of electrolytes favor the neoformation of three-layer minerals.*

— *High concentrations of alumina, low pH values, and desaturation in electrolytes will favor the neoformation of kaolinite.*

— *The appearance in the gel of crystalline nuclei with the structure of gibbsite is necessary for the development of kaolinite.*

4° Synthesis of sepiolite and kaolinite by Wey and Siffert

Wey and Siffert approached the synthesis of clay minerals by way of dilute solutions, treating them as simple chemical reactions, and setting up the most favorable conditions for obtaining the desired clay minerals.

Synthesis of magnesian montmorillonite. — We saw that the solubility of silica strongly decreases in presence of Mg^{2+} ions, between pH 10 and 12.5. At pH 11.3, a solution saturated with monomeric silica and containing an equimolar quantity of Mg, introduced in the form of $MgCl_2$, yields a crystalline precipitate. The X-ray diffraction pattern presents the diffraction peaks of a three-layer trioctahedral clay mineral. Thus, *a mineral similar to stevensite is obtained by simple chemical reaction* from dilute solutions at ordinary temperature.

Synthesis of sepiolite (Siffert and Wey, 1962 ; Siffert, 1962). — Variable volumes of more concentrated solutions of $MgCl_2$ (0.1 M) are added to identical volumes of saturated solutions of $Si(OH)_4$. The pH is adjusted by means of 0.1 N NaOH, and the solution is left for three weeks. The pH decreases in the course of the reaction; only the solutions having a final pH higher than 7.8 yielded a precipitate. This precipitate contains a magnesian montmorillonite and also another crystalline product. The greatest abundance of this second mineral is obtained with a solution in which the SiO_2/MgO ratio is 0.7, the initial pH is 11.2 and the final pH 8.73. The X-ray diffraction pattern is that of a sepiolite, differential thermal analysis curves and the infrared absorption spectra are similar to those of sepiolite, the electron photomicrograph is that of a fibrous mineral, and the periodicity along the axis of this fiber is 5.3 Å, identical to that of natural sepiolite (fig. 69).

Thus the *synthesis of sepiolite is achieved at ordinary temperature* by means of an increase of the Mg^{2+} concentration in siliceous solutions.

Synthesis of kaolinite (Siffert and Wey, 1961 ; Siffert, 1962). — If solutions saturated with monomeric silica $Si(OH)_4$ and containing Al^{3+} are made alkaline, or if solutions containing $Al(OH)_4^-$ are made acid, amorphous gels only are precipitated (Wey and Siffert, 1961). This failure is due to the tendency of the Al^{3+} and $Al(OH)_4^-$ ions to go into tetracoordination instead of hexacoordination, which is necessary for the structure of kaolinite (Caillère, Hénin, 1961; Gastuche, Fripiat, de Kimpe, 1961;

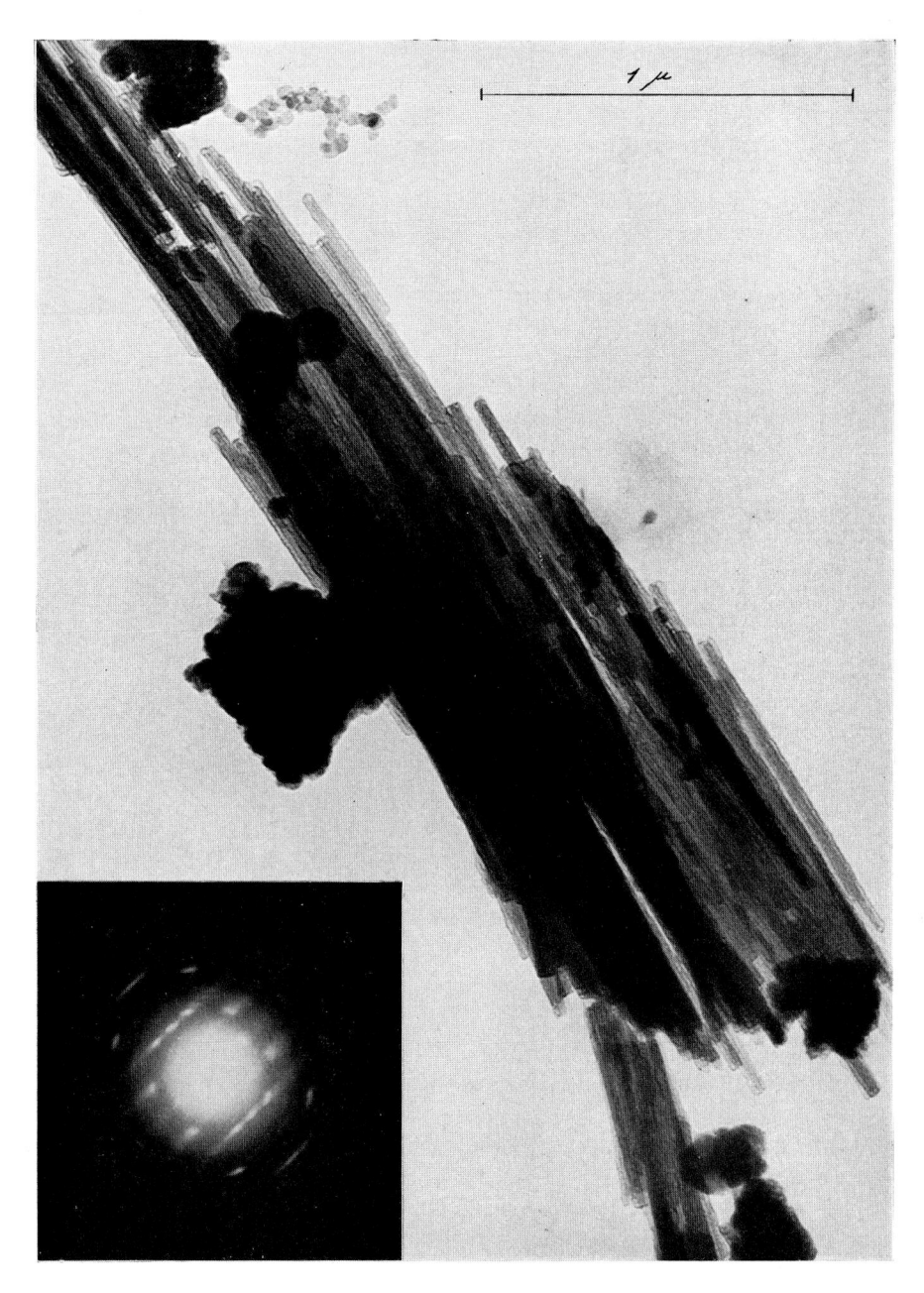

FIG. 69. — *Electron photomicrograph of the synthetic sepiolite of* SIFFERT *and* WEY *(1962), which can be compared with the natural sepiolite of the figure 3.* Corresponding electron microdiffraction pattern (photo by SIFFERT, 1962).

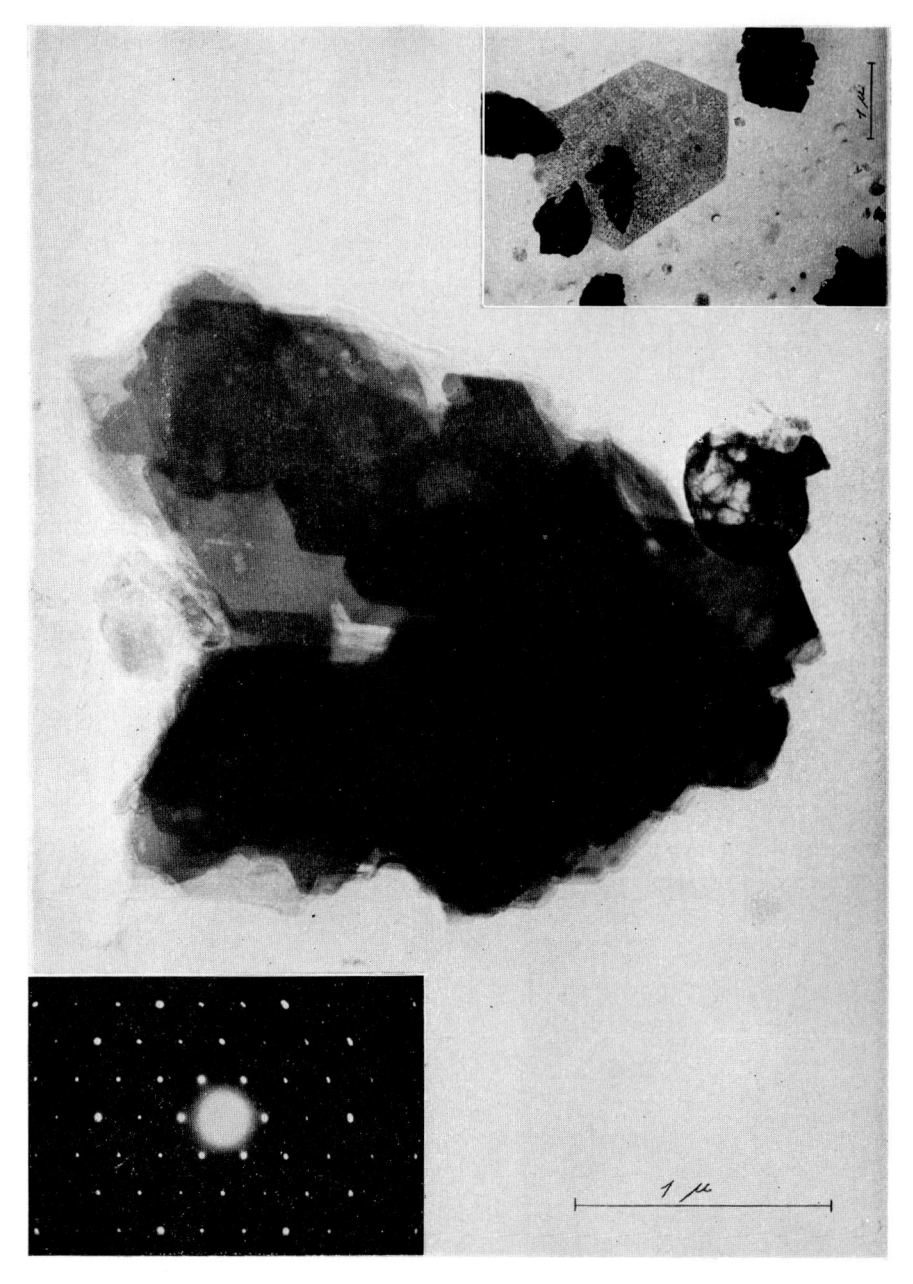

FIG. 70. — *Electron photomicrograph of the synthetic mineral similar to kaolinite. Corresponding electron microdiffraction pattern* (photo by SIFFERT, 1962).

WEY and SIFFERT, 1961). To resolve this problem, SIFFERT and WEY replaced the former ions by the complex alumino-oxalic ions, $[Al(C_2O_4)_3]^{3-}$, in which Al is hexacoordinate. The excess oxalate ions are eliminated in the form of insoluble calcium oxalate in order to lower pH to 6.75 and to retain just enough oxalate to permit slow precipitation of the hydroxide at this pH. After three weeks a precipitate is collected. The X-ray diffraction pattern reveals a crystalline product the peaks of which correspond to the strongest reflection of a two-layer clay mineral. There is no trace of hydroxide. The differential thermal analysis curve presents the exothermic peak at 950 °C, characteristic of a kaolinite lattice, and an endothermic peak at 400 °C, intermediate between that of gibbsite and that of kaolin. The chemical composition is marked by a deficit in silica, revealing an incomplete siliceous tetrahedral sheet. Finally, the electron microscope shows particles of regular outline, and the pattern of electron diffraction reveals a pseudo-hexagonal lattice (fig. 70).

The product synthesized from solutions of saturated silica and of aluminum complexed by the oxalic ion is a kaolinite with an incomplete tetrahedral sheet. The anions capable of forming hexacoordinate complexes with aluminum act, therefore, as mineralizers in the formation of kaolinite. It can be supposed that, in nature, an analogous role is played by some constituents of humus.

Lessons learned from these works. — The studies of WEY and SIFFERT are of great importance for three reasons, at least:

— *The synthesis of clay minerals is ensured directly by simple reaction from dilute solutions.*

— *Sepiolite and kaolinite, which could not be synthesized before, are now easily synthesized.*

— *The hexacoordination of aluminum is certainly the principal factor* conditioning the synthesis of kaolinite; here it is imposed by a complexing factor.

5° Syntheses at high temperature

I have no intention of presenting all that is known about syntheses of clay minerals at high temperature because my subject is the study of clay minerals of the hydrosphere and because these works are very numerous. However, as they are carried out under conditions so different from those of the hydrosphere, they are very interesting. Syntheses are obtained with ease and with the possibility to vary all the parameters in a continuous manner. The clay minerals obtained are well defined and can be studied very precisely by means of the most modern physico-chemical methods so that the crystallochemistry of these silicates can be understood.

The precursors are NOLL (1935, 1936) and NORTON (1939, 1940). NOLL worked with mixtures of silica, alumina and various cations. Kaolinite was obtained in neutral solution free of alkaline ions and in acid solution containing an alkaline ion below 400 °C. Montmorillonite formed in the presence of a moderate quantity of alkaline and alkaline-earth ions in alkaline solution, but more easily in presence of Mg. Sericite formed in alkaline solution in which the concentration of potassium was higher than in the case of montmorillonite and almost equal to the quantity required by the formula of muscovite. These results were obtained no matter what the proportion of alumina or silica in the environment was. This shows that at high temperature (300 to 400 °) the cations direct the phenomena.

NORTON started with minerals such as albite, anorthite, orthoclase, leucite, petalite (Li_2O, Al_2O_3, $3SiO_2$) and spodumene (Li_2O, Al_2O_3, $4SiO_2$). He varied pressure, temperature and carbonic gas. Among many results, the following can be emphasized: albite and anorthite give rise to montmorillonite; orthoclase, nepheline and leucite give rise to sericite; petalite and spodumene yield kaolinite. NORTON concluded that in the conditions under which he worked, the product obtained was determined by the cations present and the structure of the original mineral.

Much work has been done since then. ROY, the best specialist at present, recently made a general presentation (ROY, 1961). I shall give the features of it that are especially interesting for our subject.

Thanks to a new apparatus, ROY and his collaborators can work from 0 to 5 000 Atm. and from 0 to 1 000 °C. This permits systematic studies of the properties of synthetic clays.

Kaolinite is easily synthesized, but dickite and nacrite, appearing occasionally, evolve to kaolinite. Halloysite has never been correctly obtained. One cannot introduce more than 1 or 2 % of Fe_2O_3, Cr_2O_3, Ga_2O_3 or GeO_2 into the lattice.

In contrast, in the trioctahedral series of serpentine, the replacement of Mg by Fe or Ni and of Si by Ge, as well as the balanced replacements of Mg and Si by Al, are easily obtained.

It has been experimentally demonstrated that polymorphic relations exist between the chlorite series and the serpentine series; the compositions required for the synthesis of both series are the same, but understanding the way in which polymorphs can be obtained still requires much work.

Talc and pyrophyllite are easily synthesized, but the isomorphic replacements in these two minerals are infinitesimal.

Micas have been the subject of numerous experiments in which numerous substitutions have been made for the tetra-, hexa-, dodecacoordinate cations. Thus, ferro-ferric micas have been obtained; their composition and stability depend on water pressure and the partial pressure of oxygen.

Minerals with variable basal spacing have been prepared under good conditions and constitute three families of clay minerals, which are, moreover, better crystallized and larger than the natural surficial minerals. Dioctahedral beidellites, dioctahedral montmorillonites and trioctahedral saponites have been obtained. The study of substitutions and of the charge on the layers can be advanced with a palette of well defined products.

Mixed layers can be produced by the modification of minerals under hydrothermal conditions, but not by the coprecipitation of two different minerals. The most important of the minerals with interstratifications, illite, could be prepared by controlling the composition and especially by a deficit of potassium designed to balance the charge of the layers.

The great importance of these studies which have been made by many investigators, the principal ones being listed by ROY (1961), is of a physico-chemical and mineralogical order. But, as noted by MERING after the communication of ROY, one is impressed by the fact that synthetic neoformations at high temperature furnish an exact copy of natural minerals. This is very important for our subject. In fact, nature shows:

— prohibition of the coexistence of di- and trioctahedral types within one and the same mineral;
— the restricted range of variation of the charge of a layer within each group of minerals.

It was formerly possible to suppose that synthesis would permit the filling in of the missing members and the completing of the series of natural products. It is not so. This means that the methods of preparation in the laboratory and the modes of natural neoformation both obey strict rules imposed by thermodynamics. And this also means that the composition of natural minerals really corresponds to the greatest lattice stability.

III. — NEOFORMATIONS AND SYNTHESES BY SUBTRACTION AND ADDITION: CONCLUSIONS

There we have a general view of neoformation of clay minerals in nature and of their synthesis in the laboratory. The least that one can say is that there is a great convergence of interpretations. This is not a coincidence, but simply the fact that each of the attempts was nourished by the other. Naturalists, who follow the experimental studies, take up the results obtained in order to compare them with the information and coincidences of the natural environment. Experimentalists try ceaselessly to place their experimentation in the neighborhood of natural conditions; by varying the factors they give evidence of limits and of mechanisms.

Using here the fruit of both methods, I would like to draw some conclusions about the problem of neosynthesis of clay minerals such as it can be understood today.

1° THE SOLUBILITY OF THE SILICA OF CLAY MINERALS HAS BEEN MEASURED (WEY and SIFFERT, 1961). It is by far inferior to the solubility of amorphous silica in water and does not exceed 10 ppm or mg/l (fig. 71). *From the thermodynamic point of view, a solution containing more than 10 ppm of silica can nourish, with suitable ions, neoformations of clays.*

2° Clay minerals form by *fixation of free silica of the solution onto a layered hydroxide that is in process of development*. Suitable conditions, therefore, are necessary so that the cation or cations can precipitate *with hexacoordination in the form of layered hydroxides*.

3° *Certain cations that enter commonly into the structure of natural clay minerals precipitate easily in the form of layered hydroxides.* Thus magnesium gives rise to brucite, which is a layered hydroxide. Iron can precipitate in the form of several oxides, but ferrous iron enters a brucite structure, whereas ferric iron seems to give rise most readily to goethite, which, although its structure is different, is also layered.

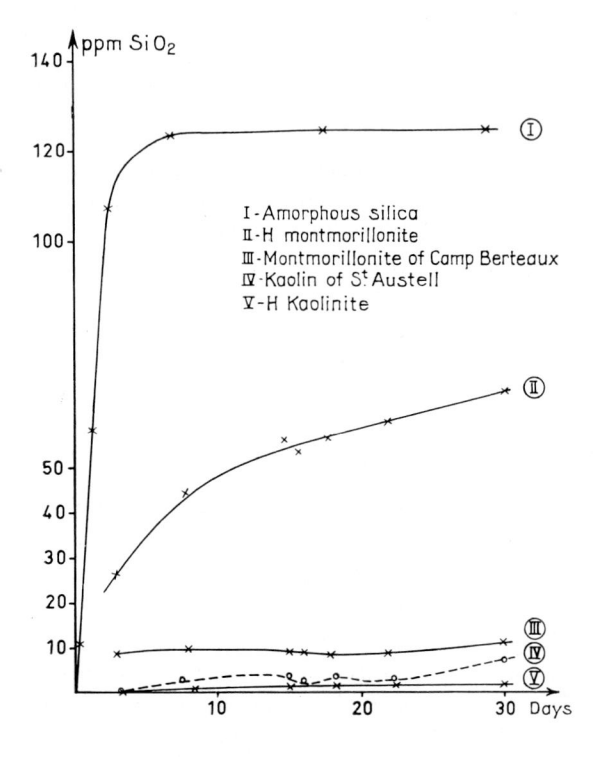

FIG. 71. — *Solubility of the silica of various clays in comparison with the curve of solubility of amorphous silica* (after WEY and SIFFERT, 1961).

I - Amorphous silica
II - H montmorillonite
III - Montmorillonite of Camp Berteaux
IV - Kaolin of St Austell
V - H Kaolinite

4° On the other hand, aluminum is characterized by a strong tendency to precipitate with tetracoordination, whereas the layered structure of gibbsite *requires hexacoordination*. Studies on syntheses show that the hexacoordination can be imposed by complexing factors. They also show that an *acid pH, desaturation in cations, and relative abundance of alumina favor this coordination.*

5° GENESIS OF KAOLINITE. — Neoformation requires only the hexacoordination. In an environment of lessivage, pH is acid, cations are removed, and alumina organizes into gibbsite, if silica is lacking; this is one of the important phenomena of lateritization. But if silica is present in solution, it fixes itself on the growing layers of gibbsite and gives rise to kaolinite. If organic matter intervenes, it can only favor this indispensable hexacoordination.

6° GENESIS OF MAGNESIAN MINERALS. — In a magnesian environment, the pH is alkaline. *Magnesia precipitates in the form of brucite layers on which silica fixes itself during their growth.* If the pH is still close to neutrality, and if silica is not very abundant, it is antigorite, a 7 Å mineral with one siliceous layer, that appears. If the pH increases and if silica is more abundant, a 14 Å mineral with two siliceous layers occurs: stevensite.

7° GENESIS OF MIXED ALUMINO-MAGNESIAN AND FERRIFEROUS MINERALS. — In a confined environment, the pH is alkaline and cations, including aluminum, magnesium and iron, are present. Magnesium precipitates and organizes into brucite, whereas aluminum comes into play by being shared between the tetrahedral and octahedral positions. If the pH is still close to neutrality, and if silica is not very abundant, 7 Å chamosites appear. If the pH and the concentration of silica increase, varieties of montmorillonites appear.

8° GENESIS OF ILLITE, SERICITE AND CHLORITE. — Direct experimental neoformation of illite, sericite and chlorite has not been obtained in the laboratory at low temperature. The conditions of formation deduced from observations in nature are the same as the preceding and depend on the quantity of cations present.

9° GENESIS OF CLAY MINERALS BY SUBTRACTION. — Confined genetic environments and those of lessivage have been defined. It has been seen that in the environments of lessivage clay minerals form by subtraction, whereas in confined environments they form by addition. Let us reexamine the first point.

A silicate rock, subject to lessivage, is hydrolysed; it loses a part of its substance. Alumina being the least soluble subsists preferentially, then silica, and finally the most soluble cations. So that as weathering progresses, a crystal becomes impoverished in cations more quickly than in silica, and in silica more quickly than in alumina. Upon being hydrolysed, a crystal loses its structure and its composition; it experiences an environment that is changing with the duration or the intensity of lessivage (temperature, discharge, etc.). In this less and less crystalline environment, the SiO_2/Al_2O_3 ratio decreases. *From the beginning, the proportions will be favorable for the genesis of sericite;* then, *after removal of potassium* and of the other cations, *favorable for kaolinite,* then, *after removal of silica, favorable for gibbsite.*

A glassy rock or a bed of ash is always rich in silicate, but the silicate is a glass. The silica and alumina tetrahedra are arranged with a rather low degree of order, so that the triply periodic structure is not realized (SAUCIER, 1960); cations occupy the holes of this initial structure. By lessivage cations are going to be released, and then silica, alumina being as always the most stable element. The mechanism is similar to the preceding one, but dissolution in the mass of the glass will lead it toward a gel progressively impoverished in alkaline ions, then in silica. In the case of mild and appropriate lessivage, montmorillonite is obtained, and here is the origin of bentonites. In the case of more accentuated lessivage, kaolinite or bauxitic horizons are the result. These latter neoformations occur at the moment when the desaturation in cations and the decrease in concentration of silica "release" aluminum with hexacoordination.

In a gel, which functions as an ion exchanger, or in a natural allophane resulting from lessivage in soils, one will encounter successively the compositions of montmorillonite, kaolinite and gibbsite. This is exactly what is observed in nature in the case of stronger and stronger lessivage. It is also the meaning of the pertinent comment of SABATIER (1961) at the C.N.R.S. Colloquium as a result of his studies on experimental weathering of feldspars at 200 °C (LAGACHE, WYART and SABATIER, 1961). It was noted in these experiments that alumina was almost insoluble. The variation of the Si/Na,K ratio as a function of 1/V was studied, V being the velocity of drainage of the feldspar. Figure 72 shows that this ratio is equal to 3 in the case of very active drainage, this value being precisely that of the Si/Na,K ratio in a feldspar; alumina alone remains in place. Then the ratio decreases toward 1 in the case of weaker drainage; at this moment silica and alumina are concentrated in the insoluble phase. Thus, in the latter, owing to weaker and weaker lessivage, one encounters successively the SiO_2/Al_2O_3 ratios of hydroxides, kaolinite, illite and montmorillonite.

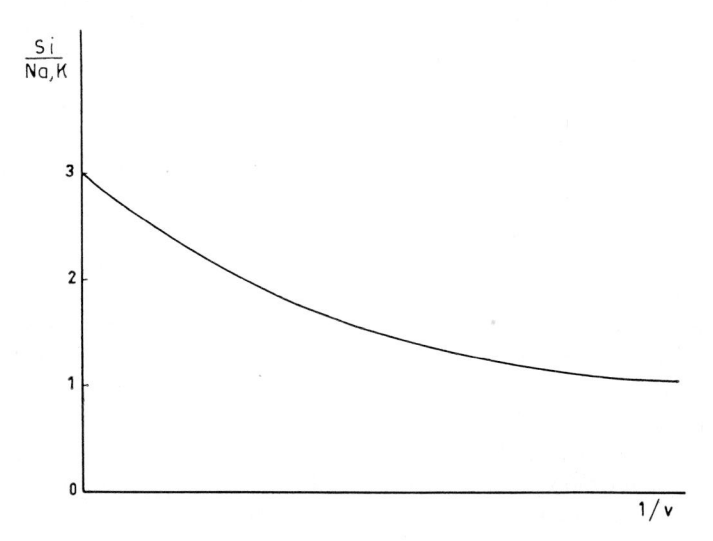

FIG. 72. — *Dissolution of alkaline feldspars in water at 200 °C. Variation in the Si/Na, K ratio as a function of 1/V, V being the velocity of drainage (after SABATIER, 1961).*

All this shows us that in neoformations by subtraction in an environment of lessivage, alumina must be considered as being an element that is practically stable. It is released by the "*removal of its companions*", and it is capable of organizing *in situ* with silica and the available cations according to their abundance.

Furthermore, this corresponds to observations in nature: bauxitization, kaolinization (acid environment) and sericitization (alkaline environment) occur in place either during weathering or during post-depositional diagenesis. *Alumina does not emigrate, and minerals rich in alumina form on the residues of the original silicates.* Chloritization is analogous, either by subtraction of potassium from biotite, or by hydrolysis of ferromagnesian minerals. The montmorillonite of bentonites formed from glass is obtained by the same mechanism. Here we find the reason for the fact that gibbsite, kaolinite, sericite and chlorite are neoformed during weathering and diagenesis and not in sediments (the chlorite of sedimentary transformation is another question).

10° GENESIS OF CLAY MINERALS BY ADDITION. — The mechanisms here are exactly complementary. Solutions emigrate and are poor in alumina. *But they contain various cations and silica. The cations and silica are concentrated in the environment where they arrive, which can be a hydromorphic soil or a sedimentary basin of chemical character.* These environments can only be alkaline, and neoformations take place under the influence of the precipitation of magnesium, iron, and accessorily aluminum, which is a poor traveller but is often present. It is obvious that leaching by salt brines, in the course of diagenesis, amounts to the same thing. And, in all these cases, various montmorillonites, chamosites, attapulgites and sepiolites form. Likewise, degraded minerals are reconstructed by aggradation. Moreover, we find here again diagenetic sericitization, but from kaolinites or illites, as well as chloritization, but from illite or sericite.

Here we see the reason why sedimentary environments cannot ensure the neosynthesis either of kaolinite or of illite. These minerals are too aluminous, and the necessary alumina is found neither in solutions nor *in situ*.

11° CLASSIFICATION OF NEOFORMATIONS. — Seen from this point of view, the experimental syntheses can be classified in two groups:

The syntheses of FRIPIAT, GASTUCHE and DE KIMPE are attempts to reproduce **neoformations by subtraction**. Gels are manipulated until they reproduce the neoformations occurring in natural environments of lessivage, such as the zones of weathering of the Congo that these authors know very well. It is obvious that the trials of experimental bauxitization of PÉDRO form a part of this group.

In contrast, the syntheses of CAILLÈRE, HÉNIN, ROBICHET and ESQUEVIN are **neoformations by addition**, achieved from solutions. That which they borrow from the glass sphere is only silica, and they reproduce natural neoformations in confined environments, such as the alkaline and ferriferous sedimentation they know very well. Likewise, PÉDRO, in his syntheses from residues of lessivage, imitates neoformations by addition in an environment that becomes concentrated. A point worthy of notice is the use of colloidal products resulting from weathering as in natural phenomena. Lastly, WEY and SIFFERT perform neoformations by addition in the pure state since, as good chemists, they use only products of the laboratory. But in the synthesis of kaolinite, they add a supplementary trick, the intervention of an organic product that imposes the hexacoordination of aluminum. Now, we know that organic matter is prevalent in lateritic profiles because it is this organic matter that ensures the mobility of iron. This research route is thus very important for the future.

I hope that *this distinction between neoformation by subtraction and neoformation by addition* will be useful.

To impoverish a crystal, a volcanic glass, or a gel until the disordered residue within it comes to the conditions and the proportions favorable to the growth of crystals is not the same thing as to send into a tropical lowland or into a basin of chemical sedimentation solutions that will become concentrated and organized.

In both cases there is crystallization of a clay mineral that is the stable form of silicates in the hydrosphere. *But in the first case, the phenomenon takes place in situ and the little soluble alumina is in command. In the second case, the phenomenon takes place far from parent rocks, and it is magnesium and iron that are imposed on the silica.* Alumina is but accessory.

Such is the teaching of Nature.

CHAPTER XII

SUPERFICIAL GEOCHEMISTRY AND THE SILICATE CYCLE

By a slow process clays have been described in their deposits and then in the mechanisms of their formation. It is instructive at the end of this process to consider the whole problem in geochemical terms. We shall do it in the four following sections:

— The geochemistry of the evolution of clays.

— The geochemistry of the constitutive elements of clays.

— The geochemistry of landscapes and sedimentary environments.

— The geochemical cycle of silicates.

I. — GEOCHEMISTRY OF CLAYS

1° *Places and mechanisms of the evolution of clays*

The three natural environments. — The complete lesson drawn from the preceding chapters shows that clays are born, evolve, and die within three principal stages of the geochemical cycle, and these are the three surficial stages of that cycle. These three stages are:

— The zone of weathering.

— The zone of sedimentation.

— The zone of diagenesis.

Problems of limits are raised; we must avoid being entangled in them. The limit between weathering and sedimentation is not clear because there exist phenomena of weathering within sedimentary basins; these are the phenomena of halmyrolysis, which take on their full importance when volcanic material falls directly into a lake or the sea. This is an interesting but a special case. It must be resolutely admitted that, on the whole, weathering is typical of the zone of lessivage of the continental crust, whereas sedimentation is characteristic of depressed zones of the crust, which are in some cases continental (lakes) but mostly marine.

Likewise, the limit between sedimentation and diagenesis is extremely vague. Equilibria that can develop rapidly are achieved during sedimentation itself, but others develop later. Next, the diagenetic environment brings its own modifications of temperature, pressure, evacuation of water, leaching by various solutions, and assumes its own particular character. But, on the whole, the later events occurring within sediments belong to diagenesis.

The four great mechanisms. — In each of the preceding zones clays have an origin that one can, as the case may be, attribute upon preliminary examination to three principal mechanisms which have been examined and which are the following:

detrital heritage	*detritisches Erbteil*	*héritage mécanique*
transformation	*Umbildungen*	*transformations*
neoformation	*Neubildungen*	*néoformations*

The problems of limits arise once more; it is necessary to solve them once and for all. The limit between heritage and transformations is vague. As a matter of fact, transformed particles are inherited before they are subjected to transformation, and the least transformed ones will be above all inherited, etc. Likewise, the limit between transformations and neoformations is difficult because it can be said that transformations also are reorganizations and growth of crystals with the collaboration of solutions.

Thus, one is caught once more in the heritage-neoformation or detrital-authigenic discussion. This discussion has already consumed much energy without arriving at clarity. This is the reason that it is necessary to use a presentation of mechanisms in three terms. It is of interest to reserve an autonomous place for the median term. Transformed clays are inherited or detrital through their structure and modified by the environment. All that is not inherited in these clays is modified by crystalline rearrangements, and growth takes place according to the same rules as for neoformations from solutions. Common sense indicates that if a clay is partially inherited, it is partially neoformed, and it is this bivalent group that is honored by the term "transformation". Let us note that it comes from the Latin translation of the Greek "metamorphism", and we find that pleasing. But then, we are going to encounter positive and negative transformations just as there is positive metamorphism and retrograde metamorphism, or retromorphism. We speak of transformations by aggradation and of transformations by degradation. If the median mechanism is divided in two, **there will be four possible origins for clays.** The important point consists in understanding the ensembles and in defining one's language; mine is as follows.

1. *Inherited clays,* coming detritally from a preceding environment.

2. *Clays transformed by degradation* of inherited particles.

3. *Clays transformed by aggradation* of degraded particles.

4. *Neoformed clays,* which are built entirely from solutions.

Places and mechanisms: 12 possibilities. — Four mechanisms taking place in three natural environments gives twelve possibilities for the evolution of clay minerals. The essentials are readily presented by means of table XV.

TABLE XV. — EVOLUTION OF CLAY MINERALS IN THE THREE NATURAL ENVIRONMENTS BY MEANS OF THE FOUR GREAT GEOCHEMICAL MECHANISMS
(after ESQUEVIN, 1958)

	Heritage	Transformation		Neoformation
		Degradation	Aggradation	
Weathering and soils	Micas from crystalline and metamorphic rocks Clay minerals from shales and sedimentary rocks	Weathering of clay minerals of rocks sericite chlorite illite chlorite vermiculite mixed layers montmorillonite	Aggradations in horizons of accumulation disordered kaolinite and halloysite → kaolinite	Neoformations during weathering sericite chlorite kaolinite gibbsite montmorillonite Neoformation in the horizons of accumulation of calcimorphic and hydromorphic soils
Sedimentation	All the clay minerals from : erosion of rocks weathering soils Amorphous material Solutions	Halmyrolysis = subaqueous alteration (volcanic material)	Sedimentary aggradations mixed layers illite corrensite chlorite glauconie	Sedimentary neoformations attapulgite sepiolite montmorillonite talc chamosite glauconie
Diagenesis	All the clay minerals from sedimentation	Diagenetic degradations kaolinite mixed layers sericite chlorite	Diagenetic aggradations sericite chlorite talc	Diagenetic neoformations kaolinite halloysite pyrophyllite sericite chlorite attapulgite sepiolite

2° The two major pathways of surficial geochemical evolution of silicates

Clay minerals are more stable than the other common silicates in the hydrosphere. Thus, they will form within the hydrosphere. As long as the conditions of the hydrosphere remain constant (pressure, temperature, concentration of solutions), clay minerals will remain stable. These conditions of stability are those of their genesis. On the other hand, if the conditions of the hydrosphere change (concentration of solutions, renewal of solutions, temperature, pressure), or if minerals are transported into a different environment, which amounts to the same thing, stability ceases and evolution is possible.

Variations of pressure will intervene only during diagenesis.

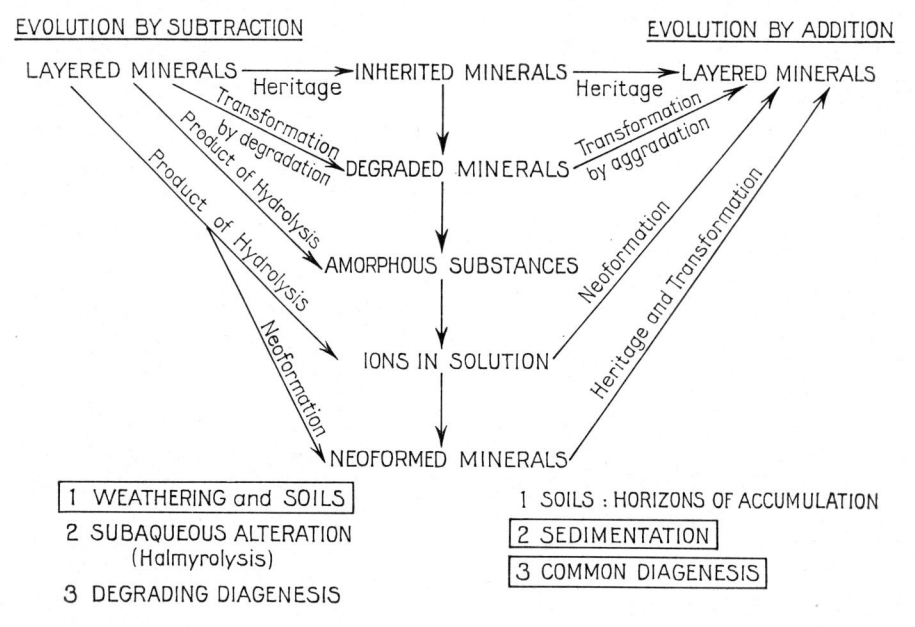

```
EVOLUTION BY SUBTRACTION                               EVOLUTION BY ADDITION
LAYERED MINERALS ────Heritage───> INHERITED MINERALS ───Heritage───> LAYERED MINERALS
         Transformation by degradation ─> DEGRADED MINERALS <─ Transformation by aggradation
         Product of Hydrolysis ↘         ↓                    ↙ Neoformation
                                 AMORPHOUS SUBSTANCES
                                         ↓                       Heritage and Transformation
                                   IONS IN SOLUTION
                  Neoformation ↘         ↓                    ↙
                                 NEOFORMED MINERALS
```

1 WEATHERING and SOILS		1 SOILS : HORIZONS OF ACCUMULATION
2 SUBAQUEOUS ALTERATION (Halmyrolysis)		2 SEDIMENTATION
3 DEGRADING DIAGENESIS		3 COMMON DIAGENESIS

FIG. 73. — *Evolution of silicates by subtraction and by addition in the hydrosphere.*

Variations of temperature will intervene in two ways, during diagenesis and by change of latitude; it must never be forgotten that the rate of reactions, and especially of hydrolysis, is approximately doubled for a difference of 10 °C in temperature, and thus quadrupled for 20 °C. This means that there is as much difference in the intensity of weathering between 10 and 20 °C as between 20 and 25.85 °C, and as much difference between 20 and 25.85 °C as between 25.85 and 30 °C.

Variations in the concentration and the renewal of solutions are still more important since they do act not only on the intensity of mechanisms but also on their direction. Concerning this point, there are two main possibilities for the hydrodynamic regime of solutions in nature. To counterbalance supplies a water body, either subaerial (surface of a lake or the sea) or subterranean (ground water and intraformational water), has two mechanisms at its disposal by which to ensure hydrodynamic equilibrium:

— loss of liquid by percolation or by downstream outlets;
— loss by evaporation.

In other words, the water of a water body can escape "toward the earth" or "toward the sky". In the first case, the environment is renewed, the escaping waters carry away material in solution; there is lessivage. *This is the environment of lessivage.* In the second case, the environment loses only water vapor and concentrates dissolved material. *This is the confined environment.*

All geochemical evolution of silicates in the hydrosphere is governed in its direction and its intensity by the activity of the environment of lessivage and of the confined environment. A graphic representation of this activity is given by the outline in figure 73.

3° *Geochemical evolution by subtraction in an environment of lessivage*

The environment of lessivage is characterized by the percolation of waters by gravity. Waters escape either vertically through the fissures, joints or pores of the underlying rocks, or laterally by the flow of ground water or by surficial flow due to the functioning of overflows. This geochemical environment is generally inhabited by rain water. *It is open, always subject to dissolution and hydrolysis; it is an environment that constantly is losing substance.*

Silicates in this environment of lessivage evolve by subtraction.

Non-layered silicates are hydrolysed and lose substance under the influence of the renewal of waters. Feldspars are impoverished in cations and silica. The conditions are rapidly assembled for the neoformation of sericite. Ferromagnesian minerals are hydrolysed and the conditions become appropriate for the neoformation of chlorite. Volcanic glass is hydrolysed and, with or without an obvious colloidal phase, the conditions for the neoformation of montmorillonite are realized. All these neoformations of clay minerals are made from the substance of silicates subjected to subtraction. If the flow of solutions is accelerated, the lessivage of cations is speeded up, the neoformed clay minerals are hydrolysed in their turn, and silica and alumina alone remain and organize into kaolinite. If lessivage is intense right from the beginning, the preceding stages are by-passed, and kaolinization prevails directly, with the release of iron. If drainage is still more intense, and if no water level subsists to maintain silica in solution at the disposal of alumina, it is gibbsite that forms. We pass from lateritization to bauxitization by subtraction.

Layered silicates and clays are not insensitive to this environment of lessivage. Although more stable in the hydrosphere than other silicates, they will not resist intense lessivage. Subtractions that were favorable to them when they were discreet will be fatal if they are intensified. Sericite, illite and chlorite will lose their interlayer ions and become "open minerals". Then some ions of the octahedral sheets migrate between the layers in an attempt to equilibrate the lattice, but they are removed by lessivage in their turn. Aluminum from the tetrahedral sheets goes into the octahedral sheets. Biotite, muscovite, sericite, and illite tend toward interstratifications in which layers of the vermiculite type and of the montmorillonite type multiply. By subtraction one comes to montmorillonite, which is merely the residue of layers from which the cations of replacement have been removed, leading to the low charge and to the qualities of expansion. But a great part of the structure has been broken down and has given rise to allophane, or amorphous gels, which dissolve slowly. If desaturation of cations increases, the environment becomes acid during the same time that alumina increases by the continual removal of silica. Conditions become appropriate for the appearance of kaolinite through the release of hexacoordinate aluminum which fixes at once the silica tetrahedra of the surrounding solutions. If these tetrahedra are lacking, bauxitization occurs.

Here then is the outline of the evolution by subtraction of silicate minerals in an environment of lessivage. It involves many different mechanisms and multiple stages corresponding to successive domains of stability. In brief, *in this environment we note progressive transformations by degradation as the layered minerals themselves are subjected to a more and more degrading lessivage.* Throughout the phenomenon, structures are degraded and structures are built, but degradation ceaselessly has the advantage over construction. Amorphous and inorganized substances appear, ephemeral in most cases, but durable in certain cases, and the products of subtraction are continually removed in solution through pores, fissures, joints, by the flow of water bodies, by natural overflow, by rain wash, springs, etc.

This evolution by subtraction can stop at any time if an equilibrium is established, and this gives rise to the variety of weathered products and soils.

At any moment, this evolution can be definitively stopped by erosion; this is the termination of weathering and the dawn of sedimentation. The materials exported from the environment of lessivage to ulterior environments are the following: inert and inherited minerals; degraded minerals; amorphous substances; solutions; neoformed minerals.

Such is the material that will be delivered to the confined environments found in soils, in sedimentary basins, or in the zone of diagenesis.

Before we come to the confined environments, it is essential to see clearly that environments of lessivage are widely developed in the zone of weathering; we shall find them again in diagenesis and in sedimentation. Certain diagenetic evolutions are similar in all ways to the preceding ones: kaolinization, sericitization of feldspars, chloritization of biotite, interstratification of the various clay minerals. It is a question of underground circulation of acid solutions that are capable of degrading the sediments. The behavior of silicates or glass of volcanic origin in the presence of water of lakes or seas is rather similar. The environment is degrading for these minerals placed into conditions of instability. It is capable of removing their most soluble ions by lessivage. All this verifies the fact that the mechanisms are independent of place but dependent on environmental conditions.

4° Geochemical evolution by addition in confined environments

As soon as evaporation exceeds the vertical or lateral loss of waters, the environment becomes confined in order to ensure the equilibrium of water bodies. This concerns soils as well as sedimentary basins. There is quite a group of soils in which water accumulates through poor drainage, while evaporation in the dry season maintains the level of the water table and its seasonal pulsations. This is also the case of lakes, lagoons, and seas in which the concentration of ions in solution increases through evaporation which counterbalances the water supply. Certain solutions in the domain of diagenesis, and even the majority of them, will play a similar role by their concentration, although they have another origin.

Here we have, therefore, confined environments in which the ions of the hydrosphere tend to accumulate and to concentrate in various proportions. Nature offers us a whole range of possibilities, calcic or hydromorphic soils, lacustrine or marine, hyposaline or hypersaline environments; and each of these environments can vary in its concentration of chlorides, sulfates, carbonates, iron oxides, phosphates, etc.

Let us suppose, for the purpose of this exposé, that the material transported into the sedimentary basin is complete, i.e., that it contains all the products of the list given above, from inherited particles to solutions.

Heritage. — Silicates, layered or not, that have resisted continental lessivage in a desaturated environment have no reason to evolve in an environment richer in cations. They remain inert and are buried in a state of continuing stability.

To these inherited minerals which never evolved are added minerals that were neoformed in environment of lessivage, foremost among which are the kaolinites of weathering, sericites, illites and chlorites. If these minerals were able to form within an environment subject to the subtractions of lessivage, they will find themselves secure in a more saturated environment. Given the frequency of these products resulting from weathering, one can understand the importance of heritage in alkaline soils and, above all, in sedimentation in general. The sedimentary environment generally constitutes a "non-agressive" milieu for well formed clay minerals born in the zone of weathering.

Transformations by aggradation. — The degraded minerals find themselves thrown into conditions exactly opposite to those of their genesis. The environment of lessivage had started their alteration and had advanced it to a greater or lesser extent. If the environment into which they arrive is little saturated, their degradation will be only interrupted. But the more the saturation of environment increases, the more intense and the more rapid the transformations by aggradation will be; first, damage to the lattice will be repaired, then modification and growth of lattice will take place. The work can begin immediately upon arrival in the marine environment, but it can be prolonged during sedimentation, then during early or late stages of diagenesis.

Potassium and magnesium ions slip into interlayer positions. From these positions magnesium tends to migrate into octahedral position, and this pushes aluminum into tetrahedral position, where it is admitted. "Open" minerals have the tendency to be charged with magnesium and iron. At the extreme, the same dioctahedral layers supplied by weathering tend to reconstruct trioctahedral minerals. On the whole, open minerals that, by means of interstratification, were previously oriented toward montmorillonite, a mineral lacking in cations, now evolve in the other direction; by new interstratifications they multiply the more regular layers, and this leads to well crystallized illite and chlorite. If the environment is potassic, illite is slowly reconstructed, but if the environment is ferriferous, that ferriferous illite called glauconite appears. If the environment is magnesian, there is evolution toward corrensite, then toward chlorite. Montmorillonite itself is not insensitive to this general tendency; neoformed in the zone of weathering, it tends to disappear very slowly but inevitably all the way through diagenesis, i.e., up to the threshold of metamorphism, where it no longer exists.

It is certainly understood that these transformations are achieved element by element, by addition of the ions of solutions to the more or less degraded lattices coming from environments of lessivage.

Neoformations. — During this time amorphous substances and solutions arrive in the horizons of accumulation in soils or in sedimentary basins. If conditions are appropriate, and studies on syntheses have shown us their essential features, neoformations begin. Clay minerals are built from scratch from solutions that may or may not be supplied with silico-aluminous debris such as amorphous gels.

Thus, neoformed montmorillonites are formed in calcic and hydromorphic soils, and, likewise, attapulgites, sepiolites, neoformed montmorillonites, neoformed glauconites, ferriferous chamosites and possibly talc form in sediments.

Here the phenomenon of addition is obvious. But it is no longer an addition of the available ions to an imperfect but already present structure. It is a question of formation of crystals on a nucleus and their growth by addition of ions in the environment to the developing structure. One sees very well to what extent there is analogy and continuity between these two processes. At the extreme, the difference concerns only the size of the nucleus; in both cases there is growth of crystals in a confined environment in which the constituent ions of silicates are added up.

The ensemble. — Here are the outlines of the geochemical evolution of silicates in the hydrosphere by addition in confined environments. This way of looking at things permits an understanding of the variety of natural phenomena:

— If the detrital supply in the sedimentary basin is composed of well crystallized clay minerals, the latter subsist without change, and evolution is minimal. The general aspect of this sedimentation is dull. It accumulates an inherited and inert material. This is commonly the case in lakes, estuaries or seas of normal chemical composition that receive well crystallized minerals. Those who have studied these cases tell us that transformations and neoformations do not exist.

— If, on the other hand, continental evolution has rather well degraded the clay minerals available to it, considerable work remains to be done downstream to reconstruct these minerals. If this work is spread out across long distances and difficult to observe, it occurs nevertheless because we can observe the end products.

— Lastly, if chemical weathering is very intense, it is mainly solutions that reach the hydromorphic lowlands, continental lakes, or epicontinental seas. In this case neoformations take place, and it is likely that, from a geochemical point of view, many resemblances will appear between the montmorillonites of soils in tropical lowlands and the sedimentary and neoformed montmorillonites of the same regions. They are the end products of the same lessivage and the echo of powerful weathering.

— These variations in geochemical evolution by addition are numerous and give rise to all the possible intermediates or mixtures; these are the variations in soils and sediments.

— This evolution continues during diagenesis. Solutions circulating through the sedimentary column are also confined in most cases, if one makes exception of the degrading solutions examined above. It is not evaporation that confines them, but imprisonment and the rise of temperature which increases solubility. Their effect is similar to that which has just been described, but thus exaggerated. Illitization, followed shortly by sericitization, becomes general, and chloritization as well.

— This progressive course toward metamorphism can be interrupted at any moment by tectonic uplift or by emergence. One has then a shortened cycle that delivers once again, prematurely, the sedimentary material to the zone of weathering.

II. — GEOCHEMISTRY OF THE CONSTITUENT ELEMENTS OF CLAYS

The major elements contained in clays are Si, Al, Fe, Mg, Na and K. Among the major elements of silicate rocks only calcium is not included in this list. But its influence on environmental conditions in nature is such that we shall examine its behavior along with that of the other elements.

1° *Alkaline ions: Na and K*

Crystalline and metamorphic rocks contain on the average as much potassium as sodium. The ratio of the numbers of atoms occurring in these rocks can be written K/Na = 1. It is not the same thing in sediments or in the sea. Thus potassium and sodium behave differently in the hydrosphere, as seen in Chapter III, and the reason for this lies in the behavior of the K^+ ion in water; it does not become hydrated, and its apparent volume is three times smaller than that of Na^+ ions.

If one considers the behavior of K^+ ions in solution, one sees that they are preferentially adsorbed by the fine-grained particles of the sediments. The K/Na ratio is equal to 2.8 for clays, 3.3 for sandstones, and 7.7 for limestones; it is close to 3 for the ensemble of sedimentary rocks. These values are obtained, of course, from samples collected from outcrops; they can certainly be criticized because samples in quarries are leached of their soluble salts.

If one considers the behavior of the K$^+$ ions in crystals, one sees that it is K$^+$ and not Na$^+$ that fulfills best the crystallochemical conditions required for the stability of mica, and thus of illite. This does not mean that sodium is not found in illites, which are irregular micas, but it is much less abundant than potassium.

For these two reasons, preferential adsorption and incorporation in silicates, potassium is much less abundant in the solutions of the hydrosphere than sodium. As a matter of fact, both mechanisms function from the beginning of weathering. In the case of adsorption, this is obvious, and in the case of incorporation in silicates, it is certain that sericitization of feldspars and illitization are the two chief phenomena of weathering. For this reason, freshwater in the zone of cementation has a K/Na ratio equal to 1/10 and sea water a ratio equal to 1/28.5. It has already been noted that rubidium and cesium accompany potassium, as had been seen by GOLDSCHMIDT (1937).

For a long time it has been explained that the sodium that characterizes sea water found a refuge upon leaving the hydrosphere in peculiar places such as saline deposits. But in sedimentary series, the latter represent only a small tonnage. The normal refuge of soluble salts after sedimentation is in connate waters that are imprisoned in the sediments. Enclosed in the sedimentary column, sodium salts are found in increasing amounts in deep bore holes. These salts are confined for the most part definitively and will be submitted to metamorphism during which they will enter crystalline structures, and especially albite, according to the persedimentary scheme of NIEUWENKAMP (1956). If this point has escaped the attention of geochemists for a long time, it is because calculations for metamorphism have been made by comparison with the composition of samples from quarries from which sodium has been removed by les-

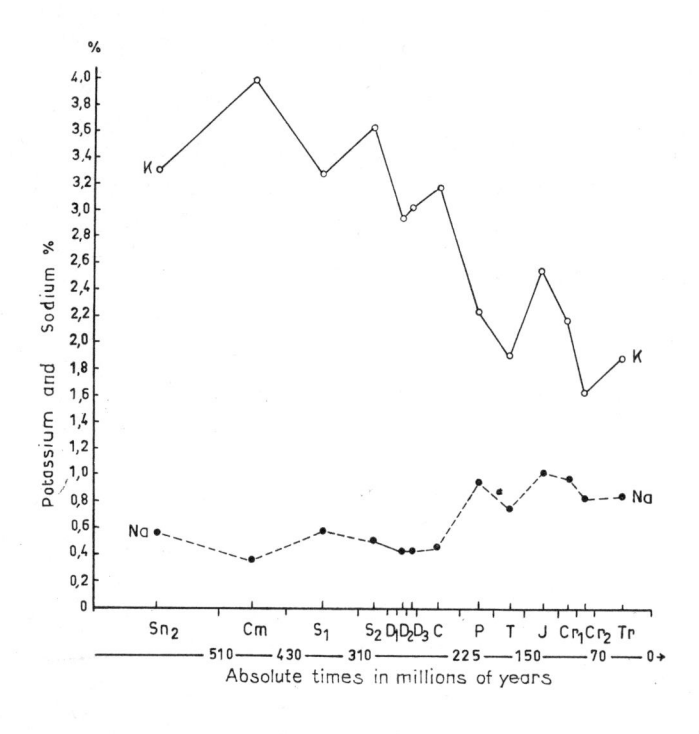

FIG. 74. — *Variation of the average content of potassium and sodium in clays of the Russian platform* (after VINOGRADOV and RONOV, 1956).

sivage in the weathering zone. Today, by analysis of core samples, we have quite different information.

From the point that interests us, we can say that *sodium remains in the solutions of sedimentary rocks whereas potassium enters to a great extent into silicates.*

EVOLUTION THROUGHOUT GEOLOGICAL TIME. — VINOGRADOV and RONOV (1956) have presented the results of analyses of the amount of K and Na in clay minerals from the Sinian (Cambrian) to the Tertiary. The amount of potassium decreases, and the amount of sodium increases (fig. 74). This is interpreted in the following way: formations exposed to weathering were mainly crystalline rocks at the beginning of the Paleozoic, and the crystalline areas subjected to weathering decreased throughout the course of time.

2° *Alkaline-earth ions: Ca and Mg*

Calcium and magnesium are considered to be two ions with similar behavior, and this is true in many phenomena of the hydrosphere. The usual occurrence of these two ions is in carbonate sediments where reduced iron accompanies them.

But carbonate rocks can be of double origin: organic or chemical. In carbonate rocks of organic origin, calcium has a clear advantage because the tests of organisms are composed chiefly of calcium carbonate. Carbonates of chemical origin are at the same time calcic, magnesian, and ferrous. Here we find the phenomenon of dolomitization, the study of which has seen great improvements thanks to the American works of GRAF and GOLDSCHMIDT (1956) and to the French works of BARON (1958, 1958, 1960). GRAF and GOLDSCHMIDT (1956) have achieved the synthesis of dolomite in the laboratory at high temperature. They have, in addition, defined protodolomite which is a double calcium and magnesium carbonate in which the amount of $CaCo_3$ is greater than half (about 55 %) and in which the superstructure X-ray reflections are weak or absent. This protodolomite is a stage in the synthesis of dolomite. BARON realized the synthesis of dolomite by precipitation from solution and an experimental dolomitization of calcite by action of a solution. The organization of dolomite occurs by stages. First, a solid solution forms; it is unstable because of the difference between the ionic radii ($Ca^{2+} = 0.99$; $Mg^{2+} = 0.66$). This solid solution evolves and the double carbonate appears, but it is poorly organized: this is the protodolomite of GRAF and GOLDSCHMIDT. By intercrystalline diffusion, the superstructure reflections appear and one ends up with dolomite.

If during organic sedimentation calcium has preference, and if during carbonate sedimentation calcium and magnesium are equally possible, except in weight, during silicate sedimentation it is magnesium that wins out and exclusively. During the reorganization of clays by transformation from open, interstratified, or degraded minerals, as well as by neoformation of magnesian clay minerals, magnesium is consumed in large quantities in the hydrosphere. Chlorite, mixed layers with chloritic layers, montmorillonite, attapulgite, sepiolite, and talc are magnesian clay minerals, and it is well known that illites, glauconites and various mixed layers contain magnesium. Calcium does not come into play.

From the point of view that interests us, we can say that *calcium remains engaged in carbonates*, whereas *magnesium is an important component of silicates.*

EVOLUTION THROUGHOUT GEOLOGICAL TIME. — VINOGRADOV and RONOV (1956) have shown an evolution of the Ca/Mg ratio in clays and shales in the course of the geological history of the deposits of the Russian platform. It is possible that the interpretation is similar to that concerning K and Na, i.e., that the surface of the crystalline basement available for weathering has decreased through time. In any case, the evolution of these geochemical ratios is a very interesting subject of study for the reconstruction of the past (fig. 75).

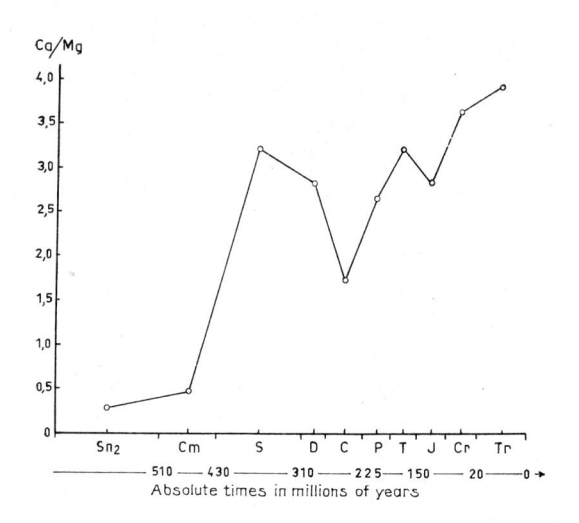

FIG. 75. — *Variation of the Ca/Mg ratio in clays of the Russian platform* (after VINOGRADOV and RONOV, 1956).

3° Iron and aluminum

Iron and aluminum are rather similar elements. They are classified in the group of "hydrolysates" of GOLDSCHMIDT, and both of them give rise to amphoteric hydroxides. But iron can present two degrees of oxidation. This gives to this pair aspects that are in some cases concordant and in other cases discordant.

From the beginning of weathering, the first kind of divergence is produced. If one takes as example the case of lateritization, which is the most striking because the liberation is the most intense, iron follows two paths. A part of the iron is mobilized in the ferrous state by the reducing waters, either in true solution or in the state of ferro-humic complexes. This iron is evacuated. The other fraction is released *in situ* and gives rise to the iron sesquioxides, predominantly goethite. The latter is associated with free alumina crystallized into gibbsite, or with alumina already engaged in lateritic or bauxitic crusts (*cuirasses*) which are chiefly ferric and aluminous.

During the same time, soluble iron has reached directly or indirectly the basins downstream, and there it is already separated from aluminum to a great extent. In fact, it can occur in four forms: in the free state as a pigment, as it occurs in the great detrital red beds; in the carbonate state as siderite, in which the iron is reduced; in the sulfide state as colloidal pyrites of sediments, such as hydrotroilite; lastly, in the silicate state. It is only in these ferriferous silicates that iron rejoins a small fraction of the aluminum, the essential part of which enters into alumino-potassic silicates.

Let us note that in sediments, and especially during diagenesis, iron has a great tendency to re-enter silicate structures. And this silication of the iron oxide results in the formation of glauconite in marine deposits, of chamosite in iron ores, and sooner or later in the development of chlorite. Whether it be at the expense of free sesquioxides, sulfides, carbonates, glauconites, chamosites, or even illites, which are always more or less ferriferous, diagenesis builds up the chlorite lattice which will be submitted to metamorphism. In this case, iron is in the octahedral position in the company of magnesium.

During this time, alumina, which hardly moves, enters silicate combinations earlier. This gives rise to sericites and illites in the zones of moderate weathering and to kaolinites in the zones of intense weathering. If one makes exception of bauxitic crusts in which alumina can maintain its geochemical liberty and which are but curiosities on the global scale, it can be said that the alumina that has just been released is captured again by silica and engaged again in silico-aluminous structures.

From this point of view, iron and aluminum witness similar fate: release in the state of an hydroxide in the hydrosphere, but this release is ephemeral and leads to engagement in micaceous lattices of the mica or chlorite type. We can say that iron is able to reside for a longer time in other isolated structures, carbonate or pyritic, but that the *resulting silicate, precocious in the case of aluminum, tardive for iron, is inevitable.*

4° *Aluminum and silicon*

If aluminum is similar to iron in certain kinds of behavior, it is also similar to silicon in others, and we have just recalled how alumina just after it is released must organize with silica in clay minerals.

During weathering in open environments, silica is more soluble than alumina. But the silica evacuated has two destinies. Either its migration is immediately interrupted in the subterranean weathering horizons, and argillaceous and siliceous neoformations take place in the weathering profiles; or migration continues and silica collects in sedimentary basins where it is organized again, and this gives rise to the argillaceous and siliceous neoformations in sediments. Each time silica combines, it is by priority with alumina; in comparison with the combinations of silica with alumina, the silica of silicifications is of little importance.

For its part, the liberated alumina can remain alone and crystallize into gibbsite, if drainage is strong enough to withdraw silica from contact with it. In the most frequent case, water bodies constantly provide to the alumina the silica necessary for the neoformation of clay minerals. Silicon and aluminum, thus, have behaviors that are similar in most cases and opposite only in a few cases. In fact, on the whole these two elements recombine as soon as they are released. But if in some cases they can remain alone, this does not occur at the same place. Alumina, which is little soluble, accumulates *in situ* in the weathering profile, and this leads to the aluminous crusts.

In contrast, silica, which is more soluble, undergoes either limited migration toward subterranean silicifications or longer migration toward sedimentary silicifications and neoformations.

Let us say that, essentially, *silicon and aluminum tend to organize together into clay minerals. If they do not, alumina stays* in situ *with iron, whereas silica is removed with lime and magnesia.*

5° Silicon and other cations in neoformations

If the line of reasoning presented for siliceous neoformations is valid, one sees that it is the foreign cations that come in to disturb progressively the crystalline edification of silica. We find macrocrystalline quartz, then silicifications of microcrystalline quartz, then chalcedony, then the cristobalite-tridymite opals, then the still more disordered opals. Disorder is seen to increase under the influence of cations, of first importance among which is calcium which, however, cannot form a silicate in the hydrosphere.

But if the other cations are present in notable proportions, order reappears. As soon as the quantities of aluminum, iron and magnesium are sufficient, layered hydroxides form and silica fixes on them to ensure the neoformations of clay minerals.

In this process, alumina has a preferential role. In a desaturated environment, it fixes silica in the form of kaolinite; in an environment charged with various cations, it associates with silica in the tetrahedral sheet, but it also fills in the octahedral sheet. The higher its concentration is in the environment, the more abundant it will be in neoformation. It exercises a kind of priority.

For historical and pedagogical reasons we have been used to emphasizing the leading role of the linkage of silica tetrahedra in silicate structures. Aluminum can replace silicon, and the other cations fill in the holes in the structure. This is the result of the fact that the periodicities accessible to X-rays are largely determined by the type of arrangement of the tetrahedra, the center of which is siliceous. But the ways in which relationships are discovered or taught does not perforce coincide with the reality of phenomena. In the case of the growth of crystals we are led more and more to give the cations an important role. This brings us to the work of BELOV (1957) who has shown the occurrence of structures more varied than the structures of silicates with which we are familiar, more varied as a function of the size of certain cations. As a matter of fact, in the presence of silica, it is the cations which orient and command the edification of the structure. We may use GOLDSZTAUB's expression: "The cations are wrapped up in the siliceous lattice like fish in a net." And it is certainly true that the net adapts to the size and the arrangement of the fish.

Let us note in addition that, as long as the cations of the environment play a minor role in the growing siliceous structure, the latter is three-dimensional, like that of quartz. As the perturbating cations increase, the three-dimensional silica structures are distributed, and we have alternating layers of cristobalite or tridymite with disorder perpendicular to their plane. These are siliceous layers that outline the rings of cristobalite, but regular development in the third dimension fails. We are allowed to

GEOCHEMISTRY OF THE CONSTITUENT ELEMENTS OF CLAYS 369

speak of the beginning of a two-dimensional structure. In clays, the latter will be organized in a way more and more suitable for the formation of clay minerals to such an extent that we can say that the imperative intervention of K, Mg and Al cations has imposed in an obligatory way the two-dimensional structure. This is the way it is on the surface of the Earth. In contrast, at depth, K, Na, Ca and Al cations are quite capable of accommodating themselves to a three-dimensional structure, that of the feldspars.

It is of extreme importance in surficial geochemistry always to keep in mind the fact that *we have much more to do with compounds of cations with silica than with silicates. The guiding role in the arrangement of silica in the hydrosphere is ensured by cations, among which aluminum is the first to intervene.*

6° Disorder, order and crystal size

In siliceous neoformations, increasing disorder results in a decreasing size of crystals, to such an extent that one arrives, in the case of opals, at the limit of the crystalline state. In other words, progressively as cations intervene in the edification of crystalline silicas in the hydrosphere, the size decreases.

When order is once again established in argillaceous neoformations, the size of crystals increases. And it increases all the more as isotypic or homeotypic substitutions decrease. The multiplicity of replacements disturbs the development of the structures, and their size remains small; this is the lot of clay minerals. By diagenesis, but mainly by metamorphism under the influence of temperature and pressure, minerals are organized; foreign ions are evacuated or regrouped, as we shall see, in order to attain a minimum of internal energy, and the size of clay minerals increases.

It is of extreme importance in surficial geochemistry to emphasize that the small size of crystals is but the visible reflection of disorder in their lattice. It is an exaggeration to pretend that independence results in disorder, but let us say that, the former structures being abandoned, a completely new effort for organization is necessary. Figure 76 shows schematically this crystallochemical behavior. In nature all lessivage of cations from a silicate lattice disturbs the order of the lattice and pulverizes it; this

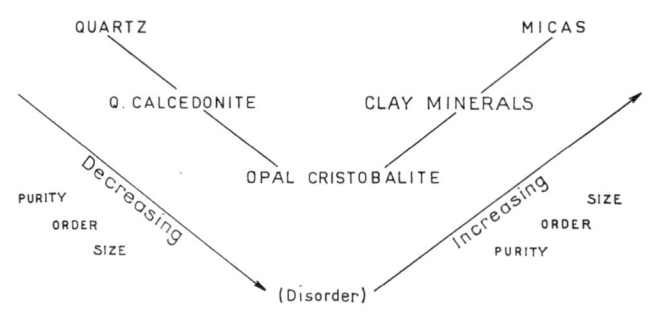

FIG. 76. — *Disorder, order and crystal size in siliceous and argillaceous neoformations.*

is one of the keys to the mechanism of weathering. On the other hand, every impurity in a growing mineral hinders the acquisition of good periodicities and restricts this mineral to "cryptocrystalline" sizes. Every purification of these impurities regularizes the structure and permits its growth.

III. — GEOCHEMISTRY OF LANDSCAPES AND OF SEDIMENTARY ENVIRONMENTS

At the end of this book one can take into account the bonds that unite tectonics, climates, and sedimentation in order to establish a "Geochemistry of landscapes". Three landscapes of extreme type have been chosen, along with some variants. They are illustrated by several simplified sketches designed to make the ideas firm. It is obvious that nature does not present only extreme cases, but a continuous variety of intermediates between these schematic types.

1° *Landscapes of physical erosion. Detrital sedimentation*

These landscapes will be dominated above all by *intense mechanical erosion which leads to detrital sedimentation*. This can be obtained by the play of active tectonics or by the play of climates that develop weathering of physical character only.

Active tectonics. Mechanical erosion. Physical sedimentation. — Sketch I of figure 77 shows the profile of a young mountain chain. The experience of geology teaches us that young mountains are also living mountains. The slopes are too steep and the energy of waters is too vigorous to permit the development of soils and slow dissolution. Mechanical erosion associated with mechanisms of disaggregation is predominant. Sedimentation is detrital. Along with the gravels and the sands, the clay minerals of crystalline and metamorphic rocks, of shales and sediments are carried away. Heritage prevails; argillaceous sedimentation itself is detrital, mechanical, and allogenic.

One variant of this landscape is the volcanic landscape (sketch II, fig. 77). In this case, tectonism is expressed by eruptions that give rise to a variety of pyroclastic rocks. If the volcanic products have been altered to bentonite or to kaolin hydrothermally, we come to deposits of bentonite or kaolin of secondary reworking. If volcanic materials, and in particular tuff and ash, fall directly into the water of sedimentary basins, they can give rise to deposits of bentonite by alteration of the glass.

Another variant is provided by tectonic rejuvenation that brings on new erosion after a long period of calm. During the calm, soils were able to develop deeply, but now they are attacked and swept away into sedimentary basins. The most typical example is that of the siderolithic facies which results from the reworking of lateritic soils (sketch III, fig. 77).

PHYSICAL EROSION – DETRITAL SEDIMENTATION

I Living tectonism
   Physical erosion
   Detrital sedimentation

II Volcanism
   Physical sedimentation
   Pyroclastic sedimentation
   Bentonites

   Pyroclastic sedimentation . Bentonites

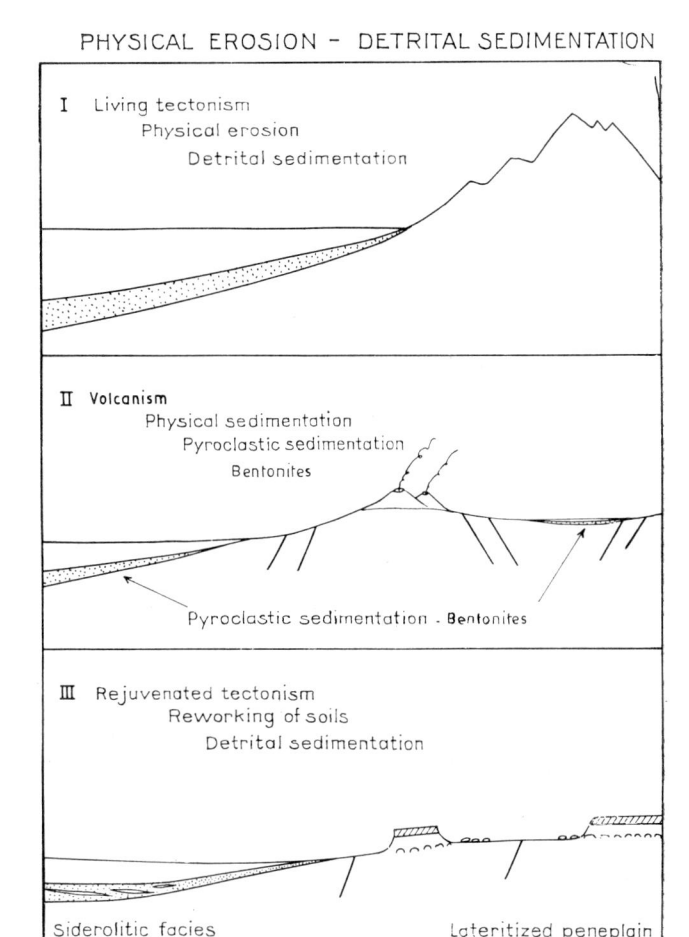

III Rejuvenated tectonism
    Reworking of soils
    Detrital sedimentation

Siderolitic facies   Lateritized peneplain

FIG. 77. — *Geochemical landscapes.* Three diagrams of terrains subject to physical erosion and detrital sedimentation.

Climates of physical weathering. Mechanical erosion. Detrital sedimentation. — The glacial landscape is again a landscape of mechanical erosion. The force driving the erosion is no longer active tectonics, but glacial and periglacial climatic activity. Immense tonnages of particulate sediments are transported into lakes, fjords, and the sea.

The desert landscape is another variant. It is again the climate that controls the granular erosion. But, in this case, dryness replaces cold in order to provide the detrital products that spread out in long flattened cones, the section of which calls up the image of an architectural pediment, whence the English term "pediment".

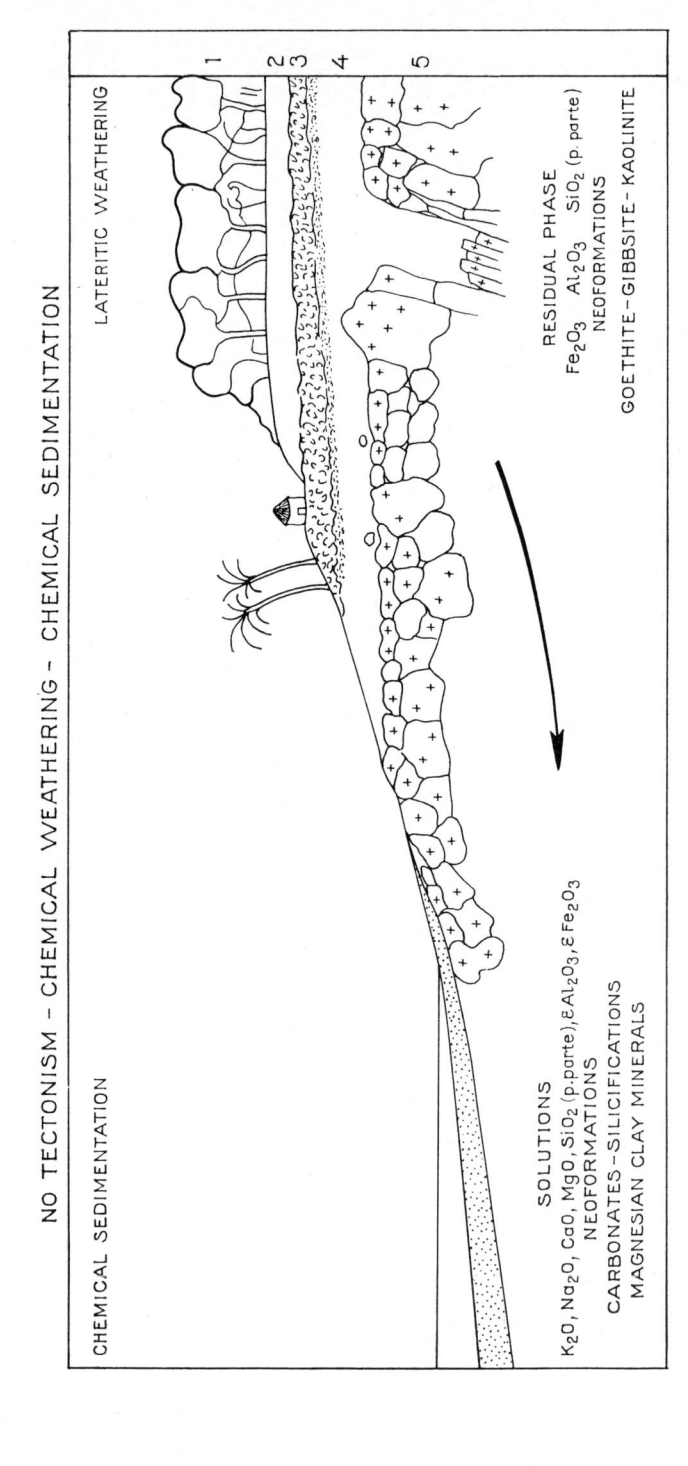

FIG. 78. — *Geochemical landscapes.* Diagram of a tropical landscape supplying chemical sediments in the absence of tectonic movements. 1. Shady tropical forest; 2. Top of the profile; 3. Iron crust; 4. Kaolinitic clays (lithomarge); 5. Parent rocks.

2° Moderate tectonics. Moderate weathering. Detrital sedimentation with transformations

This second category of landscapes is that with which we, in the zones of temperate climates, are familiar. Tectonic agitation is much weakened. Water runs more slowly and can penetrate into the surface to give rise to soils. Physical and chemical weathering are combined. The clay minerals resulting from weathering are partly degraded and partly dissolved; neoformations of clay minerals can take place in weathering profiles and in soils. All this material, together with solutions, reaches lakes and seas. The sedimentation is detrital, but a part of this detrital sedimentation is composed of altered minerals which can be repaired within the sedimentary environment under the influence of solutions. This possibility is facultative; it corresponds to transformations by "aggradation", which have been described at great length. They are capable of modifying, and in some cases completely, the products of heritage.

These landscapes represent a whole gamut of possibilities between the humid and cold landscapes that develop podzols and the steppe landscapes where montmorillonite forms. The sediments that form are also as varied; their clay fraction still contains the principal detrital minerals, illite and chlorite, and also a multitude of mixed layers which are the evidence of weathering all the way to the final stages, such as represented by vermiculite or montmorillonite of transformation and montmorillonite of neoformation. In the sedimentary environment this heritage of clay minerals will start the constructive transformations that have been evoked, if the solutions are concentrated enough to induce them. All the cases are possible.

3° Null tectonics. Strong weathering. Chemical sedimentation

This third category of landscapes (fig. 78) leads us to the immobile shield areas where tectonics has been appeased a long time ago. Moreover, humid tropical and equatorial climates prevail there. Forest develops and protects the thick chemical weatherings. Dissolutions are considerable. Only the major part of the iron and the aluminum and a fraction of the silica remain *in situ*. The residues are evacuated along with all the soluble cations and anions. The forest destroys the minerals but preserves the profiles. That which issues from such continents is almost uniquely dissolved in the ionic state in the waters of springs and rivers. Sedimentation then becomes purely chemical: carbonates, flint, chert and silexites, phosphates, and magnesian clays.

This case developed to an extreme during the lower and middle Eocene in France and in Africa.

4° The reciprocal. Reconstruction of genetic environments

When geologists began to inventory the clay minerals of sedimentary series, they were hoping that clay minerals could easily characterize the different genetic environments. This entire book shows that this is not possible. Furthermore, if often clay minerals help in the reconstruction of the past, this is not accomplished with ease but after serious labor in favorable cases, and most often with the contribution of comparison with all other methods.

The ensemble of the methods to be used in the reconstruction of genetic environments can be classified under four headings.

Paleontology. — Paleontology remains the privileged method that contributes to the definition of genetic conditions of ancient deposits. It includes the study of fossil macrofaunas, microfaunas, spores, pollen and plankton. Macrofaunas are commonly rare in some kinds of deposits and generally are destroyed by drilling. This is the reason why the study of microfossils has taken on such a great development. In addition to stratigraphic information, pollen also gives information on climate, and Foraminifera and paleoplankton give information on the characteristics of the sedimentary environment; but it seems that Ostracoda give the most precise indications of the chemical properties of solutions in which the sediments were deposited, thanks to variations in the ornamentation of their valves. These are the reasons why petroleum companies have developed their paleontological, micropaleontological and palynological sections, not only in order to establish stratigraphic correlations, but also to reconstruct variations of the former geographies which condition the position of the formed parent rocks and of the present reservoir rocks.

Analysis of the form of detrital products and of "sedimentary bodies". — Today we are well acquainted with methods of analysis of the shapes of sand grains, gravel, and pebbles within detrital formations. These methods were developed in France by CAILLEUX (1943, 1945), BERTHOIS (1950), TRICART (1950, 1959) and CAILLEUX and TRICART (1953, 1959). By means of these methods one can obtain valuable information about the original environment of the sediments and about the climates that presided over the shaping of detrital elements.

Today we also have at our disposal new methods of analysis by measurement and study of the shape of "sedimentary bodies". In fact, sedimentary deposits rarely occur as beds of regular thickness. On the contrary, they occur as lenses resulting from oblique stratification or cross-bedding. Each lens, each "fish", is a sedimentary body and its quantitative study is very demonstrative. DEBYSER and his students (BEUF, MONTADERT and DEBYSER, 1962) have reconstructed the forms of sandstone beds in the Cambro-Ordovician detrital deposits of the Tassilis Ajjers. Ribbons, spoon-shaped formations, and the elongation of arenaceous lenses lead to an approach to the reconstruction of the dynamics of detrital sedimentation. This work can be undertaken in accord with the acquired knowledge of the dynamics of sediments (GLANGEAUD, 1941) and by hydrodynamic experiments with models. PETERSON and OSMOND (1961) have presented the American studies on the dynamic and geometric analysis of arenaceous bodies.

Minerals. — The information on environmental conditions that can be drawn from minerals of sedimentary rocks is abundant; this is obvious for the minerals of limestones, dolomites, phosphates, iron ores, etc. As for clays, the significance can be either null or precise. A sedimentary series containing ubiquitous minerals such as illite and chlorite will in many cases give no indication at all. On the other hand, the presence of montmorillonite can signify so many things that one has too much to choose from: reworking of hydrothermal products; weathering within the basin of volcanic glass; weathering residues of evolved soils; reworking of calcic or hydromorphic soils; or neoformation in an alkaline environment. Lastly, glauconite will characterize the marine environment and palygorskite will be typical of precise chemical sedimentation, although they are indifferent to the salinity of the basin.

This important question can be summarized by the following points:

1° *Detrital argillaceous sedimentation is not significative of the genetic environment because the particles come from elsewhere.*

2° *On the contrary, detrital argillaceous sedimentation testifies to the types of erosion and weathering acting on the neighboring continent.* Illite and chlorite represent direct mechanical weathering, all the more so since chlorite is very sensitive to the initiation of weathering. Montmorillonite, when it has a pedological origin, characterizes soils with poor drainage and alkaline character. Kaolinite, when it has a pedological origin, testifies to strong weathering.

3° *Neoformed argillaceous sedimentation is very significative of environments of chemical sedimentation.* The best examples are glauconite, palygorskite, chamosite, and in some cases talc.

4° *Moreover, neoformed clay minerals are significative of the climatic conditions prevailing on the continent.* The example of attapulgite is the best; it corresponds to strong hydrolysis on the continent.

5° *Transformed clays share the characters of the two preceding categories. Degraded clays characterize the continental environment in which they originate. Clays transformed by aggradation are significative of an environment ensuring the reparation and then the growth of weathered minerals.* Here, the inventory has hardly begun, but it has given very beautiful results for Triassic sedimentation (LUCAS, 1962).

It is prudent, therefore, to approach the study of the clay fraction of sedimentary rocks only with discretion. In certain cases its study will throw bright light on the climates and other characteristics of the past. In other cases it will lead to nothing because the clay minerals, on the whole, will be only detrital minerals analogous to quartz grains and, like these, divested of significance.

Chemistry. Major elements. Minor elements. Isotopes. — In the study of clays, the most considerable effort for some twenty years has been applied to identifications by means of X-ray diffraction. Improvement of the methods, application of mineralogical investigations to the study of mixed layers, and automation of the apparatus has taken this much time. But now that we know more or less how to identify the types of minerals occurring in rocks, we are going back to chemistry. This is difficult because most of the clays are mixtures, and the chemical analysis of complex mixtures of minerals with variable formulas is often undecipherable. It is necessary, therefore, to perfect our techniques, and they must be rapid enough to follow the output of the apparatus of physics. The work is being advanced in three directions: variations in major elements, variations in minor elements, and the geochemistry of isotopes.

VARIATIONS IN MAJOR ELEMENTS. — We have taken the habit of designating clay minerals by the name of the structural type identified by X-ray diffraction. But it is well agreed that this type admits great variations in its chemical composition.

The mica type is represented by aluminous illite and ferric glauconite. However, from the works of JUNG (1954) and KELLER (1958) we already know intermediate types. Moreover, the illites themselves contain more or less aluminum, magnesium, or iron, and these variations are not indifferent. Transformed illites, in particular, must reflect partially the composition of the environment that repaired them and caused them to grow. The chlorite type is still more varied due to the distribution of aluminum, magnesium and iron in the tetrahedral, octahedral and brucitic sheets. Since we know that chamosites and chlorites of iron ores are rather (but not exclusively) ferriferous, and that transformed chlorites of the Triassic are magnesian and very poor in iron, a long inventory is necessary to lead to suitable systematics of sedimentary chlorites.

The montmorillonite type is extremely varied. Not only must distinctions between the dioctahedral and the trioctahedral series be used more often, but within each series the variations of homeotypic replacement are considerable. A remarkable study of GRIM and KULBICKI (1961) has already shown us that aluminous dioctahedral montmorillonites do not form a continuous series. High temperature reactions permit the distinction of two types: the Wyoming type and the Cheto type. The first is more aluminous in its octahedral layer than the second, which is more magnesian and more ferriferous. Study of the relations between the different types of montmorillonites and their genetic conditions is in progress.

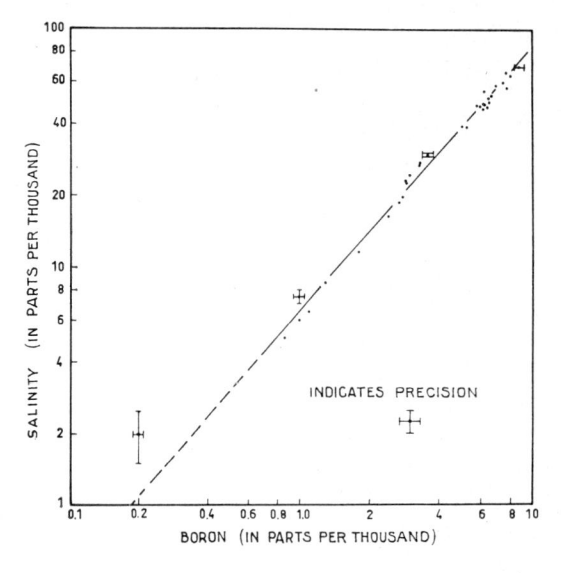

FIG. 79. — *Relationship between boron content and salinity in the Gulf of Mexico* (after FREDERICKSON and REYNOLDS, 1960).

VARIATIONS IN MINOR ELEMENTS. — The study of minor elements, or trace elements, has been under way for a long time. It takes on a new interest today now that we know the mineralogy of clays better. The case of boron will be briefly presented here.

Boron is much more abundant in marine water than in freshwater. This had already been noted by DIEULAFAIT in 1877 in a treatise which it is difficult to believe was written so many years ago. The first study of the variation of boron in sediments was made by GOLDSCHMIDT and PETERS (1932). They noted that boron was especially tied to clays and that it was two times more abundant in marine clays than in continental clays. LANDERGREN (1945, 1958, 1959) has studied the variation of the boron content in Swedish sediments. He showed that modern sediments had the same content as Cambrian sediments of the same region. He confirmed that the fixation of boron occurs essentially in clays and that the amount in modern clays varies with particle size and with the salinity of the environment.

Since 1957 studies on boron and minor elements have multiplied. First, by the continuation of the studies carried out in Sweden by LANDERGREN (1958, 1959). Then, by the work of several laboratories in the U.S.A.: in Pennsylvania by DEGENS, WILLIAMS and KEITH (1957, 1958) and KEITH and DEGENS (1959); in California by GOLDBERG and ARRHENIUS (1958); and in Oklahoma by FREDERICKSON and REYNOLDS (1960). In addition, HARDER (1959, 1960) arrived at comparable results in Germany at the same time. In France the method was systematically utilized by the research center of the Société nationale des Pétroles d'Aquitaine on the oil-bearing series of Africa and Europe (STÉVAUX, 1961; KULBICKI, STÉVAUX, ESQUEVIN, LUCAS, 1962).

In brief, the broad outlines of the knowledge obtained are the following:

1° There is a linear relation between the concentration of boron in sea water and

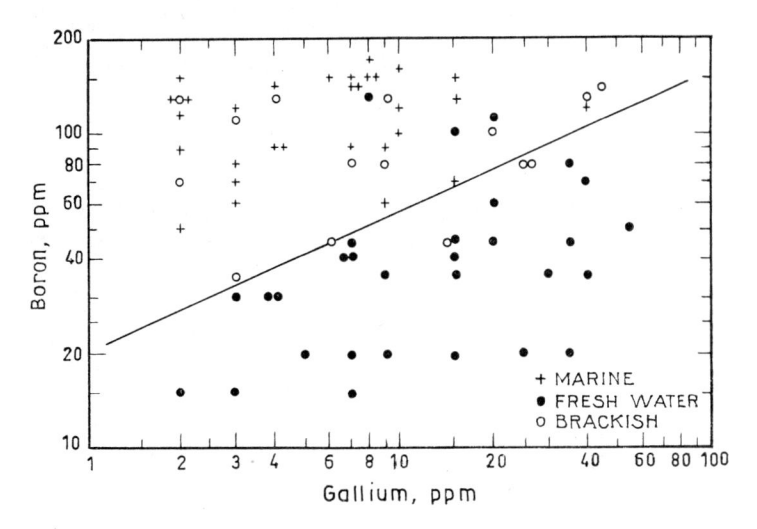

FIG. 80. — *Relationship between boron and gallium in Pennsylvanian shales of marine, freshwater and brackish origin* (after DEGENS, WILLIAMS and KEITH, 1957).

the salinity. Figure 79 shows this variation from the measurements made in the Gulf of Mexico by FREDERICKSON and REYNOLDS (1960).

2° The major part of the boron in sediments is localized in the clay fraction, principally in illites.

3° There is a relation between the salinity of the genetic environment and the boron content of clays. Figure 80 shows such a correlation for Pennsylvanian shales, according to the works of DEGENS, WILLIAMS and KEITH (1957). It shows the boron and gallium contents of marine, brackish, and lacustrine sediments, the genetic environments of which have been reconstructed from their fossils and their facies.

4° The reciprocal. — In sediments without fossils the boron content of the clay fraction can be used as an indicator of paleosalinity. The concentration may not be the same from one formation to another, but the variations between environments of different salinities are always clear within a single formation. In Chapter VIII we have seen the results obtained on the Saharan series in which the boron content designates the transgression.

5° Interpretation. — Boron seems to be fixed by modern clay minerals according to size (LANDERGREN, 1959), i.e., according to the size of the external surface. But this boron is rapidly incorporated into the structure of the mineral. In fact, chemical attack on illites allows one to think that it occurs as a substitution in the tetrahedral sheets of illite. Now, the boron content of micas turns out to be extremely low (GOLDSCHMIDT, 1954). It is indispensable, therefore, to reconstruct the phenomenon of fixation and incorporation into the illite lattice during sedimentation. This is possible because we know that detrital illites originating in weathering and soils are degraded. During their reparation and aggradation they incorporate boron in quantities that can reach several hundreds of ppm (g per ton).

Boron is thus a supplementary proof of these transformations that are contemporaneous with sedimentation (this is the "diagenesis" of American authors) and that reshape the inherited material in the likeness of the sedimentary environment. A detrital clay mineral can have two different histories:

— In freshwater, should it remain inert, or should its lattice be repaired, or should it grow, it is in contact only with waters that have a low boron content; therefore, its own boron content remains low;

— In saline or hypersaline marine environments, all rearrangement of the structure and, *a fortiori*, all growth or neoformation incorporate trace elements, including boron, available in the environment.

Vanadium is recognized by all authors as an indicator of organic matter (HUTCHINSON, 1957; BLACK and MITCHELL, 1952; GOLDBERG, 1957). During degradation of organic matter, a part of the vanadium enters organo-metallic complexes; another part is found in the form of vanadates. But it is probable that these vanadates, like the organo-metallic complexes, are fixed by clays (GOLDSCHMIDT, 1954; DEGENS et al., 1957) There is less vanadium in continental clays than in marine clays, and this is probably connected with the abundance of planktonic life in marine environments.

Gallium and chromium are more abundant in continental environments, in contrast to boron, rubidium and nickel. Correlation curves have been established, such as that in figure 80 for gallium and that in figure 81 for lithium (DEGENS, WILLIAMS and KEITH, 1957; KEITH and DEGENS, 1959).

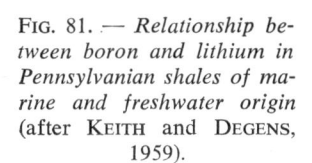

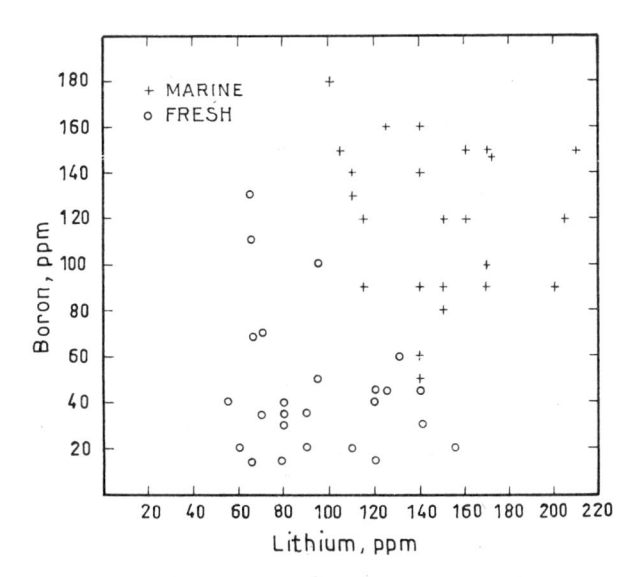

FIG. 81. — *Relationship between boron and lithium in Pennsylvanian shales of marine and freshwater origin* (after KEITH and DEGENS, 1959).

GEOCHEMISTRY OF ISOTOPES. — Many geological problems could be clarified or solved by the geochemistry of isotopes.

The measurement of paleotemperatures has been undertaken by UREY et al. (1951), EPSTEIN et al. (1951, 1953) and MACCREA (1950). The proportion of the O^{18} isotope of oxygen in a carbonate is proportional to that in the carbonic gas dissolved in the water during the formation of that carbonate. Now, this latter proportion is a function of the temperature of the water, a function that has been demonstrated experimentally. Thus we possess a method for the determination of paleotemperatures by means of the quantitative analysis of the O^{18}/O^{16} ratio in a sedimentary carbonate, be it precipitated or organic. The most spectacular results were obtained by the recording of seasonal rhythms in the concentric layers of a belemnite shell and by the determination of ocean temperatures during the Upper Cretaceous in the U.S.A. and Europe.

Attempts have been made to examine the variations in the C^{13} content as a function of the salinity of the environment (KEITH and DEGENS, 1959). Samples of marine origin have a C^{13} content higher than that of freshwater samples.

Lastly, the geochemistry of isotopes has led to very important results in geochronology and especially in the domain of crystalline rocks. This method has been extended to the glauconite of sedimentary rocks and has given some good results because glauconite is a mineral formed contemporaneously with sedimentation. The calculations are based on the measurement of Sr^{87} coming from Rb^{87} and of Ar^{40} coming from K^{40}. Studies are in progress to apply these methods to clay minerals, and especially to illites. Various answers are possible; either the clay mineral is detrital and the measured age is much older than that of the geological formation containing it, or the clay mineral, such as glauconite, is contemporaneous with sedimentation and the measured age is that of the geological formation, or the clay mineral is diagenetic and the measured age dates the diagenesis. In all cases the result is interesting, and it is particularly interesting in the case of transformations. If the transformation is rather advanced, there is reparation and growth of much degraded particles, and the lattice is practically new as far as its rubidium and potassium content is concerned. When transformations are weak, interference appears and the results can no longer have meaning.

Thus, by the chemistry of major elements, then by that of minor elements and by that of isotopes, a great range of studies is opened to us with which to reconstruct the genetic environments of the past, i.e., the paleogeography.

Conclusion

The evolution of landscapes under the influence of climates is the result of two very different mechanisms: *weathering of predominantly physical character and chemical weathering*. The first provides "whole" silicates, intact or almost intact; their lattice contains all the major or minor elements that entered therein during metamorphism or during the genesis of crystalline rocks. The second, by hydrolysis, undertakes destruction of the silicates, advances it more or less far, and releases the constituent ions of the minerals.

The results are very different in the case of argillaceous sedimentation, which represents the mass of silicates during their passage through the hydrosphere. *In the first case, nearly intact minerals are buried and prepare for a new phase of metamorphism. In the second case, the debris of silicates and solutions accumulate in sedimentary basins and reorganize.*

The reconstruction of the ancient conditions of sedimentation is difficult, therefore, in the places of sedimentation supplied with unaltered particles. In contrast, an entire phase of geochemistry appropriate to sedimentation develops in sedimentary basins where the products of hydrolysis are assembled.

The major elements are dispersed: iron, alumina and part of the silica remain on the continent; silica, lime, magnesia and alkaline elements reach lacustrine or marine basins downstream. The whole suite of trace elements has the same destiny in accordance with the rules of the behavior of ions in the hydrosphere, based largely on the ionic potential of GOLDSCHMIDT. Thus, we can try to follow the cycle of each element during this evolution. We have seen that rubidium follows potassium, that gallium follows aluminum, that nickel and cobalt follow magnesium, and that phosphorus and arsenic unite with iron ores. We have just seen that boron follows silicon and rejoins it in the tetrahedra of illite. All these activities constitute surficial geochemistry, whose mechanisms are becoming well defined.

IV. — THE SILICATE CYCLE

1° Metamorphism of clay minerals

Clay represent the essential form of silicate rocks during the surficial stage of the geochemical cycle.

Just as weathering was the work of destruction and transformation of silicate rocks, metamorphism is their regeneration. It is out of the question to begin a development of metamorphism here, but we can examine in geochemical terms the broad outlines of the metamorphic evolution of clay minerals and layered minerals.

Diagenetic evolution of kaolinite and montmorillonite.

— Kaolinite is transformed into illite and sericite prior to metamorphism during diagenesis. This is deduced from two categories of observations: first, the very general phenomenon of sericitization of kaolinites under the influence of concentrated diagenetic solutions; then, the absence of kaolinite in schists at the threshold of metamorphism.

One must find a place for sedimentary rocks rich in kaolinite and poor in potash, the evolution of which would lead to pyrophyllite. ROQUES, who is well acquainted with the metamorphic series, mentions micaschists that are so poor in potash that he was compelled to calculate pyrophyllite in them, as has in addition been reported by BARIC (1955).

— Montmorillonite is still more sensitive to the phenomena of diagenesis, and it disappears as a rule before the onset of low-grade metamorphism. Its destiny has never yet been followed step by step, but it is reasonable to think that aluminous montmorillonites give rise to illite and sericite under the influence of hypersaline solutions in diagenesis. Evolution toward chlorite can also be conceived of under the influence of magnesian solutions or from trioctahedral montmorillonites.

Thus, kaolinite and montmorillonite are transformed by diagenesis well before metamorphism.

Clay minerals in schists. — It is quite different for illite and chlorite. These are minerals that develop systematically during diagenesis. In schists (sericite schists, chlorite schists, low-grade phyllites, greenschists, *schistes lustrés*, etc.), the two fundamental constituents are sericite and chlorite. But there is quite an evolution from the clay minerals of shales, the size of which is measured in μ or in tens of μ, to the clay minerals of epimetamorphic schists, the size of which is measured in mm or in tens of mm.

The evolution of clay minerals can be followed by physical methods. FAIRBAIRN (1943) applied X-ray diffraction to foliated rocks, by which means he was able to show the nature of the layered minerals in the rock and their greater or lesser degree of orientation relative to the schistosity. BATES (1947) showed the increase in size and crystallinity of the clay minerals of schists compared with those of shales, along with amelioration of the orientation. WEAVER (1960) described studies made on the evolution of diffraction peaks of clay minerals with depth, all the way down to regions of low-grade metamorphism.

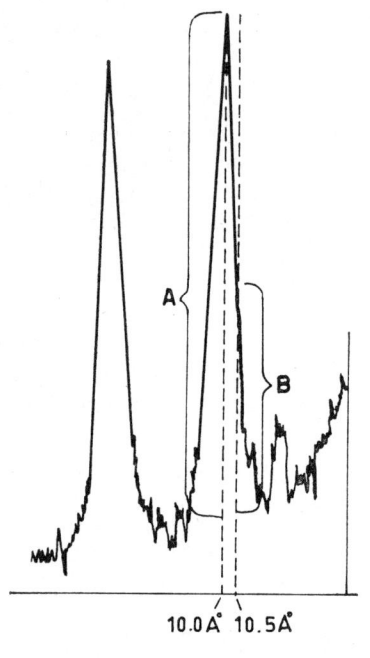

FIG. 82. — *Method of measuring the ratio A/B on an X-ray diffraction diagram* (after WEAVER, 1960). A: total height of peak; B: height of secondary effects due to the irregularity of stacking.

The 10 Å peak of illite is carefully examined (fig. 82). The total height of the peak is measured and designated A. One measures also the height of the secondary peaks on the internal flank of the peak; these correspond to the effects of opened layers, and this second height is designated B. Since metamorphism results in the closing of the expanded layers, the A/B ratio increases with the degree of metamorphism. In the Ouachita Mountain geosyncline in Oklahoma (U.S.A.) the results are the following:

Degree of evolution	A/B ratio
Unmetamorphosed Atoka	1.8
Unmetamorphosed Stanley	2.3
Incipient metamorphism	2.3
Incipient to very weak metamorphism	4.5
Very weak to weak metamorphism	6.3
Low-grade metamorphism	12.1

All these data show the suppression of interstratifications, the regularization of lattices, and the increase of size during the first stages of metamorphism.

The evolution can be followed by petrographic and chemical methods, and the most penetrating works are the oldest ones, such as those of ROSENBUSCH (1877) in

Alsace, BARROIS (1884) in Brittany, and HUTCHINGS (1890, 1894, 1896) in Great Britain. The latter showed that the sericite developing from a shale was essentially a sericite, although impure because it contained lime, magnesia, and ferrous and ferric oxides, which are foreign elements to pure muscovite. In the process of reconstruction sericite throws off its impurities, and its development is accompanied by that of a substance similar to chlorite but poorly characterized. In schists, on the other hand, chlorite is well separated from the mica, which is definitively muscovite. BRAMMAL (1921) also found this evolution. His point of departure was a mass in which very small flakes of mica and chlorite and some obscure "chloritic matter" were identified. In the process of metamorphic reconstruction, the latter decreased and disappeared. The crystallized chlorite increased, forming well individualized crystals which had affinities with clinochlore and pennine. White mica was intercalated in obvious fashion, and the length of the blades was ten to fifteen times larger than that of the clay minerals of the matrix. All these evolutions lead to schists in which clay minerals attain the order of a millimeter in size and belong to the two great families of sericites and of chlorites. Based on mineralogical and chemical studies, these large clay minerals must be classified among varieties of micas (FOSTER, 1956), sericites (SCHALLER, 1950) and chlorites (ORCEL, CAILLÈRE and HÉNIN, 1950). It must be noted that the sericites are in some cases magnesian, and that they are in many cases rather far from the composition of muscovite, as shown by MICHEL (1953) for the schistes lustrés of the Grand Paradis in the Alps, in which the predominant clay mineral is phengite.

All these works permit us to present the transition from the clay minerals of shales to the clay minerals of schists according to the mechanism that I proposed in 1949. *One must not at all consider that illite gives rise to sericite or that the chlorite of shales gives rise to the chlorite of schists. One must examine the evolution in common of the illite-chlorite mixture toward the sericite-chlorite association by a new distribution of the constituents.* In fact, sedimentary illites and chlorites are clay minerals with many isomorphic substitutions, and they have an extremely impure chemical composition compared to the large clay minerals of metamorphic rocks. As metamorphism develops in response to the increase of temperature and pressure, the minerals acquire a better organization by the tendency for a minimum of internal energy. Illite expels iron and commonly magnesium from its lattice, and these will nourish the neighboring flakes of chlorite. Reciprocally, chlorite expels its foreign ions, and aluminum in particular, which will nourish the growing micaceous lattice. This new distribution of elements is accompanied by the growth of crystals, i.e., by the increase in size of the clay minerals. *For clay minerals owe their small size to the intensity of substitutions and as metamorphism "purifies" the lattices, the size increases* (MILLOT, 1949).

We must understand well that, during this evolution from shales to schists, *there is no filiation of a large-size clay mineral from a small-size clay mineral, but there is sorting, redistribution, and exchange of elements between the clay minerals in order to arrive at the most regular arrangements, therefore the largest sizes.* The result is a schist in which the clay minerals are an association of aluminous sericite, of siliceous and magnesian sericite, and of chlorite. The abundance of these minerals depends on the initial composition.

The micas of micaschists. — Metamorphism continues and one passes from the low-grade phyllites to the true micaschists characterized by the two typical micas, muscovite and biotite. The evolution of sericite is clearly understood by the continuation of the previous mechanism: removal of the elements that were still tolerated by the lattice, especially magnesium, and regularization of the crystal, which loses water and gains potassium. At the same time an important transformation takes place, that of chlorite to biotite; in the case of this transformation one can only report that magnesium leaves the brucitic layers of chlorite to be installed in the octahedral layers along with iron, whereas it is potassium that ensures the interlayer bonds. A redistribution of elements occurs in the white mica - black mica ensemble, which represents the stable structures of the micaschist zone, by a mechanism inverse to that which we have seen functioning during weathering. Let us note that if abundant sodium has been brought into this low-grade metamorphism, it is unable to hide any longer in micaceous lattices, and albite appears, as is the case in numerous micaschists. This seems to be much more frequent than the development of paragonite, which still remains dubious.

Regeneration of feldspars. — Sodium and calcium of the rocks undergoing metamorphism tend, for their part, to give rise to soda-lime feldspars. As for the series of clay minerals, potassic feldspathization will occur sooner or later. First muscovite is feldspathized by tetracoordination of aluminum and consumption of silica. Biotite cannot be feldspathized because magnesium and iron cannot find a place in a feldspathic lattice. Thus, muscovite and biotite have very different behaviors in the mesozone of metamorphism because muscovite, which is essentially aluminous, possesses the possibilities of rearrangement, which is not the case for biotite. Therefore, silicate rocks of the mesozone are no longer essentially constituted of layered minerals, as in schists and micaschists, but are feldspathic, and biotite is the only witness of the former state. The new state is also ephemeral. Upon reaching the catazone, biotite will be feldspathized in its turn, giving rise to orthoclase feldspar and to ferromagnesian elements devoid of alumina, such as the orthorhombic pyroxenes. Thus, the clay minerals that were prevalent in the hydrosphere disappear definitively during highest-grade metamosphism, such as that of the granulite facies, for instance, in which aluminum is feldspathic, while magnesium and part of the iron are pyroxenic. Let us note that the tolerances for substitution of iron and titanium in orthorhombic pyroxenes are limited. Therefore, we see that these elements, which diagenesis and incipient metamorphism attempt to silicate, are released again in the form of titaniferous iron which is typical of the granulite facies.

The study, in terms of crystallochemistry, of the stages of metamorphism is a fascinating work, but it is not appropriate here. The sole aim of this short exposé is to show how the layered clay minerals, which were formed in great abundance during weathering and sedimentation, resist metamorphism for a little while in the form of micas. But thermodynamic conditions at depth become so imperious that silicates leave this layered organization to take on other styles that are more compact and non-hydrated.

2° Convergence and divergence of the geochemical cycles of elements
(MILLOT, 1949)

Between the surfaces and depth, the constituent elements of silicates are engaged progressively in different structures, layered in the hydrosphere and essentially feldspathic in metamorphic and crystalline rocks. Each element works its way through the cycle after its own fashion; the geochemical cycles of the elements can be described as follows:

Alkaline elements and calcium (fig. 83). — At depth, sodium, potassium and calcium ions are essentially feldspathic. All three occupy holes in the three-dimensional structures of feldspars. This equality of function corresponds to an equality of proportion because the silico-aluminous crystalline and metamorphic rocks are, on the average, as rich in potassium as in sodium and in calcium.

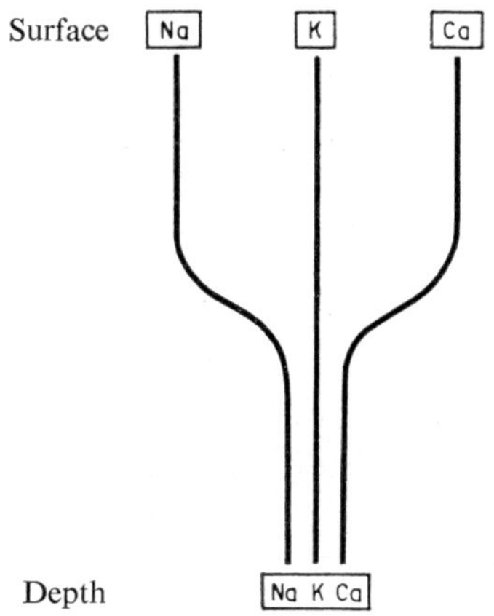

FIG. 83. — *Convergence and divergence of the geochemical cycles of* Na, K *and* Ca.

Upon reaching the surface, feldspars are weathered, and the three associates have different fates. Potassium is immediately engaged in the clay minerals of weathering, the sericites and illites; also it is adsorbed preferentially by mineral particles in which it is always ready to become engaged. Sodium, for its part, is at home in solution; it populates the waters of the sea. Then it will in some cases find refuge in saline deposits, but the largest part will enter sediments with connate waters, of which it is the chief component. Calcium is also soluble, but, if it is not admitted into layered silicates, it is hardly stable in solution and precipitates in the form or carbonates, either in limestone beds or mixed with clays to become marnes.

Restricting ourselves to broad outlines, we can say that *the fate of the* Na, K *and* Ca *elements converged at depth, but that it diverges on the surface.* Potassium enters silicates, calcium enters carbonates, and sodium remains in solution in the form of chlorides. But when metamorphism sets in, the calcium of marnes and the sodium confined in sediments rejoin potassium in feldspathic structures.

Magnesium and iron (fig. 84). — At depth magnesium and iron are associated. They are the main components of those silicates that petrographers call, appropriately, ferromagnesian. In these minerals, micas, amphiboles, pyroxenes and peridots, iron and magnesium have found refuge because the intense development of feldspathic structures does not tolerate them.

Upon reaching the surface, these two companions are liberated and in most cases they separate. Magnesium divides itself between two principal pathways. It rejoins, on one hand, potassium in clay minerals; or, on the other hand, it rejoins calcium in carbonates. For its part, iron has rather numerous modes of occurrence, as already noted, because of its polyvalency and its easy bonding with organic matter. Either it chooses liberty in the form of sesquioxides, rejoining aluminum, or it enters carbonates, rejoining calcium, or it enters silicates, rejoining potassium. One sees that the first choice corresponds to the oxidized form of iron which is trivalent, whereas the two others bring more readily into play the bivalency of ferrous iron. One sees also that

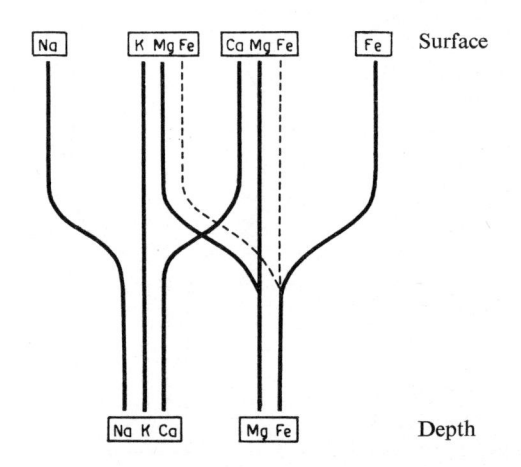

Fig. 84. — *Convergence and divergence of the geochemical cycles of* Na, K, Ca, Mg *and* Fe.

when it is trivalent, surficial iron is associated with aluminum, whereas it accompanies magnesium when it is bivalent. We have noted how free iron, during diagenesis and incipient metamorphism, tends to enter silicates; as a result *ferromagnesian minerals will be reconstructed by the time* the iron reaches the *middle zone of metamorphism*. In the catazone, with the development of the granulite facies, a new separation (from silica) will be imposed on them.

Aluminum and silicon (fig. 85). — At depth aluminum is essentially feldspathic and consequently tetracoordinate. This behavior is identical to that of silicon. In alkaline feldspars, one out of four tetracoordinate positions is occupied by aluminum and three out of four by silicon, whereas at the calcic pole of plagioclases, silicon and aluminum are on equal terms. Thus, at depth, aluminum functions in the same way as feldspathic silicon, which it replaces in a proportion that varies between 1/4 and 1/2. It was J. DE LAPPARENT (1941) in his *Logique des Minéraux du Granite* who pointed out this behavior, which seems obvious to us today, by contrasting it to the hexacoordinate activity of aluminum on the surface. This contrast is fundamental and summarizes the whole evolution of silicates.

As a matter of fact, this evolution takes place by small steps. During ordinary weathering, aluminum is engaged in sericite in which it is one-third tetracoordinate and two-third hexacoordinate. But during more vigorous weathering, which releases cations, aluminum combines with silica to give rise to kaolinite in which it is hexacoordinate. An extreme degree of weathering leaches even silica and leads to gibbsite, in which aluminum is hexacoordinate and free. One sees that the release of alumina bestows on it graduated possibilities, but, once free of cations, alumina comes into hexacoordination.

During metamorphism the reverse process takes place. Oxides become silicated, kaolinite is transformed into mica, and all the silico-aluminous material becomes

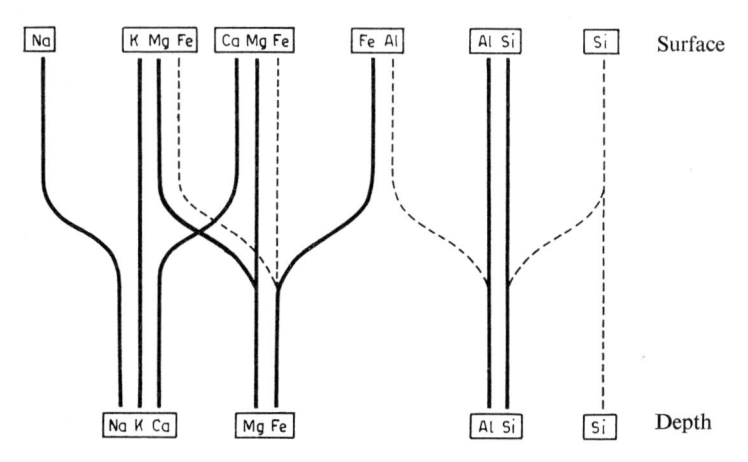

FIG. 85. — *Convergence and divergence of the geochemical cycles of* Na, K, Ca, Mg, Fe, Al *and* Si.

micaceous in epizone metamorphism. Therefore, aluminum is already one-third tetracoordinate. During feldspathization, the tetracoordination becomes general.

At depth, the fate of aluminum and that of silicon are closely convergent within feldspathic structures. Upon reaching the surface, aluminum and silicon are released by hydrolysis. It happens in some cases that they make use of their liberty. In this case, aluminum functions like trivalent iron, and its occurrence in lateritic crusts (*cuirasses*) is the best example. For its part, isolated silica is oriented toward silicifications. But in most cases, as soon as they are released, silicon and aluminum are nearly always associated again in the suite of clay minerals. This association will be more or less close as long as aluminum occupies the position and the function of silicon in the tetrahedral sheets or the "cation" position in the octahedral sheets. Let us note well that the more the other alkaline or alkaline-earth cations are absent, the more aluminum will tend to be the cation of the silicic acid, and the more we shall be able to speak properly of aluminum silicate, the model of which is kaolinite. On the other hand, the more abundant mono- and bivalent cations are, the more aluminum will be thrown back toward silicon, and we shall be able to speak of silico-aluminates. And this activity which begins on the surface can only increase in amplitude with depth. *And it can be verified that during the geochemical cycle aluminum passes from complete chemical liberty where it is hexacoordinate to geochemical dependence in feldspars in which it is tetracoordinate* (MILLOT, 1949).

Conclusions. — One of the most interesting studies for a geochemist is to follow the itinerary of elements through the geochemical cycle. There is agreement that this cycle through its succession of weathering, sedimentation, metamorphism and the genesis of the crystalline rocks summarizes the evolution of the Earth's crust.

Now, the first two stages take place in water and the following two in the crystalline state. And the behavior of elements is not the same in solution as within crystals, which are essentially silicates. During the cycle, therefore, the elements behave differently in their itinerary on the surface and their itinerary at depth.

What is curious is the fact that for certain elements these different behaviors result in common itineraries. But this community of trajectory is often ephemeral. Two elements that played similar roles at depth, where the crystalline state prevails, are separated in the hydrosphere where they are in solution. It is in this manner that the convergences and the divergences of the geochemical cycles of elements can be drawn. Here they are drawn for the major elements. Those who would like to can complicate this business by the examination of the itineraries of minor elements.

But such an analysis is only one means of understanding the behavior of the silicates of the Earth's crust. The silicates of the lithosphere make up metamorphic and crystalline rocks; they are essentially feldspathic. The silicates of the hydrosphere, for their part, are clays which are essentially layered. We all know of the patient work that was and is still necessary in order to reconstruct the history of the rocks at depth. When the lithosphere crops out, it undergoes a new transformation of incomparable seriousness relative to its previous history; the lithosphere is transformed into clays. But this is not a simple process that is accomplished all at once. On the contrary, this transformation is modest, progressive and varying in its evolution through weathering, sedimentation and diagenesis. The evolution of silicates in the hydrosphere is as much alive as that which takes place in the lithosphere; here we call it "the geology of clays".

BIBLIOGRAPHY

ABICH, H. (1856). — Untersuchungen über die Zusammensetzung des kaspischen Meerwassers. *Mem. Acad. Imp. Sci.*, 7, p. 12.

AGARD, J., DESTOMBES, J. et JEANNETTE, A. (1953). — Les gisements de vermiculite du Haut Atlas de Midelt. *Notes et Mém. Serv. Géol. Maroc*, 117, pp. 275-280.

AHRENS, L. H. (1952). — The use of ionization potentials. I. Ionic radii of the elements. *Geochim. Cosmochim. Acta*, 2, pp. 155-169.

AHRENS, L. H. (1953). — The use of ionization potentials. II. Anion affinity and geochemistry. *Geochim. Cosmochim. Acta*, 3, pp. 1-29.

ALEXANDER, G. B. (1953). — Preparation of monosilicic acid. *Journ. Amer. Chem. Soc.*, 75, pp. 2887-2888.

ALEXANDER, G. B. (1953). — Reaction of low molecular weight silicic acid with molybdic acid. *Journ. Amer. Chem. Soc.*, 75, pp. 5655-5657.

ALEXANDER, G. B., HESTON, W. M. and ILER, H. K. (1954). — The solubility of amorphous silica in water. *Journ. Phys. Chem.*, 58, pp. 453-455.

ALEXANDER, L. T., HENDRICKS, S. B. and NELSON, R. A. (1939). — Minerals present in soil colloïds. *Soil Sci.*, 48, p. 273.

ALIMEN, H. (1936). — Etude sur le Stampien du Bassin de Paris. *Mém. Soc. Géol. Fr.*, 14, n° 31, 304 pages.

ALIMEN, H. (1944). — Roches gréseuses à ciment calcaire du Stampien, étude pétrographique. *Bull. Soc. Géol. Fr.*, 14, pp. 307-328.

ALIMEN, H. (1954). — Colorimétrie des sédiments quaternaires et paléoclimats. Premiers résultats. *Bull. Soc. Géol. Fr.*, 4, pp. 609-620.

ALIMEN, H. et DEICHA, G. (1958). — Observations pétrographiques sur les meulières pliocènes. *Bull. Soc. Géol. Fr.*, 8, pp. 77-90.

ALLENS, V. T. and JOHNS, W. D. (1960). — Clays and clay minerals of New England and eastern Canada. *Bull. Geol. Soc. Amer.*, 71, pp. 75-86.

ANDREATTA, A. C. (1950). — Sull'alterabilità dei minerali delle rocce. *XVIIIth Intern. Geol. Congr. Great Britain*, 2, pp. 5-8.

ANDRUSSOV, N. (1897). — Der Atschi-Daria oder Kara-Bugas Busen. *Peterm. Mitt.*, 43, pp. 25-34.

AOMINE, S. et YOSHINAGA, N. (1955). — Clay minerals of some well-drained volcanic ash soils in Japan. *Soil Sci.*, 79, pp. 349-358.

ARBEY, F. et LE FOURNIER, J. (1962). — Note préliminaire sur une altération de la glauconie à la base de la série ordovicienne du Sahara. *C. R. Acad. Sci. Fr.*, 254, pp. 143-145.

ARNEMAN, H. F. and WRIGHT, H. E. Jr. (1959). — Petrography of some Minnesota tills. *Journ. Sedim. Petrol.*, 29, pp. 540-555.

ATAMAN, G. et WEY, R. (1961). — Sur les phyllites du bassin potassique d'Alsace. *Bull. Serv. Carte Géol. Als. Lor.*, 14, pp. 129-137.

AUBERT, G. et MONJAUZE, A. (1946). — Observations sur quelques sols de l'Oranie nord-occidentale. *C. R. Séances Soc. Biogéogr.*, 23, n° 199, pp. 44-51.

AUBERT, G. (1951). — Les sols et l'aménagement agricole de l'Afrique occidentale française. *Cahiers des Ing. Agron.*, 11 pages.

AUBERT, G. (1954). — Les sols latéritiques. — Ve *Congr. Intern. Sci. Sol*, 1, pp. 103-118.

AUBERT, G. et DUCHAUFOUR, Ph. (1956). — Projet de classification des sols. VIe *Congr. Intern. Sci. Sol*, 5, E, pp. 597-604.

AUBERT, G. (1958). — Classification des sols. *Bull. ORSTOM*, 8, pp. 1-3.

AUBRÉVILLE, A. (1949). — *Climats, forêts et désertification de l'Afrique tropicale.* Soc. Edit. Marit. Col., Paris.
AUBRÉVILLE, A. (1949). — *Contribution à la paléohistoire des forêts de l'Afrique tropicale.* Soc. Edit. Marit. Col., Paris, 98 pages.
AUFRÈRE, L. (1930). — Les formations continentales éocènes du Berry oriental et méridional (calcaire lacustre et sidérolithique des auteurs) et leur signification morphologique. *Ass. Fr. Av. Sci., Congr. Alger, C. R.*, pp. 152-155.
AUFRÈRE, L. (1937). — L'origine de la silice de l'eau de mer. *Bull. Soc. Océan. Fr.*, 95, pp. 1645-1648.
AUZEL, M. et CAILLEUX, A. (1949). — Silicifications nord-sahariennes. *Bull. Soc. Géol. Fr.*, 5, pp. 553-559.
BANNISTER, F. (1943). — Brammallite (sodium-illite) a new mineral from Llandebie, South Wales. *Miner. Mag.*, 26, pp. 304-307.
BARDOSSY, G. (1958, 1959). — The geochemistry of hungarian bauxites. *Acta Geol. Acad. Sci. Hungar.*, 5, fasc. 3/4, parts 1-2; 6, fasc. 1/2, parts 3-4.
BARDOSSY, G. (1961). — *A magyar bauxit geokemiai vizsgalata.* Müszaki Könyvkiadò, Budapest.
BAREN, J. VAN (1928). — Microscopical, physical and chemical studies of limestones and limestone soils from the East Indian archipelago. *Med. Landbouwk. Wageningen*, 32, Verh. 7, 186 pages.
BARIC, L. (1955). — Pyrophyllitschiefer von Parsovici in der Herzegowina. *Bull. Sci. Yougoslavie*, 2, n° 3, p. 91.
BARON, G. (1958). — Précipitation de la giobertite et de la dolomie à partir des solutions de chlorures de magnésium et de calcium. *C. R. Acad. Sci. Fr.*, 247, n° 19, pp. 1606-1608.
BARON, G. et FAVRE, J. (1958). — Etat actuel des recherches en direction de la synthèse de la dolomie. *Rev. Inst. Fr. Pétrole*, 13, p. 1067.
BARON, G. (1960). — Sur la synthèse de la dolomite. Application au phénomène de la dolomitisation. *Rev. Inst. Fr. Pétrole*, 15, pp. 3-69.
BARRELL, J. (1908). — Relations between climate and terrestrial deposits. *Journ. Geol.*, 16, pp. 159-190, 255-295, 363-384.
BARROIS, C. (1884). — Le granite de Rostrennen, ses apophyses et ses contacts. *Ann. Soc. Géol. Nord*, 12, pp. 1-119.
BARSHAD, I. (1948). — Vermiculite and its relation to biotite as revealed by base exchange reactions, X-ray analyses, differential thermal curves, and water content. *Amer. Min.*, 33, pp. 655-678.
BARSHAD, I. (1949). — The nature of lattice expansion and its relation to hydratation in montmorillonite and vermiculite. *Amer. Min.*, 34, pp. 675-684.
BARSHAD, I. — (1950). — The effect of the interlayer cations on the expansion of the mica type of crystal lattice. *Amer. Min.*, 35, pp. 225-238.
BASSETT, W. A. (1960). — Role of hydroxyl orientation in mica alteration. *Bull. Geol. Soc. Amer.*, 71, n° 4, pp. 449-456.
BATES, T. F. (1947). — Investigations on the micaceous minerals in slate. *Amer. Min.*, 32, pp. 625-636.
BATES, T. F., SAND, L. B. and MINK, J. F. (1950). — Tubular crystals of chrysotile asbestos. *Science*, U.S.A., 3, n° 2889, pp. 512-513.
BEAVERS, A. H., JOHNS, W. D., GRIM, R. E. and ODEL, R. T. (1955). — Clay minerals in some Illinois soils developed from loess and till under grass vegetation. *Clays and clay minerals* (3rd Nat. Conf., 1954), pp. 356-372.
BEHNE, W. und HOENES, D. (1955). — Die Kaolinlagerstätte von Geisenheim (Rheingau). *Heidelberg. Beitr., Min. Petr.*, 4, pp. 412-433.
BELOV, N. V. (1957). — Nouvelles structures de silicates. *Cristallographie, Acad. Sci. URSS*, 2, p. 366.
BENSON, W. N. (1918). — The origin of serpentine. *Amer. Journ. Sci.*, 46, pp. 693-731.
BENTOR, Y. K. (1952). — Relations entre la tectonique et les dépôts de phosphates dans le Neguev israélien. *XIXe Congr. Intern. Géol. Alger*, fasc. 9, pp. 93-102.

BENTOR, Y. K. (1957). — Genetic classification of clays and shales in Israel. *Bull. Research Council Israel*, Sec. B, pp. 279-280.
BERGOUNIOUX, F. M. (1947). — Sur la genèse des argiles sidérolithiques. *C. R. Soc. Géol. Fr.*, 17, pp. 167-168.
BERNAL, J. D. and FOWLES, R. H. (1933). — Note on the pseudo-crystalline nature of water. *Faraday Soc. Trans.*, 29, pp. 1049-1056.
BERNAL, J. D. et MEGAW, H. D. (1935). — Le rôle de l'hydrogène dans les forces intermoléculaires. *Proc. Royal Soc.*, Sér. A, pp. 384-420.
BERNARD, A. (1958). — Contribution à l'étude de la province métallifère sous-cévenole. *Thèse Sci. Nancy*, et *Sciences de la Terre* (1959), 5, pp. 123-403.
BERSIER, A. (1953). — La sédimentation cyclique des faciès détritiques molasse et houiller, signification et causes. *Rev. Inst. Fr. Pétrole*, 8, numéro spécial, pp. 51-55.
BERSIER, A. (1958). — Séquences détritiques et divagations fluviales. *Eclogae Geol. Helv.*, 51, pp. 854-893.
BERTHIER, P. (1820). — Analyse des nodules de chaux phosphatée qui se trouvent dans la craie du cap La Hève (près Le Havre). *Ann. Mines*, pp. 197-204.
BERTHIER, P. (1821). — Analyse de plusieurs minéraux que l'on rapporte ordinairement à l'espèce chlorite. *Ann. Mines*, pp. 459-464.
BERTHIER, P. (1826). — Analyse de l'halloysite. *Ann. Chim. Phys.*, 32, pp. 332-335.
BERTHOIS, L. (1950). — Méthode d'étude des galets. Applications à l'étude de l'évolution des galets marins actuels. *Rev. Géomorph. Dyn.*, 5, pp. 199-225.
BERTRAND, L. et LANQUINE, A. (1919). — Les roches siliceuses envisagées au point de vue de la fabrication des briques de silice. *Bull. Off. Rech. Invent.*, 1, pp. 55-61; 2, pp. 121-127.
BERTRAND, L. et LANQUINE, A. (1919). — Sur les relations entre la composition chimique, la structure microscopique et les qualités céramiques des argiles. *C. R. Acad. Sci. Fr.*, 169, pp. 1171-1174.
BERTRAND, L. et LANQUINE, A. (1922). — Sur la composition et la structure microscopique des argiles. Leur fusibilité et les transformations qu'elles subissent à haute température. *Bull. Off. Rech. Invent.*, 27.
BÉTREMIEUX, R. (1951). — Etude expérimentale de l'évolution du fer et du manganèse dans les sols. *Ann. Agron.*, 3, pp. 193-295.
BEUF, S., MONTADERT, L. et DEBYSER, J. (1962). — Sur des structures sédimentaires dénommées « cordon » dans les grès de l'unité IV, Cambro-Ordovicien du Tassili des Ajjers entre l'oued Tassed et Djanet. *C. R. Acad. Sci. Fr.*, 254, pp. 892-894.
BICHELONNE, J. et ANGOT, P. (1938). — *Le bassin ferrifère de Lorraine*. Berger-Levrault, Nancy-Strasbourg, 483 pages.
BIEN, G. S., CONTOIS, D. E. et THOMAS, W. H. (1958, 1959). — The removal of soluble silica from fresh water entering the sea. *Geochim. Cosmochim. Acta*, 14, pp. 35-54 (1958) et *Soc. Econ. Pal. Miner., Spec. Publ.*, 7, pp. 20-35 (1959).
BIGOTTE, G. (1956). — Contribution à la géologie du bassin du Niari. *Thèse Sci., Nancy*, et *Bull. Direct. Mines, Géol. A.E.F.* (1959), 9, 188 pages.
BIGOTTE, G., BONIFAS, M. et MILLOT, G. (1957). — Présence du talc dans les roches sédimentaires infracambriennes du bassin du Niari, Congo français (A.E.F.). *Bull. Serv. Carte Géol. Als. Lor.*, 10, fasc. 2, pp. 3-7.
BILLIET, V. (1944). — Quelques aspects de la minéralogie moderne. *Bull. Soc. belge Géol.*, 53, n° 3, pp. 179-198.
BIROT, P. (1947). — Résultats de quelques expériences sur la désagrégation des roches cristallines. *C. R. Acad. Sc. Fr.*, 225, p. 745.
BIROT, P. (1954). — Désagrégation des roches cristallines sous l'action des sels. *C. R. Acad. Sci. Fr.*, 238, p. 1145.
BIROT, P., CAPOT-REY, R. et DRESCH, J. (1955). — Recherches morphologiques dans le Sahara central. *Trav. Inst. Rech. Sahar.*, 13, pp. 13-74.
BIROT, P., CAILLÈRE, S. et HÉNIN, S. (1959). — Etude du problème des premiers stades de l'altération de quelques roches. *Ann. Agron.*, Sér. A, pp. 257-265.

BIROT, P. et PINTA, J. (1961). — Données sur les variations de l'altération argileuse dans le Hoggar. *C. R. Soc. Géol. Fr.*, pp. 21-22.

BIROT, P., HÉNIN, S., GUILLIEN, Y. et DELVERT, J. (1961). — *Contribution à l'étude de la désagrégation des roches.* Centre Docum. Universit., Paris, 231 pages.

BISQUE, R. E. (1962). — Clay polymerization in carbonate rocks: a silicification reaction defined. *Clays and clay minerals* (9th Nat. Conf., 1960), pp. 365-374.

BLACK, W. A. P. and MITCHELL, R. L. (1952). — Trace elements in the common brown algae and in sea water. *Journ. Marine Biol. Assoc. United Kingd.*, 30, p. 575.

BLOOMFIELD, C. (1952). — Translocation of iron in podzol formation. *Nature*, 170, p. 540.

BLOOMFIELD, C. (1955). — The movement of sesquioxides and clay in the soil. *African Soils*, 3, pp. 488-506.

BÖHM, J. (1925). — Über Aluminium und Eisenhydroxyde. *Zeitschr. anorg. allg. Chemie*, 11, p. 203.

BÖHM, J. (1928). — Röntgenographische Untersuchung der mikrokristallinen Eisenhydroxydminerale. *Zeitschr. Krist.*, 68, pp. 567-585.

BOILLOT, G. et MILLOT, G. (1962). — Sur une formation « sidérolithique » en place sous le Lutétien au large de Roscoff. *C. R. Acad. Sci. Fr.*, 254, pp. 3008-3010.

BOLGER, R. C. and WEITZ, J. H. (1952). — Mineralogy and origin of the mercer fireclay of North-Central Pennsylvania. *Prob. Clay Laterite Genesis, A.I.M.M.E.*, New York, pp. 81-93.

BONIFAS, M. (1959). — Contribution à l'étude géochimique de l'altération latéritique. *Thèse Sci. Strasbourg et Mém. Serv. Carte Géol. Als. Lor.*, 17, 159 pages.

BONTE, A. (1958). — Réflexion sur l'origine des bauxites et sur l'altération superficielle des calcaires. *LXXXIIIe Congr. Soc. sav.*, pp. 147-165.

BONYTHON, C. W. (1956). — The salt of lake Eyre. Its occurrence in Madigan Gulf and its possible origin. *Trans. Roy. Soc. South Australia*, 79, pp. 66-92.

BORDET, P. (1951). — Présence de latérites fossiles dans l'Atakor du Hoggar. *C. R. Soc. Géol. Fr.*, 1, p. 97.

BORDET, P. (1955). — La série de Sérouenout (Continental intercalaire) et son substratum précambrien en Ahaggar oriental. *Serv. Carte Géol. Algérie*, N.S., 5, Trav. Collab. (1954), pp. 7-42.

BOSAZZA, V. L. (1948). — The petrography and petrology of south african clays. *Thèse, Johannesbourg*, 313 pages.

BOURCART, J. (1936). — Le gisement d'argiles smectiques de Camp Berteaux (Maroc oriental). *C. R. Soc. Géol. Fr.*, 6, pp. 94-95.

BOURCART, J. (1937). — *Les pénéplaines du Maroc et du Sahara.* Recueil offert à E. F. Gautier. Arthaud, Tours, 75 pages.

BOURCART, J. (1938). — La marge continentale. Essai sur les régressions et transgressions marines. *Bull. Soc. Géol. Fr.*, 8, pp. 393-474.

BOUREAU, E. et FREULON, J. M. (1959). — Sur les flores jurassiques du Continental intercalaire saharien. *C. R. Soc. Géol. Fr.*, 3, p. 53.

BOUROZ, A., DOLLE, P. et PUIRABAUD, C. (1958). — La série stratigraphique et les tonstein du Westphalien C du Sud-Ouest de la concession de Noeux. *Ann. Soc. Géol. Nord*, 78, pp. 108-123.

BRADLEY, W. F. (1940). — The structural scheme of attapulgite. *Amer. Min.*, 25, pp. 204-205.

BRADLEY, W. F. (1945). — Diagnostic criteria for clay minerals. *Amer. Min.*, 30, pp. 704-713.

BRADLEY, W. F. (1950). — The alterning layer sequence of rectorite. *Amer. Min.*, 35, pp. 590-595.

BRADLEY, W. F. (1953). — Analysis of mixed-layer clay mineral structures. *Analytical Chem.*, 25, pp. 727-730.

BRADLEY, W. F. (1955). — Structural irregularities in hydrous magnesium silicates. *Clays and clay minerals* (3rd Nat. Conf., 1954), pp. 94-102.

BRADLEY, W. F. and WEAVER, C. E. (1956). — A regularly interstratified chlorite vermiculite clay mineral. *Amer. Min.*, 41, pp. 497-504.

BRADLEY, W. H. (1929). — The occurrence and origin of analcite and meerschaum beds in the Green River formation of Utah, Colorado, and Wyoming. *U.S. Geol. Survey Prof. Paper*, 158-A, pp. 1-9.

BRAGG, W. L. and WARREN, B. E. (1930). — The structure of chrysotile $H_4Mg_3Si_2O_9$. *Zeitschr. Krist.*, 76, pp. 201-210.

BRAITSCH, O. (1957). — Über die natürlichen Faser- u. Aggregationstypen beim SiO_2, ihre Verwachsungsformen, Richtungsstatistik u. Doppelbrechung. *Heidelberg. Beitr. Min. Petr.*, 5, pp. 331-372.

BRAJNIKOV, B. (1937). — Recherches sur la formation appelée « argile à silex » dans le Bassin de Paris. *Rev. Géogr. Phys. Géol. Dyn.*, 10, pp. 1-90 et 109-129.

BRAJNIKOV, B. et MALYCHEFF, V. (1938). — Sur un constituant de néoformation et sur le milieu physico-chimique de quelques formations superficielles. *Rev. Géogr. Phys. Géol. Dyn.*, 11, pp. 249-253.

BRAJNIKOV, B. (1942). — Quelques considérations sur l'évolution des limons. *Bull. Soc. Géol. Fr.*, 12, pp. 91-96.

BRAMMAL, A. (1921). — Reconstitution process in shales, slates and phyllites. *Miner. Mag.*, 19, pp. 211-225.

BRAMMAL, A., LEECH, J. G. C. and BANNISTER, F. A. (1937). — The paragenesis of cookeite and hydromuscovite associated with gold at Ogofau, Carmarthenshire. *Miner. Mag.*, 24, pp. 507-520.

BRAUNER, K. und PREISINGER, A. (1956). — Struktur und Entstehung des Sepioliths. *Tchermaks Min. Petr. Mitt.*, 1-2, pp. 120-140.

BRIGGS, L. I. (1958). — Evaporit facies. *Journ. Sedim. Petrol.*, 28, p. 46.

BRINDLEY, G. W. and ROBINSON, K. (1946). — The structure of kaolinite. *Miner. Mag.*, 27, pp. 242-253.

BRINDLEY, G. W., ROBINSON, K. and GOODYEAR, J. (1948). — X-ray studies of halloysite and metahalloysite. *Miner. Mag.*, 28, pp. 393-428.

BRINDLEY, G. W. (1949). — Mineralogy and crystal structure of chamosite. *Nature*, 164, pp. 319-320.

BRINDLEY, G. W. (1951). — The crystal structure of some chamosite minerals. *Miner. Mag.*, 29, pp. 502-525.

BRINDLEY, G. W. (1951). — X-ray identification and crystal structures of clay minerals. *Miner. Soc. Great Britain*, Monograph, London, 345 pages.

BRINDLEY, G. W. (1952). — Structural relationships in the kaolin group of minerals. *XVIIIth Intern. Geol. Congr., Great Britain* (1948), Part 13, p. 305.

BRINDLEY, G. W. and YOUELL, R. F. (1953). — Ferrous chamosite and ferric chamosite. *Miner. Mag.*, 30, 220, pp. 57-70.

BRINDLEY, G. W. and GILLERY, F. H. (1954). — A mixed-layer kaolin-chlorite structure. *Clays and clay minerals* (2nd Nat. Conf., 1953), pp. 349-353.

BRINDLEY, G. W. and GILLERY, F. H. (1956). — X-ray identification of chlorite species. *Amer. Min.*, 41, pp. 169-186.

BRINDLEY, G. W. (1957). — Fuller's earth from near Dry Branch, Georgia, a montmorillonite-cristobalite clay. *Clay Min. Bull.*, 3, 18, p. 167.

BRINDLEY, G. W. (1959). — X-ray and electron diffraction data for sepiolite. *Amer. Min.*, 44, pp. 459-500.

BROECK, E. VAN DEN (1878). — Du rôle de l'infiltration des eaux météoriques dans l'altération des dépôts superficiels. *C. R. Congr. Géol., Paris*.

BROECK, E. VAN DEN (1898). — Les coupes du gisement de Bernissart. Caractères et dispositions sédimentaires de l'argile ossifère du Cran aux Iguanodons. *Bull. Soc. belge Géol.*, 12, p. 216.

BRONGNIART, A. (1807). — *Traité élémentaire de minéralogie*. Paris, 2 vol.

BRONGNIART, A. (1822). — Notice sur la magnésie du Bassin de Paris et sur le gisement de cette roche dans divers lieux. *Ann. Mines*, 7, pp. 291-318.

BRONGNIART, A. (1823). — *Mémoire sur les terrains de sédiments supérieurs calcaréo-trappéens du Vicentin*. Levrault, Paris, 85 pages.

BRONGNIART, A. (1828). — Notice sur les brèches osseuses et les minerais de fer pisiformes de même position géognostique. *Ann. Sci. Nat.*, 14, p. 410.

BROPHY, J. A. (1959). — Heavy mineral ratios of Sangamon weathering profiles in Illinois. *Illinois State Geol. Surv.*, 273, 22 pages.

BROWN, B. E. and JACKSON, M. L. (1958). — Clay mineral distribution in the Hiawatha sandy soils of Northern Wisconsin. *Clays and clay minerals* (5th Nat. Conf., 1956), pp. 213-226.

BROWN, G. and MACEWAN, D. M. C. (1950). — The interpretation of X-ray diagrams of soil clays. II. Structures with random interstratification. *Journ. Soil Sci.*, 1, pp. 239-253.

BROWN, G. and NORRISH, K. (1952). — Hydrous micas. *Min. Mag.*, 29, pp. 929-932.

BROWN, G. (1953). — The dioctahedral analogue of vermiculite. *Clay Min. Bull.*, 10, pp. 64-69.

BROWN, G. (1955). — Report of the clay minerals group subcommittee on nomenclature of clay minerals. *Clay Min. Bull.*, 2, pp. 294-302.

BROWN, G. (1961). — The X-ray identification and crystal structures of clay minerals. *Min. Soc.* (Clay Min. Group), London, 544 pages.

BROWN, L. G. (1953). — Geological aspects of the St Austell granite. *Clay Min. Bull.*, 9, pp. 17-21.

BUBENICEK, L. (1960). — Développement diagénétique des chlorites de la minette lorraine. *C. R. Acad. Sci. Fr.*, 251, p. 765.

BUBENICEK, L. (1960). — Recherches sur la constitution et la répartition du minerai de fer dans l'Aalénien de Lorraine. *Thèse Sci., Nancy*, 177 pages.

BUBENICEK, L. (1961). — Conditions paléogéographiques de formation de la minette lorraine. *C. R. Acad. Sci. Fr.*, 253, pp. 1468-1469.

BUERGER, M. J. (1954). — The stuffed derivatives of the silica structures. *Amer. Min.*, 39, pp. 600-614.

BURST, J. F. (1956). — Mineralogic variability in glauconitelike pellets and its application in stratigraphic studies. *Geol. Soc. Amer. Bull.*, 67, p. 1678.

BURST, J. F. (1958). — "Glauconite"-pellets: their mineral nature and applications to stratigraphic interpretations. *Bull. Am. Ass. Petrol. Geol.*, 42, pp. 310-327.

BURST, J. F. (1958). — Mineral heterogenity in "Glauconite"-pellets. *Amer. Min.*, 43, pp. 481-497.

BURST, J. F. (1959). — Postdiagenetic clay mineral environmental relationships in the Gulf Coast Eocene. *Clays and clay minerals* (6th Nat. Conf., 1957), pp. 327-341.

BYSTRÖM, A. M. (1954). — Mineralogy of the ordovician bentonite beds at Kinnekulle (Sweden). *Sver. Geol. Unders.* Sér. C, 540, 62 pages.

BYSTRÖM, A. M. (1957). — The clay minerals in the Ordovician bentonite beds in Billingen, Southwest Sweden. *Geol. Fören. Förh.*, 79, pp. 52-56.

CAILLÈRE, S. (1936). — Contribution à l'étude des minéraux des serpentines. *Bull. Soc. Fr. Minér.*, 59, pp. 163-326.

CAILLÈRE, S. (1936). — Etude de quelques silicates magnésiens à faciès asbestiforme ou papyracé n'appartenant pas au groupe de l'antigorite. *Bull. Soc. Fr. Minér.*, 59, pp. 352-373.

CAILLÈRE, S. (1939). — Sur quelques palygorskites du Sahara occidental. *C. R. Congr. Soc. Sav.*, Sect. Sci., pp. 147-150.

CAILLÈRE, S., HÉNIN, S. et MERING, J. (1947). — Passage expérimental de la montmorillonite à une pyhllite à équidistance stable de 14 Å. *C. R. Acad. Sci. Fr.*, 224, pp. 842-843.

CAILLÈRE, S., HÉNIN, S. et MERIAUX, S. (1948). — Transformation expérimentale d'une montmorillonite en une phyllite à 10 Å type illite. *C. R. Acad. Sci. Fr.*, 226, pp. 680-681.

CAILLÈRE, S., HÉNIN, S. et GUENNELON, G. (1949). — Transformation expérimentale du mica en divers types de minéraux argileux par séparation des feuillets. *C. R. Acad. Sci. Fr.*, 228, pp. 1741-1742.

CAILLÈRE, S., PERRIN-BONNET, I. et KRAUT, F. (1950). — Etude minéralogique des schistes de la mine Amélie II (Bassin de Mulhouse). *Mémor. Serv. Chim. Etat Fr.*, 35, 2, pp. 75-86.

CAILLÈRE, S. (1951). — Sur la présence d'une palygorskite à Tafraout (Maroc). *C. R. Acad. Sci. Fr.*, 233, pp. 697-698.

CAILLÈRE, S. et HÉNIN, S. (1951). — Etude de quelques altérations de la phlogopite à Madagascar. *C. R. Acad. Sci. Fr.*, 233, pp. 1383-1385.

CAILLÈRE, S. (1952). — Sur la présence de la sépiolite à Bouazer (Maroc). *Bull. Soc. Fr. Min. Crist.*, pp. 305-307.

CAILLÈRE, S. et HÉNIN, S. (1952). — Extraction et étude minéralogique des phyllites des minerais de fer. *XIXe Congr. Géol. Intern. Alger,* CIPEA, pp. 57-64.

CAILLÈRE, S., HÉNIN, S. et ESQUEVIN, J. (1953). — Recherches sur la synthèse des minéraux argileux. *Bull. Soc. Fr. Min. Crist.*, 76, pp. 300-314.

CAILLÈRE, S., HÉNIN, S. et ESQUEVIN, J. (1953). — Synthèse à basse température de phyllites ferrifères. *C. R. Acad. Sci. Fr.*, 237, pp. 1724-1726.

CAILLÈRE, S., HÉNIN, S. et ESQUEVIN, J. (1954). — Synthèse de quelques phyllites nickélifères. *C. R. Acad. Sci. Fr.*, 239, pp. 1535-1537.

CAILLÈRE, S. et KRAUT, F. (1954). — Les gisements de fer du bassin lorrain. *Mém. Mus. Hist. Nat. Fr.* Sér. C, 1re éd., 4, 2e éd. (1957), 7, 175 pages.

CAILLÈRE, S., BIROT, P. et HÉNIN, S. (1954). — Etude expérimentale du mécanisme de la désagrégation de quelques roches éruptives et métamorphiques. *Actes et C. R. Ve Congr. Intern. Sci. Sol*, 2, pp. 18-22.

CAILLÈRE, S., HÉNIN, S. et ESQUEVIN, J. (1955). — Synthèse à basse température de quelques minéraux ferrifères (silicates et oxydes). *Bull. Soc. Fr. Min. Crist.*, 78, pp. 227-242.

CAILLÈRE, S., HÉNIN, S. et ESQUEVIN, J. (1956). — Etude de quelques silicates nickélifères naturels et de synthèse. *Bull. Groupe Fr. Arg.*, 7, pp. 21-31.

CAILLÈRE, S. et JOURDAIN, M. A. (1956). — Sur quelques particularités de la région de Montguyon (Charente-Maritime). *Bull. Groupe Fr. Arg.*, 7, pp. 31-39.

CAILLÈRE, S., HÉNIN, S. et BIROT, P. (1957). — Sur la formation transitoire de montmorillonite dans certaines altérations latéritiques. *C. R. Acad. Sci. Fr.*, 244, pp. 788-791.

CAILLÈRE, S., HÉNIN, S. et ESQUEVIN, J. (1957). — Synthèse des minéraux argileux. Etude des réactions du silicate de sodium et des cations Al et Mg. *Bull. Groupe Fr. Arg.*, 9, pp. 67-77.

CAILLÈRE, S. et MALYCHEFF, V. (1957). — Etude de la fraction argileuse des loess du Bassin de Paris. *C. R. Acad. Sci. Fr.*, 245, pp. 1446-1448.

CAILLÈRE, S., HÉNIN, S. et ESQUEVIN, J. (1958). — Synthèse des argiles cobaltifères. *Clay Min. Bull.*, 3, pp. 232-238.

CAILLÈRE, S. et ROUAIX, S. (1958). — Sur la présence de palygorskite dans la région de Taguenout-Hagueret (A.O.F.). *C. R. Acad. Sci. Fr.*, 246, pp. 1442-1444.

CAILLÈRE, S. et HÉNIN, S. (1959). — La classification des argiles. *XXe Congr. Géol. Intern. Mexico* (1956), pp. 1-8.

CAILLÈRE, S. et HÉNIN, S. (1959). — Les phyllites des minerais de fer sédimentaires et leurs conditions de genèse. *XXe Congr. Géol. Intern. Mexico* (1956), pp. 9-20.

CAILLÈRE, S., HÉNIN, S. et ESQUEVIN, J. (1959). — Essai de synthèse de phyllites manganésifères. *Bull. Groupe Fr. Arg.*, 11, pp. 53-57.

CAILLÈRE, S. (1960). — Compte rendu de la réunion du CIPEA à Copenhague en 1960. *Bull. Groupe Fr. Arg.*, 12, pp. 97-98.

CAILLÈRE, S., GATINEAU, L. et HÉNIN, S. (1960). — Préparation à basse température d'hématite alumineuse. *C. R. Acad. Sci. Fr.*, 250, pp. 3677-3679.

CAILLÈRE, S. et HÉNIN, S. (1960). — Propriétés des ions et conditions de synthèse des minéraux argileux : le cas du fer. *XXIst Geol. Intern. Congr.*, Norden, 24, pp. 73-79.

CAILLÈRE, S. et POBEGUIN, T. (1961). — Sur les minéraux ferrifères des bauxites. *C. R. Acad. Sci. Fr.*, 253, pp. 288-291.

CAILLÈRE, S. et HÉNIN, S. (1961). — Vues d'ensemble sur le problème de la synthèse des minéraux phylliteux à basse température. *Coll. Intern. C.N.R.S.*, 105, pp. 31-43.

CAILLÈRE, S. et HÉNIN, S. (1961). — Préparation d'hydroxydes mixtes d'aluminium et de fer. *C. R. Acad. Sci. Fr.*, 253, pp. 690-692.

CAILLÈRE, S., ESTÉOULE, J. et HÉNIN, S. (1962). — Préparation de silicates alumino-magnésiens à partir de gels. *C. R. Acad. Sci. Fr.*, 254, pp. 2380-2383.

CAILLÈRE, S., HÉNIN, S. et POBEGUIN, T. (1962). — Présence d'un nouveau type de chlorite dans les « bauxites » de Saint-Paul-de-Fenouillet (Pyrénées-Orientales). *C. R. Acad. Sci. Fr.*, 254, pp. 1657-1659.

CAILLEUX, A. (1945). — Distinction des sables marins et fluviatiles. *Bull. Soc. Géol. Fr.*, 13, pp. 125-138.

CAILLEUX, A. (1945). — Distinction des galets marins et fluviatiles. *Bull. Soc. Géol. Fr.*, 15, pp. 375-404.

CAILLEUX, A. (1947). — Concrétions quartzeuses d'origine pédologique. *Bull. Soc. Géol. Fr.*, 17, pp. 475-483.

CAILLEUX, A. (1953). — Graviers du Pliocène et du Quaternaire inférieur européens et nord-américains. *Geol. Bavarica*, 19, pp. 307-314.

CAILLEUX, A. et TAYLOR, G. (1954). — Cryopédologie. Etude des sols gelés. *Actualités Scient. Industr.*, 1203, Paris, 218 pages.

CAILLEUX, A. et TRICART, J. (1959). — *Initiation à l'étude des sables et galets*. Centre Docum. Universit., Paris, 3 vol.

CAILLEUX, A. (1961). — Sur une poussière transportée par le vent en mer Rouge. *C. R. Acad. Sci. Fr.*, 252, pp. 905-908.

CAMEZ, T. et ROTH, C. (1957). — Evolution des minéraux argileux des lehms des environs de Strasbourg. *Bull. Serv. Carte Géol. Als. Lor.*, 10, pp. 21-25.

CAMEZ, T., LUCAS, J. et MILLOT, G. (1959). — Minéraux argileux interstratifiés dans certains sols et leur évolution. *Bull. Groupe Fr. Arg.*, 11, pp. 43-47.

CAMEZ, T., FRANC DE FERRIÈRE, P. J. J., LUCAS, J. et MILLOT, G. (1960). — Sur l'importance de la vermiculite dans certains sols tempérés et dans les dépôts du Quaternaire ancien. *C. R. Acad. Sci. Fr.*, 250, pp. 3038-3041.

CAMEZ, T. (1962). — Etude sur l'évolution des minéraux argileux dans les sols des régions tempérées. *Thèse Sci., Strasbourg et Mém. Serv. Carte Géol. Als. Lor.*, 20, 90 pages.

CAPDECOMME, L. (1952). — Sur les phosphates alumineux de la région de Thiès (Sénégal). *C. R. Acad. Sci. Fr.*, 235, p. 187.

CAPDECOMME, L. et KULBICKI, G. (1954). — Argiles des gîtes phosphatés de la région de Thiès (Sénégal). *Bull. Soc. Fr. Min. Crist.*, 77, pp. 500-518.

CAROZZI, A. (1951). — Glauconitisation de la biotite dans le Crétacé moyen des chaînes subalpines et du Jura. *Cahiers géol. Thoiry*, 4, pp. 33-37.

CAROZZI, A. (1953). — *Pétrographie des roches sédimentaires*. Lausanne, 250 pages.

CAROZZI, A. (1960). — *Microscopic sedimentary petrography*. London and New York, 485 pages.

CARROLL, D. and STARKEY, H. C. (1960). — Effect of sea-water on clay minerals. *Clays and clay minerals* (7th Nat. Conf., 1958), pp. 80-102.

CARTLEDGE, G. H. (1928). — Studies on the periodic system. I. The ionic potential as a periodic function. *Journ. Amer. Chem. Soc.*, 50, pp. 2855-2863.

CAVANAUGH, R. J. and KNUTSON, C. F. (1960). — Laboratory technique for plastic saturation of porous rocks. *Bull. Amer. Assoc. Petr. Geol.*, 44, 5, pp. 628-630.

CAYEUX, L. (1897). — Contribution à l'étude micrographique des terrains sédimentaires. *Mém. Soc. Géol. Nord*, 4, fasc. 2, pp. 1-589.

CAYEUX, L. (1913). — Les minerais de fer sédimentaires considérés dans leurs rapports avec la destruction des chaînes de montagnes. *C. R. Acad. Sci. Fr.*, 156, p. 1185.

CAYEUX, L. (1916). — Introduction à l'étude pétrographique des roches sédimentaires. *Mém. Carte Géol. Fr.*, 2 vol., 524 pages.

CAYEUX, L. (1922). — *Les minerais de fer oolithiques de France, minerai de fer secondaire*, II, Imp. Nat., Paris, 1051 pages.

CAYEUX, L. (1929). — Les roches siliceuses. *Mém. Carte Géol. Fr.*, 774 pages.

CAYEUX, L. (1932). — Les manières d'être de la glauconie en milieu calcaire. *C. R. Acad. Sci. Fr.*, 195, pp. 1050-1052.

CHALARD, J. (1951). — Tonstein du bassin houiller du Nord de la France. *IIIe Congr. Intern. Stratigr. et Géol. du Carbonifère*, Heerlen, pp. 73-77.

CHAMBRE SYNDICALE de la Recherche et de la Production du Pétrole et du Gaz naturel (1961). — *Essai de nomenclature des roches sédimentaires*, Paris, 80 pages.

CHARLES, G. (1949). — Sur le phénomène de rubéfaction et ses conditions climatiques. *C. R. Acad. Sci. Fr.*, 228, p. 589.

CHARLOT, G. (1949). — *Théorie et méthode nouvelle d'analyse qualitative*. Masson et Cie, Paris.

CHEPILOV, K. R., ERMOLOVA, E. P. et ORLOVA, N. A. (1959). — Epigenetic minerals as indicators of the time of arrival of petroleum into commercial sand reservoirs. *Dokl. Akad. Nauk. SSSR*, 125, n° 5, pp. 1097-1099. (Trad.: Assoc. Techn. Serv., Inc., East-Orange, New Jersey, U.S.A.)

CHILINGAR, G. V. (1956). — Distribution and abundance of chert and flint as related to the Ca/Mg ratio of limestones. *Geol. Soc. Amer. Bull.*, 67, pp. 1559-1561.

CHOLLEY, A. (1943). — Recherches sur les surfaces d'érosion et la morphologie de la région parisienne. *Ann. Géogr.*, 52, n°s 289, 290, 291, pp. 81-97 et 161-189.

CHOUBERT, G. (1945). — Note préliminaire sur le Pontien au Maroc. Essai de synthèse orogénique du Maroc atlasique. *Bull. Soc. Géol. Fr.*, 15, pp. 677-764.

CHOUBERT, G. (1950). — Réflexions au sujet du Pliocène continental. *Notes et Mém. Serv. Géol. Maroc.*, 76-3, pp. 13-92.

CHOUBERT, G. (1953). — Les rapports entre les formations marines et continentales quaternaires. *Actes IVe Congr. INQUA*, pp. 576-591.

CHOUBERT, G., JOLY, F., GIGOUT, M., MARÇAIS, J., MARGAT, J. et RAYNAL, R. (1956). — Essai de classification du Quaternaire continental au Maroc. *C. R. Acad. Sci. Fr.*, 243, p. 504.

CHOUBERT, G. (1957). — L'Adoudounien et le Précambrien III dans l'Anti-Atlas. *LXXVIe Coll. Intern. C.N.R.S., Paris*, pp. 143-162.

CHOUBERT, G. (1959). — Coup d'œil sur la fin du Précambrien et le début du Cambrien dans le Sud marocain. *Notes et Mém. Serv. Géol. Maroc*, 17, pp. 7-34.

CLABAUGH, S. E. and BARNES, V. E. (1959). — Vermiculite in Central Texas. *Bureau Econ. Geol. Repart. Invest.*, 40, 32 pages.

CLOUD, P. E. Jr. (1955). — Physical limits of glauconite formation. *Bull. Amer. Assoc. Petrol. Geol.*, 39, pp. 484-493.

COLE, N. F. (1942). — X-ray analysis (by the powder method) and microscopic examination of the product of weathering of the Gingin upper Greensand. *Journ. Roy. Soc. West Australia*, 27, pp. 229-243.

COLLET, L. W. (1908). — *Les dépôts marins*. Doin, Paris, 325 pages.

COLLIER, D. (1951). — Contribution à l'étude de l'altération des granites et de l'évolution des sols sur les plateaux granitiques de l'Auvergne. *Bull. Ass. Fr. Et. Sol*, pp. 1-12.

COLLIER, D. (1951). — Sur l'altération du granite à gros grain en Auvergne. *C. R. Acad. Sci. Fr.*, 233, p. 96.

COLLIER, D. (1961). — Mise au point sur les processus de l'altération des granites en pays tempéré. *Ann. Agron.*, 12, pp. 273-331.

COLLINI, B. (1950). — Om våra kvartära lerors mineralogiska sammansättning. *Geol. Fören, Stock. Förh.*, 72, pp. 192-206.

COLLINI, B. (1956). — On the origin and formation of the fennoscandian Quaternary clays. *Geol. Fören. Stock. Förh.*, 78, pp. 528-534.

COMMITTEE ON SEDIMENTATION (1936). — Report on the committee on sedimentation. *Nat. Res. Council, Washington*, 1935-1936.

CORRENS, C. W. (1929). — Bestimmung der Brechungsexponenten in Gemengen feinkörniger Minerale und von Kolloiden. *Fortschr. Min. Kryst. Petr.*, 14, pp. 26-27.

CORRENS, C. W. und NAGELSCHMIDT, G. (1933). — Über Faserbau und optische Eigenschaften von Chalzedon. *Zeitschr. Krist.*, 85, pp. 199-213.

CORRENS, C. W. (1937). — *Scientific results of the German Atlantic Expedition on the exploration ship "Meteor"*. Deutsche Atl. Exp. Meteor. 1925-1927.

CORRENS, C. W. und ENGELHARDT, W. VON (1938). — Neue Untersuchungen über die Verwitterung des Kalifeldspates. *Chemie der Erde*, 12, pp. 1-22.

CORRENS, C. W. (1938). — Zur Frage der Neubildung von Glimmer in jungen Sedimenten. *Geol. Rundsch.*, 29, pp. 220-222.

CORRENS, C. W. (1939). — *Die Sedimentgesteine*. I. Verwitterung, in BARTH, CORRENS and ESKOLA, Springer, Berlin, pp. 116-130.
CORRENS, C. W. (1939). — Pelagic sediments of the North Atlantic Ocean. Recent marine sediments. *Symposium SEPM, Sp. Publ.*, 4, pp. 373-395.
CORRENS, C. W. (1940). — Die chemische Verwitterung der Silicate. *Die Naturwissensch.*, 28, pp. 369-376.
CORRENS, C. W. (1941). — Über die Löslichkeit von Kieselsäure in schwachsauren und alkalischen Lösungen. *Chemie der Erde*, 13, pp. 92-96.
CORRENS, C. W. (1947). — Über die Bildung der sedimentaren Eisenerze. *Forsch. u. Fortschr.*, 4-6, pp. 21-23.
CORRENS, C. W. (1948). — Probleme der Sedimentpetrographie. *Zeitschr. Deutsch. Geol. Gesellsch.*, 100, pp. 158-163.
CORRENS, C. W. (1950). — Zur Geochemie der Diagenese. I. Das Verhalten von CO_3Ca und SiO_2. *Geochim. Cosmochim. Acta*, 1, pp. 49-54.
CORRENS, C. W. und PILLER, H. (1954). — Mikroskopie der feinkörnigen Silikatminerale. *Handb. Mikrosk. i.d. Techn.*, 4, pp. 699-780.
CORRENS, C. W. (1961). — The experimental chemical weathering of silicates. *Clay Min. Bull.*, 4, pp. 249-265.
CUTHBERT, F. L. (1944). — Clay minerals in Lake Erie sediments. *Amer. Min.*, 29, pp. 378-388.
DAMOUR, A. A. et SALVETAT, D. (1847). — Notice et analyses sur un hydrosilicate d'alumine trouvé à Montmorillon (Vienne). *Ann. Chim. Phys.*, Sér. 3, 21, pp. 376-383.
DAPPLES, E. C. (1959). — The behaviour of silica in diagenesis. *Soc. Econ. Pal. Miner. Spec. Publ.*, 7, pp. 36-54.
DARMOIS, E. (1946). — *Mémorial des Sciences Physiques*, fasc. 48.
DARS, R. (1957). — Sur l'existence du Continental intercalaire au nord-est de Nara (Afrique Occidentale française). *C. R. Soc. Géol. Fr.*, 12, p. 248.
DARS, R. (1960). — Les formations sédimentaires et les dolérites du Soudan occidental (Afrique de l'Ouest). *Thèse Sci., Paris et Mém. B.R.G.M.*, 12 (1961), 327 pages.
DAUBRÉE, A. (1857). — Recherches expérimentales sur le striage des roches dû au phénomène erratique et sur les décompositions chimiques produites dans les réactions mécaniques. *C. R. Acad. Sci. Fr.*, 44, p. 997.
DEBRENNE, F. (1958). — Un cas d'épigénie de fossiles par la chlorite. *C. R. Acad. Sci. Fr.*, 247, pp. 2023-2025.
DEBYSER, J. (1957). — Note sur un procédé de préparation des plaques minces dans les sédiments fins actuels. *Rev. Inst. Fr. Pétrole*, 12, pp. 489-492.
DEBYSER, J. (1959). — Contribution à l'étude géochimique des vases marines. *Thèse Sci., Paris*, 210 pages.
DEFFEYES, K. S. (1959). — Zeolites in sedimentary rocks. *Journ. Sedim. Petrol.*, 29, pp. 602-610.
DÉFOSSEZ, M. (1958). — Contribution à l'étude géologique et hydrogéologique de la boucle du Niger. *Thèse Sci. Strasbourg*, 227 pages.
DEGENS, E. T., WILLIAMS, E. G. and KEITH, M. L. (1957). — Environmental studies of carboniferous sediments. I. Geochemical criteria for differentiating marine and freshwater shales. *Bull. Amer. Assoc. Petrol. Geol.*, 41, pp. 2427-2455.
DEGENS, E. T., WILLIAMS, E. G. and KEITH, M. L. (1958). — Environmental studies of carboniferous sediments. II. Application of geochemical criteria. *Bull. Amer. Assoc. Petrol. Geol.*, 42, p. 981.
DEGENS, E. T., KEITH, M. L., WILLIAMS, E. G. and KANEHIRO, Y. (1959). — Environmental study of some recent sediments from Hawaï. *Journ. Geol.*
DEICHA, G. (1942). — Sur les conditions de dépôt dans le golfe du gypse parisien. *C. R. Acad. Sci. Fr.*, 214, p. 863.
DEICHA, G. (1942). — Zones bimensuelles, saisonnières et annuelles dans le gypse parisien. *C. R. Soc. Géol. Fr.*, 9, p. 83.
DEICHA, G. (1945). — Le quartz et le problème de la silicification. *C. R. Soc. Géol. Fr.*, 8, pp. 93-95.

DEICHA, G. (1946). — Individualité pétrographique et faciès cristallographique. *C. R. Soc. Géol. Fr.*, 16, pp. 220-222.

DEJOU, J. (1959). — Etude comparative des phénomènes d'altération sur granite porphyroïde de Lormes et sur anatexites à cordiérite du Morvan Nord et les sols qui en dérivent. *Ann. Inst. Nat. Rech. Agron.*, Sér. A, pp. 101-123.

DEJOU, J. (1961). — Les anatexites à cordiérite de Neuvic d'Ussel (Corrèze). Leur altération comparée à celle du Morvan Nord. *Bull. Ass. Fr. Et. Sol*, numéro spécial, pp. 54-85.

DEKEYSER, W. et AMELINCKX, S. (1955). — *Les dislocations et la croissance des cristaux*. Masson et Cie, Paris, 185 pages.

DEKEYSER, W., HOEBEKE, F. et VAN KEYMEULEN, J. (1955). — Les minéraux argileux des polders et de quelques autres régions naturelles. *I.R.S.I.A., C. R. Rech.*, 14, pp. 123-161.

DEKEYSER, W., VAN KEYMEULEN, J., HOEBEKE, F. et VAN RYSSEN, A. (1955). — L'altération et l'évolution des minéraux micacés et argileux. *I.R.S.I.A., C. R. Rech.*, 14, pp. 91-103.

DEMANGEON, P. et SALVAYRE, H. (1961). — Sur la genèse de palygorskite dans un calcaire dolomitique. *Bull. Soc. Fr. Min. Crist.*, 84, 2, pp. 201-202.

DEMOLON, A. et BASTISSE, E. (1936). — Genèse des colloïdes argileux dans l'altération du granite en cases lysimétriques. *C. R. Acad. Sci. Fr.*, 203, p. 736.

DEMOLON, A. (1939). — *Dynamique du sol*. Dunod, Paris (4e édition en 1948).

DEMOLON, A. et BASTISSE, E. (1946). — Observations sur les premiers stades de l'altération spontanée d'un granite et la genèse des colloïdes argileux. *C. R. Acad. Sci. Fr.*, 223, p. 115.

DENISOFF, I. (1957). — Un type particulier de concrétionnement en cuvette centrale congolaise. *Pédol. Gand*, 7, pp. 119-123.

DERRUAU, M. (1960). — Quel était le climat du Massif Central français pendant la seconde moitié de l'ère tertiaire ? *Rev. d'Auvergne*, 74, pp. 179-184.

DESCHAMPS, M. (1957). — Le conglomérat sidérolithique de Doyet-Montvic (Allier). *C. R. Acad. Sci. Fr.*, 244, pp. 637-639.

DESCHAMPS, M. (1958). — Les relations de l'arkose de Cosne avec les grès rouges sidérolithiques (feuille de Moulins, Allier). *C. R. Acad. Sci. Fr.*, 246, pp. 1444-1447.

DESCHAMPS, M. (1960). — Les rapports du Sidérolithique du Cher, daté Sannoisien supérieur, avec le petit bassin tongrien de Gouzon (Creuse). *C. R. Acad. Sci. Fr.*, 250, pp. 730-732.

DESCHAMPS, M. (1962). — L'altération des anatexites dans le Sidérolithique du Lembron (Puy-de-Dôme). *C. R. Acad. Sci. Fr.*, 254, pp. 1831-1834.

DEVIGNE, J. P. et REYRE, D. (1957). — Notice explicative sur la feuille Mayoumba-Ouest. *Carte géologique de reconnaissance au 1/500 000*. Gouvernement Général de l'A.E.F.

DEVORE, G. W. (1956). — Surface chemistry as a chemical control on mineral association. *Journ. Geol.*, 64, pp. 31-55.

D'HOORE, J. (1954). — L'accumulation des sesquioxydes libres dans les sols tropicaux. *I.N.E.A.C.*, Sér. Sci., 62, 131 pages.

DIADTCHENKO, M. G. et HATUNTCEVA, A. I. (1955). — Au sujet de la genèse de la glauconie. *Bull. Acad. Sci. URSS*, 101.

DIENERT, F. et WANDENBULCKE, F. (1923). — Sur le dosage de la silice dans les eaux. *C. R. Acad. Sci. Fr.*, 176, pp. 1478-1480.

DIETZ, R. S. (1941). — Clay minerals in recent marine sediments. *Thèse Univ. Illinois*. Résumé in *Amer. Min.*, 1942, 27.

DIEULAFAIT, L. (1877). — L'acide borique ; méthodes de recherches; son existence normale dans les eaux de mers modernes et dans celles de tous les âges... *Ann. Chim. Phys.*, 12, pp. 318-354.

DIEULAFAIT, L. (1884). — Origine et mode de formation des phosphates de chaux en amas dans les terrains sédimentaires. Leur liaison avec les minerais de fer et les argiles des terrains sidérolithiques. *Ann. Chim. Phys.*, 6e sér., 5, p. 204.

DONNAY, J. D. H. (1936). — La biréfringence de forme dans la calcédoine. *Ann. Soc. Géol. Belg.*, 59, pp. 289-302.

DORSEY, G. E. (1926). — The origin of the color of red beds. *Journ. Geol.*, 34, pp. 131-143.

DOUVILLÉ, H. (1936). — Les sables et les argiles granitiques, leur distribution et leur origine. *Bull. Soc. Géol. Fr.*, 6, pp. 17-40.

DRENCK, K. (1959). — *X-ray particlesize determination and its application to flint*. X-ray and crystal structure Labor., Pennsylv. Univ., 158 pages.
DROSTE, J. B. (1956). — Alteration of clay minerals by weathering in Wisconsin tills. *Bull. Geol. Soc. Amer.*, 67, pp. 911-916.
DROSTE, J. B. et THARIN, J. C. (1958). — Alteration of clay minerals in Illinoian tills by weathering. *Bull. Geol. Soc. Amer.*, 69, pp. 61-68.
DROSTE, J. B. (1959). — Clay minerals in Playas of the Mojave desert (California). *Science*, 130, p. 100.
DROSTE, J. B., BHATTACHARYA, H. and SUNDERMAN, J. A. (1962). — Clay mineral alteration in some Indiana soils. *Clays and clay minerals* (9th Nat. Conf., 1960), pp. 329-343.
DUCHAUFOUR, P., MICHAUD, R. et MILLOT, G. (1952). — Etude des éléments fins de quelques types de sol de climat atlantique. *Géol. Appl. Prosp. Min., Nancy*, 4, 3, pp. 31-62.
DUCHAUFOUR, P. (1956). — *Pédologie. Applications forestières et agricoles*. Ecole Nat. Eaux Forêts, Nancy, 438 pages.
DUCHAUFOUR, P. (1960). — *Précis de pédologie*. Masson et Cie, Paris, 438 pages.
DUNCAN HERON, S. (1960). — Clay minerals of the outcroping basal Cretaceous beds between the Cape Fear river, North Carolina, and Lynches river, South Carolina. *Clays and clay minerals* (7th Nat. Conf., 1958), pp. 148-161.
DUNHAM, K. C. (1952). — Red coloration in desert formations of permian and triassic age in Britain. *XIXth Congr. Intern. Alger*, Sect. 7, fasc. 7, pp. 25-32.
DUNOYER DE SEGONZAC, G. et MILLOT, G. (1962). — Pyrophyllite de diagenèse dans le Dévonien inférieur du synclinal de Laval (Massif Armoricain). *C. R. Acad. Sci. Fr.*, 255, pp. 3438-3440.
DUPARQUE, A. (1946). — Remarques préliminaires sur les caractères pétrographiques des grès et des schistes houillers du Nord de la France. *Ann. Soc. Géol. Nord*, pp. 137-157.
DUPLAIX, S., DUPUIS, J., CAMEZ, T., LUCAS, J. et MILLOT, G. (1960). — Sur la nature des minéraux argileux inclus dans les calcaires sénoniens de l'Angoumois. *Bull. Serv. Carte Géol. Als. Lor.*, 13, pp. 157-162.
DURAND, J. G. (1960). — Utilisation d'une réserve synthétique polymérisable dans la confection des lames minces sur échantillon calcaire ou argileux. *Rev. Inst. Fr. Pétrole*, 15, pp. 673-679.
DURAND, J. H. (1959). — *Les sols rouges et les croûtes en Algérie*. Serv. Et. Sci., Clairbois-Birmandreis, 187 pages.
DURAND, S. et MILON, Y. (1959). — Relations des sédimentations crétacées et tertiaires en Bretagne. *LXXXIVe Congr. Soc. Sav.*, pp. 146-162.
DURAND, S. (1960). — Le Tertiaire de Bretagne. *Mém. Soc. Géol. Bret.*, 12, 378 pages.
DWYER, F. P. and MELLOR, D. P. (1934). — An X-ray study of opals. *Journ. Roy. Soc. New South Wales*, 68, pp. 47-50.
DZENS-LITOVSKY, A. I. (1952-1954). — Le Kara-Bogaz-Gol dans le passé géologique et actuellement. Dokl. ezhegod. Chten. Pam. L.S. Berga S.S.S.R., 1-3, pp. 131-169 (*Trad. B.R.G.G.M., n° 2044, 1956*).
ECKHARDT, F. J. (1958). — Über Chlorite in Sedimenten. *Geol. Jb.*, 75, pp. 437-474.
ECKHARDT, F. J. (1960). — Die Veränderungen eines devonischen Tonschiefers durch die Mineralienwandlungen infolge der tertiären Zersetzung. *Zeitschr. deutsch. geol. Ges.*, 112, pp. 188-196.
EDELMAN, C. H., BAREN, F. A. VAN et FAVEJEE, J. C. L. (1939). — *Overdruk Meded. Landbouwk.*, 43, pp. 1-39.
EDELMAN, C. H. (1947). — Relations entre les propriétés et la structure de quelques minéraux argileux. *Verres et silicates*, 12-6, pp. 3-6.
EDMUNDS, F. H. (1948). — The Wealdien district. *Geol. Surv. and Mus. and Mus. Brit. reg. Geol.*, p. 88.
ELLENBERGER, F. (1948). — Métamorphisme, silicification et pédogenèse en Bohême méridionale. *Ann. Sci. Franche-Comté*, pp. 122-125.
ELOUARD, P. (1959). — Etude géologique et hydrogéologique des formations sédimentaires du Guebla mauritanien et de la vallée du Sénégal. *Thèse Sci., Paris*, 372 pages.

ELOUARD, P. et MILLOT, G. (1959). — Observations sur les silicifications du Lutétien en Mauritanie et dans la vallée du Sénégal. *Bull. Serv. Carte Géol. Als. Lor.*, 12, pp. 15-21.
ENDELL, J. (1948). — Röntgenographischer Nachweis kristalliner Zwischenzustände bei der Bildung von Cristobalit aus Kieselgur beim Erhitzen. *Koll. Zeit.*, 111, pp. 19-22.
ENDELL, J. (1955). — Clay minerals in coals and in their ashes. *Clay Min. Bull.*, 2, pp. 289-294.
ENDELL, K., HOFMANN, U. und MÄGDEFRAU, E. (1935). — Über die Natur des Tonanteils in Rohstoffen der deutschen Zementindustrie. *Zement*, 24, pp. 625-632.
EPSTEIN, S., BUCHSBAUM, R., LOWENSTAM, H. and UREY, H. C. (1951). — Carbonate-water isotopic temperature scale. *Bull. Geol. Soc. Amer.*, 62, pp. 417-426.
EPSTEIN, S., BUCHSBAUM, R., LOWENSTAM, H. and UREY, H. C. (1953). — Revised carbonate-water isotopic temperature scale. *Bull. Geol. Soc. Amer.*, 64, pp. 1315-1325.
ERHART, H. (1940). — Sur l'altération des basaltes miocènes du Cantal et la nature des sols qui en dérivent. *C. R. Acad. Sci. Fr.*, 210, pp. 537-539.
ERHART, H. (1943). — Les latérites du Moyen-Niger et leur signification paléo-climatique. *C. R. Acad. Sci. Fr.*, 217, pp. 323-325.
ERHART, H. (1953). — Sur la nature minéralogique et la genèse des sédiments de la cuvette tchadienne. *C. R. Acad. Sci. Fr.*, 237, pp. 401-403.
ERHART, H. (1955). — « Biostasie » et « Rhexistasie ». Esquisse d'une théorie sur le rôle de la pédogenèse en tant que phénomène géologique. *C. R. Acad. Sci. Fr.*, 241, pp. 1218-1220.
ERHART, H. (1956). — *La genèse des sols en tant que phénomène géologique*, Masson et Cie, Paris, 83 pages.
ERHART, H. (1961). — Sur la genèse de certains gîtes sédimentaires de fer. *C. R. Acad. Sci. Fr.*, 252, p. 3307.
ERHART, H. (1962). — Témoins pédogénétiques de l'époque permo-carbonifère. *C. R. Soc. Biogéogr.*, 335-336-337, pp. 23-53.
ERMOLOVA, E. P. (1955). — Analcime et mordenite dans les dépôts oligocènes et miocènes du Caucase de l'Ouest. *Acad. Sci. U.R.S.S. Trav. Musée Minér.*, 7, pp. 76-82.
ESQUEVIN, J. (1956). — Synthèse des phyllites zincifères. *Bull. Gr. Fr. Argiles*, 8, 3, pp. 23-27.
ESQUEVIN, J. (1958). — Les silicates de zinc. Etude de produits de synthèse et des minéraux naturels. *Thèse Sci. Paris*, 1958, et *Ann. Inst. Nat. Rech. agron.*, A 11, 1960, pp. 497-556.
ESTÉOULE-CHOUX, J. (1962). — Etude comparée de la sédimentation argileuse dans les bassins tertiaires de Campbon et Saffre (Loire-Atlantique). Premier Coll. Intern. Strat. Paléogène, Bordeaux, 1962.
FABRE, J. et FEYS, R. (1962). — Réflexions sur la genèse des bassins houillers et la théorie bio-rhexistasique. *C. R. Soc. Biogéogr.*, 335-336-337, pp. 6-15.
FAIRBAIRN, H. W. (1943). — Packing in ionic minerals. *Bull. Geol. Soc. Amer.*, 54, pp. 1305-1374.
FAIRBAIRN, H. W. (1943). — X-ray petrology of some fine grained foliated rocks. *Amer. Min.*, 28, pp. 246-256.
FALKE, H. (1961). — La question des conditions probables du climat de l'Autunien et du Saxonien de l'Europe centrale et occidentale. *Bull. Soc. Géol. Fr.*, 3, pp. 463-467.
FAURE, H. (1961). — Intérêt de la paléogéographie pour la prospection des substances utiles au Niger. *Note destinée au Bull. du B.R.G.M. Dakar*.
FAURE, H. (1962). — Reconnaissance géologique des formations postpaléozoïques du Niger oriental. *Thèse Sci., Paris*, 483 pages.
FAUST, G. T., HATHAWAY, J. C. and MILLOT, G. (1959). — A restudy of stevensite and allied minerals. *Amer. Min.*, 44, pp. 342-370.
FERSMANN, A. (1908). — *Mém. Acad. Sci. Saint-Pétersbourg*, 6, p. 645.
FERSMANN, A. (1913). — Recherches sur les silicates de magnésie. *Mém. Acad. Sci. Saint-Pétersbourg*, 32, pp. 377-392.
FERSMANN, A. (1926). — Über die Erscheinung der Silicifizierung in der Mittelasiatischen Wüste Karakum. *C. R. Acad. Sci. U.R.S.S.*, p. 145.
FIELDES, M. (1955). — Clay mineralogy of New-Zealand soils. *Journ. Sci. and Techn. N. Zeal.*, 37, pp. 336-350.

FINATON, Ch. (1934). — Les dépôts lagunaires et le gypse du Bassin Parisien. *Rev. Géog. Phys. Géol. Dynam.*, 7, p. 357.
FINATON, Ch. (1935). — La formation des gisements salifères. *Rev. Géog. Phys. Géol. Dynam.*, 8, pp. 285-303.
FINATON, Ch. (1937). — Essai d'interprétation mathématique de quelques phénomènes de sédimentation. *Rev. Géog. Phys. Géol. Dynam.*, 10, pp. 255-263.
FLAWN, P. T. (1953). — Petrographic classification of argillaceous sedimentary and low-grade metamorphic rocks in subsurface. *Bull. Am. Ass. Petrol. Geol.*, 37, pp. 560-565.
FLEISHER, M. (1943). — Discredited species: illite — bravaisite. *Amer. Min.*, 28, pp. 214 and 470.
FLEURY, E. (1909). — Le Sidérolithique suisse. Contribution à la connaissance des phénomènes d'altération superficielle des sédiments. *Mém. Soc. Fribourgeoise Sci. Nat.*, 6, 260 pages.
FLÖRKE, O. W. (1955). — Zur Frage des „Hoch"-Cristobalit in Opalen, Bentoniten und Gläsern. *Neues Jb. Min., Monatsh.*, 10, pp. 217-223.
FLÖRKE, O. W. (1961). — A discussion of the tridymite-cristobalite problem. *Silicates industr.*, 26, pp. 415-418.
FLÖRKE, O. W. (1961). — Untersuchungen an amorphem und mikrokristallinem SiO_2. *Silicates industr.*, 26, pp. 424-427.
FOLK, R. L. and WEAVER, C. E. (1952). — A study of the texture and composition of chert. *Amer. Journ. Sci.*, 250, pp. 498-510.
FORESTIER, H. et CHAUDRON, G. (1925). — Points de transformation des solutions d'alumine ou de sesquioxyde de chrome dans le sesquioxyde de fer. *C. R. Acad. Sci. Fr.*, 180, pp. 1264-1266.
FOSTER, M. D. (1953). — Geochemical studies of clay minerals. *Geochim. Cosmochim. Acta*, 3, pp. 143-154.
FOSTER, M. D. (1954). — The relation between "illite", beidellite and montmorillonite. *Clays and clay minerals* (2nd Nat. Conf., 1953), pp. 386-397.
FOSTER, M. D. (1955). — The relation between composition and swelling in clays. *Clays and clay minerals* (3rd Nat. Conf., 1954), pp. 205-220.
FOSTER, M. D. (1956). — Correlation of dioctahedral potassium micas on the basis of their charge relations. *U. S. Geol. Surv. Bull.*, 1036-D, pp. 57-67.
FRANC DE FERRIÈRE, P. J. J. (1956). — Un problème pédologique : la consommation en potasse des départements français. *VIe Congr. Sci. Sol, Paris*, IV, 66, pp. 449-456.
FRANC DE FERRIÈRE, P. J. J., CAMEZ, T. et MILLOT, G. (1957). — Nutrition potassique de la pomme de terre et du blé en fonction des types d'argile des sols et des pluviosités. *Bull. Serv. Carte Géol. Als. Lor.*, 10, 2, pp. 11-17.
FRANC DE FERRIÈRE, P. J. J., CAMEZ, T. et MILLOT, G. (1958). — Influence des types d'argile des sols et de la pluviosité sur la nutrition potassique de la pomme de terre. *Ann. Agron.*, Série A, pp. 51-70.
FRANC DE FERRIÈRE, P. J. J., MILLOT, G. et CAMEZ, T. (1959). — Argile des sols des formations tertiaires et quaternaires d'Aquitaine. *Bull. Ass. Fr. Et. Sol*, pp. 512-526.
FRANC DE FERRIÈRE, P. J. J., BLANCHET, R., MILLOT, G. et CAMEZ, T. (1960). — Influence de la nature des argiles sur la dynamique du potassium dans les sols de quelques champs d'essais. *Ann. Agron.*, A, pp. 163-175.
FRANC DE FERRIÈRE, P. J. J. (1961). — Etude des minéraux argileux des sols du périmètre syndical d'irrigation de Montbreton à Pessac-sur-Dordogne. *Bull. Ass. Fr. Et. Sol*, pp. 54-68.
FRANCIS, E. H. (1961). — Thin beds of graded kaolinized tuff and tuffaceous siltstone in the carboniferous of Fife. *Bull. Geol. Surv. of Great Brit.*, 17, pp. 191-214.
FRANK-KAMENETSKY, V. A. (1960). — A crystallochemical classification of simple and interstratified clay minerals. *Clay Min. Bull.*, 4, pp. 161-172.
FREDERICKSON, A. F. (1951). — Mécanisme de l'altération. *Bull. Geol. Soc. Amer.*, 62, pp. 221-231.
FREDERICKSON, A. F. and REYNOLDS, R. C. (1960). — Geochemical method for determining paleosalinity. *Clays and clay minerals* (8th Nat. Conf., 1959), pp. 203-213.

FRIDMAN, R. (1953). — Les minéraux argileux des vases côtières de l'Atlantique. *C. R. Acad. Sci. Fr.*, 236, pp. 2095 2097.
FRIPIAT, J. J. (1960). — Application de la spectroscopie infra-rouge à l'étude des minéraux argileux. *Bull. Gr. Fr. Argiles*, 12, pp. 25-43.
FÜCHTBAUER, H. (1950). — Die nichtkarbonatischen Bestandteile des Göttinger Muschelkalkes mit besonderer Berück. *Heidelberg. Beitr.*, 2, pp. 235-254.
FÜCHTBAUER, H. (1955). — Zur Petrographie des Bentheimer Sandsteins im Emsland. *Erdöl u. Kohle*, 8, pp. 616-617.
FÜCHTBAUER, H. und GOLDSCHMIDT, H. (1956). — Ein Zechsteinanhydrit-Profil mit Einlagerungen von Montmorillonit und einer abweichenden Serpentinvarietät. *Heidelberg. Beitr.*, 5, pp. 187-203.
FÜCHTBAUER, H. und GOLDSCHMIDT, H. (1959). — Die Tonminerale der Zechsteinformation. *Beitr. z. Miner. u. Petrogr.*, 6, pp. 320-345.
FÜCHTBAUER, H. (1961). — Zur Quartzneubildung in Erdöllagerstätten. *Erdöl u. Kohle*, 14, pp. 169-173.
GABERT, P. (1961). — Un problème de l'altération quaternaire : le ferretto de l'Italie du Nord. *C. R. Soc. Géol. Fr.*, pp. 22-24.
GABIS, V. (1958). — Etude préliminaire des argiles oligocènes du Puy-en-Velay (Haute-Loire). *Bull. Soc. Fr. Min. Crist.*, 81, pp. 183-186.
GABIS, V. (1959). — Note sur les argiles oligocènes du bassin du Puy-en-Velay (Haute-Loire). *C. R. Acad. Sci. Fr.*, 248, pp. 3583-3584.
GALLIHER, E. W. (1935). — Glauconite genesis. *Bull. Geol. Soc. Amer.*, 46, pp. 1351-1366.
GALLITELLI, P. (1961). — Remarques sur la genèse de quelques minéraux argileux des Apennins de l'Italie du Nord. *Coll. Intern. C.N.R.S.*, 105, pp. 191-195.
GALPIN, S. L. (1912). — Studies on Flint Clays and their associates. *Trans. Amer. Ceramic Soc.*, 14, pp. 301-346.
GARRELS, R. M. and HOWARD, P. (1959). — Reactions of feldspar and mica with water at low temperature and pressure. *Clays and clay minerals* (6th Nat. Conf., 1957), pp. 68-88.
GASTUCHE, M. C., DELVIGNE, J. et FRIPIAT, J. J. (1954). — Altération chimique des kaolinites. *Ve Congr. Intern. Sci. Sol*, 2, pp. 439-449.
GASTUCHE, M. C., FRIPIAT, J. J. et DE KIMPE, C. (1961). — La genèse des minéraux argileux de la famille du kaolin. I. Aspect colloïdal. *Coll. Intern. C.N.R.S.*, 105, pp. 57-65.
GASTUCHE, M. C. et DE KIMPE, C. (1961). — La genèse des minéraux argileux de la famille du kaolin. II. Aspect cristallin. *Coll. Intern. C.N.R.S.*, 105, pp. 67-81.
GAUBERT, R. (1925). — Sur l'identité de la limonite fibreuse avec la goethite. *C. R. Acad. Sci. Fr.*, 181, pp. 869-872.
GÈZE, B. (1947). — Paléosols et sols dus à l'évolution actuelle. Importance relative en pédologie théorique et appliquée. *Ann. Ec. Nat. Agric. Montpellier*, 27, pp. 263-288.
GIESEKING, J. E. (1949). — *Adv. Agron.*, 1, p. 159.
GIESEKING, J. E. and MORTLAND, M. M. (1951). — Influence of the silicate ion on potassium fixation. *Soil Sci.*, 71, pp. 381-385.
GIESLER (1875). — Das oolithische Eisensteinvorkommen in Deutsch-Lothringen. *Zeitschr. f. Berghütten- u. Salinenwesen im preuß. Staat.*
GILLIS, E. et DEKEYSER, W. (1961). — Expériences avec des gels de silice et d'alumine. *Coll. Intern. C.N.R.S.*, 105, pp. 25-29.
GINSBURG, I. I. (1912-1913). — Kaolin i ego genesis. *Isv. CPb. polytech. Inst.*, 17 et 18.
GIOT, R. P. (1944). — Contribution à l'étude des terrains tertiaires du Royans (Isère et Drôme). *Trav. Lab. Géol. Univ. Grenoble*, 24, pp. 49-68.
GJEMS, O. (1960). — Some notes on clay minerals in podzol profiles in Fennoscandia. *Clay Min. Bull.*, 4, pp. 208-211.
GLANGEAUD, L. (1941). — Corrélation statistique, classification et hiérarchie des facteurs intervenant dans la formation des sédiments. *Bull. Soc. Géol. Fr.*, 11, pp. 371-388.
GLASS, H. D., POTTER, P. E. and SIEVER, R. (1956). — Clay mineralogy of some basal pennsylvanian sandstones. clays and shales. *Bull. Am. Ass. Petrol. Geol.*, 40, pp. 750-754.
GLASS, H. D. (1958). — Clay mineralogy of Pennsylvanian sediments in southern Illinois. *Clays and clay minerals* (5th Nat. Conf., 1956), pp. 227-241.

GLENN, R. C., JACKSON, M. L., HOLE, F. D. and LEE, G. B. (1960). — Chemical weathering of layer silicate clays in loess-derived Tama silt loam of southern Wisconsin. *Clays and clay minerals* (8th Nat. Conf., 1959), pp. 63-83.

GODARD, A., PAQUET, H. et MILLOT, G. (1961). — Contribution à l'étude de quelques paléosols du Nord de l'Ecosse. *Bull. Serv. Carte Géol. Als. Lor.*, 14, pp. 111-128.

GOLDBERG, E. D. (1957). — Biochemistry of trace metals. *Mem. Geol. Soc. Amer.*, 67, pp. 345-357.

GOLDBERG, E. D. and ARRHENIUS, G. O. S. (1958). — Chemistry of Pacific pelagic sediments. *Geochim. Cosmochim. Acta*, 13, pp. 153-212.

GOLDICH, S. S. (1938). — A study in rock-weathering. *Journ. Geol.*, 46, pp. 17-23.

GOLDICH, S. S. (1948). — Origin and development of aluminous laterites and bauxites. *Bull. Geol. Soc. Amer.*, 59, p. 1326.

GOLDSCHMIDT, V. M., BARTH, T. W. F., LUNDE, G. and ZACHARIASEN, W. (1926). — Geochemische Verteilungsgesetze der Elemente. *Skr. Norske Vid. Acad. Oslo*, I, Math-Nat. Kl. 2.

GOLDSCHMIDT, V. M. und PETERS, Cl. (1932). — Zur Geochemie des Bors. *Nachr. Ges. u. Wiss. Math. Phys. Kl. Göttingen*.

GOLDSCHMIDT, V. M. (1934). — Drei Vorträge über Geochemie. *Geol. Fören. Förhandl.*, 56, pp. 385-427.

GOLDSCHMIDT, V. M. (1937). — Les principes de la répartition des éléments chimiques dans les minéraux et les roches. *Journ. Chem. Soc.*, p. 655.

GOLDSCHMIDT, V. M. (1945). — Fondements géochimiques de la répartition des oligo-éléments. *Soil Sci.*, 60, pp. 1-7.

GOLDSCHMIDT, V. M. (1954). — *Geochemistry*. Muir A., Oxford, 730 pages.

GOLDSZTAUB, S. (1935). — Etude de quelques dérivés de l'oxyde ferrique. Détermination de leur structure. *Bull. Soc. Fr. Min.*, 58, pp. 6-76.

GORDON, M. and TRACEY, J. I. (1952). — Origin of the Arkansas bauxite deposits. *Problems of clay and laterite genesis*, A.I.M.M.E. Symposium vol., pp. 12-34.

GOSSELET, J. (1888). — L'Ardenne. *Mém. Carte Géol. Fr.*, 881 pages.

GOTO, K., OKURA, T. and KAYAMA, I. (1953). — *Kagaku*, 23, p. 428 (cité dans l'article Okamoto *et al.*, 1957).

GRABAU, A. W. (1904). — On the classification of sedimentary rocks. *Amer. Geol.*, 33, pp. 225-247.

GRAF, D. and GOLDSMITH, J. R. (1956). — Some hydrothermal syntheses of dolomite and protodolomite. *Journ. Geol.*, 64, pp. 173-186.

GRAFF-PETERSEN, P. (1961). — *Lermineralogien i de limniske jura-sedimenter pa Bornholm*. Kobenhavn, 149 pages.

GRAHAM, E. R. (1941). — Les argiles acides : un agent de l'érosion chimique. *Journ. Geol.*, 49, pp. 392-401.

GRAHAM, E. R. (1950). — The plagioclases feldspars as an index to soil weathering. *Soil Sci. Soc. Amer. Proc.*, 14, p. 300.

GRANGEON, P. (1959). — Sur une couche d'altération climatique miocène de la région du Coiron (Ardèche). *C. R. Acad. Sci. Fr.*, 248, pp. 1370-1372.

GRANGEON, P. (1960). — Contribution à l'étude des terrains tertiaires de la tectonique et du volcanisme du Massif du Coiron (Sud-Est du Massif Central français). *Trav. Lab. Géol. Univ. Grenoble*, 36, pp. 143-284.

GREBER, Ch. (1962). — Réflexions sur la flore houillère et les problèmes qu'elle pose. *C. R. Soc. Biogéogr.*, 335-336-337, pp. 16-22.

GREIGERT, J. (1961). — Notice explicative sur la carte géologique de la feuille Dosso, République du Niger. *B.R.G.M. Dakar*.

GRESSLY, A. (1838-1841). — Observations géologiques sur le Jura soleurois. *Mém. Soc. Helv.*, II, III, V, 549 pages.

GRIFFIN, G. M. and INGRAM, R. L. (1955). — Clay minerals of the Neuse river estuary. *Journ. Sedim. Petrol.*, 25, pp. 194-200.

GRIFFIN, J. J. and JOHNS, W. D. (1958). — Clay mineral composition of sediments of the Mississippi river and major tributaries. *Bull. Soc. Geol. Amer.*, 69, p. 1574.

GRIFFITHS, J. C. (1952). — Reaction relation in the finer-grained rocks. *Clay Min. Bull.*, 1, pp. 251-256.
GRIFFITHS, J. C., BATES, T. F. and SHADLE, H. W. (1956). — Guide to field trip of fourth national clay conference: clay minerals in sedimentary rocks. *Clays and clay minerals* (4th Nat. Conf., 1955), pp. 1-20.
GRIGGS, D. T. (1936). — The factor of fatigue in rock exfoliation. *Journ. Geol.*, 44, pp. 783-796.
GRIM, R. E. (1935). — Petrology of Pennsylvanian shales and non calcareous underclays associated with Illinois coals. *Bull. Amer. Ceram. Soc.*, 14, pp. 113-119, 129-134, 170-176.
GRIM, R. E., LAMAR, J. E. and BRADLEY, W. F. (1937). — The clay minerals in Illinois limestones and dolomites. *Journ. Geol.*, 45, pp. 829-843.
GRIM, R. E., BRAY, R. H. and BRADLEY, W. F. (1937). — The mica in argillaceous sediments. *Amer. Min.*, 22, pp. 813-829.
GRIM, R. E. and ROWLAND, R. A. (1942). — Differential thermal analysis of clay mineral and other hydrous materials. *Amer. Min.*, 27, pp. 746, 761-801, 818.
GRIM, R. E. (1947). — Relation of clay mineralogy to origin and recovery of petroleum. *Bull. Am. Ass. Petrol. Geol.*, 31, pp. 1491-1499.
GRIM, R. E. and BRADLEY, W. F. (1948). — The illite clay minerals. *XVIIIth Intern. Geol. Congr. Great Britain*, pp. 127-128.
GRIM, R. E., DIETZ, R. S. and BRADLEY, W. F. (1949). — Clay mineral composition of some sediments from the Pacific Ocean of the California Coast and the gulf of California. *Bull. Geol. Soc. Amer.*, 60, pp. 1785-1808.
GRIM, R. E. (1951). — The depositional environment of red and green shales. *Journ. Sedim. Petrol.*, 21, pp. 226-232.
GRIM, R. E. (1953). — *Clay mineralogy*. MacGraw-Hill, N. Y., 384 pages.
GRIM, R. E. and JOHNS, W. D. (1954). — Clay mineral investigations of sediments in the northern gulf of Mexico. *Clays and clay minerals* (2nd Nat. Conf., 1953), pp. 81-103.
GRIM, R. E. and BRADLEY, W. F. (1955). — Structural implications in diagenesis. *Geol. Rundsch.*, 43, pp. 469-474.
GRIM, R. E., BRADLEY, W. F. and WHITE, W. A. (1957). — Petrology of the paleozoic shales of Illinois. *Illinois St. Geol. Surv.*, 203, 35 pages.
GRIM, R. E. (1958). — Concept of diagenesis in argillaceous sediments. *Bull. Am. Ass. Petrol. Geol.*, 42, pp. 246-253.
GRIM, R. E., DROSTE, J. B. and BRADLEY, W. F. (1960). — A mixed layer clay mineral associated with an evaporite. *Clays and clay minerals* (8th Nat. Conf., 1959), pp. 228-236.
GRIM, R. E., KULBICKI, G. and CAROZZI, A. V. (1960). — Clay mineralogy of the sediments of the Great Salt Lake, Utah. *Bull. Geol. Soc. Amer.*, 71, pp. 515-520.
GRIM, R. E. et VERNET, J. P. (1961). — Etude par diffraction des minéraux argileux de vases méditerranéennes. *Schweiz. Min. Petrog. Mitt.*, 41, pp. 65-70.
GRIM, R. E. and KULBICKI, G. (1961). — Montmorillonite: High temperature reactions and classification. *Amer. Min.*, 46, pp. 1329-1369.
GRÜNER, J. W. (1922). — Organic matter and the origin of Biwabik iron bearing formation of the Mesabi range. *Econ. Geol.*, 17, pp. 407-460.
GRÜNER, J. W. (1932). — The crystal structure of kaolinite. *Zeitschr. Krist.*, pp. 75-88.
GRÜNER, J. W. (1934). — The structures of vermiculites and their collapse by dehydration. *Amer. Min.*, 19, pp. 557-575.
GRÜNER, J. W. (1934). — The crystal structures of talc and pyrophyllite. *Zeitschr. f. Krist.*, Abt. A, 88, pp. 412-419.
GRÜNER, J. W. (1935). — The structural relationship of glauconite and mica. *Amer. Min.*, 20, pp. 699-714.
GRÜNER, J. W. (1937). — Notes on the structure of serpentines. *Amer. Min.*, 22, pp. 97-103.
GRÜNER, J. W. (1940). — Abundance and significance of cristobalite in bentonites and fuller's earths. *Econ. Geol.*, 35, pp. 869-875.
GUEIRARD, S. (1957). — Sur l'origine de la « collobriérite » du Massif des Maures (Var). *C. R. Acad. Sci. Fr.*, 245, pp. 2339-2341.

HALLIMOND, A. F. (1925). — Iron ores bedded ores of England and Wales. *Spec. Rep. Min. Ressourc. Gr. Britain*, 29, pp. 26-27.

HARDER, H. (1951). — Über den Mineralbestand und die Entstehung einiger sedimentären Eisenerze des Lias (gamma). *Heidelberg. Beitr.*, 2, pp. 455-476.

HARDER, H. (1956). — Untersuchungen an Paragoniten und an natriumhaltigen Muskoviten. *Heidelberg. Beitr.*, 5, pp. 227-271.

HARDER, H. (1957). — Zum Chemismus der Bildung einiger sedimentären Eisenerze. *Zeitschr. deutsch. Geol. Ges.*, 109, pp. 69-72.

HARDER, H. (1959, 1960). — Beitrag zur Geochemie des Bors. *Nachr. Akad. Wiss. Göttingen Math. Phys. Kl.*, 5 und 6, pp. 67-183.

HARDON, H. J. and FAVEJEE, J. Ch. L. (1939). — Qualitative X-ray analysis of clay fraction of the principal soil types of Java. *Meded. Landbouwk. Hogeschool*, 43, pp. 53-59.

HARMAN, R. W. (1925, 1926, 1927). — Aqueous solutions of sodium silicates. *Journ. Phys. Chem.*, 29, pp. 1155-1168 ; 30, pp. 359-368, 917-924, 1100-1111 ; 31, pp. 355-373, 511-518, 616-625.

HARRISON, J. B. (1933). — *The catamorphism of igneous rocks under humid tropical conditions.* Imp. Bur. Soil Sci., Harpenden, 79 pages.

HARRISON, J. L. and DROSTE, J. B. (1960). — Clay partings in gypsum deposits in southwestern Indiana. *Clays and clay minerals* (7th Nat. Conf., 1958), pp. 195-200.

HARVEY, H. W. (1949). — *Chimie et biologie de l'eau de mer.* Presses univ., Paris.

HATHAWAY, J. C. (1955). — Studies of some vermiculite-type clay minerals. *Clays and clay minerals* (3rd Nat. Conf., 1954), pp. 74-86.

HEEZEN, B. C., NESTEROFF, W. D. et SABATIER, G. (1960). — Répartition des minéraux argileux dans les sédiments profonds de l'Atlantique nord et équatorial. *C. R. Acad. Sci. Fr.*, 251, pp. 410-413.

HEIM, D. (1957). — Über die mineralischen, nicht karbonitischen Bestandteile des Cenoman und Turon der mitteldeutschen Kreidemulden und ihre Verteilung. *Heidelberg. Beitr.*, 5, pp. 302-330.

HENDRICKS, S. B. and FRY, W. H. (1929). — The results of X-ray and microscopical examinations of soil colloïds. *Bull. Am. Soil Surv. Ass.*, 11, pp. 194-195.

HENDRICKS, S. B. (1938). — On the structure of the clay minerals: dickite, halloysite and hydrated halloysite. *Amer. Min.*, 23, pp. 295-301.

HENDRICKS, S. B. and JEFFERSON, M. E. (1938). — Crystal structure of vermiculites and mixed vermiculite-chlorites. *Amer. Min.*, 23, pp. 851-862.

HENDRICKS, S. B. and ROSS, C. S. (1941). — Chemical composition and genesis of glauconite and celadonite. *Amer. Min.*, 26, pp. 683-708.

HENDRICKS, S. B. and TELLER, E. (1942). — X-ray interference in partially ordered layer lattices. *Journ. Chem. Phys.*, 10, pp. 147-167.

HÉNIN, S. (1947). — La formation des argiles et la pédologie. *C. R. Conf. Pédologie méditer. Alger-Montpellier*, pp. 97-108.

HÉNIN, S. et CAILLÈRE, S. (1947). — Sur une synthèse de l'antigorite à basse température. *C. R. Acad. Sci. Fr.*, 224, pp. 1439-1440.

HÉNIN, S. et CAILLÈRE, S. (1952). — Sur l'évolution de la phlogopite à Madagascar. *XIXe Congr. Géol. Intern. Alger*, 18, pp. 139-147.

HÉNIN, S. et ROBICHET, O. (1953). — Sur les conditions de formation des minéraux argileux par voie expérimentale, à basse température. *C. R. Acad. Sci. Fr.*, 236, pp. 517-519.

HÉNIN, S., ROBICHET, O. et DU ROUCHET, M. J. (1953). — Altération expérimentale d'un schiste, comparaison avec un granite. *C. R. Acad. Sci. Fr.*, 237, pp. 1437-1439.

HÉNIN, S. et ROBICHET, O. (1954). — Nouveaux résultats concernant la préparation de minéraux argileux au laboratoire ; synthèse de l'antigorite. *C. R. Acad. Sci. Fr.*, 238, pp. 2554-2556.

HÉNIN, S. et ROBICHET, O. (1954). — A study of the synthesis of clay minerals. *Clay Min. Bull.*, 2, pp. 110-115.

HÉNIN, S. et ROBICHET, O. (1955). — Résultats obtenus au cours de nouveaux essais de synthèse de minéraux argileux. *Bull. Gr. Fr. Argiles*, 6, pp. 19-22.

HÉNIN, S. (1956). — Classification des minéraux argileux. *Bull. Gr. Fr. Argiles*, 8, pp. 29-37.
HÉNIN, S. (1957). — Essais de synthèse des minéraux argileux à partir de gels de silice. *C. R. Acad. Sci. Fr.*, 244, pp. 225-227.
HÉNIN, S. et PÉDRO, G. (1957). — Mise en évidence d'un effet de dislocation du granite à biotite par traitement à l'eau oxygénée. *C. R. Acad. Sci. Fr.*, 245, pp. 1451-1454.
HERBILLON, A. et GASTUCHE, M. C. (1962). — Synthèse et genèse de l'hydrargillite. *C. R. Acad. Sci. Fr.*, 254, pp. 1105-1107.
HEYSTECK, H. and SCHMIDT, E. R. (1954). — The mineralogy of the attapulgite montmorillonite deposit in the Springbok Flats, Transvaal. *Trans. Geol. Soc. South Africa*, 56, pp. 99-115.
HILLY, J. (1957). — Etude géologique du Massif de l'Edough et du Cap de Fer (Est Constantinois). *Bull. Serv. Carte Géol. Algérie.*
HOFMANN, F., GEIGER, Th. und SCHWARZACHER, W. (1949). — Über ein Vorkommen von Montmorillonit in der ostschweizerischen Molasse. *Schweiz. Min. Petrogr. Mitt.*, 29, pp. 43-49.
HOFMANN, F. (1958). — Zusammenhänge zwischen Entstehungsbedingungen und Beschaffenheit toniger Sedimente mit gleichartigem Ausgangsmaterial an einem Beispiel aus dem Tertiär des Kantons Schaffhausen (Schweiz). *Eclogae geol. Helv.*, 51, pp. 981-989.
HOFMANN, U., ENDELL, K. und WILM, D. (1933). — Kristallstruktur und Quellung von Montmorillonit. *Zeitschr. Krist.*, 86, pp. 340-348.
HOLMS, R. S. and HEARN, W. E. (1942). — Chemical and physical properties of some of the important alluvial soils of Mississippi drainage Basin. *US Dept Agri. Tech. Bull.*, 833.
HONEYBORNE, D. B. (1951). — The clay minerals in the Keuper marl. *Clay Min. Bull.*, 1, 5, pp. 150-157.
HOSKING, J. S. (1940). — The soil clay mineralogy of some Australian soils developed on granitic and basaltic parent materials. *Journ. Coun. Sc. Ind. Res. Australia*, 13, pp. 206-216.
HOSS, H. (1959). — Nachtrag zu den Untersuchungen über die Petrographie kulmischer Kieselschiefer. *Beitr. Min. Petr.*, 6, pp. 248-260.
HOUGH, J. L. (1958). — Fresh water environment of deposition of Precambrian banded iron formations. *Journ. Sedim. Petrol.*, 28, pp. 414-431.
HOURCQ, V. et REYRE, D. (1956). — Les recherches pétrolifères dans la zone côtière du Gabon. *XXe Congr. Intern. Mexico*, 1, pp. 113-141.
HOUTEN, F. B. VAN (1948). — Origin of red banded early Cenozoïc deposits in Rocky Mountains region. *Bull. Am. Ass. Petrol. Geol.*, 32, pp. 2083-2126.
HOUTEN, F. B. VAN (1961). — Ferric oxides in red beds as paleomagnetic data. *Journ. Sedim. Petrol.*, 31, pp. 296-301.
HUMBERT, R. P. and SHAW, B. (1941). — Studies of clay particles with the electron microscope. *Soil Sci.*, 52, pp. 481-487.
HUME, W. F. (1925). — *Geology of Egypt*. Surv. of Egypt. Le Caire, 408 pages.
HUMMEL, K. (1922). — Die Entstehung eisenreicher Gesteine durch Halmyrolyse (= submarine Gesteinszersetzung). *Geol. Rundsch.*, 13, pp. 40-81 et 97-136.
HUTCHINGS, W. M. (1890). — Notes on the probable origin of some slates. *Geol. Mag.*, 7, pp. 264-273.
HUTCHINGS, W. M. (1894). — Notes on the composition of clays, slates, etc., and on some points in their contact metamorphism. *Geol. Mag.*, 1, pp. 36-45.
HUTCHINGS, W. M. (1896). — Clays, shales and slates. *Geol. Mag.*, 3, pp. 309-317.
HUTCHINSON, G. E. (1957). — *A treatise on limnology*. Wiley & Sons, N. Y.
ILER, R. K. (1955). — *The colloïd chemistry of silica and silicates*. Cornell Univ. Press, N. Y., 324 pages.
JACKSON, M. L., TYLER, S. A., WILLIS, A. L., BOURBEAU, G. A. and PENNINGTON, R. P. (1948). — Weathering sequence of clay-size minerals in soils and sediments. *Journ. Phys. Coll. Chem.*, 52, pp. 1237-1260.

JACKSON, M. L., HSEUNG, V., COREY, R. B., EVANS, E. J. and VAN DEN HEUVEL, R. C. (1952). — Weathering sequence of clay-size minerals in soils and sediments. *Proc. Soil Sci. Soc. Amer.*, 16, pp. 3-6.

JACKSON, M. L., WHITTIG, L. D., VAN DEN HEUVEL, R. C., KAUFMAN, A. and BROWN, B. E. (1954). — Some analyses of soil montmorin, vermiculite, mica, chlorite, and interstratified layer silicates. *Clays and clay minerals* (2nd Nat. Conf., 1953), pp. 218-240.

JACKSON, M. L. (1959). — Frequency distribution of clay minerals in major great soil group as related to the factors of soil formation. *Clays and clay minerals* (6th Nat. Conf., 1957), pp. 133-143.

JASMUND, K. (1951, 1955). — *Die silicatischen Tonminerale.* Verlag Chemie, Weinheim, 1re éd., 142 pages, 2e éd., 190 pages.

JEANNETTE, A. (1952). — Argiles smectiques et Rhassoul. *XIXe Congr. Géol. Intern. Alger*, 1, pp. 371-384.

JEANNETTE, A. et LUCAS, J. (1955). — Sur l'extension au Maroc des niveaux à chlorite dans les argiles du Permo-Trias. *Notes Serv. Géol. Maroc*, 125, pp. 129-134.

JEANNETTE, A., MONITION, A., ORTELLI, I. et SALVAN, H. (1959). — Premiers résultats de l'étude des argiles de la série phosphatée du Bassin de Khouribga (Maroc). *XXe Congr. Géol. Intern. Mexico*, 1956, pp. 53-62.

JOHNS, W. D. and GRIM, R. E. (1958). — Clay mineral composition of recent sediments from the Mississippi River delta. *Journ. Sedim. Petrol.*, 28, pp. 186-199.

JOHNSON, S. W. and BLAKE, J. M. (1867). — On kaolinite and pholerite. *Amer. Journ. Sci.*, 43, pp. 351-361.

JOHNSTONE, A. (1889). — On the action of pure water and of water saturated with carbonic acid gas, on the minerals of the mica family. *Quart. Journ. Geol. Soc.*, 45, p. 363.

JOLLES, H. and NEURATH, F. (1898). — Colorimetric estimation of silica in water. *Zeitschr. Angew. Chem.*, 11, pp. 315-316.

JONAS, E. C. and BROWN, T. E. (1959). — Analysis of interlayer mixtures of three clay mineral types by X-ray diffraction. *Journ. Sedim. Petrol.*, 29, pp. 77-86.

JONAS, E. C. and ROBERSON, H. E. (1960). — Particle size as a factor influencing expansion of the three layer clay minerals. *Amer. Min.*, 45, pp. 828-838.

JONAS, E. C. and THOMAS, G. L. (1960). — Hydratation properties of potassium deficient clay micas. *Clays and clay minerals* (8th Nat. Conf., 1959), pp. 181-192.

JONAS, E. C. and BROWN, T. E. (1961). — The effects of particle size on micaceous clay mineral composition and related properties. *Clays and clay minerals* (10th Nat. Conf.).

JOULIA, F., BONIFAS, M., CAMEZ, T., MILLOT, G. et WEIL, R. (1958). — Analcimolites sédimentaires dans le Continental intercalaire du Sahara central, Bassin du Niger, A.O.F. *Bull. Serv. Carte Géol. Als. Lor.*, 11, 2, pp. 67-70.

JOULIA, F., BONIFAS, M., CAMEZ, T., MILLOT, G. et WEIL, R. (1959). — Découverte d'un important niveau d'analcimolite gréseuse dans le Continental intercalaire de l'Ouest de l'Aïr (Sahara central). *Notes Serv. Géol. Prosp. Min., Dakar*, 4, 40 pages.

JUNG, J. (1954). — Les illites du bassin oligocène de Salins (Cantal). *Bull. Soc. Fr. Min. Crist.*, 77, pp. 1231-1249.

KAHLENBERG, L. and LINCOLN, A. T. (1898). — Solutions of silicates of the alkalies. *Journ. Phys. Chem.*, 2, pp. 77-90.

KAISER, E. (1926). — *Die Diamantenwüste Südwest-Afrikas.* Berlin, 853 pages.

KAISIN, F. Jr. (1956). — Le rôle de la substitution dans l'altération météorique des roches sédimentaires, spécialement des calcaires. *Mém. Inst. Géol. Univ. Louvain*, 20, pp. 50-163.

KALKOWSKY, E. (1901). — Die Verkieselung der Gesteine in der nördlichen Kalahari. *Abh. Naturwiss. Ges. Isis. Dresden*, pp. 55-107.

KEELING, P. S. (1956). — Sepiolite at a locality in the Keuper marl of the Middlands. *Min. Mag.*, 31, pp. 328-332.

KEFERSTEIN, Ch. (1828). — *Deutsch. Geognotisch. Geol. dargestellt*, 5, 3, p. 510.

KEITH, M. L. and DEGENS, E. T. (1959). — Geochemical indicators of marine and freshwater sediments. *Research in Geochemistry*, Abelson, pp. 38-61.

KELLER, W. D. (1941). — Petrography and origin of the Rex chert. *Bull. Geol. Soc. Amer.*, 52, pp. 1279-1298.

Keller, W. D. and Foley, R. (1949). — Missouri river sediments in river water, ocean water and sodium oxalate solution. *Journ. Sedim. Petrol.*, 19, pp. 78-81.
Keller, W. D. (1952). — Analcime in the Popo Agie member of Chugwater formation. *Journ. Sedim. Petrol.*, 22, pp. 70-82.
Keller, W. D. and Frederickson, A. F. (1952). — Role of plants and colloïdal acids in the mechanism of weathering. *Amer. Journ. Sci.*, 25, pp. 594-608.
Keller, W. D., Wescott, J. F. and Bledsoe, A. O. (1954). — The origin of Missouri fireclays. *Clays and clay minerals* (2nd Nat. Conf., 1953), pp. 7-46.
Keller, W. D. (1956). — Clay minerals as influenced by environments of their formation. *Bull. Am. Ass. Petrol. Geol.*, 40, pp. 2689-2710.
Keller, W. D. (1958). — Argillation and direct bauxitisation in terms of concentrations of hydrogen and metal cations at surface of hydrolysing aluminium silicates. *Bull. Am. Ass. Petrol. Geol.*, 42, pp. 233-245.
Keller, W. D. (1958). — Glauconitic mica in the Morrison Formation in Colorado. *Clays and clay minerals* (5th Nat. Conf., 1956), pp. 120-128.
Kelley, W. P. (1929). — The nature of the base exchange material of soils. *Journ. Amer. Soc. Agron.*, 21, p. 1210.
Kelley, W. P., Dore, W. H. and Brown, S. M. (1931). — The nature of the base exchange material of bentonite soils and zeolites as revealed by chemical investigations and X-ray analysis. *Soil Sci.*, 31, pp. 25-55.
Kelley, W. P., Woodford, A. O., Dore, W. H. and Brown, S. M. (1939). — Comparative study of the colloïds of a Cecil and Susquehanna soil profiles. *Soil Sci.*, 47, pp. 175-193.
Kelley, W. P., Dore, W. H. and Page, J. B. (1941). — The colloïdal constituents of American alkali soils. *Soil Sci.*, 51, pp. 101-124.
Kelley, W. P. (1942). — Modern clay researches in relation to agriculture. *Journ. Geol.*, 50, pp. 307-319.
Kern, R. (1953). — Etude du faciès de quelques cristaux ioniques à structure simple. *Bull. Soc. Fr. Min. Crist.*, 76, pp. 325-364 et 391-414.
Kern, R. (1961). — Sur la formation des macles de croissance. *Bull. Soc. Fr. Min. Crist.*, 84, pp. 292-311.
Kerr, P. F. and Cameron, E. N. (1936). — Fuller's earth of bentonitic origin from Tehachapi, California. *Amer. Min.*, 21, pp. 230-237.
Kerr, P. F. (1955). — Hydrothermal alteration and weathering. *Geol. Soc. Amer. spec. Paper*, 62, pp. 525-544.
Kesler, T. L. (1952). — Occurrence and exploration of Georgia's kaolin deposits, in *Problems of clay and laterite genesis*, A.I.M.M.E. Symposium vol., pp. 162-177.
Kieslinger, A. (1954). — Studien über Verkieselung. *Tschermaks min. petrogr. Mitt.*, 5, pp. 70-84.
Kikoine, J. et Radier, H. (1949). — Quartzites d'altération au Soudan oriental. *C. R. Soc. Géol. Fr.*, 19, pp. 339-340.
Kikoine, J. et Radier, H. (1950). — Silicifications au Soudan oriental ; le calcaire à silex du Continental terminal. *C. R. Soc. Géol. Fr.*, 20, p. 168.
Kilian, C. (1931). — Des principaux complexes continentaux du Sahara. *C. R. Soc. Géol. Fr.*, 1, pp. 109-111.
Klein, Cl. (1961). — A propos du « Sidérolithique » sous-vendéen. *C. R. Acad. Sci. Fr.*, 253, p. 151.
Klein, Cl. (1961). — Sur la « formation de la Brenne » et ses extensions en Montmillonais et en Châtelleraudais. *C. R. Acad. Sci. Fr.*, 253, pp. 2087-2089.
Klingebiel, A. (1961). — Lithostratigraphie comparée des formations éocènes de la basse vallée de la Dordogne. *Thèse 3e cycle, Bordeaux*, 81 pages.
Konta, J. (1955). — Clay minerals and free silica in the carbonate sediments of the Silurian of Bohemia. *Acta Univ. Carolinae Geol.*, 1, pp. 29-70.
Konta, J. (1960). — Clay minerals from cassiterite-wolframite and molybdenite ore veins in Krupka (Graupen). *Acta Univ. Carolinae Geol.*, 1, pp. 23-50.
Konta, J. et Pouba, Z. (1961). — *Excursion guide. Second conference on clay mineralogy and petrography, Prague.* Karlovy Univ., Prague, 1961, 50 pages.

KORZHINSKY, D. S. (1940). — Liquid inclusions as the cause of imaginary pelitization of feldspars. *C. R. Acad. Sci. U.R.S.S.*, 29, pp. 112-114.

KOSSOVSKAIA, A. G. and SHUTOV, V. D. (1958). — Zonality in the structure of terrigene deposits in platform and geosynclinal regions. *Eclogae geol. Helv.*, 51, pp. 656-666.

KRAUSKOPF, K. B. (1956). — Dissolution and precipitation of silica at low temperatures. *Geochim. Cosmochim. Acta*, 10, pp. 1-27.

KRAUSKOPF, K. B. (1959). — The geochemistry of silica in sedimentary environments. *Soc. Econ. Pal. Miner. Spec. Publ.*, 7, pp. 4-19.

KRAUT, F. et VATAN, A. (1938). — Sur l'origine des roches argileuses des environs de Confolens (Charente) attribuées au Sidérolithique. *C. R. Acad. Sci. Fr.*, 206, pp. 443-445.

KRUMBEIN, W. C. and GARRELS, R. M. (1952). — Origin and classification of chemical sediments in terms of pH and oxidation-reduction potentials. *Journ. Geol.*, 60, pp. 1-34.

KRUSEMAN, G. P. (1962). — Etude paléomagnétique et sédimentologique du Permien de Lodève (Hérault). *Thèse Utrecht et Geologia Ultraiectina.*

KRYNINE, P. D. (1945). — Sediments and the search for oil. *Producer's Monthly*, 9, pp. 12-22.

KRYNINE, P. D. (1949). — The origin of red beds. *Trans. N. Y. Acad. Sci.*, 2, 2, pp. 60-68.

KÜBLER, B. (1959). — Etude de l'Oehningien (Tortonien) du Locle (Neuchâtel, Suisse). *Thèse Sci. Neuchâtel* et *Bull. Soc. Neuchâteloise Sci. Nat.*, 25, 1962, pp. 6-42, et *Beitr. z. Miner. und Petrogr.*, 8, 1962, pp. 267-314.

KULBICKI, G. (1949). — Sur la consolidation des roches argileuses en vue de la confection de plaques minces pétrographiques. *Bull. Soc. Fr. Min. Crist.*, 72, pp. 367-373.

KULBICKI, G. (1953). — Constitution et genèse des sédiments argileux sidérolithiques et lacustres du Nord de l'Aquitaine. *Thèse Sci., Toulouse* et *Sci. Terre*, 4, 1956, pp. 5-101.

KULBICKI, G. (1953). — Sur les conditions de cristallisation des minéraux kaoliniques dans le sidérolithique d'Aquitaine. *C. R. Acad. Sci. Fr.*, 237, pp. 194-196.

KULBICKI, G. (1954). — Phénomènes de diagenèse dans les sédiments argileux. *Clay Min. Bull.*, 2, pp. 183-188.

KULBICKI, G. et VETTER, P. (1955). — Etude des roches argileuses de quelques bassins houillers de la bordure occidentale du Massif Central. *Bull. Soc. Géol. Fr.*, 5, pp. 645-653.

KULBICKI, G. et VETTER, P. (1955). — Sur la présence d'argiles bauxitiques dans le Stéphanien de Decazeville. *C. R. Acad. Sci. Fr.*, 240, pp. 104-106.

KULBICKI, G. et MILLOT, G. (1960). — L'évolution de la fraction argileuse des grès pétroliers cambro-ordoviciens du Sahara central. *Bull. Serv. Carte Géol. Als. Lor.*, 13, pp. 147-156.

KULBICKI, G. and MILLOT, G. (1961). — Diagenesis of clays in sedimentary and petrol series. *Clays and clay minerals* (10th Nat. Conf.).

KULBICKI, G., STÉVAUX, J., ESQUEVIN, J. et LUCAS, J. (1962). — Les oligo-éléments de la série argilo-gréseuse de l'Ordovicien du Mésozoïque inférieur d'Hassi-Messaoud (Sahara). *Bull. Serv. Carte Géol. Als. Lor.*, 15, 4, pp. 171-178.

LACROIX, A. (1893-1913). — *Minéralogie de la France et de ses colonies*. Baudry et Cie, Paris, 5 vol.

LACROIX, A. (1913). — Les latérites de la Guinée et les produits d'altération qui leur sont associés. *Nouv. Arch. Museum*, 5, pp. 255-356.

LAFFITTE, R. (1949). — Sédimentation et orogenèse. *Ann. Hébert Haug*, 7, pp. 239-289.

LAFOND, R. (1961). — Etude minéralogique des argiles actuelles du bassin de la Vilaine. *C. R. Acad. Sci. Fr.*, 252, pp. 3614-3616.

LAFOND, R., RIVIÈRE, A. et VERNHET, S. (1961). — Etude de la composition minéralogique de quelques argiles glaciaires. *C. R. Acad. Sci. Fr.*, 252, p. 3310.

LAGACHE, M., WYART, J. et SABATIER, G. (1961). — Dissolution des feldspaths alcalins dans l'eau pure ou chargée de CO_2 à 200 °C. *C. R. Acad. Sci. Fr.*, 253, pp. 2019-2022.

LAGACHE, M., WYART, J. et SABATIER, G. (1961). — Mécanisme de la dissolution des feldspaths dans l'eau pure ou chargée de CO_2 à 200 °C. *C. R. Acad. Sci. Fr.*, 253, pp. 2296-2299.

LAJOINIE, J. P. et BONIFAS, M. (1961). — Les dolérites du Konkouré et leur altération latéritique (Guinée, Afrique occidentale). *Bull. B.R.G.M., Paris*, 2, pp. 1-34.

LANDERGREN, St. (1945). — Contribution to the geochemistry of boron. *Arkiv. Kemi. Min. Geol.*, 19 A, 25, pp. 1-7 and 26, pp. 1-31.

LANDERGREN, St. (1958). — Distribution of boron in different size classes in marine clay sediments. *Geol. Fören. Förhandl.*, 80, pp. 104-107.
LANDERGREN, St. (1959). — Sur la distribution du bore dans les sédiments marins argileux. *Coll. Intern. C.N.R.S.*, 83, pp. 29-32.
LANQUINE, A. et CUVILLIER, J. (1941). — Sur les faciès siliceux du Sparnacien dans l'Est et le Sud-Est du Bassin de Paris. *Bull. Soc. Géol. Fr.*, 11, pp. 195-206.
LANQUINE, A. et HALM, L. (1951). — Sur la présence et les conditions de gisement d'argiles mégalumineuses dans le Sidérolithique du Bassin d'Aquitaine. *C. R. Acad. Sci. Fr.*, 232, pp. 1570-1572.
LAPPARENT, A. F. DE (1952). — Etat actuel de nos connaissances sur la stratigraphie, la paléontologie et la tectonique des grès de Nubie du Sahara central. *XIXe Congr. Géol. Intern. Alger*, 21, pp. 113-125.
LAPPARENT, J. DE (1909). — Etude comparative de quelques porphyroïdes françaises. *Bull. Soc. Fr. Min.*, 32, pp. 174-304.
LAPPARENT, J. DE (1923). — *Leçons de pétrographie.* Masson et Cie, Paris, 501 pages.
LAPPARENT, J. DE (1924). — Roches à radiolaires du Dévonien de la vallée de la Bruche. *Bull. Serv. Carte Géol. Als. Lor.*, 1, pp. 47-64.
LAPPARENT, J. DE (1927). — L'alumine hydratée des bauxites. *C. R. Acad. Sci. Fr.*, 184, p. 1661.
LAPPARENT, J. DE (1930). — Comportement minéralogique et chimique des produits d'altération élaborés aux dépens des gneiss du Massif Central français avant l'établissement des dépôts sédimentaires de l'Oligocène. *C. R. Acad. Sci. Fr.*, 190, pp. 1062-1064.
LAPPARENT, J. DE (1930). — Les bauxites de la France méridionale. *Mém. Carte Géol. Fr.*, 187 pages.
LAPPARENT, J. DE (1934). — Constitution et origine de la leverriérite. *C. R. Acad. Sci. Fr.*, 198, pp. 669.
LAPPARENT, J. DE (1934). — La boehmite et le diaspore dans les fireclays de l'Ayrshire (Ecosse). *C. R. Acad. Sci. Fr.*, 199, pp. 1629-1631.
LAPPARENT, J. DE (1935). — Sur un constituant essentiel des terres à foulon. *C. R. Acad. Sci. Fr.*, 201, p. 481.
LAPPARENT, J. DE (1937). — Formules structurales et classification des argiles. *Zeitschr. Krist.*, 98, pp. 233-258.
LAPPARENT, J. DE (1937). — Les argiles d'El Goléa (Sahara algérien). Etude pétrographique. *Serv. Carte Géol. Algérie*, 5e sér., 3, 53 pages.
LAPPARENT, J. DE (1937). — Les phénomènes anciens de rubéfaction dans le Sahara central. *C. R. Acad. Sci. Fr.*, 205, p. 196.
LAPPARENT, J. DE (1937). — Structure et origine des terres naturelles susceptibles d'être utilisées pour la décoloration des huiles minérales. *IIe Congr. Mond. Pétrole, Paris.*
LAPPARENT, J. DE (1937). — Sur l'origine des bentonites de l'Afrique du Nord. *C. R. Soc. Géol. Fr.*, 7, pp. 126-128.
LAPPARENT, J. DE (1938). — Défense de l'attapulgite. *Bull. Soc. Fr. Min.*, 61, pp. 253-283.
LAPPARENT, J. DE et HOCART, R. (1939). — La leverriérite des formations latéritiques de l'Afrique Occidentale française. *C. R. Acad. Sci. Fr.*, 208, pp. 1465-1467.
LAPPARENT, J. DE (1941). — Logique des minéraux du granite. *Rev. Scientif.*, pp. 285-292.
LAPPARENT, J. DE (1945). — L'épisode du dépôt des argiles smectiques de l'Afrique du Nord. *C. R. Acad. Sci. Fr.*, 221, pp. 335-337.
LAPPARENT, J. DE (1947). — Sur la séricite. *C. R. Géol. Fr.*, 16, p. 349.
LAUNAY, L. DE (1913). — *Gîtes minéraux et métallifères*, Paris.
LAVEZARD (1906). — *Contribution à l'étude des argiles de France*, in *Contribution à l'étude des argiles de la céramique*, Paris.
LE CHATELIER, H. (1894). — Constitution des calcaires marneux. *C. R. Acad. Sci. Fr.*, 118, p. 262.
LE CHATELIER, H. (1914). — *La silice et les silicates*, Paris, Hermann.
LEFRANC, J. Ph. (1958). — Stratigraphie des séries continentales intercalaires au Fezzan Nord-occidental (Libye). *C. R. Acad. Sci. Fr.*, 247, pp. 1360-1363.

LEFRANC, J. Ph. (1959). — Existence au Fezzan Nord-occidental (Libye) de lacunes et discordances dans les séries du Continental intercalaire. *C. R. Acad. Sci. Fr.*, 249, pp. 2345-2348.

LEMCKE, K., ENGELHARDT, W. VON und FÜCHTBAUER, H. (1953). — Geologische und sedimentpetrographische Untersuchungen im Westteil der ungefalteten Molasse des süddeutschen Alpenvorlandes. *Beitr. Geol. Jb.*, 11, 109 pages.

LENEUF, N. (1959). — L'altération des granites calco-alcalins et des granodiorites en Côte-d'Ivoire forestière et les sols qui en sont dérivés. *Thèse Sci., Paris*, 210 pages.

LEVIN, I. and OTT, E. (1932). — The cristallinity of opals and the existence of high temperature cristobalite at room temperature. *Amer. Chem. Journ.*, 54, pp. 828-829.

LEVIN, I. and OTT, E. (1933). — X-ray study of opals, silica glass and silica gel. *Zeitschr. Krist.*, 81, pp. 305-318.

LIENHARDT, G. (1960). — Caractères généraux des tonstein du bassin stéphanien de Lons-le-Saunier (Jura). *Bull. Soc. Géol. Fr.*, 2, pp. 661-666.

LIENHARDT, G. (1961). — Causes et genèse des colorations rouges et vertes du Saxonien et du Trias de la région de Lons-le-Saunier (Jura). *Coll. sur le Trias, Montpellier*, mars 1961.

LIENHARDT, G. (1961). — Subsidence et énallaxie, phénomènes fondamentaux régissant les dépôts du Stéphanien de Lons-le-Saunier (Jura). *C. R. Acad. Sci. Fr.*, 252, pp. 2572-2574.

LINDGREN, W. (1933). — *Mineral Deposits*, New York, 1049 pages.

LIPPI-BONCAMBI, C., MACKENZIE, R. C. and MITCHELL, W. A. (1955). — The mineralogy of some soils from central Italy. *Clay Min. Bull.*, 2, pp. 280-288.

LIPPMANN, F. (1954). — Über einen Keuperton von Zaisersweiher bei Maulbronn. *Heidelberg. Beitr.*, 4, pp. 130-134.

LIPPMANN, F. (1956). — Clay minerals from the röth member of the triassic near Göttingen, Germany. *Journ. Sedim. Petrol.*, 26, pp. 125-129.

LIPPMANN, F. (1959). — Corrensit. Hintze-Chudoba „Handbuch der Mineralogie". Erg., 2, pp. 688-691.

LOCHER, F. W. (1952). — Kurzer Bericht über sedimentpetrographische Untersuchung zweier Lotkerne der Albatross-Expedition. *Heidelberg. Beitr.*, 3, pp. 193-218.

LOMBARD, A. (1953). — Les rythmes sédimentaires et la sédimentation générale. *Rev. Inst. Fr. Pétrole*, 8, numéro spécial, pp. 9-45.

LOMBARD, A. (1956). — *Géologie sédimentaire*, Masson et Cie, Paris, 722 pages.

LONGCHAMBON, H. et MOURGUES, F. (1927). — Sur le gisement de magnésite de Salinelles (Gard). *Bull. Soc. Min.*, 50, pp. 66-74.

LONGCHAMBON, H. (1936). — Sur les propriétés caractéristiques des palygorskites. *C. R. Acad. Sci. Fr.*, 203, p. 672.

LONGCHAMBON, H. (1937). — Sur certaines caractéristiques de la sépiolite d'Ampandandrava et la formule des sépiolites. *Bull. Soc. Fr. Min.*, 60, pp. 232-276.

LONGUET-ESCART, J. (1950). — Fixation des hydroxydes par la montmorillonite. *Trans. IVth Congr. Soil Sc.*, 3, p. 40.

LOSSAINT, P. (1959). — Etude expérimentale de la mobilisation du fer des sols sous l'influence des litières forestières. *Ann. Agron.*, 1re part., pp. 369-414, 2e part., pp. 493-452.

LOVERING, T. S. (1941). — The origin of the tungsten ores of Boulder County, Colorado. *Econ. Geol.*, 36, pp. 229-279.

LOVERING, T. S. (1950). — The geochemistry of argillic and related types of rock alteration. *Quart. Colorado Sch. Mines*, 45, pp. 231-260.

LUCAS, G. (1942). — Description géologique et pétrographique des Monts de Ghar-Rouban et du Sidi el Abed. *Bull. Serv. Carte Géol. Algérie*, 2e sér., 16.

LUCAS, G. (1955). — Caractères géochimiques et mécaniques du milieu générateur des calcaires noduleux à faciès amonitico-rosso. *C. R. Acad. Sci. Fr.*, 240, pp. 2000-2002.

LUCAS, J., CAMEZ, T. et MILLOT, G. (1959). — Détermination pratique aux rayons X des minéraux argileux simples et interstratifiés. *Bull. Serv. Carte Géol. Als. Lor.*, 12, pp. 21-33.

LUCAS, J. et BRONNER, A. M. (1961). — Evolution des argiles sédimentaires dans le bassin triasique du Jura français. *Bull. Serv. Carte Géol. Als. Lor.*, 14, pp. 137-149.

Lucas, J. et Jehl, G. (1961). — Etude de l'action de la chaleur sur la chlorite et la kaolinite par diffraction des rayons X. Application à la distinction de ces minéraux. *Bull. Serv. Carte Géol. Als. Lor.*, 14, pp. 159-173.

Lucas, J. (1962). — La transformation des minéraux argileux dans la sédimentation. Etudes sur les argiles du Trias. *Mém. Serv. Carte Géol. Als. Lor.*, 23, 202 pages.

MacCrea, J. M. (1950). — On the isotopic chemistry of carbonates and a paleo-temperature scale. *Journ. Chem. Phys.*, 18, pp. 849-857.

MacEwan, D. M. C. (1948). — A trioctahedral montmorillonite derived from biotite. *XVIIIth Intern. Geol. Congr. Great Britain*, Abstr., p. 128.

MacEwan, D. M. C. (1948). — Les minéraux argileux de quelques sols écossais. *Verres et Silic. industr.*, 13, pp. 41-46.

MacEwan, D. M. C. (1949). — Some notes on the recording and interpretation of X-ray diagrams of soil clays. *Journ. Soil Science*, 1, pp. 90-103.

MacEwan, D. M. C. (1956). — A study of interstratified illite-montmorillonite clay from Worcestershire, England. *Clays and clay minerals* (4th Nat. Conf., 1955), pp. 166-172.

Mackenzie, R. C. (1959). — The classification and nomenclature of clay-minerals. *Clay min. Bull.*, 4, pp. 52-66.

MacMillan, N. J. (1956). — Petrology of the Nodaway underclay (Pennsylvanian) Kansas. *Kansas Geol. Surv. Bull.*, 119, 6, pp. 187-249.

MacMurchy, R. C. (1934). — The crystal structure of the chlorite minerals. *Zeitschr. Krist.*, 88, pp. 421-432.

Mägdefrau, E. und Hofmann, U. (1937). — Glimmerartige Mineralien als Tonsubstanzen. *Zeitschr. Krist.*, 98, pp. 31-59.

Maignien, R. (1954). — Différents processus de cuirassement en A.O.F. *C. R. Conf. Interafr. Sols*, 116, pp. 1469-1486.

Maignien, R. (1958). — Contribution à l'étude du cuirassement des sols en Guinée française. *Mém. Serv. Carte Géol. Als. Lor.*, 16, 235 pages.

Maignien, R. (1959). — Les sols subarides au Sénégal. *L'Agron. trop.*, 5, pp. 536-571.

Maignien, R. (1961). — Sur les sols d'argiles noires tropicales d'Afrique occidentale. *Bull. Ass. Fr. Et. Sol*, numéro spécial, pp. 131-144.

Maikovsky, V. (1941). — Contribution à l'étude paléontologique et stratigraphique du Bassin potassique d'Alsace. *Mém. Serv. Carte Géol. Als. Lor.*, 6, 184 pages.

Mallard, E. (1878). — Sur la bravaisite. *Bull. Soc. Fr. Min.*, 1, p. 5.

Mallard, E. (1890). — Sur la lussatite, nouvelle variété minérale cristallisée de silice. *Bull. Soc. Fr. Min.*, 13, pp. 63-66.

Marel, H. W. van der (1954). — Potassium fixation in Dutch soils: mineralogical analyses. *Soil Sci.*, 78, pp. 163-179.

Marel, H. W. van der (1959). — La fixation du potassium, un caractère des sols favorable aux plantes cultivées. *Zeitschr. Pflznähr., Düng., Bodenk.*, 84, pp. 51-62.

Marel, H. W. van der (1960). — Relation between the morphological characteristics of clay minerals and their physical properties. *L.G.M. Meded.*, 4, pp. 1-27.

Marlière, R. (1946). — Deltas wealdiens du Hainaut, sables et graviers de Thieu ; argiles réfractaires d'Hautrage. *Bull. Soc. belge Géol.*, 55, pp. 69-101.

Marlière, R. (1947). — Les argiles réfractaires d'âge wealdien du Hainaut. *Verres et Silicates*, 12, p. 32.

Marshall, C. E. (1935). — Layer lattices and the base-exchange clays. *Zeitschr. Krist.*, 91, p. 433.

Marshall, C. E. (1935). — Mineralogical methods for the study of silts and clays. *Zeitschr. Krist.*, 90, p. 8.

Martin-Vivaldi, J. L. and Cano-Ruiz, J. (1956). — Contribution to the study of sepiolite: II. Some considerations regarding the mineralogical formula. *Clays and clay minerals* (4th Nat. Conf., 1955), pp. 173-176.

Martin-Vivaldi, J. L., Fontbote, J. M., Raussell-Colom, J. A. y Truyols, Y. J. (1957). — Sobre la composición mineralogica de las arcillas del Mioceno der Valles-Penedes. *Estud. Geol.*, 14, 35-36, pp. 305-321.

MARTIN-VIVALDI, J. L. and MACEWAN, D. M. C. (1957). — Triassic chlorites from the Jura and the Catalan coastal range. *Clay Min. Bull.*, 3, pp. 177-184.
MARTIN-VIVALDI, J. L. and MACEWAN, D. M. C. (1960). — Corrensite and swelling chlorite. *Clay Min. Bull.*, 4, pp. 173-181.
MARTIN-VIVALDI, J. L. and SANCHEZ CAMAZANO, M. (1961). — A dioctahedral clay vermiculite in a soil from Saucelle, Salamanca. *Clay Min. Bull.*, 4, pp. 299-306.
MASON, B. (1949). — Oxidation and reduction in geochemistry. *Journ. Geol.*, 57, pp. 62-72.
MASSEPORT, J. (1959). — Premiers résultats d'expériences au laboratoire sur les roches. *Rev. Géogr. Alpine*, 47, pp. 531-538.
MATTSON, S. (1928). — The electrokinetic and chemical behaviour of colloïdal material. *Journ. Phys. Chem.*, 32, pp. 1532-1552.
MATTSON, S. and HESTER, J. B. (1935). — The laws of soil colloïdal behaviour. *Soil Sci.*, 39, pp. 75-84.
MAUGUIN, Ch. (1928). — Etude des micas au moyen des rayons X. *Bull. Soc. Fr. Min. Crist.*, 51, pp. 285-332.
MAUGUIN, Ch. (1930). — La maille cristalline des chlorites. *Bull. Soc. Fr. Min. Crist.*, 53, pp. 279-299.
MAUREL, P. (1959). — Etude minéralogique de quelques marnes oxfordiennes. *Bull. Soc. Fr. Min. Crist.*, 82, pp. 276-284.
MAUREL, P. et BROUSSE, R. (1959). — Sur les phyllites contenues dans quelques arkoses. *Bull. Soc. Fr. Min. Crist.*, 82, pp. 87-90.
MAUREL, P. (1962). — Sur la présence d'albite dans le Permien supérieur des environs de Saint-Affrique (Aveyron) et de Lodève (Hérault). *C. R. Acad. Sci. Fr.*, 254, pp. 3003-3006.
MEHMEL, M. (1935). — Über die Struktur von Halloysit und Metalhalloysit. *Zeitschr. Krist.*, 90, pp. 35-43.
MENSCHING, H. (1956). — Les problèmes de la terre rouge méditerranéenne. *Phot. und Forsch.*, 7, 2, pp. 33-39.
MERING, J. (1946). — On the hydratation of montmorillonite. *Trans. of Faraday Soc.*, 42 B, pp. 205-219.
MERING, J. (1949). — L'interférence des rayons X dans les systèmes à stratification désordonnée. *Acta Cryst.*, 2, pp. 371-377.
MERRELL, H. W., JONES, D. J. and SAND, L. B. (1957). — Sedimentation features in Paradox shales, southeastern Utah. *Bull. Geol. Soc. Amer.*, 68, p. 1766.
MERRILL, G. P. (1902). — What constitutes a clay. *Amer. Geologist*, 30, p. 318.
MEUGY (1869). — Sur le Lias. *Bull. Soc. Géol. Fr.*, 26, pp. 484-513.
MEYER, C. (1946). — Notes on the cutting and polishing of thin sections. *Econ. Geol.*, 41, pp. 166-172.
MICHEL, R. (1953). — Les schistes cristallins des massifs du Grand Paradis et de Sesia-Lanzo (Alpes franco-italiennes). *Thèse Sci. Clermont* et *Sciences de la Terre*, 1, n° 3-4, 290 pages.
MIDGLEY, H. G. (1951). — Chalcedony and Flint. *Geol. Mag.*, 88, pp. 179-184.
MILLOT, G. (1949). — Relations entre la constitution et la genèse des roches sédimentaires argileuses. *Thèse Sci. Nancy* et *Géol. Appl. Prospec. Min.*, 2, n°s 2, 3, 4, pp. 1-352.
MILLOT, G. (1950). — Comparaison des indications de la pédologie et de la géologie au sujet de la genèse des minéraux argileux. *Trans. Intern. Congr. Soil Sci. Amsterdam*, 3, pp. 37-40.
MILLOT, G. (1953). — Héritage et néoformation dans la sédimentation argileuse. *XIXe Congr. Intern. Alger*, 1952. CIPEA, pp. 163-177.
MILLOT, G. (1953). — Minéraux argileux et leurs relations avec la géologie. *Rev. Inst. Fr. Pétrole*, 8, numéro spécial, pp. 75-86.
MILLOT, G. (1954). — La Ghassoulite, pôle magnésien de la série des montmorillonites. *C. R. Acad. Sci. Fr.*, 238, pp. 257-259.
MILLOT, G. (1954). — Observations de M. Millot sur la communication de MM. RIVIÈRE et VISSE. 12 janvier 1954. *Bull. Gr. Fr. Argiles*, 6, pp. 41-42.
MILLOT, G. (1954). — Sur l'existence de roches argileuses composées essentiellement de chlorites. *C. R. Acad. Sci. Fr.*, 238, pp. 363-364.

MILLOT, G. et BONIFAS, M. (1955). — Transformations isovolumétriques dans les phénomènes de latéritisation et de bauxitisation. *Bull. Serv. Carte Géol. Als. Lor.*, 8, pp. 3-10.
MILLOT, G. (1957). — Des cycles sédimentaires et de trois modes de sédimentation argileuse. *C. R. Acad. Sci. Fr.*, 244, pp. 2536-2539.
MILLOT, G., CAMEZ, T. et BONTE, A. (1957). — Sur la montmorillonite dans les craies. *Bull. Serv. Carte Géol. Als. Lor.*, 10, n° 2, pp. 25-26.
MILLOT, G., CAMEZ, T. et WERNERT, P. (1957). — Evolution des minéraux argileux dans les lœss et les lehms d'Achenheim. *Bull. Serv. Carte Géol. Als. Lor.*, 10, fasc. 2, pp. 17-21.
MILLOT, G., RADIER, H. et BONIFAS, M. (1957). — La sédimentation argileuse à attapulgite et montmorillonite. *Bull. Soc. Géol. Fr.*, 7, pp. 425-435.
MILLOT, G. et PALAUSI, G. (1959). — Sur un talc d'origine sédimentaire. *C. R. Soc. Géol. Fr.*, pp. 45-47.
MILLOT, G., RADIER, H., MULLER-FEUGA, R., DÉFOSSEZ, M. et WEY, R. (1959). — Sur la géochimie de la silice et les silicifications sahariennes. *Bull. Serv. Carte Géol. Als. Lor.*, 12, fasc. 2, pp. 3-15.
MILLOT, G. (1960). — Silice, silex, silicifications et croissance des cristaux. *Bull. Serv. Carte Géol. Als. Lor.*, 13, pp. 129-146.
MILLOT, G., ELOUARD, P., LUCAS, J. et SLANSKY, M. (1960). — Une séquence sédimentaire et géochimique des minéraux argileux : montmorillonite, attapulgite, sépiolite. *Bull. Gr. Fr. Argiles*, 12, pp. 77-82.
MILLOT, G. (1961). — Silicifications et néoformations argileuses : problèmes de genèse. *Coll. Intern. C.N.R.S.*, 105, pp. 167-173.
MILLOT, G., LUCAS, J. et WEY, R. (1963). — Some researches on the evolution of clay minerals and argillaceous and siliceous neoformations. *Clays and clay minerals* (10th Nat. Conf., 1961).
MILLOT, G., PERRIAUX, J. et LUCAS, J. (1961). — Signification climatique de la couleur rouge des grès permo-triasiques des Vosges et des grandes séries détritiques rouges. *Bull. Serv. Carte Géol. Als. Lor.*, 14, pp. 91-101.
MILLOT, G. (1962). — Crystalline neoformations of clays and silica. *Physical Sci. Some recent advances in France and the United States*. N. Y. Univ. Press, pp. 180-194.
MILLOT, G. (1962). — Some geochemical aspects of weathering and sedimentation. *Physical Sci. Some recent advances in France and the United States*. N. Y. Univ. Press, pp. 159-169.
MILNE, I. H. and EARLY, J. W. (1958). — Effect of source and environment on clay minerals. *Bull. Am. Ass. Petrol. Geol.*, 42, pp. 328-338.
MILON, Y. (1930). — L'extension des formations sidérolithiques éocènes dans le centre de la Bretagne. *C. R. Acad. Sci. Fr.*, 194, pp. 1360-1362.
MITCHELL, B. D. and MITCHELL, W. A. (1956). — The clay mineralogy of Ayrshire soils and their parent rocks. *Clay Min. Bull.*, 3, pp. 91-97.
MITCHELL, B. D. (1961). — The influence of soil forming factors on clay genesis. *Coll. Intern. C.N.R.S.*, 105, pp. 139-147.
MITCHELL, W. A. (1955). — Review of the mineralogy of Scottish soil clays. *Journ. Soil Science*, 16, pp. 94-98.
MOHR, E. C. J. (1944). — *The soils of equatorial regions, with special reference to the Netherlands East Indies*. Ann Arbor, Michigan, 766 pages.
MOHR, E. C. J. and BAREN, F. A. VAN (1959). — *Tropical soils*. La Haye, 498 pages.
MORTLAND, M. M., LAWTON, K. and UEHARA, G. (1956). — Alteration of biotite to vermiculite by plant growth. *Soil Sci.*, 82, pp. 477-481.
MUIR, A. (1951). — Notes on the soils of Syria. *Journ. Soil Science*, 2, pp. 163-181.
MULLER, G. (1961). — Die rezenten Sedimente im Golf von Neapel. *Beitr. Miner. Petrogr.*, 8, pp. 1-20.
MULLER-FEUGA, R. (1952). — Contribution à l'étude de la géologie, de la pétrographie et des ressources hydrauliques et minérales du Fezzan. *Thèse Sci.* Nancy et *Mém. 12 des Ann. Min. et Géol. Tunisie* (1954).
MURATA, K. J. (1942-1945). — The significance of internal structure in gelatinizing silicate minerals. *U. S. Geol. Surv. Bull.*, 950, pp. 25-33.

MURRAY, H. H. and SAYYAB, A. S. (1955). — Clay mineral studies of some recent marine sediments of the North Carolina Coast. *Clays and clay minerals* (3rd Nat. Conf., 1954), pp. 430-441.

MURRAY, H. H. and LEININGER, R. K. (1956). — Effect of weathering on clay minerals. *Clays and clay minerals* (4th Nat. Conf., 1955), pp. 340-347.

MURRAY, J. and RENARD, A. F. (1891). — *Deep sea deposits: Scientific results of the voyage of HMS "Challenger" 1873-1876*, London, pp. 400-411.

NAGELSCHMIDT, G., DESAI, A. D. and MUIR, A. (1940). — The minerals in the clay fractions of a black cotton soil and a red earth from Hyderabad. *Journ. Agric. Sci.*, 30, pp. 639-653.

NAGELSCHMIDT, G. and HICKS, D. (1943). — The mica of certain coal-measure shales in South Wales. *Min. Mag.*, 26, pp. 297-303.

NAGELSCHMIDT, G. (1944). — X-ray diffraction experiments on illite and bravaisite. *Min. Mag.*, 27, pp. 59-61.

NAGELSCHMIDT, G. (1944). — The mineralogy of soil colloïds. *Imp. Bur. Soil, Sci. Tech. Comm.*, 42, 22 pages.

NAGY, B. and BRADLEY, W. F. (1955). — The structural scheme of sepiolite. *Amer. Min.*, 40, pp. 885-892.

NASH, V. E. and MARSHALL, C. E. (1956). — The surface reactions of silicate minerals. II. Reactions of feldspar surfaces with salt solutions. *Missouri Univ. Agr. Ex. Stat. Research, Bull.*, 614, 36 pages.

NELSON, B. W. (1960). — Clay mineralogy of the bottom sediments, Rappahannock River, Virginia. *Clays and clay minerals* (7th Nat. Conf., 1958), pp. 135-148.

NESTEROFF, W. D. et SABATIER, G. (1961). — « Apport » et « Néogenèse » dans la formation des argiles des grands fonds marins. *Coll. Intern. C.N.R.S.*, 105, pp. 149-158.

NICOLAS, J. (1956). — Contribution à l'étude géologique et minéralogique de quelques gisements de kaolins bretons. *Thèse Sci. Paris*, 254 pages.

NICOLAS, J. (1961). — Sur la présence de « Glauconie » en Bretagne Centrale. *Coll. Intern. C.N.R.S.*, 105, pp. 197-207.

NIEUWENKAMP, W. (1956). — Géochimie classique et transformiste. *Bull. Soc. Géol. Fr.*, 6, pp. 407-431.

NIGGLI, P. (1938). — Zusammensetzung und Klassifikation der Lockergesteine. *Schw. Arch. f. angew. Wissensch. u. Tech.*, 4 pages.

NOLL, W. (1931). — Über die geochemische Rolle der Sorption. *Chem. d. Erde.*, 6, p. 352.

NOLL, W. (1935). — Mineralbildung im System Al_2O_3, SiO_2, H_2O. *Neues Jahrb. Min. Geol. Beil. Bd.*, 70, pp. 65-115.

NOLL, W. (1936). — Über die Bildungsbedingungen von Kaolin, Montmorillonit, Pyrophyllit und Analcim. *Min. Petr. Mitt.*, 48, pp. 210-247.

NORIN, E. (1953). — Occurrence of authigenic illitic mica in sediments of the Central Tyrrhenian Sea. *Bull. Geol. Inst. Uppsala*, 34, pp. 279-280.

NORTON, F. H. (1939). — Hydrothermal formation of clay minerals in the laboratory. Part I. *Amer. Min.*, 24, pp. 1-17.

NORTON, F. H. (1941). — Hydrothermal formation of clay minerals in the laboratory. Part II. *Amer. Min.*, 26, pp. 1-17.

OBERLIN, A. (1957). — Altération des cristaux de kaolinite ; détermination par microdiffraction électronique de la structure des produits altérés. *C. R. Acad. Sci. Fr.*, 244, pp. 1658-1661.

OBERLIN, A. et TCHOUBAR, C. (1957). — Etude en microscopie électronique de l'altération des cristaux de kaolinite. *C. R. Acad. Sci. Fr.*, 244, pp. 1524-1526.

OBERLIN, A., FREULON, J. M. et LEFRANC, J. Ph. (1958). — Etude minéralogique de quelques argiles des grès de Nubie du Fezzan. *Bull. Soc. Fr. Min. Crist.*, 81, pp. 1-4.

OBERLIN, A., HÉNIN, S. et PÉDRO, G. (1958). — Recherches sur l'altération expérimentale du granite par épuisement continu à l'eau. *C. R. Acad. Sci. Fr.*, 246, pp. 2006-2008.

OBERLIN, A. et FREULON, J. M. (1958). — Etude minéralogique de quelques argiles des séries primaires du Tassili N'Ajjer et du Fezzan (Sahara central). *Bull. Soc. Fr. Min. Crist.*, 81, pp. 186-189.

OBERLIN, A. et TCHOUBAR, C. (1959). — Reconnaissance de certains minéraux du groupe de la kaolinite par microdiffraction électronique. *C. R. Acad. Sci.*, 248, pp. 3184-3186.

OBERLIN, A., TCHOUBAR, C., SCHILLER, C., PÉZERAT, H. et KOVACEVIK, S. (1961). — Etude du fireclay produit par altération de la kaolinite et de quelques fireclays naturels. *Coll. Intern. C.N.R.S.*, 105, pp. 45-57.

OCHSENIUS, C. (1877). — *Die Bildung der Steinsalzlager und ihrer Mutterlaugensalze*. Halle, 172 pages.

OKAMOTO, G., OKURA, T. and GOTO, K. (1957). — Properties of silica in water. *Geochim. Cosmochim. Acta*, 12, pp. 123-132.

ORCEL, J. (1927). — Recherches sur la composition chimique des chlorites. *Bull. Soc. Fr. Min. Crist.*, 50, pp. 75-426.

ORCEL, J., HÉNIN, S. et CAILLÈRE, S. (1949). — Sur les silicates phylliteux des minerais de fer oolitiques. *C. R. Acad. Sci. Fr.*, 229, pp. 134-135.

ORCEL, J., CAILLÈRE, S. et HÉNIN, S. (1950). — Nouvel essai de classification des chlorites. *Min. Mag.*, 29, pp. 329-340.

PACKHAM, R. F., ROSAMAN, D. and MIDGLEY, H. G. (1961). — A mineralogical examination of suspended solids from nine English rivers. *Clay Min. Bull.*, 4, pp. 239-242.

PALAUSI, G. (1959). — Contribution à l'étude géologique et hydrogéologique des formations primaires au Soudan et en Haute-Volta. *Bull. Serv. Géol. Prosp. Min.*, 33, Dakar, 209 pages.

PALLMANN, H. (1947). — Pédologie et Phytosociologie. *Congr. Intern. Pédol. Méd. Montpellier*, pp. 1-36.

PAQUET, H. (1961). — Etude de la fraction argileuse de quelques sols d'Afrique. *Thèse 3ᵉ Cycle, Strasbourg*.

PAQUET, H., MAIGNIEN, R. et MILLOT, G. (1961). — Les argiles des sols des régions tropicales semi-humides d'Afrique occidentale. *Bull. Serv. Carte Géol. Als. Lor.*, 14, pp. 111-129.

PARJADIS DE LA RIVIÈRE, N. (1959). — Contribution à l'étude minéralogique des calcaires argileux du Mésozoïque des environs de Grenoble (« couches à ciment » du Valbonnais, Genevray-de-Vif, Voreppe, Grenoble, Sassenage). *Trav. Labo. Géol. Grenoble*, 35, pp. 161-190.

PATTERSON, S. H. and HOSTERMAN, J. W. (1960). — Geology of the clay deposits in the Olive Hill district, Kentucky. *Clays and clay minerals* (7th Nat. Conf., 1958), pp. 178-195.

PAULING, L. (1927). — The sizes of ions and the structure of ionic crystals. *Journ. Amer. Chem. Soc.*, 49, p. 763.

PAULING, L. (1930). — The structure of the micas and related minerals. *Proc. Nat. Acad. Sci.*, 16, pp. 123-129.

PAULING, L. (1930). — The structure of the chlorites. *Proc. Nat. Acad. Sc.*, 16, p. 578.

PAVLOVITCH, St. (1937). — Les roches éruptives de Zlatibor. *Bull. Soc. Fr. Min. Crist.*, 60, pp. 5-137.

PÉDRO, G. (1957). — Mécanisme de la désagrégation du granite et de la lave de Volvic sous l'influence des sels de cristallisation. *C. R. Acad. Sci. Fr.*, 245, pp. 333-335.

PÉDRO, G. (1957). — Nouvelles recherches sur l'influence des sels dans la désagrégation des roches. *C. R. Acad. Sci. Fr.*, 244, pp. 2822-2824.

PÉDRO, G. (1958). — Etude par voie expérimentale du processus de formation d'éléments fins par altération des roches à l'aide d'eau pure. *Bull. Gr. Fr. Argiles*, 10, pp. 45-61.

PÉDRO, G. (1958). — Premiers résultats concernant la réalisation expérimentale d'un processus de latéritisation. *C. R. Acad. Sci. Fr.*, 247, pp. 1217-1220.

PÉDRO, G. (1959). — Considérations sur une forme de l'altération des roches : l'arénisation. *C. R. Acad. Sci. Fr.*, 248, pp. 993-996.

PÉDRO, G. (1960). — Altération expérimentale des roches par l'eau sous atmosphère de CO_2. *C. R. Acad, Sci. Fr.*, 250, pp. 2035-2038.

PÉDRO, G. (1960). — Genèse de minéraux argileux par évolution des matériaux amorphes provenant de l'altération de diverses roches. *C. R. Acad. Sci. Fr.*, 250, pp. 1697-1700.

PÉDRO, G. (1961). — An experimental study on the geochemical weathering of crystalline rocks by water. *Clay Min. Bull.*, 4, 26, pp. 266-281.

PÉDRO, G. (1961). — Genèse des minéraux argileux par lessivage des roches cristallines au laboratoire. *Coll. Intern. C.N.R.S.*, 105, pp. 99-107.

PÉDRO, G. (1961). — Sur l'altération spontanée du granite en milieu naturel : résultats obtenus au bout de trente ans dans l'expérience lysimétrique de Versailles. *C. R. Acad. Sci. Fr.*, 253, pp. 2242-2245.

PELTO, C. R. (1956). — A study of chalcedony. *Amer. Journ. Sci.*, 254, pp. 32-50.

PERELMAN, A. I. (1950). — Palygorskite dans les fossiles et reliques des solonetz désertiques d'Asie centrale (en russe). *Dokl. Akad. Nauk. SSSR*, 71, pp. 541-543.

PERRIAUX, J. (1961). — Contribution à la géologie des Vosges gréseuses. *Mém. Serv. Carte Géol. Als. Lor.*, 18, 236 pages.

PERRIN, R. M. S. (1957). — The clay mineralogy of some tills in the Cambridge district. *Clay Min. Bull.*, 3, pp. 193-205.

PETERSON, J. A. and OSMOND, J. C. (1961). — *Geometry and sandstone bodies.* Publ. Am. Ass. Petrol. Geol., 240 pages.

PETERSON, M. N. A. (1961). — Mineral assemblages from evaporitic rocks, in the systems MgO, CaO,Al_2O_3, SiO_2, H_2O. *Abst. Prog. Ann. Meetgs*, p. 121 A.

PETERSON, M. N. A. (1962). — The mineralogy and petrology of upper Mississippian carbonate rocks of the Cumberland plateau in Tennessee. *Journ. Geol.*, 70, pp. 1-31.

PETROV, V. P. (1958). — Genetic types of white clays in the U.R.S.S. and laws governing their distribution. *Clay Min. Bull.*, 3, pp. 287-297.

PETTIJOHN, F. J. (1956). — *Sedimentary rocks.* Harper and Brothers, N. Y., 718 pages.

PÉZERAT, H. (1961). — Discussion in Genèse et synthèse des argiles. *Coll. Intern. C.N.R.S.*, 105, pp. 175-176.

PINSAK, A. P., MURRAY, H. H. (1960). — Regional clay mineral patterns in the Gulf of Mexico. *Clays and clay minerals* (7th Nat. Conf., 1958), pp. 162-178.

POWERS, M. C. (1954). — Clay diagenesis in the Chesapeake bay area. *Clays and clay minerals* (2nd Nat. Conf., 1953), pp. 68-80.

POWERS, M. C. (1957). — Adjustment of land derived clays to the marine environment. *Journ. Sedim. Petrol.*, 27, pp. 355-372.

POWERS, M. C. (1959). — Adjustment of clays to chemical change and the concept of the equivalence level. *Clays and clay minerals* (6th Nat. Conf., 1957), pp. 309-326.

PRUVOST, P. (1934). — Bassin houiller de la Sarre et de la Lorraine. *Etude des Gîtes minér. de la France*, 172 pages.

PUSTOWALOFF, L. W. (1955). — Über sekundäre Veränderungen der Sedimentgesteine. *Geol. Rundsch.*, 43, pp. 535-550.

QUAIDE, W. (1958). — Clay minerals from salt concentration ponds. *Amer. Journ. Sci.*, 256, pp. 431-437.

QUIÉVREUX, F. (1935). — Esquisse du monde vivant sur les rives de la lagune potassique. *Bull. Soc. Ind. Mulhouse*, 101, pp. 161-187.

RADIER, H. (1953). — Contribution à l'étude stratigraphique et structurale du détroit soudanais. *Bull. Soc. Géol. Fr.*, 3, pp. 677-695.

RADIER, H. (1957). — Le Précambrien saharien au Sud de l'Adrar des Iforas. Le bassin crétacé et tertiaire de Gao. Contribution à l'étude géologique du Soudan oriental. *Thèse Sci. Strasbourg* et *Bull. Serv. Géol. Prosp. Min. A.O.F.*, 1959, 26, 556 pages.

RAMAN, Sir C. V. and JAYARAMAN, A. (1953). — The structure of opal and the origin of its iridescence. *Proc. Indian Acad. Sci.*, A, 38, pp. 101-108.

RAT, P. (1960). — Sur l'âge et la nature des couches de base du Wealdien dans la province de Santander et à ses abords (Espagne). *C. R. Acad. Sci. Fr.*, 251, pp. 2207-2209.

RAYMOND, P. E. (1927). — The significance of red color in sediments. *Amer. Journ. of Sc.*, Vth ser., 13, pp. 234-251.

RAYNAL, P. (1953). — Etude pétrographique des calcaires concrétionnés de la Limagne. *Rev. Sci. Nat. Auvergne*, pp. 29-37.

REIFENBERG, A. (1947). — *Soils of Palestine.* Th. Murby, London, 174 pages.

RICH, C. I. (1958). — Muscovite weathering in a soil developed in the Virginia piedmont. *Clays and clay minerals* (5th Nat. Conf., 1956), pp. 202-212.

Ricour, J. (1960). — La genèse des niveaux salifères ; cas du Trias français. *Rev. Géogr. phys. Géol. dynam.*, 3, pp. 133-139.

Ricour, J. (1962). — Contribution à une révision du Trias français. *Thèse Sci., Paris* et *Mém. Carte Géol. Fr.*, 471 pages.

Rinne, F. (1924). — Röntgenographische Untersuchung an einigen feinzerteilten Mineralien, Kunstprodukten und dichten Gesteinen. *Zeitschr. Krist.*, 60, pp. 55-69.

Rivière, A. (1946). — Contribution à l'étude des sédiments argileux. *Bull. Soc. Géol. Fr.*, 16, pp. 43-55.

Rivière, A. (1946). — Sur les argiles micacées (Illites). *C. R. Acad. Sci. Fr.*, 222, pp. 1445-1446.

Rivière, A. (1946). — Sur l'identité structurale des illites et de la bravaisite de Noyant (Allier). *C. R. Acad. Sci. Fr.*, 223, p. 95.

Rivière, A. et Vernhet, S. (1951). — Sur la sédimentation des minéraux argileux en milieu marin en présence des matières humiques. Conséquences géologiques. *C. R. Acad. Sci. Fr.*, 233, pp. 807-808.

Rivière, A. (1953). — Sur l'origine des argiles sédimentaires. XIXe *Congr. Géol. Intern. Alger*, 18, pp. 177-180.

Rivière, A. et Visse, L. (1954). — L'origine des minéraux des sédiments marins. *Bull. Soc. Géol. Fr.*, 4, pp. 467-474.

Robb, G. L. (1949). — Red bed coloration. *Journ. Sedim. Petrol.*, 19, pp. 99-103.

Robbins, C. and Keller, W. D. (1952). — Clays and other non carbonate minerals in some limestones. *Journ. Sedim. Petrol.*, 22, pp. 146-152.

Roberts, A. L. (1958). — Minéralogie des argiles réfractaires : la livesite. *Bull. Soc. Fr. Céramique*, 41, pp. 29-33.

Robinson, K. and Brindley, G. W. (1949). — A note on the crystal structure of the chlorite minerals. *Proc. Leeds Philos. Soc.*, 5, pp. 102-108.

Rogers, L. E., Martin, A. E. and Norrish, K. (1954). — The occurrence of palygorskite near Ipswich, Queensland (Australia). *Min. Mag.*, 30, pp. 534-540.

Rolfe, B. N. and Jeffries, C. D. (1953). — Mica weathering in three soils in Central New York, U.S.A. *Clay Min. Bull.*, 2, pp. 85-94.

Rolfe, B. N. (1957). — Surficial sediment in lake mead. *Journ. Sedim. Petrol.*, 27, pp. 378-386.

Rondeau, A. (1958). — Géomorphologie et Géochimie. *Bull. Ass. Géogr. Fr.*, 271, pp. 17-23. *C. R. Soc. Géol. Fr.*, 13, pp. 288-290.

Rosenbusch, H. (1877). — Die Steiger Schiefer und ihre Contactzone von Barr-Andlau und Hohwald. *Abd. geol. Specialk. Els. Lothr.*, 1, H. 2, 383 pages.

Rosenqvist, I. T. (1955). — Investigations in the clay electrolyte water system. *Norg. Geotekn. Inst. Pub.*, 9, pp. 9-30.

Rosenqvist, I. T. (1961). — What is the origin of the hydrous micas of Fennoscandia? *Bull. Geol. Inst. Univ. Uppsala*, 40, pp. 265-268.

Ross, C. S. and Shannon, E. V. (1926). — Minerals of bentonite and related clays and their physical properties. *Journ. Am. Ceram. Soc.*, 9, pp. 77-96.

Ross, C. S. (1927). — The mineralogy of clays. *Proc. Intern. Congr. Soil Sci.*, 4, pp. 555-561.

Ross, C. S. (1928). — Sedimentary analcite. *Amer. Min.*, 13, pp. 195-197.

Ross, C. S. and Kerr, P. F. (1931). — The clay minerals and their identity. *Journ. Sedim. Petrol.*, 1, p. 59.

Ross, C. S. and Kerr, P. F. (1931). — The kaolin minerals. *U. S. Geol. Surv. Prof. Pap.*, 165 E., pp. 151-170.

Ross, C. S. and Kerr, P. F. (1934). — Halloysite and allophane. *U. S. Geol. Surv. Prof. Pap.*, 185 G., pp. 135-148.

Ross, C. S. (1941). — Sedimentary analcite. *Amer. Min.*, 26, pp. 627-629.

Ross, C. S. and Hendricks, S. B. (1943-1944). — Minerals of the montmorillonite group. *U. S. Geol. Surv. Prof. Pap.*, 205 B.

Ross, C. S. (1945). — Minerals and mineral relationships of the clay minerals. *Journ. Am. Ceram. Soc.*, 28, pp. 173-183.

ROUGERIE, G. (1960). — Le façonnement actuel des modelés en Côte-d'Ivoire forestière. *Mém. Inst. Fr. Afrique Noire, Dakar*, 542 pages.

ROUGERIE, G. (1961). — Etude comparative de l'évacuation de la silice en milieux cristallins tropical humide et tempéré humide (premiers résultats). *Ann. Géogr.*, 70, pp. 45-70.

ROY, C. J. (1945). — Silica in natural waters. *Amer. Journ. Sci.*, 243, pp. 393-403.

ROY, R. (1961). — The preparation and properties of synthetic clay minerals. *Coll. Intern. C.N.R.S.*, 105, pp. 85-98.

RUELLAN, F. (1931). — La décomposition et la désagrégation du granite à biotite au Japon et en Corée et les formes du modelé qui en résultent. *C. R. Congr. Intern. Géogr. Paris*, pp. 670-684.

RUTHRUFF, R. F. (1941). — Vermiculite and hydrobiotite. *Amer. Min.*, 26, pp. 478-484.

SABATIER, G. (1949). — Recherches sur la glauconie. *Bull. Soc. Fr. Min.*, 10, pp. 475-541.

SABATIER, G. (1961). — Discussion in *Coll. Intern. C.N.R.S.*, p. 75.

SADRAN, G., MILLOT, G. et BONIFAS, M. (1955). — Sur l'origine des gisements de bentonites de Lalla Maghnia. *Pub. Serv. Carte Algérie, Trav. Coll.*, 5, pp. 213-234.

SAKAMOTO (1954). — Zonal arrangement of residual clays. *Journ. Fac. Sci. Imp. Univ. Tokyo*, II, 9, pp. 301-324.

SARROT-REYNAULD DE CRESSENEUIL, J. (1953). — Observations sur la nature physico-chimique des argiles d'Eybens (Isère). *Trav. Labo. Géol. Grenoble*, 31, pp. 243-246.

SAUCIER, H. (1960). — Rhéologie des verres minéraux, in PERSOZ, *Intr. à l'étude de la rhéologie*, chap. VII, pp. 131-145.

SCHALLER, W. T. (1950). — An interpretation of the composition of highsilica sericites. *Min. Mag.*, 29, pp. 406-415.

SCHEERE, J. (1959). — La kaolinite du Houiller belge. *Silicates Industr.*, p. 475.

SCHLOCKER, J. and VAN HORN, R. (1958). — Alteration of volcanic ash near Denver, Colorado. *Journ. Sedim. Petrol.*, 28, p. 31.

SCHOELLER, H. (1941). — Etude sur le Sidérolithique du Lot et du Lot-et-Garonne. *Bull. Serv. Carte Géol. Fr.*, 43, pp. 1-20.

SCHOELLER, H. (1941). — Les conditions de formation des molasses et du Sidérolithique de la bordure Nord-Est du Bassin d'Aquitaine. *C. R. Soc. Géol. Fr.*, 7, pp. 32-34.

SCHOELLER, H. (1942). — La diorite d'Anglars (Lot). *Soc. Linn. Bordeaux. Proc. Verb.*, pp. 88-93.

SCHROEDER, D. (1955). — Mineralogische Untersuchungen an Lößprofilen. *Heidelberg. Beitr. Min.*, 4, pp. 443-463.

SCHULLER, A. (1951). — Zur Nomenclatur und Genese der Tonsteine. *Neues Jahrb. Min. Mon.*, pp. 97-109.

SCHULTZ, L. G. (1958). — Petrology of underclays. *Bull. Geol. Soc. Amer.*, 69, pp. 363-402.

SEDLETZKY, I. O. (1939). — Absorbing complex of soil, a paragenetic system of colloïdal minerals. *C. R. Acad. Sci. URSS*, 23, pp. 258-263.

SÉGALEN, P. (1957). — Etude des sols dérivés de roches volcaniques basiques à Madagascar. *Mém. Inst. Sci., Madagascar*, D, 8, 182 pages (Thèse).

SERRES, M. DE (1818). — Mémoire sur les terrains d'eau douce. *Journ. Phys.*, 87, pp. 134-135.

SIEFFERMANN, G. (1959). — Premières déterminations des minéraux argileux des sols du Cameroun. *Rapp. ORSTOM*, n° 99, pp. 1-10.

SIEVER, R. (1957). — The silica budget in the sedimentary cycle. *Amer. Min.*, 42, pp. 821-841.

SIEVER, R. (1959). — Petrology and geochemistry of silica cementation in some Pennsylvanian sandstones. *Soc. Econ. Pal. Min. Spec. Publ.*, 7, pp. 55-79.

SIEVER, R. (1962). — Silica solubility, 0-200 °C, and the diagenesis of siliceous sediments. *Journ. of Geol.*, 70, pp. 127-150.

SIFFERT, B. et WEY, R. (1961). — Sur la synthèse de la kaolinite à la température ordinaire. *C. R. Acad. Sci. Fr.*, 253, pp. 142-145.

SIFFERT, B. (1962). — Quelques réactions de la silice en solution. La formation des argiles. *Mém. Serv. Carte Géol. Als. Lor.*, n° 21.

SIFFERT, B. et WEY, R. (1962). — Synthèse d'une sépiolite à température ordinaire. *C. R. Acad. Sci. Fr.*, 254, pp. 1460-1463.

SITTLER, C. (1962). — Relations sédimentologiques entre divers bassins oligocènes de l'Est de la France. *Ier Coll. Intern. Strat. Paléogène, Bordeaux*, 1962.
SLANSKY, E. (1956). — The mineral composition of painter's earth and ochres from some localities in western Bohemia. *Rozpr. Cesk. Akad. ved. rada matem.*, 68, 1.
SLANSKY, M. (1958). — Vue d'ensemble sur le bassin sédimentaire côtier du Dahomey-Togo. *Bull. Soc. Géol. Fr.*, 8, pp. 555-580.
SLANSKY, M. (1959). — Contribution à l'étude géologique du bassin sédimentaire côtier du Dahomey et du Togo. *Thèse Sci. Nancy*, 355 pages.
SLANSKY, M., CAMEZ, T. et MILLOT, G. (1959). — Sédimentation argileuse et phosphatée au Dahomey. *Bull. Soc. Géol. Fr.*, 1, pp. 150-155.
SLAUGHTER, M. and MILNE, I. H. (1960). — The formation of chlorite-like structures from montmorillonite. *Clays and clay minerals* (7th Nat. Conf., 1958), pp. 114-125.
SLOSS, L. L. (1953). — The significance of evaporites. *Journ. Sedim. Petrol.*, 23, pp. 143-161.
SMITH, J. V. and YODER, H. S. (1956). — Experimental and theorical studies of the mica polymorphs. *Min. Mag.*, 31, pp. 209-235.
SMOOT, T. W. (1960). — Clay mineralogy of pre-Pennsylvanian sandstones and shales of the Illinois Basin, Part I. — Relation of permeability to clay mineral suites. *Illinois State Geol. Surv.*, 286, 20 pages.
SMOOT, T. W. and NARAIN, K. (1960). — Clay mineralogy of pre-Pennsylvanian sandstones and shales of the Illinois Basin, Part II. — Relation between clay mineral suites of oil-bearing and non-oil-bearing rocks. *Illinois State Geol. Surv.*, 287, pp. 1-14.
SMOOT, T. W. (1960). — Clay mineralogy of pre-Pennsylvanian sandstones and shales of the Illinois Basin, Part III. — Clay minerals of various facies of some Chester formations. *Illinois State Geol. Surv.*, 293, pp. 1-19.
SOSMAN, R. B. (1927). — *The properties of silica*. New York. The Chemical Catalog.
SOUGY, J. (1959). — Les formations crétacées du Zemmour Noir (Mauritanie septentrionale). *Bull. Soc. Géol. Fr.*, 1, pp. 166-182.
SOVERI, U. (1950). — Differential thermal analyses of some quaternary clays of Fennoscandia. *Ann. Acad. Sci. Fenn.*, A, 23, pp. 1-97.
STACH, E. (1950). — Vulkanische Aschenregen über dem Steinkohlenmoor. *Glückauf*, 86, pp. 41-50.
STEINBERG, M. (1961). — Nature minéralogique de la fraction argileuse des sédiments sidéro-lithiques et des argiles de décalcification de la feuille de Poitiers. *C. R. Acad. Sci. Fr.*, 253, pp. 148-151.
STEINWEHR, H. E. VON (1954). — Über das Pigment roter Gesteine. *Neues Jahrb. Geol. Pal. Abh.*, 99, pp. 355-360.
STEPHEN, I. and MACEWAN, D. M. C. (1951). — Some chlorite clay minerals of unusual type. *Clay Min. Bull.*, pp. 157-162.
STEPHEN, I. (1952). — *Journ. of Soil Sci.*, 3, pp. 219-237.
STEPHEN, I. (1954). — An occurrence of palygorskite in the Shetland Isle. *Min. Mag.*, 30, pp. 471-479.
STÉVAUX, J. (1961). — Etude sédimentologique des formations ordoviciennes et mésozoïques inférieures des forages au Sud du Champ d'Hassi-Messaoud. *Thèse 3e Cycle, Strasbourg*.
STEVENS, R. E. (1942-1945). — A system for calculating analysis of micas and related minerals to end members. *Geol. Surv. Bull.*, 950, pp. 101-113.
STEVENS, R. E. and CARRON, M. K. (1948). — Simple field test for distinguishing minerals by abrasion pH. *Amer. Min.*, 33, pp. 31-49.
STORZ, M. (1928). — Die sekundären authigenen Kieselsäuren in ihrer petrogenetisch-geologischen Bedeutung. *Monogr. z. Geologie und Pal.*, Ser. II, 481 pages.
STRAKHOV, N. M. (1957). — *Méthodes d'étude des roches sédimentaires*. 2 vol. Moscou. Traduc. B.R.G.M. : Ann. 35.
SUDO, T. (1954). — Clay mineralogical aspects of the alteration of volcanic glass in Japan. *Clay Min. Bull.*, 2, pp. 96-106.
SUDO, T., HAYASHI, H. and SHIMODA, S. (1962). — Mineralogical problems of intermediate minerals. *Clays and clay minerals* (9th Nat. Conf., 1960), pp. 378-392.

SUTRA, G. (1946). — Sur la dimension des ions électrolytiques. *Journ. Chim. Phys.*, pp. 190-326.
SUTRA, G. (1949). — Résultats récents concernant la constitution des ions électrolytiques. *Rev. Gén. Sci. pures et appl.*, pp. 54-64.
SWINEFORD, A. and FRYE, J. C. (1955). — Petrographic comparison of some loess samples from western Europe with Kansas loess. *Journ. Sedim. Petrol.*, 25, pp. 3-23.
SWINEFORD, A. and FRANKS, P. C. (1959). — Opal in the ogallala formation in Kansas. *Soc. Econ. Pal. Min. Spec. Publ.*, 7, pp. 110-120.
TAGGART, M. S. Jr. and KAISER, A. D. Jr. (1960). — Clay mineralogy of Mississippi river deltaic sediments. *Bull. Geol. Soc. Amer.*, 71, pp. 521-530.
TAKAHASI, J. et YAGI, T. (1929). — The peculiar mud grains in the recent littoral and estuarine deposits, with special reference to the origin of glauconite. *Econ. Geol.*, 24, pp. 838-852.
TAMM, O. (1924). — Experimental studies on chemical process in the formation of glacial clays. *Sver. Geol. Unders.*, C, 333.
TAYLOR, J. H. (1949). — Petrology of the Northampton and ironstone formation. *Mem. Geol. Surv. Great Britain*, London.
TCHOUKOV, F. V. (1961). — Sur la genèse des minéraux argileux dans la zone d'altération superficielle des gîtes métallifères. *Coll. Intern. C.N.R.S.*, 105, pp. 159-166.
TERMIER, H. et G. (1952). — *Histoire géologique de la biosphère.* Masson et Cie, Paris, 721 pages.
TERMIER, H. et G. (1956). — A propos de la théorie bio-rhexistasique. *Bull. Soc. Géol. Fr.*, 6, pp. 451-453.
TERMIER, H. et G. (1961). — *L'évolution de la lithosphère. III. Glyptogenèse.* Masson et Cie, Paris, 471 pages.
TERMIER, P. (1890). — Note sur la leverriérite. *Bull. Soc. Fr. Min.*, 13, pp. 325-330 et *Ann. Mines*, 17, p. 372.
TESSIER, F. (1950). — Age des phosphates et latéritoïdes phosphatés de l'Ouest du plateau de Thiès (Sénégal). *C. R. Acad. Sci. Fr.*, 230, pp. 981-983.
TESSIER, F. (1952). — Contributions à la stratigraphie et à la paléontologie de la partie ouest du Sénégal. *Bull. Direct. Mines, Gouv. Gén. de l'A.O.F.*, n° 14, 2 tomes.
TESSIER, F. (1954). — Oolithes ferrugineuses et fausses latérites dans l'Est de l'Afrique Occidentale française. *Ann. Ecol. Sup. Sci. Inst. Hautes Etudes, Dakar.*
THA HLA (1945). — Electrodialysis of mineral silicates: an experimental study of rock-weathering. *Min. Mag.*, pp. 137-145.
THEODOROVITCH, G. I. (1954). — Faciès géochimiques sédimentaires selon le profil du potentiel d'oxydo-réduction et leurs types pétrolifères probables. *Bull. Acad. Sci. URSS*, 96, 3.
THIÉBAUT, J. (1951). — Etude géologique des terrains métamorphiques de la grande Kabylie. *Bull. Serv. Carte Géol. Algérie*, 6, pp. 1-175.
THIÉBAUT, L. (1925). — Contribution à l'étude des sédiments argilo-calcaires du Bassin de Paris. *Thèse Sci. Nancy*, 170 pages.
THURMANN, J. (1838). — Son opinion sur diverses époques de dépôts qu'on observe dans le Bohnerz, *in* Réunion extraordinaire de la Société à Porrentruy, du 5 au 12 septembre 1938. *Bull. Soc. Géol. Fr.*, 9, pp. 376-379.
TOMLINSON, C. W. (1916). — The origin of red beds. *Journ. Geol.*, 24, pp. 153-179 et 238-253.
TOOKER, E. W. (1962). — Clay minerals in rocks of the lower part of the Oquirrh formation, Utah. *Clays and clay minerals* (9th Nat. Conf., 1960), pp. 355-365.
TOPKAYA, M. (1950). — Recherches sur les silicates authigènes dans les roches sédimentaires. *Bull. Labo. Géol. Min. Lausanne*, 97, 132 pages.
TOTSCHILIN, M. S. (1952). — Sur l'origine de l'hydrogoethite oolitique dans le minerai de fer à sidérite et chamosite. *Bull. Acad. Sci. URSS*, pp. 269-271.
TOURTELOT, H. A. (1961). — Thin sections of clay and shale. *Jour. Sedim. Petrol.*, 31, pp. 131-133.
TOVBORG JENSEN, A., WØHLK, C. J., DRENCK, K. and KROGH ANDERSEN, E. (1957). — A classification of danish flints, etc. based on X-ray diffractometry. *Danish Nat. Inst. Build. Res., Prog. Rep.*, D, 1. Copenhagen.

TRICART, J. et SCHAEFFER, R. (1950). — L'indice d'émoussé des galets. — Moyen d'étude des systèmes d'érosion. *Rev. Géomorph. Dyn.*, 4, pp. 151-179.

TRICART, J. (1953). — Résultats préliminaires d'expériences sur la désagrégation des roches sédimentaires par le gel. *C. R. Acad. Sci. Fr.*, 236, p. 1296.

TRICART, P. (1956). — Etude expérimentale du problème de la gélivation. *Bull. périglac. de Lodz*, pp. 285-318.

TRICART, J. (1959). — Les rapports entre la composition pétrographique des alluvions et la nature lithologique du bassin. *Rev. Géomorph. Dyn.*, pp. 2-13.

TRICART, J. (1960). — Désagrégation du granite par la cristallisation du sel marin. *Cahiers Océanogr. du C.O.E.C.*, 12, pp. 302-318.

TWENHOFEL, W. H. (1939). — *Principles of sedimentation.* N. Y. (2° édit., 1950). 673 pages.

TYRRELL, G. W. and PEACOCK, M. A. (1926). — The petrology of Iceland. *Roy. Soc. Edimb. Trans.*, 55, pp. 62-63, 71.

URBAIN, P. (1933). — Les sciences géologiques et la notion d'état colloïdal. *Act. Sci. Indust.*, 69, Paris.

URBAIN, P. (1937). — *Introduction à l'étude pétrographique et géochimique des roches argileuses.* Paris, Hermann, 2 vol., 60 et 80 pages.

URBAIN, P. (1937). — Texture microscopique des roches argileuses. *Bull. Soc. Géol. Fr.*, 7, pp. 341-352.

URBAIN, P. (1941). — Sur quelques particularités minéralogiques des argiles smectiques du bassin pliocène de la Moulouya (Maroc). *C. R. Soc. Géol. Fr.*, 11, p. 17.

URBAIN, P. (1942). — Logique des roches argileuses. *Bull. Soc. Géol. Fr.*, 12, pp. 97-112.

URBAIN, P. (1951). — Recherches pétrographiques et géochimiques sur deux séries de roches argileuses. *Mém. p. servir à l'explic. Carte Géol. Fr.*, pp. 1-278.

UREY, H. C., LOWENSTAM, H. A., EPSTEIN, S. and MACKINNEY, C. R. (1951). — Measurement of paleotemperatures and temperatures of the upper Cretaceous of England, Denmark and the southeastern United States. *Bull. Geol. Soc. Amer.*, 62, pp. 399-416.

VALETON, I. (1957). — Lateritische Verwitterungsböden zur Zeit der jungkimmerischen Gebirgsbildung im nördlichen Harzvorland. *Geol. Jahrb.*, 73, pp. 149-164.

VALETON, I. (1958). — Der Glaukonit und seine Begleitminerale aus dem Tertiär von Walsrode. *Mitt. Geol. Staatsinst.* Hamburg, 27, pp. 88-131.

VANDERSTAPPEN, R. et CORNIL, J. (1958). — Contribution à l'étude des minéraux argileux du type « Mixed Layers ». *Bull. Soc. Belge Géol.*, 67, pp. 91-103.

VANDERSTAPPEN, R. et VERBEEK, T. (1959). — Présence d'analcime d'origine sédimentaire dans le Mésozoïque du bassin du Congo. *Bull. Soc. belge Géol.*, 68, pp. 417-421.

VARLEY, E. R. (1952). — Vermiculite. *Publ. Colonial Geol. Surv. London.*

VATAN, A. (1939). — Observations 1. sur l'existence dans certaines argiles de sels alcalins solubles, 2. sur la structure des sables argileux. *C. R. Soc. Géol. Fr.*, 9, pp. 145-146.

VATAN, A. (1947). — La sédimentation continentale tertiaire dans le bassin de Paris méridional. *Thèse Sci.* Edit. Toulousaines de l'Ingénieur, 215 pages.

VATAN, A. (1947). — Remarques sur la silicification. *C. R. Soc. Géol. Fr.*, 5, pp. 99-101.

VATAN, A. (1948). — La sédimentation détritique en Aquitaine aux temps tertiaires. *C. R. Soc. Géol. Fr.*, 3, p. 48.

VATAN, A., ROUGE, P. E. et BOYER, F. (1957). — Etudes sédimentologiques et pétrographiques dans le Tertiaire subalpin et jurassien de Savoie et des régions limitrophes. *Rev. Inst. Fr. Pétrole*, 12, pp. 467-480.

VATAN, A. (1962). — Les grès et leur milieu. *C. R. Acad. Sci. Fr.*, 254, pp. 2026-2028.

VENZO, S. (1952). — Geomorphologische Aufnahme des Pleistozäns (Villafrachian-Würm) im Bergamasker Gebiet und in der östlichen Brianza: Stratigraphie, Paläontologie und Klima. *Geol. Rundsch.*, 41, pp. 109-125.

VERNADSKY (1924). — *La géochimie.* F. Alcan, Paris, 403 pages.

VERNET, J. P. (1959). — Etudes sédimentologiques et pétrographiques des formations tertiaires et quaternaires de la partie occidentale du Plateau suisse. *Eclogae geol. Helv.*, 51, pp. 1115-1152.

VERNET, J. P. (1961). — Concerning the association Montmorillonite-Analcime in the series of Stanleyville, Congo. *Journ. Sedim. Petrol.*, 31, pp. 293-295.

VERNET, J. P. (1962). — L'halloysite bleue du mont Vuache (Savoie). *C. R. Acad. Sci. Fr.*, 254, pp. 2377-2380.
VERNHET, S. (1956). — Etude chimique et minéralogique de quelques sédiments méditerranéens de moyenne et grande profondeurs. *C. R. Acad. Sci. Fr.*, 242, pp. 1049-1052.
VINOGRADOV, A. P. and RONOV, A. B. (1956). — Evolution of the chemical composition of clays of the Russian Platform. Geochemistry. *The Geoch. Soc. Ann Arbor, Mich.*, 2, pp. 123-139.
VISSE, L. (1953). — Les faciès phosphatés. *Rev. Inst. Fr. Pétrole*, 8, numéro spécial, pp. 87-98.
VISSE, L. (1954). — La sédimentation argileuse des dépôts marins tertiaires de l'Ouest sénégalais. *C. R. Soc. Géol. Fr.*, 4, pp. 27-29.
VISSE, L. (1954). — Présence de palygorskite dans les sédiments marins de la Lama (Dahomey). *C. R. Soc. Géol. Fr.*, 3, pp. 59-61.
VOGT, G. (1906). — De la composition des argiles. In *Contrib. à l'étude des argiles et de la céramique*, Paris.
VOLK, N. J. (1933). — Formation of muscovite in soils. *Amer. Journ. Sci.*, 26, pp. 114-129.
VOLTZ, L. (1834). — Réunion extraordinaire à Strasbourg. Discussion sur le bohnerz, p. 35. *Bull. Soc. Géol. Fr.*, 1re sér., 6, pp. 5-59.
WALDMANN, L. (1938). — Über weitere Begehungen im Raume der Kartenblätter Zwettl-Weitra, Ottenschlag und Ybbs. *Verh. geol. Bundesanstalt*, pp. 115-119.
WALKER, G. F. (1949). — The decomposition of biotite in the soil. *Min. Mag.*, 28, pp. 693-703.
WALKER, G. F. (1949). — Water layers in vermiculite. *Nature*, 163, p. 726.
WALKER, G. F. (1950). — Trioctahedral minerals in the soil clays of North East Scotland. *Min. Mag.*, 29, pp. 72-84.
WALKER, G. F. (1951). — *Vermiculite and some related mixed layers in Brindley "X-ray identification and structure of the clay minerals"*, Min. Soc. Great Britain, Monogr. London, 345 pages.
WALKER, G. F. (1957). — On the differenciation of vermiculites and smectites in clays. *Clay Min. Bull.*, 3, pp. 154-164.
WALKER, G. F. (1958). — Reactions of expanding-lattice clay minerals with glycerol and ethylene glycol. *Clay Min. Bull.*, 3, pp. 302-313.
WALTHER, J. (1900). — *Das Gesetz der Wüstenbildung in Gegenwart und Vorzeit.* Berlin (2e éd., Leipzig, 1924).
WARREN, B. E. and BISCOE, J. (1938). — The structure of silica glass by X-ray diffraction studies. *Journ. Amer. Ceram. Soc.*, 21, pp. 49-54.
WARREN, B. E. (1942). — X-ray study of chrysotile asbestos. *Amer. Min.*, 27, p. 235.
WARSHAW, C. M. and ROY, R. (1961). — Classification and a scheme for the identification of layer silicates. *Bull. Geol. Soc. Amer.*, 72, pp. 1455-1492.
WASHBURN, E. W. and NAVIAS, L. (1922). — Relations of calcedony to other forms of silica. *Proc. Nat. Acad. Sci.*, 8, pp. 1-16.
WEAVER, C. E. (1953). — Mineralogy and petrology of some Ordovician K-bentonites and related limestones. *Bull. Soc. Geol. Amer.*, 64, pp. 921-944.
WEAVER, C. E. (1956). — The distribution and identification of mixed-layer clays in sedimentary rocks. *Amer. Min.*, 41, pp. 202-221.
WEAVER, C. E. (1958). — A discussion on the origin of clay minerals in sedimentary rocks. *Clays and clay minerals* (5th Nat. Conf., 1956), pp. 159-173.
WEAVER, C. E. (1958). — Geologic interpretation of argillaceous sediments. Part I. Origin and significance of clay minerals in sedimentary rocks. Part II. Clay petrology of upper Mississippian-lower Pennsylvanian sediments of central United States. *Bull. Am. Ass. Petrol. Geol.*, 42, pp. 254-309.
WEAVER, C. E. (1958). — The effects and geologic significance of potassium "fixation" by expandable clay minerals derived from muscovite, biotite, chlorite and volcanic material. *Amer. Min.*, 43, pp. 839-889.
WEAVER, C. E. (1959). — The clay petrology of sediments. *Clays and clay mineral* (6th Nat. Conf., 1957), pp. 154-187.

WEAVER, C. E. (1960). — Possible uses of clay minerals in the search for oil. *Bull. Am. Ass. Petrol. Geol.*, 44, pp. 1505-1518.

WEAVER, C. E. (1960). — Possible uses of clay minerals in the search for oil. *Clays and clay minerals* (8th Nat. Conf., 1958), pp. 214-227.

WEIL, R. (1926). — Influence des impuretés sur la température de transformation paramorphique de la cristobalite. *C. R. Acad. Sci. Fr.*, 183, pp. 753-755.

WEISSE, J. G. DE (1948). — Les bauxites d'Europe centrale (Province dinarique et Hongrie). *Bull. Labo. Géol. Min. Géoph. Univ. Lausanne*, 87, p. 162.

WELBY, C. W. (1958). — Occurrence of alkali metals in some Gulf of Mexico sediments. *Journ. Sedim. Petrol.*, 28, pp. 431-453.

WEY, R. et SIFFERT, B. (1961). — Réactions de la silice monomoléculaire en solution avec les ions Al^{3+} et Mg^{2+}. Genèse et synthèse des argiles. *Coll. Intern. C.N.R.S.*, 105, pp. 11-23.

WEZEL FORESE, C. (1960). — Ricerche sedimentologiche su una serie argillitico-marnosa del « Macigno » della Regione di Bobbio (Appennino Piacentino). *Boll. Soc. Geol. Italiana*, 79, pp. 143-165.

WHITE, D. E., BRANNOCK, W. W. and MURATA, K. J. (1956). — Silica in hot spring waters. *Geochim. Cosmochim. Acta*, 10, pp. 27-59.

WHITEHOUSE, U. G. and MACCARTER, R. S. (1958). — Diagenetic modification of clay mineral types in artificial sea water. *Clays and clay minerals* (5th Nat. Conf., 1956), pp. 81-119.

WHITEHOUSE, U. G., JEFFREY, L. M. and DEBRECHT, J. D. (1960). — Differential settling tendencies of clay minerals in saline waters. *Clays and clay minerals* (7th Nat. Conf., 1958), pp. 1-79.

WHITTIG, L. D. and JACKSON, M. L. (1955). — Interstratified layer silicates in some soils of northern Wisconsin. *Clays and clay minerals* (3rd Nat. Conf., 1954), pp. 322-336.

WHITTIG, L. D. and JACKSON, M. L. (1956). — Mineral content and distribution as indexes of weathering in the Omega and Ahmeck soils of northern Wisconsin. *Clays and clay minerals* (4th Nat. Conf., 1955), pp. 362-371.

WICKMAN, F. E. (1945). — Some notes on the geochemistry of the elements in sedimentary rocks. *Ark. f. Kemi, min. och geol.*, 19, pp. 1-7.

WILHELM, E. (1962). — Echange de cations d'une vermiculite, étude thermodynamique. *Thèse 3e Cycle, Strasbourg.*

WIMMENAUER, W. (1959). — *In Erläuterungen zur geologischen Excursionskarte des Kaiserstuhls. Geol. Landesamt i. Baden-Württ.*, 139 pages.

WINCHEL, A. N. (1927). — Further studies in the mica group. *Amer. Min.*, 12, pp. 267-279.

WINKLER, H. G. (1955). — *Struktur und Eigenschaften der Kristalle.* Berlin, 314 pages.

WOLLAST, R. (1961). — Aspect chimique du mode de formation des bauxites dans le Bas-Congo. *Acad. Roy. Sci. O.-M.*, nouv. sér., 7, pp. 468-489.

WÜLFING, E. A. (1900). — Untersuchung des bunten Mergels der Keuperformation auf seine chemischen und mineralogischen Bestandteile. *Jahresh. Ver. f. Vaterl. Naturk. Württ.*, 56, pp. 1-45.

YAALON (1955). — Clays and some non-carbonate minerals in limestones and associated soils of Israël. *Bull. Res. Counc. Israël*, 5 B, 2, Sect. B, pp. 161-173.

YODER, H. S. and EUGSTER, H. P. (1954). — Phlogopite synthesis and stability range. *Geochim. Cosmochim. Acta*, 6, pp. 157-185.

YOUELL, R. F. (1960). — An electrolytic method for producing chlorite-like substances from montmorillonite. *Clay Min. Bull.*, 4, pp. 191-195.

ZANS, V. A. (1952). — Bauxites resources of Jamaïca and their development. *Col. Geol. min. Res.*, 3, pp. 307-333.

ZANS, V. A., LEMOINE, R. C. et ROCH, E. (1961). — Genèse des bauxites caraïbes. *C. R. Acad. Sci. Fr.*, 252, pp. 3302-3304.

ZUSSMAN, J., BRINDLEY, G. W. and COMER, J. J. (1957). — Electron diffraction studies of serpentine minerals. *Amer. Min.*, 42, pp. 133-153.

INDEX

Bold-face numbers indicate pages of principal interest

Addition, **322**, 349, 361
Aggradation, 307, **315**, 361
Allevardite, **28**
Allophane, 42, 47, **96**, **299**, 315, 340
Alumina-Aluminum, 22, **59**, 118, 130, 272, 280, **334**, 341, 366, **386**
Alunite, 42, 152, 273
Amesite, 26
Ampelite, 38
Analcimolite, 175
Anauxite, 28
Antigorite, 4, 26, 336
Arenization, **92**, 160
Argillite, 35, 186, 193, 199
Attapulgite, **18**, 26, 172, **199**, 257, **262**, 305, 329

Batavite, 27
Bauxite-Bauxitization, 47, 69, **130**, 147, 254, 273, 324, 331, 360
Beidellite, **11**, 27, 308, 311, 338
Bentonite, 37, **45**
Biorhexistasy, **77**, 264, 274
Biotite, 6, 27, 89, **310**, 383
Boehmite, **23**, 118, 147, 254, 336
Bohnerz, 142, 149
Bowlingite, 27
Brammalite, 8, 27
Bravaisite, **8**, 16, 30, 31, 217
Brucite, 350

Calcium, 50, 64, 119, 193, 199, 284, **332**, **365**, 385
Celadonite, 206
Chalcedony, 146, **283**, 370
Chalk, 200, 263
Chamosite, 4, 13, 211
Chernozem, 111, 325
Chert, 55, 173, 199, **277**, 289, 293, 326, 373
Chlorite, **12**, 28, 44, 108, **211**, 218, 257, 305, 311, 319, 327
Chlorite, swelling, **13**, **211**, 219, 257
Chloritization, 42, 103, 198, 210, 310, **323**, 352, 361, **383**
Chromocre, 27

Chrysotile, 5, 26
Claystone, 36
Climates, **80**, 94, 100, 159, 164, 216, 245, 289, 371
Corrensite, **15**, 28, 195, **219**, 258, 318
Corundum, 23
Cristobalite, 201, 279, **285**, 326, **370**
Cronstedtite, 5, 26
Crust, lateritic, (see "cuirassement")
Crystal growth, 209, **277**, 296, 335, 353, 369
"Cuirassement", 70, 77, **114**, 126, 146, 214, 274, 373

Damourite, 6, 27, 92
Degradation, **307**, 309, 359
Deserts, 102, 154
Diagenesis, 82, 147, 182, 189, 197, 212, 234, 247, 275, 304, **327**, 355, 381
Diaspore, **23**, 43, 152, **254**
Dickite, **3**, 26, 43, 147, 348
Dioctahedry, 4, 26
Disorder, 3, 15, 284, **301**, 369
Dombassite, 26

Environment, confined, 123, 296, **331**, 361
Environment of lessivage, 88, 96, 107, **121**, 145, **330**, 359
Eolian, 137
Estuaries, 139, 178, 316

Feldspars, 41, **62**, 88, 92, 97, 120, 156, 323, 368, **384**
Ferrallite (see Laterite)
Fireclay, **4**, 26, **254**, 306
Flint (see Chert)
Flysch, 187
Fuschite, 27

Garnierite, 5
Gels, 42, 55, 64, 277, **289**, **341**
Geochemical cycle, **68**, 273, 355, 381, **384**
Ghassoulite, 11
Gibbsite, **23**, 118, **130**, 146, **324**, 343, 359
Glacial deposits - Freeze-Thaw, 85, 102, 136
Glauconie, **9**, 27, 171, 199, **204**, 263, 305

Glauconite, 9, 205, 326
Glimmerton, 8, 30, 217
Goethite, 22, 117, **126**, 131, 151, **210**, 350, 366
Greenalite, 5, 26
Greywacke, 187
Grovesite, 26

Halloysite, 3, 26, 43, 47, 146, 306, 325
Halmyrolysis, **48**, 180, **310**, 358
Hectorite, 12, 27, 44
Hematite, **22**, 117, 156
Heritage, 32, **82**, 100, 177, 185, 192, 198, 233, 275, **302**, **356**
Homeotypy, 4
Hydrobiotite, 15, 102, 309
Hydrodynamics, 71, 123, 335, 357
Hydrolysates, 49, 69
Hydrolysis, **62**, **88**, 98, 160, 330, 357
Hydromica, 8, 28, 126, 238, 246
Hydromuscovite, 8
Hydronium, 7
Hydrothermal, 40, 44
Hypersaline, 71, 173, **215**, 257, 317, 363

Illite, 7, 27, 108, 164, 170, 208, 239, **305**, 325
Inheritance (see Heritage)
Ionic potential, 49, 69
Ionic radii, 51, 69
Ionization potential, 66
Iron, 22, 50, 69, 87, 109, **126**, 205, **213**, 311, 338, **366**, 385
Isomorphy, 4
Isotopes, 375
Isotypy, 4
Isovolume, 118, 130, 214

Jefferisite, 27
Jenkinsite, 26

Kaolin, 41
Kaolinite, **2**, 26, 41, 145, 239, 304, **325**, 336, **344**
Kaolinization, 41, **98**, 166, 246, 254, **324**, 341, 344, 352

Lakes, 146, **169**, 281
Landscapes, 202, 291, **370**
Lassalite, 20
Laterite-Lateritization, 71, **112**, 126, 143, 157, 201, 213, 254, 272, 340, 359, 372
Lateritic crusts (see "cuirassement")
Ledikite, 27
Lepidocrocite, 22
Lepidomelane, 27

Leptochlorites, 28, 211
Lessivage (see Environment of lessivage)
Leverrierite, 166, 205
Limonite, 22
Livesite, 4
Loess-Lehm, 137
Lutite, 34

Maghemite, 22
Magnesium, 44, 50, 61, 119, 196, **271**, 284, 318, 332, 344, **365**, 385
"Marne", 36, 171, **193**, 199
Mellorite, 4
"Meulière", 172, 199, 289
Micas, **5**, 27, 43, 89, **311**, **383**
Minnesotaite, 27
Mixed-layer minerals, **15**, 42, **101**, 139, 189, 218, 304, 310
Molasse, 187
Montmorillonite, **10**, 27, **45**, 108, 124, **199**, 263, 305, **324**, 336
Mudstone, 36
Muscovite, **5**, 9, 89, 219, **310**, 383

Nacrite, **3**, 26, 43, 348
Neoformation, 32, 82, **100**, 177, 201, 210, 233, 276, **323**, **356**
Nepouite, 5
Nontronite, **11**, 27, 44, 338
Noumeite, 5, 26

Oil, 240, 251
Oligiste (see Hematite)
Oolithe, 71, 151, **210**, 264
Opal, 44, 58, 279, **285**
Orthoantigorite, 26
Orthoserpentine, 5
Oxidates, 69
Oxidation-reduction, 87, 208, 211, 333
Oxonium, 8

Palygorskites, **18**, 44, 172
Paragonite, 6, 27, 383
Pelite, 34, 37
pH, 56, 62, 88, 331, 335
Phengite, 6
Phlogopite, 6, 27, 90
Phosphates, **199**, 208, 263, 375
Phyllite, 36
Piedmont, **140**, 149, 160, 237
Pilolite, 20, 29
Pinite, 6
Pisolite, 142
Podzol, 105, 141
Polytypy, 9
Potash-bearing clay, 8

Potassium, **50**, 89, 102, 179, 215, 284, 312, 318, 332, **363**, 385
Psammite, 34
Pseudochlorites, 211, 338
Pyrophyllite, **6**, 27, 43, 139, 219, 310, **329**

Quartz, 42, 69, 89, 108, 113. 156, 186, 236, **277**, **369**
Quartzification, 237, 246, 288, **297**

Rectorite, 15, 44
Regur, 111
Rendzina, 103
Rivers, **139**, 178, 316

Sandstone, 37, 69, 154, **186**, 235, **289**
Sandstones, red, **154**
Saponite, **11**, 27, 44, 336
Sauconite, 27, 44, 338
Schist, 36, 381
Sepiolite, **18**, 26, 172, 199, 219, 257, **262**, 305, 329, **344**
Sequence, series, 68, **73**, 203, 235, **271**
Sericite, 5, 27, 92, 312, **383**
Sericite-like mineral, 8, 30, 217
Sericitization-Illitization, 41, 92, 103, 137, 162, **239**, 309, **323**, 352, 360, 383
Serpentine, **4**, 26, 44, 114, 219
Shale, 36, 186
Siderolithique, **142**, 157, 215, 264.
Silica-Silicon, **55**, **277**, 334, 367, 386
Silicification, 55, 146, 158, 246, 274, **277**, 367

Silt, 37, 239
Slate, 35
Sodium, **50**, 63, 175, 215, 284, 332, **363**, 385
Soil, brown, 104
Soil, calcimorphic, **111**, 325, 361
Soil, ferruginous, **109**, 157
Soil, tropical black, 111, 325
Stevensite, **11**, 27, 339, 344
Subtraction, **322**, 349, 359
Synthesis, 336

Talc, **6**, 27, 44, 218, **329**
Temperature, 42, 85, **88**, 347, 357
Terra rossa, 110
Texture, 38
Tirs, 111, 325
Tonstein, **166**, 325
Trace elements, 214, 247, 319, **375**, 380
Transformation, 33, 43, 82, 101, 108, 174, 184, 233, 256, 275, **306**, **356**
Tridymite, 279, 285
Trioctahedry, 4, 26
Tropics, 76, **122**, 143, **154**, 273

Underclay, 163, 254

Vermiculite, **13**, 24, 27, 43, 101, 105, 140, **308**
Volcanism, **45**, 168, 176, 181, 370

Weathering, 81, **85**, 95, 114, 133, 303, **309**, 322, 355, **370**
Wolchonskoite, 27

Imprimé en Belgique
par Georges Thone à Liège